FORAGE CROPS

Properly managed forage crops combined with row crops make an excellent farming enterprise.

FORAGE CROPS

Darrell A. Miller
Professor of Agronomy
College of Agriculture
University of Illinois, Urbana–Champaign

2 8536

McGRAW-HILL BOOK COMPANY

New York   St. Louis   San Francisco   Auckland   Bogotá
Hamburg   Johannesburg   London   Madrid   Mexico   Montreal   New Delhi
Panama   Paris   São Paulo   Singapore   Sydney   Tokyo   Toronto

This book was set in Times Roman by Black Dot, Inc. (ECU).
The editors were John J. Corrigan and Mary Jane Martin;
the production supervisor was Leroy A. Young.
The cover was designed by Carla Bauer.
Project supervision was done by The Total Book.
Halliday Lithograph Corporation was printer and binder.

FORAGE CROPS

1234567890HALHAL8987654

ISBN 0-07-041980-9

Library of Congress Cataloging in Publication Data

Miller, Darrell A.
  Forage crops.

  Includes bibliographies and index.
  1. Forage plants. I. Title.
SB193.M45 1984   633.2   83-14949
ISBN 0-07-041980-9

To my wife, Norma,
who has given so much of her time,
love, and support over the years.

CONTENTS

PREFACE

Forage Crops was written in response to the need for a complete textbook on forage crops and their profitable management. This book is the result of many years of teaching forage production at the University of Illinois, but it could be used for similar forage courses almost anywhere in the United States. It is based on research conducted at various experiment stations. I hope it will serve as a reference for secondary schools, community colleges, and others interested in forage production.

This book presents forage production practices in a contemporary manner: from the perspective of soil preparation, soil fertility, seedbed preparation, species selection, seeding, management, and harvesting methods. The physiological aspects of forage production are presented in detail, with strong emphasis on applications.

Forage crops are not usually thought of as cash crops, but profit can and should be expected from raising them. Proper management of forages grown on productive soils will provide profits. At present, the number of hay markets is increasing, thus stimulating interest in forage production. In addition, with the recent energy crisis, the costs of nitrogen fertilizer have become more important, and, therefore, so has nitrogen fixation. The book presents data to convince professionals and producers of the significance of legume production, the culture of legumes with grasses, and nitrogen contribution from legumes.

This book covers most of the important legume and grass species grown in the United States, where they are grown, how they grow, and the characteristics of each. This will enable the student to determine their merits, as well as to make decisions regarding where, when, and how to grow each. Advantages and disadvantages are presented at the end of those chapters covering the various forage species. Each chapter has a summary, with questions covering the material presented.

There is also a list of references at the end of each chapter, including works other than those cited in the text, so that students can pursue the literature beyond that presented in this book.

Many people contributed their advice and counsel to this project. The author is indebted to Annette Bodzin, the project supervisor. In addition, copy editors

Debbie L. Rosenberg and Marcy Stamper greatly assisted in preparing the final version. I appreciate and wish to thank the following forage crop researchers who reviewed the text for its technical content and made many valuable suggestions: Professor John R. Campbell, University of Illinois, Urbana-Champaign; Professor William A. Householder, Eastern Kentucky University; Professor Wiley C. Johnson, Auburn University; Professor Bill Melton, New Mexico State University; and Professor Gerry L. Posler, Kansas State University.

I also want to express my appreciation to the typists: Elaine Apple, Kay Schelling, and my wife, Norma.

Darrell A. Miller

METRIC CONVERSION TABLE

Metric unit	English unit
Length	
millimeter (mm)	0.039 inch (in)
centimeter (cm)	0.394 inch (in)
meter (m)	3.281 foot (ft)
meter (m)	1.094 yard (yd)
kilometer (km)	0.621 mile (mi)
Area	
hectare (ha)	2.471 acre (a)
kilometer^2 (km^2)	247.1 acre (a)
Volume	
liter (l)	1.057 quart (qt) (U.S. liquid)
hectoliter (hl)	2.838 bushel (bu) (U.S.)
Weight	
gram (g)	0.00221 pound (lb)
kilogram (kg)	2.205 pound (lb)
quintal (q)	220.5 pound (lb)
ton (t)	1.102 ton (English)
Yield	
quintal/hectare (q/ha)	0.892 hundredweight/acre
quintal/hectare (q/ha)	89.24 pounds/acre
kilogram/hectare (kg/ha)	0.892 pound/acre
ton/hectare (t/ha)	0.446 ton/acre
Temperature	
degree Celsius (°C)	1.80°C + 32 = degree Fahrenheit (°F)

Light

$$1 \text{ lux (lx)} = 0.0929 \text{ foot-candle (ft-c)}$$
$$1 \text{ candela (cd)} = 1 \text{ lumen (lm)}$$
$$1 \text{ lux (lx)} = 1 \text{ lm/m}^2$$
$$1 \text{ lambert (L)} = 1 \text{ lm/cm}^2 = 205.7 \text{ candle/foot}^2$$

Energy

$$1 \text{ calorie (cal)} = 3.97 \times 10^{-3} \text{ British thermal unit (Btu)}$$
$$1 \text{ kilocalorie (kcal)} = 1000 \text{ cal}$$
$$1 \text{ megacalorie (Mcal)} = 1000 \text{ kcal}$$
$$1 \text{ erg (erg)} = 2.39 \times 10^{-8} \text{ cal}$$

FORAGE
CROPS

PART ONE

FORAGES IN OUR SOCIETY

CHAPTER 1

ROLE OF FORAGES

It is difficult to assess the importance of forages in our national economy. The primary reason is that forages must be consumed by an animal before they are generally considered usable for human beings. Improving animal production requires improving the quantity and quality of forage.

In recent years, forages have gained a great deal of interest as cash crops in the United States. Maintaining a productive agriculture in this country is dependent upon the utilization of forage crops in overall farming operations. In the central and midwestern states, most of the area is devoted to row crops or grain crops, and livestock and forages are raised where the topography is unsuited for row crops. Moving westward, into the dryland area, and in the southwest, rangeland becomes integral to an economical agriculture. In the far west, under highly intensified agriculture, irrigated pastures and forage seed fields are very important, as they are in the southwest and west, where irrigated forages such as alfalfa are considered cash crops. In the midsouth and south, semitropical plants come into prominence, and in this region's very diverse agriculture, forages play a significant role. Forages recently have increased in importance in the southeast. And the northeast, with its rolling topography, forages are important in relation to dairy and beef production.

Forage production is a complex mechanism that takes the energy of sunlight and transforms it into plant proteins, carbohydrates, and other compounds within the forage plant, and further involves the animal's conversion of these plant products into milk, meat, or wool. In this complex system, the plants are called the primary producers and the animals are called the secondary producers.

The ruminant livestock, cattle and sheep, can utilize most of the forage

FIGURE 1-1
Proper pasture management with the beef cow-calf herd will provide the needed red meat in our diet.

product, including the fibrous portion of the plant. Monogastric livestock, such as swine and poultry, can utilize only a portion of the forage product that they consume. Whole plants of some forages can be utilized more effectively by nonruminant animals when the plant is ground into meal.

Forages provide most of the feed for cattle and sheep in the United States (Table 1-1). Forages provide 63, 84, and 90% of the feed units consumed by dairy cattle, beef cattle, and sheep and goats, respectively (12). The overall percentage of forages consumed by all ruminants makes up 80% of the total feed unit. The remaining balance is supplied by grains, concentrates, or by product feeds. Since the ruminant livestock consume most of the primary production that man cannot utilize directly, they are considered a highly complementary component of the overall food system.

Americans consume about 645 kg (1420 lb) of food per person per year.

TABLE 1-1
FEEDS CONSUMED BY RUMINANT LIVESTOCK IN MILLION METRIC TONS
OF FEED UNITS
(One Feed Unit Equals One Pound of Corn)

Livestock	Concen-trates	Forages	Total feed	Percent forage
Dairy	23.8	41.6	65.4	63%
Beef	33.2	171.8	205.0	84
Sheep and goats	0.7	6.1	6.8	90
All ruminants	57.7	219.6	277.3	80
All livestock	157.8	240.3	398.1	60
Percent by ruminants	36	91	70	

About 42% of this is of animal origin. About 212 kg (466 lb), or one-third of our food supply, is provided by the ruminant animals (11). Foods that these animals produce are important sources of energy, protein and other nutrients, plus minerals. About 23% of the total energy consumed and 45% of the protein supply come from beef and dairy products (11). Since forages provide about 80% of the feed units consumed by the ruminant animals and since these animals contribute 23% of the energy and 45% of the protein in our diets, we could calculate that 18% of the energy and 36% of the protein in the American diet originate in forages (Fig. 1-1). No other commodity approaches forages as a contributor to the human diet in the United States.

FORAGE REGIONS

One can divide the United States into three major forage-producing regions (Fig. 1-2). These regions can be defined by drawing an imaginary line extending

FIGURE 1-2
Forage regions of the United States.

from the Canadian border south to Dallas, Texas and then southwest to San Antonio, and another line extending eastward from the first, along approximately the 35th parallel. West of the north-south line lie the arid and semiarid lands. Northeast of the 35th parallel is the humid, cool-temperate region, while south of it is the humid, subtropical region. These regions have unique climatological characteristics and soil conditions (Fig. 1-3). The distribution of the various forage-producing areas is presented in Table 1-2.

Forage production is much below its potential in the two humid regions. Improved management practices, including the use of improved cultivars, better fertility, pest management, renovation of old stands, timely harvests, and rotational grazing could increase hay and pasture production in these regions severalfold. Approximately 66% of the beef cattle and 86% of the dairy cattle are grown in the two humid regions (13). This indicates the importance of the livestock demands for a more consistent, higher-quality forage that can be produced in these humid regions. Most of the rangeland in the west is fully stocked, and improvement would be marginal in this area.

FUTURE WORLD FOOD NEEDS

No one would disagree with the statement that world food needs will increase in the future. It has been stated that this increase may be as high as 50% by 2000 A.D. (13). The increase in human population, plus the added affluence in many of the developing countries, will dictate a greater food supply. In addition to a greater demand for food products, there will be an increasing demand for animal protein.

Energy may be the key word for the future expansion of production of food in the world. Fossil energies are becoming more expensive and limited. As energy demands increase, alternative sources of energy will become much more important. At that point, forages will definitely contribute to converting marginal land into productive areas. There will also be a greater demand to increase the grain production per hectare in the highly productive regions of the United States. Increased forage production is the result of proper species selection, which is often a grass plus a legume that may provide some of the additional nitrogen for the grass component of the forage. Increased forage production means greater carrying capacity per land unit and longer-grazing pastures per animal unit, with a larger percent of red meat and less fat produced per animal carcass. Overall this will result in greater production efficiency of protein and food products for the human population. Feed grains will continue to play an important role in overall livestock production, but eventually it will be the by-products of grain production and other cellulose crops that dictate the role of future ruminant-livestock production practices.

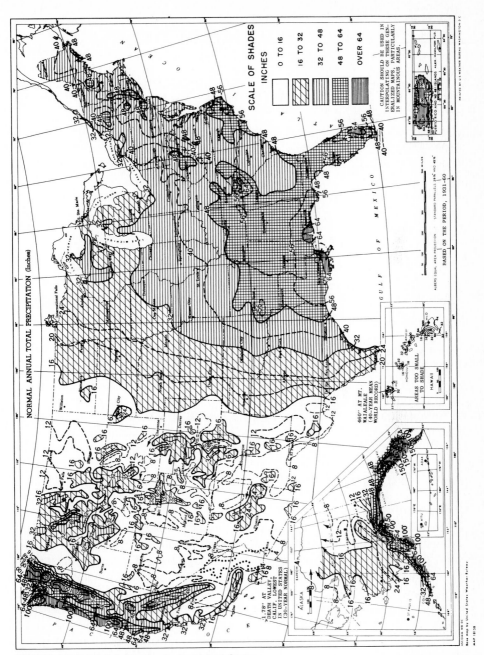

FIGURE 1-3 Rainfall distribution in the United States.

TABLE 1-2
APPROXIMATE AREA OF LAND DEVOTED TO VARIOUS TYPES OF FORAGE
PRODUCTION ON PRIVATELY OWNED LAND AS DISTRIBUTED IN THE THREE MAJOR
FORAGE REGIONS OF THE UNITED STATES
(Millions of Ha)

Forage region	Pasture-land	Range-land	Grazed forest	Hay and pasture	Total forage	%	Total U.S. land
Humid, cool-temperate	24	4	15	21	64	23	245
Humid, subtropical	9	2	13	3	27	9	105
Arid, semiarid	9	149*	27	9	194	68	301
Total	42	155	55	33	285	100	651

*There is an additional 106 million ha of federally owned rangeland available for grazing.

SIGNIFICANCE OF GRASSLANDS

Forages or grasslands are of great concern directly or indirectly to all people, whether they live in urban, suburban, or rural areas. Grasslands are one of our natural resources and a part of our national heritage that must be managed wisely. Some of the significant uses of grasslands are:

1 Support and provisions for the livestock industry
2 Use in soil and water conservation
3 Water resource development and protection of the soil
4 A balanced program of wildlife conservation and protection
5 Plant life for our outdoor recreation and leisure-time pleasure
6 A natural and effective ecological use of our land that will provide a balanced and profitable management of our total environment
7 Production of food on land that is too steep, rough, wet, or dry for row crops

Grazing Lands

Permanent pastures are vital in forage production. If properly managed, a permanent pasture can supply a considerable portion of livestock needs. The following chapters will describe proper management practices as related to fertility, reseeding, grazing, use of specific species, and the overall practices involved in efficient production of forages. A good permanent pasture will not only reduce runoff sediment, minimizing mineral leaching, but will also provide cover for wildlife.

Approximately 46% of our pasture can be improved by proper fertility, grazing, and weed control. About 33% needs reseeding or reestablishment to more productive grasses or legumes (1).

The objective of managing pastureland is to provide a full season of grazing by utilizing certain grass and legume species that are most productive at various times during a grazing season. Well-managed pastures are basic to good soil

FIGURE 1-4
Proper crop management includes forage crops in rotation on rolling topography.

conservation programs on livestock farms in the United States (Fig. 1-4). Soil conservation is a very important part of the forage program on many of the ranges in the far west.

Cropland for Hay

It is estimated that the cropland used for hay and pasture represents about 11% of the total forage resources that supply nearly all of the winter forage for beef and sheep, and a considerable portion of that which is needed for dairy cattle (16). Three-fourths of the cropland used for hay occupies some of the best soils in the United States (5). If properly fertilized, seeded with proper species, and well-managed, the carrying capacity can be as high as 4 animals per hectare (2 animals per acre). For this reason alone the forage crop is a very important component of the livestock farm's economy.

Grazed Forests

Thickly canopied forests reduce forage production tremendously, whereas forests whose tree stands are at only 10% of their capacity will allow considerable forage production. Recently the Bureau of Land Management and the Forest Service have developed a permit system, whereby the forest lands can be used for various purposes such as recreation, mining, grazing, and so on. Grazing is allowed particularly on the eastern slopes of the western mountains.

The Forest Service and the Bureau of Land Management determine the number of animal units allowed to graze, the length of grazing time, and fees. All types of livestock use such lands. In some cases livestock are trucked in for the grazing season.

Rangeland

The rangeland is dominated primarily by native vegetation, grasses, grasslike plants, and shrubs that are suitable for grazing. Rangeland includes the natural grasslands, savannas, many wetlands, tundra, and other shrub communities. Some rangelands have been reseeded to a more productive, long-lived, perennial grass, of either natural or introduced forage species.

It is difficult to increase overall production on the rangelands of the United States because low precipitation limits the total forage production. Rangeland needs to be protected from overgrazing, and it must provide sufficient groundcover to ensure proper erosion control. Weeds and brush also must be controlled so that a productive, quality forage can be grown. The cattle carrying capacity of these areas is much smaller than that of cropland areas. Improvement of the rangeland can sometimes be combined with improvement of the wildlife habitat. It is important to maintain and protect the soil and water resources within the rangelands and restore them to their potential production level. Proper grazing of the rangeland includes controls on the number and kinds of livestock and the duration of the grazing period. All of these are important aspects in the management of rangeland.

BENEFITS FROM FORAGES

A few of the more evident and most important benefits will be listed. Forages are adapted to a wide range of soil and climatic environments throughout the United States. This diversity of adaptation in our forage species makes it possible to intensify use of our overall land resources.

Forage grasses along with legumes tend to penetrate the subsoil and improve the overall drainage and soil structure. The fibrous root systems of grass will uniformly penetrate the plow layer more readily, whereas the roots of legumes may penetrate the subsoil and allow for greater penetration of water. A high-quality grass-legume sod will disperse rain drops, retain water on the field, and increase the water percolation.

One of the greatest benefits of adding forage production to cultivated land is forages' ability to prevent soil losses. Compared to a clean-cultivated crop or fallow, forages are from 200 to 2000 times more effective in preventing soil loss. This range in effectiveness results from different soil types, forage management practices, and amounts of rainfall (4). Throughout the growing season, forages protect the soil surface from the beating action of rain more than do row crops. The grass roots, and to a certain extent the legumes, when plowed for the following row crop, will add organic matter and furnish additional protection to the soil surface by binding and holding together the soil particles and increasing the percolation; a continuous row crop system does not provide these benefits (4). Forages will play an even greater role in the future as a means of preventing soil erosion. Public policy will require land currently in row crop production to be returned to forage production in order to reduce soil erosion. This is a serious matter of immense public interest.

Another benefit of forage production is the direct income, or the economics, of growing forages. It is somewhat difficult to put a market value on forage itself except when it is a cash crop. One way to estimate the economic worth of forage production would be on the basis of the cost of purchasing protein and energy. One hundred lb (45.4 kg) of alfalfa hay is equal to 33 lb (15 kg) of corn plus 22 lb (10 kg) soybean oil meal (Fig. 1-5). This will be balanced in both protein and energy. To determine the value of a ton of alfalfa hay, calculate or determine the price of corn and soybean oil meal and equilibrate to the cost of a ton of hay. In recent years, this has varied from $60 per ton to $100 per ton depending upon the current prices of corn and soybean oil meal.

The production cost can be compared to the yearly production cost of corn grown on a comparable soil type, minus the cost of nitrogen. These costs will include the total variable and nonvariable cost of producing the forage crop and harvesting at least three times. In most situations, the annual cost of producing a forage crop will be approximately $30 to $45 less per acre than the cost that would be involved in producing an acre of corn.

There are several reasons why growers seldom look at the economic advantage of growing forage: (1) it is often fed directly to livestock so no monetary value is placed upon it; (2) the producer does not know how to determine the value of a pound of forage; (3) forages are often grown on poor land; (4) forages are not considered a priority item in the overall farming operation; and (5) the producer feels that the labor, handling, and storage of the forage outweigh any economic advantage that forage might have. Recently the

FIGURE 1-5
One hundred pounds (45.5 kg) early bloom alfalfa is equal to the energy and protein in 33 lb (15 kg) of corn plus 22 lb (10 kg) of soybean oil meal.

economic advantages have been presented on the basis of net energy and protein returned by using forages (9).

Hay markets have been developed to help facilitate and distribute hay throughout the United States. These hay markets will continue to develop and improve in the near future. The cost of transportation is probably the greatest factor involved in the marketing of hay.

TRENDS IN FORAGE PRODUCTION

Forage crops and their utilization will become more important in the southeastern and southern United States. The trend in forage production will be toward less total grazing in the humid areas, and more utilization of silage and haylage. More forage hectareage will be harvested by these two methods than has been the case in recent years because they involve less labor than making hay and then removing it from the field. There will be a greater reduction in workforce devoted to harvesting, storing, and feeding forages in the future. At present silage and haylage require more energy than does hay harvest. In order of the energy required to produce the storage material, loose hay takes the least, followed by big bales, conventional square bales, haylage, and silage with the most. The conventional hay baler will decrease in use. Hay is being stored in larger packages than formerly, whether in upright stacks, compact stacks, high-density bales, or large bales. Also, more cubing of forages may occur because this facilitates transportation; cost, however, is a major limitation. More hay markets will be developed in the future. Cattle will be fed longer, utilizing more forage and reducing consumption of grain, which in turn will tie up more of the money being invested. Cattle will be held longer before marketing. The greater demand for red meat with less fat has led to higher marketing weights for cattle, which in turn necessitates the use of more forages in feeding operations (17).

A greater emphasis is being put on year-round grazing in snowfall areas, as well as the utilization of cornstalks and more stubble grazing than has been practiced in the past few years.

Minimum or conservation tillage for the establishment of forage will replace ordinary or customary methods of establishment. There will be more direct seeding of forage crops without the use of companion crops. Emphasis will be placed on higher-yielding cultivars utilizing multipest resistance.

In recent years the energy crisis has created much concern for the average grower, and so more legumes are being utilized with grass forages since the legumes reduce some of the need for nitrogen (9). The future will see much greater emphasis on energy efficiency. There is a great possibility that one will see commercial nitrogen fertilizer being used only on grasses that are grown in marginal areas where legumes are difficult to grow in combination with them, such as reed canarygrass or bermudagrass.

The grower can help reduce the cost of producing animal products by improving forage quality. This will reduce the amount of purchased protein and energy supplements that are being fed directly to livestock.

Forages will play a very important role in soil conservation and water control (4). In recent years, with more regulations being issued by the Environmental Protection Agency, and greater concern being placed upon soil conservation and soil improvement, there will be an added use of forages to help reduce soil runoff, improve water control, and increase overall soil productivity.

SUMMARY

The principal use of forage crops is as feed for ruminant livestock. Forages provide approximately 75% of all the feed units consumed by these animals. Through foods of ruminant livestock origin, forages account for 26% of the food consumed by humans in the United States, making forages by far the most important supplier of food products. Forages are produced on much land that is not suited for the production of cultivated crops, and in areas where erosion control must be implemented to provide the greatest dollar return for the investment through ruminant livestock operations. Forages are labor-intensive, though recent inventions and improvements in machinery and techniques have relieved the labor situation and increased the efficiency of operations.

The contribution of forages to the United States food-supply system is not well appreciated. The production of food by means of the forage-ruminant complex is energy-efficient. Many high-fiber crops, byproducts, and residues are utilized by forage animals who convert them into a human food. A greater emphasis is being placed upon forage quality for growing dairy and beef animals. There appears to be great promise in the area of increasing our overall food productivity through forages. The use of legumes with grasses will reduce some of the requirement for nitrogen fertilizer, the manufacture of which requires fossil fuels. In this way, forages may permit us to increase our overall productivity, perhaps by twofold or more, and still be largely independent of fossil energy.

The potential for producing forages and expanding our food production is immense and largely unrecognized. Forages and forage-producing land probably represent the greatest resources available for increasing food production in the United States. The potential for forage production is greatest in the humid, cool-temperature and the humid, subtropical regions, and less in the arid regions.

QUESTIONS

1 Why is the forage-livestock system considered to be a complex food system?
2 What is the role of forages in the overall feed source for livestock?
3 Of the food consumed by the average U.S. citizen, what is the percentage that has animal origin? How much of the food is provided by ruminants? What percentage of

the total energy and protein consumed by Americans comes from beef or dairy products?

4 Where are the various areas of forage production in the United States? Which region or regions have the greatest potential for increasing the overall production of forages?

5 What is the role of forage legumes in reducing some of the fossil energy used in food production?

6 List and discuss at least five trends related to forage production that are occurring or will occur in the future.

7 Why are forages poorly appreciated in the overall food system?

REFERENCES

1 Blakely, B. D and R. E. Williams. Our grazing land resources. In *Grasslands of the United States,* Iowa State Univ. Press, Ames, IA, pp. 6–14, 1974.

2 Box, T. W. Potential of arid and semi-arid rangelands. In *Potential of the World's Forages for Ruminant Animal Production. Winrock Report,* Winrock International Livestock Research and Training Center, Morrilton, AR, 1977.

3 Brink, R. A., J. W. Densmore, and G. A. Hill. Soil deterioration and the growing world demand for food. *Science,* 197:625–630, 1977.

4 Browning, G. M. Forages and soil conservation. In *Forages,* Iowa State Univ. Press, Ames, IA, 3d ed., pp. 30–43, 1973.

5 Bula, R. J., V. L. Lechtenberg, and D. A. Holt. Potential of temperate zone cultivated forages. In *Potential of the World's Forages for Ruminant Animal Production. Winrock Report,* Winrock International Livestock Research and Training Center, Morrilton, AR, 1977.

6 Council for Agricultural Science and Technology (CAST). *Energy Use in Agriculture. Rep. No. 68,* Ames, IA, 1977.

7 Hanson, A. A. The importance of forages to agriculture. In *Forage Fertilization,* Amer. Soc. Agron., pp. 1–16, 1974.

8 Heath, M. E. Grassland agriculture. In *Forages,* Iowa State Univ. Press, Ames, IA, 3d ed., pp. 13–20, 1973.

9 Heichel, G. H. Agricultural production and energy resources. *Am. Sci.,* 64:64–72, 1976.

10 Hibbs, J. W. and H. R. Conrad. Minimum concentrate feeding for efficient milk production. *World Anim. Rev.,* 15:33–38, 1975.

11 Hodgson, H. J. Food from plant products–forage. *Proceedings of a Symposium on Complementary Roles of Plant and Animal Products in the U.S. Food System,* Nat. Acad. Sci., Washington, DC, Nov. 29–30, 1978.

12 Hodgson, H. J. Forage crops. *Sci. Amer.,* February 1976, pp. 61–75.

13 Hodgson, H. J. Forages, ruminant livestock, and food. *BioScience, 26:625–630, 1976.*

14 Hodgson, H. J. Gaps in Knowledge and technology for finishing cattle of forages. *J. Anim. Sci.,* 44:896–900, 1977.

15 Jacobs, V. E. Forage production economics. In *Forages,* Iowa State Univ. Press, Ames, IA, 3d ed., pp. 21–29, 1973.

16 Reid, J. T. Potential for increased use of forages in dairy and beef rations. *Proc. 10th Research Industry Conference,* Amer. Forage and Grassland Council, Columbia, MO, Feb. 14–16, 1977.

17 Wedin, W. F., H. J. Hodgson, and N. L. Jacobson. Utilizing plant and animal resources in producing human food. *J. Am. Sci.,* 41:667–686, 1975.

CHAPTER 2

BOTANY OF LEGUMES AND GRASSES

Forage crops are primarily divided into two botanical families, the legumes, Leguminosae (Fabaceae) and the grasses, Gramineae (Poaceae) (1, 2).

LEGUMES

There are approximately 500 genera and 11,000 species of legumes; 4000 of them are grown in North America.

Description

The legume derives its name from the type of fruit it produces: a seed-containing pod (Fig. 2-1). Legumes are used for pasture, hay, green manure, ground cover, nectar, and human food.

Legumes are dicotylendons; that is, they have two seed leaves. Their life cycle may be either annual, biennial, or perennial.

Botanical Characteristics

Roots Within the Leguminosae family, most of the species have a taproot system, with lateral roots branching from a main root. The root system is the site for N fixation. Symbiotic bacteria are responsible for the formation of nodules on the roots. These bacteria use N_2 from the soil air and in turn multiply within the nodule. Bacteria in the nodules metabolize N_2 to NH_3 (ammonia) via an enzyme, nitrogenase, which reduces N_2 in the absence of O_2. The ammonia is then combined with organic acid. Once the N is fixed in this way, it is

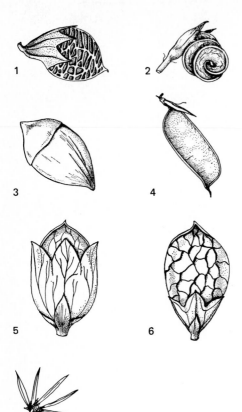

FIGURE 2-1
Different types of legume seed pods:
(1) yellow sweet clover, (2) alfalfa, (3) red
clover, (4) hairy vetch, (5) common
lespedeza, (6) Korean lespedeza,
(7) birdsfoot trefoil.

transported rapidly from the nodule to the rest of the plant. Greater nodule numbers are associated with finer root systems; that is, the more root branches, the more nodules.

Root depth varies within the Leguminosae family, as shown in Fig. 2-2. Rooting depth affects the ability of a legume species to withstand drought or tolerate a high water table. Root pattern influences susceptibility to "heaving"; the more branched the root system, the less heaving.

Stems The stems of legumes vary greatly from species to species in length, size, degree of branching, and succulence. Some stems are very woody whereas others are quite succulent. Rhizomes, or below-ground stems, are found in some strains of alfalfa and other forage species. Stolons, or horizontal stems lying on the soil surface, can be found in white clover and other forage species (Fig. 2-3).

Leaves Legume leaves may be either simple or compound in structure. A simple leaf is composed of one complete structure (Fig. 2-4-*a*1), while a

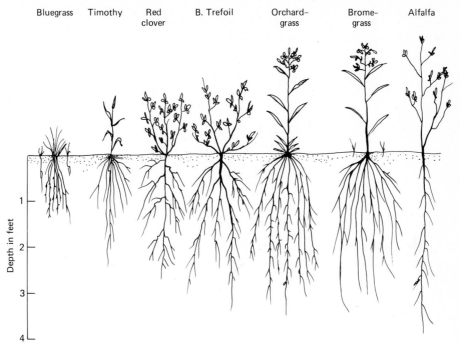

FIGURE 2-2
Rooting depth of various legumes and grasses.

compound leaf has more than one leaflet (Fig. 2-4-*b*2). The leaves are arranged alternately on the stem and may have stipules.

Flowers   A typical legume flower consists of five petals—one standard, two wings, and a keel (Fig. 2-5). The keel consists of two petals, which are more or less united. Within the keel there are 10 stamens; 9 are united and one is separate. The keel also contains the pistil, which consists of the stigma, style, and ovary (Fig. 2-6). Nectar glands are found at the base of the ovary or corolla tube. The corolla tube is formed by the overall joining of the five petals.

The inflorescence, or flowers, are arranged as racemes, as in alfalfa, or as heads, as in red clover. The length of the corolla tube is one factor in determining bee visitations, which ultimately result in pollination, or seed set (Fig. 2-7). Since honeybees are visiting the flowers primarily for honey or nectar, if the corolla is rather long, they become discouraged in their effort to collect the nectar. Generally the flowers of red clover produced early in the year or before the heat of summer are longer, exceeding 12 mm, than those flowers produced later in the year. For this reason, more red clover seed is produced from the second or third growth cycle than from the first; the honeybee can collect nectar

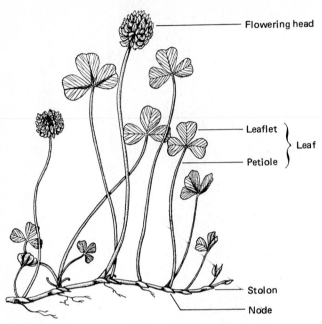

FIGURE 2-3
White clover spreads by means of stolons. Roots may develop at each node of the stolon (aboveground stem). Leaves and flowering heads are the only erect growth.

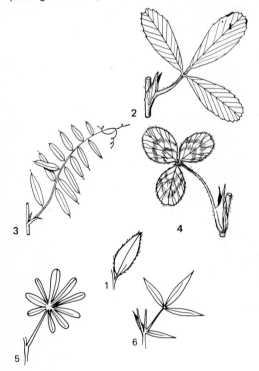

FIGURE 2-4
Types of legume leaves: (1) crotolaria, simple leaf; (2) alfalfa, trifoliate or pinnately compound; (3) hairy vetch, pinnately compound; (4) red clover, true trifoliate; (5) lupine, palmately compound; and (6) birdsfoot trefoil, compound leaf.

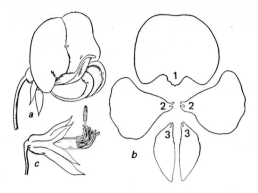

FIGURE 2-5
(a) Typical legume flower. (b) The five petals separated—(1) standard, (2) two wings, (3) two united petals forming the keel. (c) Sepals and ten stamens (nine partially united and one free) and pistil showing the stigma.

more easily from the later flowers. Pollination of red clover occurs when the honeybee visits a flower and, with its head, exerts pressure on the keel petals; the stigma and the anthers are exerted, or "tripped," and make contact with the bee, usually on the posterior part of its head. When the bee removes itself and visits another flower the process is repeated, but in addition, the pollen from the first flower is rubbed off and deposited on the second flower's stigma. Thus pollination occurs.

Pollen must be placed upon the stigma before pollination, or seed set, can occur. Some species, such as red clover, are self-sterile; they must receive pollen from some other plant or no seed will be set. Other species, which may pollinate themselves, are self-fertile.

FIGURE 2-6
Cutaway drawing of an alfalfa flower showing the various parts.

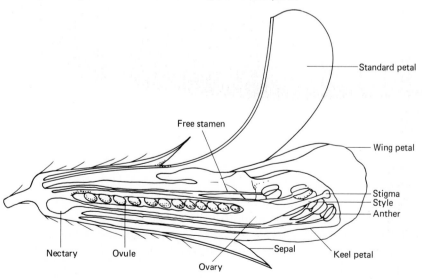

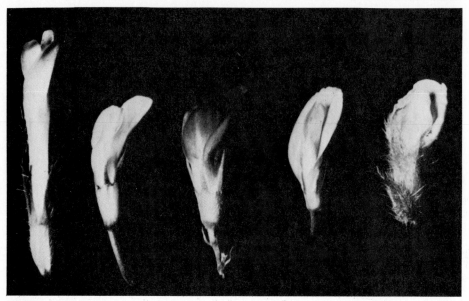

FIGURE 2-7
Close-up of various legume florets: left to right—red clover, white dutch clover, alfalfa, alsike clover, and soybeans.

It is believed that within some species, such as alfalfa, a stigmatic membrane must be ruptured before pollen will germinate on the stigma and grow down the style. These flowers must be tripped so that the stigma membrane is broken. This phenomenon will be explained in Chapter 16, which discusses alfalfa.

Fruit The fruit of a legume is called a pod, which contains one to several seeds (Fig. 2-1). The seeds are attached to the ovary wall. The hilum is the scar where the seed has been detached from the pod. The seed has two cotyledons. The cotyledons store reserve food, which will be used in germination and early seedling growth. Between the cotyledons are the plumule and radicle (Fig. 2-8). The radicle develops into the root system and the plumule develops into the above-ground portion of the plant. Each seed is enclosed in a seed coat.

GRASSES

Approximately three-fourths of the cultivated forage cropland includes grass species. Grasses make up a major portion of the native rangeland. There are approximately 600 genera and almost 5000 species worldwide. In the United States, approximately 150 genera and 1500 grass species are grown (3).

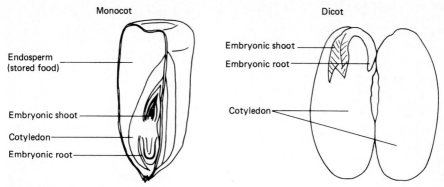

FIGURE 2-8
Legume and grass seed showing the two cotyledons of the legume seed and the one cotyledon of the grass seed, along with the plumule and radicle.

Description

Grasses are monocotyledons. Most of the grasses are herbaceous plants. Grasses are used for pasture, hay, human food, wildlife management, soil conservation, turf, and shelter (bamboo, for example) (4).

Grasses may be either annuals, winter annuals, or perennials. There are several ways to classify grasses. One common method is to classify them by type of growth: bunch type and sod-forming type. Sod-forming grasses spread by forming stolons and rhizomes, whereas bunch-type grasses do not readily spread. Several sod-forming grasses are smooth bromegrass, reed canarygrass, and bermudagrass. Timothy and orchardgrass are examples of bunch grasses.

Bunch-forming grasses usually have more rapid regrowth than do the sod-forming ones. The buds and tillers on the bunch-forming grasses are generated at a higher point in the crownal area than is the case in the sod-forming grasses. In the bunch formers, the tillers, which grow between the leaf sheath and the culm, grow erect. By contrast, the sod-forming grasses produce many procumbent shoots. In these grasses, the tiller grows horizontally through the leaf sheath, thus slowing its growth and ensuring its spreading, sod-forming characteristics. The horizontal tillers eventually elongate into upright shoots.

Another method of classifying grasses is on the basis of their adaptation or response to temperatures. The two main types are the cool-season and warm-season grasses.

The cool-season grasses require a cool, moist climate. In general they are not extremely drought-resistant. They require cool temperatures and long days for floral initiation. Examples of cool-season grasses are smooth bromegrass, orchardgrass, the fescues, the bluegrasses, timothy, and reed canarygrass.

The warm-season grasses require warm temperatures. In general these grasses are more drought-resistant than the cool-season grasses, and can tolerate high temperatures. Many require warm, short days to induce flowering. Several examples of warm-season grasses are bermudagrass, bluestems, johnsongrass, sudangrasses, the millets, and dallisgrass.

Botanical Characteristics

Roots   Grasses, in general, have fibrous root systems consisting of both seminal and adventitious roots. The seminal, or primary, root system develops when the seed germinates, and may persist for a short or extended period of time. These roots are formed at the depth at which the seed is planted. The adventitious root systems form at the lower nodes of the stem. The adventitious root system comprises the major portion of the permanent root system (2).

Stems   Grass stems are made up of nodes and internodes. The node is a solid, enlarged joint on the stem. The internode is the area of the stem between nodes (Fig. 2-9). Stems may be either hollow, pithy, or solid. Leaves arise and have their vascular connections at the node. Lateral buds arise at the axils of the leaves. These lateral buds may become either vegetative branches or flower shoots (1, 4).

Under favorable conditions, a grass plant may have several stems. Most of these stems arise from lateral buds at the lower internodes, and so develop a crownlike area. Tillers are basal, secondary culms.

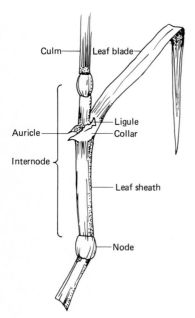

FIGURE 2-9
Parts of grass stem and leaf area.

   Modified stems can form horizontally at the crownal area, creeping along the soil surface. These above-ground creeping stems are called stolons (Fig. 2-10). They have definite nodes, internodes, and meristematic regions, which in turn may give rise to leaves, roots, or upright branches. In addition, some grasses form below-ground horizontal stems called rhizomes (Fig. 2-10). Such species as Kentucky bluegrass, quackgrass, bermudagrass, and johnsongrass possess rhizomes.

   Both stolons and rhizomes enable the grass plant to spread or form a sod. The rhizome in most cases is the overwintering portion that creates a species' perennial nature.

Leaves   A grass leaf consists of a sheath, blade, and ligule (Fig. 2-10). The sheath surrounds the stem above the node. At a particular height or sheath length, there is a region called a collar. This is the junction of the sheath and blade. The leaf blade is parallel-veined and generally flat, narrow, and sessile. At the collar some grasses have earlike appendages projecting from the leaf margin; these are called auricles. A ligule is the thin membranelike appendage at the juncture of the leaf blade and the sheath (Fig. 2-9).

Flowers   Grasses usually have perfect flowers arranged in a spikelet (Fig. 2-10). The number of flowers per spikelet varies from one to many depending on the species.

   The axis of a spikelet is called a rachilla. There are two glumes at the base of a spikelet on opposite sides of the rachilla. These glumes enclose the flowers of the spikelet (Fig. 2-10). The outer glume, the larger one, is called the lemma while the inside one, the smaller one, is called the palea (Fig. 2-10). The lemma usually encloses the palea. Generally there are three stamens but the number may vary. The pistil has a single ovary. The style is branched, forming two feathery stigmas (Fig. 2-10).

   Many of the grasses are cross-pollinated by wind. Only a few are self-pollinated; these include barley, oats, and wheat.

   The spikelets are arranged in several manners. The most common arrangement is the panicle. Within the panicle, the spikelets are borne on branched pedicels (Fig. 2-10). A spike is the simplest arrangement. Spikelets may be borne along an unbranched axis, or rachis. A simple type of inflorescence in which the elongated axis bears flowers on short stems in succession toward the apex is called a raceme; bluestems have this arrangement. A spike has sessile spikelets, as in wheatgrass (Fig. 2-10). Examples of grasses with a panicle arrangement are smooth bromegrass and Kentucky bluegrass.

Fruit   The fruit of a grass is called a caryopsis or kernel (Fig. 2-10). The caryopsis may be permanently enclosed by the lemma and palea as in oats, or it may be free as in wheat.

   The embryo is on the side of the caryopsis next to the lemma. The endosperm comprises the rest of the caryopsis, where the food for germination is stored.

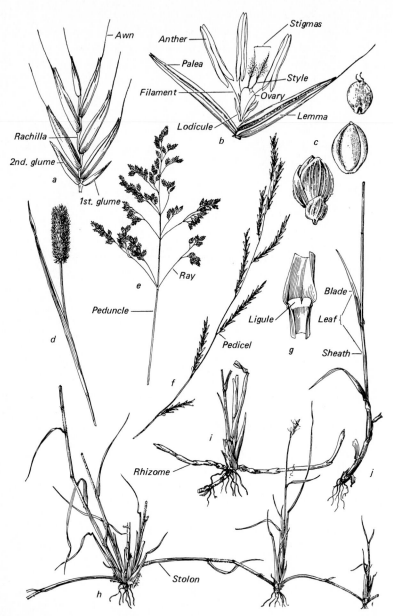

FIGURE 2-10
Characteristic growth of grass plant parts: *(a)* flowers in a spikelet arranged on a central axis enclosed in two empty glumes or bracts. *(b)* The different parts of a grass flower. *(c)* The developed fruit or seed, a caryopsis. The caryopsis is shown successively enclosed in the outer glumes with the lemma and palea both closely adhering and free. *(d)* Spikelets arranged in a terminal spike. *(e)* Spikelets arranged in a panicle. *(f)* Spikelets in a raceme. *(g)* A ligule at the junction of the leaf blade and leaf shoot. *(h, i, j)* Means of propagating or spreading: stolon, rhizome, and bulb, respectively. (*USDA Yearbook, 1948*)

The embryo has a plumule, radicle, and scutellum, or cotyledon (Fig. 2-8). Upon germination the radicle develops into the primary root system, while the plumule develops into the above-ground portion of the plant. The scutellum releases enzymes from its outer area of cells; the enzymes dissolve the food stored in the endosperm.

QUESTIONS

1 Where does the legume gets its name?
2 Discuss the differences between the grasses and legumes.
3 Draw a legume flower and label its parts.
4 Draw a grass flower and label its parts.
5 How are legumes and grass flowers pollinated?
6 Describe the regrowth differences between bunch- and sod-forming grasses.
7 What is the difference between a stolon and a rhizome?
8 Describe the development and placement of the two different root systems of a grass.
9 What is the difference between a spike, raceme, and panicle arrangement of grass inflorescence?

REFERENCES

1 Fernald, M. L. *Gray's Manual of Botany,* American Book, New York, 1950.
2 Gould, F. W. *Grass Systematics,* McGraw-Hill, New York, 1968.
3 Hanson, A. A. Grass varieties in the United States. *USDA Agr. Handbook 170,* rev. 1972.
4 Heath, M. E., D. S. Metcalf, and R. E. Barnes. *Forages,* Iowa State Univ. Press, Ames, IA, 3d, ed., 1973.
5 Hitchcock, A. S. Manual of the grasses of the United States. *USDA Misc. Publ. 200* rev. 1951.
6 Isley, D. The Leguminosae of the north central United States: Loteae and Trifolieae. *Iowa State Coll. J. Sci.,* 25:439–482, 1951.
7 Larkin, R. A. and H. O. Groumann. Anatomical structure of the alfalfa flower and an explanation of the tripping mechanism. *Bot. Gaz.,* 116:40–52, 1954.
8 Leithead, H. L., L. L. Yarlett, and T. N. Shiflet. 100 native forage grasses in 11 southern states. *USDA Agr. Handbook 389,* 1971.
9 *USDA Grass Yearbook Agr.,* 1948.

CHAPTER 3

BIOLOGICAL NITROGEN FIXATION

SYMBIOSIS

The continuous removal of nitrogen resources from the soil and the necessity for higher crop yields have led to an ever-increasing emphasis on means of conserving the limited supply of this element. Approximately 50% of the total agricultural need for nitrogen comes from synthetic and natural fertilizers; the remaining portion must be supplied from the soil reserves and through the biological fixation of atmospheric N_2. A limited number of free-living microorganisms can assimilate molecular nitrogen, whereas no higher plant or animal has the enzyme needed to catalyze the reaction. However, in some instances a symbiosis can become established in which one of the more permanent effects is the acquisition of nitrogen from the atmosphere. The association requires two members, a plant and a microorganism. Such a symbiosis exists between leguminous plants and bacteria of the genus *Rhizobium*.

INFECTION AND NODULATION

Before N_2 fixation can occur, the legume must become associated with the bacterium *Rhizobium*. The plant and the bacteria then live in association with each other. The plant provides an energy source and an ecological niche for the bacteria, while the bacteria provide a source of usable nitrogen for the plant.

There is a sequence of events that occurs before nitrogen is fixed. Bacterial growth seems to be stimulated in the vicinity of legume root hairs. Infection occurs on the root hairs, followed by nodule development and then N_2 fixation.

Stimulation of bacterial growth is probably the result of the exudation of compounds like amino and other organic acids, vitamins, and sugars from legume roots (1). These compounds can be utilized as sources of energy or as growth stimulators.

It has been shown that birdsfoot trefoil exudes a protein that can specifically attract *Rhizobium* bacteria (5). Attraction of bacteria to plant roots, or chemotaxis, has been found in alfalfa, though the specificity is not known (3). If legumes exude proteins that attract *Rhizobium* bacteria, this could be a useful tool to plant breeders for selecting plants that would interact with superior nitrogen-fixing bacterial strains.

Curling of root hairs often occurs before penetration of rhizobia. The amount of curling is a function of the *Rhizobium* strain, the number of infection points, and the plant cultivar (1, 3, 6, 7). Curling is not always required for infection; many infections become established through straight root hairs. After penetration, infection threads develop and the bacteria migrate through the threads to the root cortex. Cell division is stimulated by the bacteria, then the cortical cells enlarge, resulting in nodule formation. The bacteria multiply in the nodule tissue and finally develop the capacity to fix nitrogen (3).

An effective nodule contains a red pigment, leghemoglobin. The effectiveness of a nodule can be determined easily by observing the color of a nodule that has been dissected with a knife or fingernail. Small and numerous nodules, and nodules whose insides are white, green, or hollow, are nonfunctional.

The size and shape of the nodules are different on different legumes. Some nodules are round, whereas others may be lobed. Different legume species of the same genus may have different nodule shapes.

The number of rhizobia in the soil may be reduced as a result of drought or a lack of certain elements. Sufficient calcium, phosphorous, and potassium and a pH between 5 and 8 are essential. Alfalfa has a higher optimum pH for nodulation than does red clover.

Parts of the Nodule

There are two types of nodules, exogenous and endogenous. The exogenous type is found on clover, alfalfa, beans, and peas. These nodules are formed within the root cortex and are connected to the vascular system. They are located primarily on root hairs and the smaller roots. Endogenous nodules are found on peanuts. They are formed within the pericyclic cell layer in the root. Bacteria enter through broken tissue on the root during secondary root emergence. Thus, these nodules are located in the root axis and not on the root hairs.

A nodule consists of four parts (Fig. 3-1). The outermost spongy area is called the cortex and is composed of 4 to 10 layers of thin-walled cells. The cortex is the point of attachment of the nodule to the root. Because this point of attachment is so thin, nodules are easily broken off when the plant is removed from the soil (Figs. 3-2, 3-3).

Just inside the cortex is the vascular system, which surrounds the inner infected area and serves as the transport system into the host plant for the nitrogenous substances and by-products of the rhizobia. The host plant also supplies the rhizobia with nutrients through this same vascular system.

Inside the vascular system, at the furthermost tip of the nodule, is the meristem, or growing point. A terminal meristem produces a slender, long, or

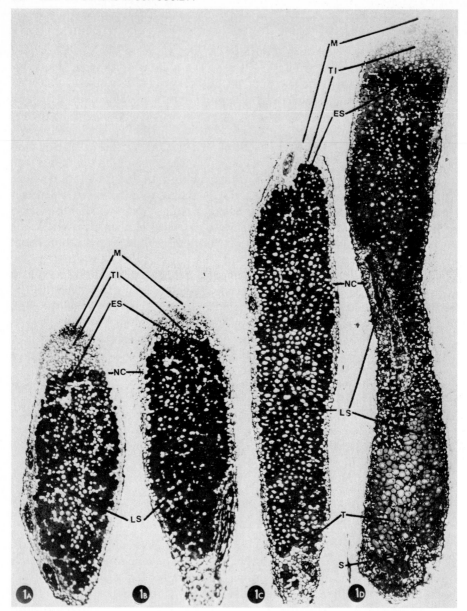

FIGURE 3-1

Light micrograph montage of median longitudinal sections of alfalfa nodules collected from unharvested control plots. Zones illustrated are: Meristem (M); threat invasion (TI); early symbiotic (ES); late symbiotic (LS); transition (T), an area of both mature and senescent bacteroid-containing cells and senescent zone (S). The nodules have vascular tissue (V) outside the central mass of bacteroid-containing cells and are enclosed by a layer of nodule cortex (NC) cells. X64. 1a. Nodule collected on day 1. 1b. Nodules collected on day 4 are increasing in length and bacteroid-containing cells. 1c. Nodule collected on day 10. Nodules continued to increase in length with increases in bacteroid-containing cells. A transition zone with bacteroid-containing cells in early stages of senescence is in evidence at the base of the nodule. 1d. Nodule collected on day 18. Nodules continued to increase in length, and transition zones increased in size. A cluster of senescent bacteroid-containing cells can be seen at the base of the nodule. *(Courtesy C. P. Vance and L. E. B. Johnson USDA, University of Minnesota)*

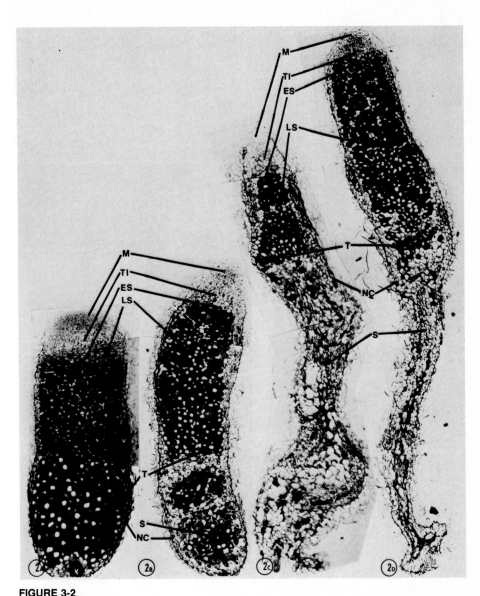

FIGURE 3-2
Longitudinal sections of alfalfa nodules collected from foliage harvested plants. X64. 3a. Nodule collected 1 day after foliage harvest, bacteroid-containing cells in late symbiotic stage of development. Compare to Fig. 3-1a. 3b. Nodule collected 4 days after a foliage harvest. Nodules increased in length, but fewer bacteroid-containing cells. Compare to Fig. 3-1b. 3c. Nodules collected 10 days after foliage harvest. Nodules increased in length, fewer mature bacteroid-containing cells are found in the late symbiotic zones. Compare to Fig. 3-1c. 3d. Nodule collected 18 days after foliage harvest. Nodules increased in length and mature bacteroid-containing cells in late symbiotic stage are increased as senescence is slowed. Transition and senescent zone occupied areas equal to those on day 10. Senescent bacteroid-containing cells have collapsed. Compare to Fig. 3-1d. *(Courtesy C. P. Vance and L. E. B. Johnson, USDA, University of Minnesota)*

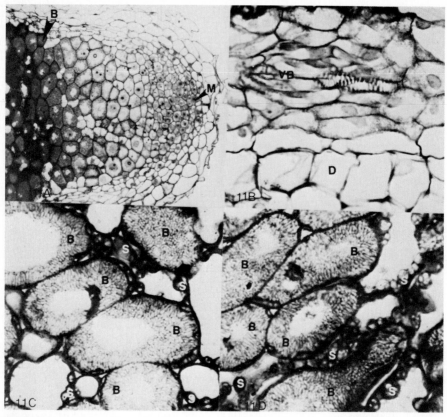

FIGURE 3-3
(a) Alfalfa apical meristem from a nodule 13 days after harvest. *(b)* Vascular bundle adjacent to disorganized and collapsed cortical cells from a nodule 13 days after harvesting. Note both apical meristem and the vascular bundle are still intact and apparently not senescent. *(c)* Bacteroid-containing cells from nodules 4 days after harvest. *(d)* Bacteroid-containing cells from nodules 10 days after harvest. X1200. *(Courtesy C. P. Vance and L. E. B. Johnson, USDA, University of Minnesota)*

multilobed nodule. Within a round nodule the meristem is in a semicircular formation.

The innermost area of the nodule is the bacteroid area. This may be the most important part of the nodule. It is a spongy area, comprising infected cells and vacuoles, where the leghemoglobin and the red color are found.

How the Nodule Functions

With the use of ^{15}N-labeled N_2, N_2 reduction, or fixation, has been traced within the nodule. Nodules metabolizing $^{15}N-N_2$ convert the substrate to $^{15}N-NH_3$, with nearly all of the newly acquired nitrogen being in the form of NH_3. Ammonia is the first stable intermediate (7). The NH_3 is then combined with organic molecules that give rise to amino acids, such as glutamic acid, glutamine, alanine, and aspartic acid. An enzyme, nitrogenase, is found in the rhizobial bacteroids. This enzyme reduces N_2 in the absence of O_2. Nitrogenase must be provided with ATP and a reducing agent in order to function. Leghemoglobin helps convert the free nitrogen to ammonia (7). Fixed nitrogen is transported rapidly from the nodule to the remainder of the plant.

INOCULATION

Inoculation is the addition of infecting rhizobia to legume seed prior to planting in order to achieve N_2 fixation. In the 1930s, farmers inoculated legume seed by adding soil from a field that earlier had the same legume growing in it. They were really adding the rhizobia to the seed by means of the soil. Bacteria will generally live in most soils for two to three years, but not in sufficient numbers to properly cause nodulation (12a).

Rhizobium may live in alkaline soils almost indefinitely. Later, laboratories effectively grew isolated rhizobia on agar, in liquids, or in peat. Such material is now sold dry, in plastic packets. The farmer adds the prepared rhizobia to the seed prior to seeding.

Inoculants are purchased from a seed dealer. One must be certain to buy the right kind of bacteria for a specific legume and to use it before the expiration date. The inoculant must be kept in a cool place prior to use. Follow instructions on the package. Most of the time water or a sugar solution is used when mixing the bacteria with the seed; this lets the inoculant stick to the seed. Inoculated seed should be seeded within 24 hours.

In recent years the seed industry has developed a process to manufacture "preinoculated seed." The legume seed is treated with rhizobia by means of a liquid, then the seed is subjected to an intense vacuum treatment. The rhizobia penetrate the seed coat and are pushed into the seed. With this method the rhizobia are protected and remain viable for a number of months. This seed costs a few cents more per unit of weight than noninoculated seed. Mixing a peat-based slurry of the rhizobia with the seed is another method industry has used. Various sugars, gums, and polysaccharides provide protection and adhesive properties. Recently, a lime-coating treatment has been added to legume seed along with the rhizobia; this increases the seed size and allows for more accurate seeding.

INOCULATION GROUPS

Not all rhizobia infect all legume species. There are seven inoculation groups and several specific strains:

Group

    1 Alfalfa—sweet clover
    2 True clovers
    3 Peas—vetch

  4 Beans
  5 Cowpeas—lespedeza
  6 Soybeans
  7 Lupines

Specific Strains

  1 Birdsfoot trefoil
  2 Sainfoin
  3 Crownvetch

By inoculating legumes, one ensures that the proper *Rhizobium* are present for nodulation and N_2 fixation. If effective *Rhizobium* strains have been used, the nodules that develop from the infection points of the rhibozia provide sufficient N for proper forage growth, and release it when the legume is plowed down. Inoculation is a simple, inexpensive method of obtaining N for both the legume and the subsequent crop.

AMOUNT OF N FIXED

Properly inoculated and nodulated legumes will fix considerable amounts of N. Table 3-1 shows an estimate of the amount of nitrogen fixed annually by several legume species.

TABLE 3-1
NITROGEN FIXED ANNUALLY BY RHIZOBIA IN LEGUMES

Legume	kg/ha	lb/A
Alfalfa	217	194
Ladino clover	201	179
Lupines	169	151
Sweet clover	133	119
Alsike clover	133	119
Red clover	128	114
Kudzu	120	107
White clover	115	103
Lentils	115	103
Crimson clover	105	94
Cowpeas	101	90
Lespedeza (annual)	95	85
Vetch	90	80
Burclover	87	78
Peas	81	72
Soybeans	65	58
Winterpeas	56	50
Peanuts	47	42
Beans	45	40

Source: R. A. Date and J. Brockwell, "Rhizobium strain competitions and host interactions for nodulation," in J. R. Wilson (ed.), *Plant Relations in Pastures*, Commonwealth Scientific and Industrial Research Organization, pp. 202–216, 1978.

TABLE 3-2
NITROGEN FIXATION BY VARIOUS LEGUME SPECIES

Species	N_2 fixed/ha/yr* (kg)		Equivalent cu.ft. natural gas†	% Plant N from atmosphere
	Range	Average		
Pea	52–77	65	3640	23
Soybeans	1–168	103	5768	60
Clover	45–673	183	10248	26
Alfalfa	56–463	197	11144	58

*Multiply by 0.892 to convert to lb/A.
†1 kg N is equivalent to 56 cu.ft. natural gas.
Source: G. H. Heichel. Personal communications. University of Minnesota. 1981.

The N_2-fixing capacity of a nodule is a measure of its effectiveness. Nodule formation does not ensure that N_2 fixation will occur. Plant cultivar-bacterial interaction studies have shown that the N_2 fixation in nodulated alfalfa plants may vary from less than 2.25 mg N/plant up to more than 5 mg N/plant (5).

On a seasonal basis, alfalfa consistently shows greater amounts of N_2 fixation than any other legume species (Table 3-2). The variation in the amount of nitrogen fixed is caused by several factors: (1) differences between individual plants; (2) variations in management practices; (3) climatic differences; and (4) differences in *Rhizobium* strains. As shown in Table 3-2, soybeans and alfalfa are very efficient in the N_2 fixation process: 58 to 60% of their total N is obtained through fixation (5). Recent research has shown that the N_2 fixed by the host plant is highly efficient in N_2 fixation (8a, 9, 10, 11).

Large amounts of energy are required for production of N fertilizer. It is worthwhile to determine how much energy would be required for production of an amount of N fertilizer equivalent to that produced through fixation (Table 3-2). One can see from these data the amount of energy saved when alfalfa is used as a green manure crop.

FACTORS AFFECTING N_2 FIXATION

Several factors affect the amount of N_2 fixed in forage legumes. The primary factors are pH, K levels, nitrate, and management (1a).

Soil pH affects nodulation. As the soil pH decreases, the amounts of nodulation and infection also decrease (1). Increasing soil K levels results in a greater assimilation of N_2 and transport for the synthesis of amino acid (1a). Increasing the level of nitrate decreases nodulation, infection, nodule growth, and N_2 fixation. Increasing levels of soil nitrogen in the form of nitrate causes a decrease in the amount of N_2 fixed, as well as the number and size of nodules (1). There is very little evidence supporting the theory that following cutting or grazing, alfalfa loses its nodules and new ones are formed. It has been found that shading, defoliation, and lodging have an adverse effect on nodules in clover and birdsfoot trefoil (13). Shading and defoliation reduce photosynthesis, thereby reducing the flow of nutrients for nodular growth. Shading and defoliation of red clover and birdsfoot trefoil cause a severe reduction in nodule numbers, while in white clover defoliation does not appear to affect nodule number and may in fact stimulate nodulation (13).

SUMMARY

N_2 fixation in legumes is the result of a complex interaction between the plant and the bacteria *Rhizobia*. The efficiency of this interaction is a function of both the host plant and the bacteria. Recent research has shown that N_2 fixation efficiency in alfalfa may be between 50 to 60%, which means that alfalfa must obtain 40 to 50% of its own N from the soil reservoir (9, 10, 11). It should be possible to increase N_2 fixation and yield in legumes with the interaction of plant breeders, physiologists, agronomists, and pathologists. New, efficient N_2 fixing cultivars are being developed through the cooperative research of plant breeders and physiologists.

QUESTIONS

1 What does symbiosis mean?
2 What is meant by an exogenous nodule?
3 What is the red pigment inside a nodule?
4 Name the parts of a nodule and the function of each.
5 How does N_2 fixation occur?
6 Rank at least five legumes and the relative amounts of N fixed by these legumes if properly nodulated.
7 What is the importance of biological nitrogen in agriculture?
8 Determine the value of an alfalfa green manure crop as it relates to nitrogen fixation.

REFERENCES

1 Allen, O. N. Rhizobiaceae. In *McGraw-Hill Encyclopedia of Science and Technology,* McGraw-Hill, New York, 3d ed., 11:578, 1971.
1 a Barta, A. L. Response of symbiotic N_2 fixation and assimilate partitioning to K supply in alfalfa. *Crop Sci.,* 22:89–92, 1982.
2 Bergersen, F. J. and G. L. Turner. Nitrogen fixation by the bacteroid fraction of breis of soybean root nodules. *Biochem. Biophys. Acta,* 141:507–515, 1967.
3 Burton, J. C. Nodulation and symbiotic nitrogen fixation. In *Alfalfa Sci. and Tech.,* Amer. Soc. Agron., Madison, WI, pp. 229–246, 1972.
4 Burton, J. C. *Rhizobium* culture and use. In H. J. Peppler (ed.), *Microbiol Technology,* Reinhold, New York, pp. 1–33, 1967.
5 Date, R. A. and J. Brockwell. Rhizobium strain competitions and host interactions for nodulation. In J. R. Wilson (ed.), *Plant Relations in Pastures,* Commonwealth Scientific and Industrial Research Organization, pp. 202–216, 1978.
6 Erdman, L. W. *USDA Farmers Bull. 2003,* rev. 1967.
7 Evans, H. J. How legumes fix nitrogen. In *How Crops Grow—A Century Later.* Conn. Agr. Exp. Sta. Bull. 708, pp. 110–127, 1969.

7 a Evers, G. W. Seedling growth and nodulation of arrowleaf, crimson, and subterranean clovers. *Agron. J.,* 74:629–632, 1982.

7 b Fishbeck, K. A. and D. A. Phillips. Host plant and rhizobium effects on acetylene reduction in alfalfa during regrowth. *Crop Sci.,* 22:251–254, 1982.

8 Fottrell, P. F. Recent advances in biological nitrogen fixation. *Sci. Progr.,* Oxford, 56:541–555, 1968.

8 a Hardarson, G., G. H. Heichel, P. K. Barnes, and C. P. Vance. Rhizobial strain preference of alfalfa populations selected for characteristics associated with N_2 fixation. *Crop Sci.,* 22:55–58, 1982.

9 Hardarson, G., G. H. Heichel, C. P. Vance, and D. K. Barnes. Evaluation of alfalfa and *Rhizobium meliloti* for compatability in nodulation and nodule effectiveness. *Crop Sci,* 21:562–567, 1981.

10 Heichel, G. H., D. K. Barnes, and C. P. Vance. Nitrogen fixation of alfalfa in the seedling year. *Crop Sci.,* 21:330–335, 1981.

11 Hoffman, D. and B. Milton. Variation among alfalfa cultivars for indices of nitrogen-fixation. *Crop Sci.,* 21:8–10, 1981.

12 Jensen, H. L and D. Frith. Production of nitrate from roots and root nodules of lucerne and subterranean clover. *Proc. Linnean Soc. N. S. Wales,* 69:210–214, 1944.

12 a Mahler, R. L. and A. G. Wollum, II. Seasonal variation of *Rhizobium meliloti* in alfalfa hay and cultivated fields in North Carolina. *Agron. J.,* 74:428–431, 1982.

12 b Rogers, D. D., R. D. Warren, Jr., and D. S. Chamblee. Remedial postemergence legume inoculation with *Rhizobium. Agron. J.,* 74:613–619, 1982.

13 Vincent, J. M. Environmental factors in the fixation of nitrogen by the legume. In W. V. Bartholomew and F. E. Clark (eds)., *Soil Nitrogen.* Amer. Soc. Agron., Madison, WI, pp. 384–435, 1965.

14 Vincent, J. M. Survival of the root-nodule bacteria. In E. G. Hallsworth (ed.), *Nutrition of the Legumes,* Academic Press, New York, pp. 100–123, 1958.

CHAPTER 4

ECOLOGICAL AND PHYSIOLOGICAL FACTORS DETERMINING FORAGE PRODUCTIVITY

Forages have played an important role in American agriculture since it began. Before animals were domesticated, our lands were used for grazing by wild animals. Early in history, we found that we could cut green forage, cure it, and store it for use during the winter months, or in periods of drought or inclement weather. In fact, forages have helped establish a stabilized agriculture wherever progressive civilizations have developed.

More than half the land area in the United States has forages growing on it. More than 80% of all feed units consumed by ruminants in the United States come from forages (51).

Forages serve many important functions. Besides providing a source of nutrients for ruminants, one of the more important roles for forages is probably soil conservation. Forages protect the soil from both wind and rain erosion. The forage crop serves as a canopy that reduces wind velocity. It intercepts raindrops and distributes rain gently over the soil surface. Forages also catch and hold snow, which in turn serves as a blanket to insulate the forage and as a source of water when the snow melts in the spring. Forages are used in waterways to help reduce water runoff and soil erosion. When forages are plowed down they add organic matter and improve the tilth of the soil. The improved physical condition of the soil helps reduce erosion and increases water infiltration (Fig. 4-1).

ROLE OF LEGUMES

Legumes are important not only in providing food for human and livestock consumption, but also because of other aspects they possess. They provide their own nitrogen when properly inoculated and provide a considerable amount of nitrogen for a companion grass when grown together. In fact, early in history

FIGURE 4-1
An effective forage program involves an efficient utilization of existing ecological and physiological factors.

this was probably the major reason legumes were grown—for a source of nitrogen. When grasses were grown without the addition of manure or some other source of nitrogen, they soon became unproductive. When legume stand densities in sward become less than one-third, we consider them too low to provide enough nitrogen for the companion grass and would then fertilize with nitrogen.

Legumes tend to be higher in protein and minerals than grasses (Table 4-1). For this reason alone, livestock producers have added legumes in their forage mixtures to increase the protein level and reduce the cost of purchasing protein supplement. Many livestock producers feed pure legume forages so that no protein supplement is needed.

Seasonal production is usually more uniform in legumes than in grasses. This is especially true for alfalfa grown in areas of adequate rainfall.

Legumes are usually more productive than grasses (Table 4-2). The exception is when wet conditions exist and high rates of nitrogen fertilizer are added; then some grasses will have higher yields than do legumes. When N is not applied, productiveness of the mixture tends to be correlated with the proportion and kind of legume in a mixture. The problem of management is to maintain the legume as long as possible. Winter hardiness is a greater problem in legumes than in grasses. Phosphorus has both an indirect and a direct effect upon the host legume's growth and metabolic functions related to N_2 fixation. Also, it has direct effects on early root development and on the formation and activity of the nodules. Therefore, phosphorus plays a very important role in maintaining a healthy legume stand and N_2 fixation. Nitrogen fixation demands much readily available photosynthate in the form of sugar, which is vital for photosynthesis, energy transfer, and formation of more sugars.

TABLE 4-1
PROTEIN AND MINERAL CONTENT OF VARIOUS LEGUMES AND GRASSES
(University of Illinois, Urbana, IL)

Species	Stage of growth	Protein (%)	Ca (%)	P (%)	K (%)	Fe (ppm)	Mn (ppm)	Cu (ppm)
						Minerals		
Alfalfa	Bud	23.3	1.47	0.34	1.61	184	50	6.8
	1⁄10 bloom	17.9	1.41	0.28	1.42	154	52	5.8
Red clover	Late bud	20.5	1.57	0.28	2.01	164	50	9.2
	1⁄10 bloom	15.9	1.55	0.22	1.36	173	47	9.5
Ladino	1⁄10 bloom	21.8	1.55	0.28	1.96	239	48	7.4
Smooth bromegrass	Boot	15.2	0.40	0.38	2.60	238	123	9.7
	Headed	10.1	0.20	0.17	2.08	40	76	3.0
Orchardgrass	Boot	14.2	0.35	0.27	2.60	185	155	6.0
	Headed	9.2	0.22	0.20	2.50	35	118	1.0
Timothy	Boot	14.5	0.34	0.27	2.40	170	120	7.4
	Headed	7.4	0.20	0.15	1.90	65	38	1.0

In comparison to grasses, legumes usually have more seedling vigor, which enhances rapid stand establishment (Table 4-3). Legumes tend to dominate grass-legume mixtures during the seeding year and the first year following establishment (Table 4-4). Since livestock selectively graze legumes, one should be alert for a potential bloat problem when grazing a legume-grass mixture for the first two years. With a higher percentage of legumes in the sward, cattle may

TABLE 4-2
THIRD HARVEST DRY MATTER YIELDS OF VARIOUS LEGUME-GRASS
MIXTURES AT END OF 3 YEARS OF PRODUCTION
(Northern Illinois Experiment Field, Shabbona, IL)

	3d harvest total yield (kg/ha)	% legume or grass	Yield (kg/ha)	
			Legume	Grass
Legume in mix*				
Alfalfa	1121	86.4	964	157
Red clover	326	58.5	168	158
Ladino	673	77.2	516	157
Grass in mix†				
Timothy	818	8.4	751	67
Smooth brome	751	10.0	673	78
Orchardgrass	796	37.0	505	291

*Each legume grown with timothy, smooth bromegrass, and orchardgrass.
†Each grass grown with alfalfa, red clover, and ladino.

TABLE 4-3
SEEDLING VIGOR RATINGS OF VARIOUS FORAGE
SPECIES
(University of Illinois, Urbana, IL)

Species	Seedling vigor*
Alfalfa	1.2
Red clover	1.3
Ladino clover	1.9
Timothy	3.1
Orchardgrass	3.3
Tall fescue	3.4
Smooth bromegrass	3.6
Reed canarygrass	4.1

*1=vigorous, 5=poor vigor

graze the more succulent legume species, and skip some of the grasses, especially during the warmer seasons.

ROLE OF GRASSES

Grasses, like legumes, have specific functions. As discussed earlier, soil conservation is a major role of forages. Grasses, due to their fibrous root system, are more effective than legumes in reducing soil erosion.

In a grass-legume sward, grasses reduce bloat hazard. This is a definite advantage when grazing is the means of harvest. In some years, problems with bloat may occur when a sward shifts to legume predominance. For example, the addition of manure or lime to a sward, plus considerable rainfall, could shift the sward stand to a high percentage of ladino clover almost immediately. However,

TABLE 4-4
PERCENT LEGUMES ON SEPT. 30 IN A STAND SEEDED
AS A 50% LEGUME/ 50% GRASS MIXTURE
(University of Illinois, Urbana, IL)

| Species* | Year | | | | |
	1	2	3	4	5
Alfalfa/orchardgrass	62	57	50	49	42
Alfalfa/smooth brome	65	61	56	43	40
Red clover/timothy	61	56	52	31	22
Red clover/smooth brome	62	51	31	20	18
Ladino/orchardgrass	60	55	48	47	42

*P_2O_5 and K_2O were applied annually in the fall according to the amount removed by the harvest.

under continual high rainfall and high harvest height, a shift to grasses would occur.

Grasses aid in the curing of hay. Because of their hollow stems and long fiber cells, grasses tend to make a more fluffy windrow when cut and crimped than do legumes. Legumes have more solid stems and short fiber cells, and they tend to mat down when cut, crimped, and windrowed. This reduces the air flow so the hay takes a longer period to cure.

Grasses make better silage than do legumes alone. Since grasses have a lower mineral content than legumes do, the silage process is aided because a desired pH can be obtained more easily. It is very easy to obtain a pH of 4.0 to 3.6 in grass silage, while in pure legume silage it is difficult to get a pH of 4.2 or lower.

Grasses and legumes help spread the risk in establishment, but when wet conditions prevail at the time of seeding, the grasses tend to germinate faster and quickly develop a fibrous root system. This alone may save soil and the companion legume seed from washing away. Grass seedlings are also more tolerant of frost than are legume seedlings.

As the stand becomes older, grasses tend to dominate. Legumes are generally not as long-lived as grasses because legumes are more susceptible to crown and root diseases and winter injury (Table 4-4).

ECOLOGICAL FACTORS

Ecological factors determine the productiveness of forage crops and pastures. Ecology is the branch of biology that deals with the mutual relationship between organisms and their environment. Mixtures of grasses and legumes are usually grown for several years and harvested by animals or man. The overall system of living organisms within the nonliving environment is called an ecosystem (51, 52).

The ecosystem comprises four components. These are: (1) abiotic substances; (2) producer organisms; (3) consumer organisms; and (4) decomposer organisms (30).

The nonliving components of our environment—water, sunlight, oxygen, carbon dioxide, soil, organic matter, and plant nutrients—comprise the abiotic substances. These components may be the most important, or limiting, factor in the ecosystem.

The producer component is the green plant community. Among the forages, this includes primarily grasses, legumes, and grassy and broadleaf weeds. The green plants, through photosynthesis, manufacture their own food, which in turn is consumed as food by the consumer organisms.

Animals are the consumer organisms. Those that feed entirely upon green plants are called the primary consumers, or herbivores, and those that feed upon the primary consumers are called the secondary consumers, or carnivores. Examples of the primary consumers are cattle, sheep, horses, rabbits, mice, birds, and insects, and the secondary consumers include dogs, cats, spiders, and snakes, to name only a few. Humans may be considered both primary and secondary consumers.

The decomposer organisms include two groups—microflora and fauna. The final step of the cycle is death and breakdown of the plant tissue. Probably the most important decomposers are dung beetles and earthworms. Upon their death and decay, along with the death and decay of plant life, chemicals are released and may then be recycled. The decomposers play a very important role in our ecosystem; without them, the earth would soon be covered with waste.

Forage management involves all aspects of the ecosystem: its nonliving components as well as its living means of production, consumption, and eventual decomposition. People must manage each phase of the ecosystem so that it is maintained at a very productive level. Nutrients must be applied to ensure excellent plant growth. Also, the addition of a legume into a grass sod generally increases production and reduces the need for large amounts of N fertilizer (54). The selection of highly productive cultivars is very important. Some cultivars are more productive under grazing conditions than when harvested for hay. Therefore, by combining the proper species, the proper cultivars, and the proper grazing animal one can develop a very efficient, productive ecosystem.

PHYSIOLOGICAL FACTORS

The management of forages varies with the region or area of the country in which the crop is grown. The environment plays a very important role in determining plant growth. Environmental factors influence both physiological reactions and growth responses. The major environmental factors influencing plant growth are light, temperature, moisture, and CO_2.

Light

Light affects plant growth through its influence on energy production by means of photosynthesis. It also influences structural development and seed production. One can discuss three aspects of light: (1) quality; (2) intensity; and (3) duration.

Quality Light quality refers to the wavelengths of light rays. The entire spectrum of radiant energy ranges from very short cosmic waves to very long electric and radio waves (Table 4-5). Most of the wavelengths reaching earth from the sun are from 300 nm to 2600 nm. Plant development is best under the full spectrum of sunlight rather than just one portion of the spectrum. Ultraviolet wavelengths may kill plants or retard their growth, whereas plants under infrared waves usually grow continuously, as in darkness.

There is a phytochrome system present in most of the higher plants, which allows them to respond to light quality. This is often referred to as the red ⟷ far-red reaction. It was first observed in the germination of lettuce seed (23, 37).

TABLE 4-5
SPECTRUM OF ELECTROMAGNETIC WAVELENGTHS

Light wavelengths (nm)	Classification
<0.0001	Cosmic
0.0001–0.01	Gamma
0.01–10	X-rays
10–390	Ultraviolet
390–760	Visible
760–100,000	Infrared
>100,000	Electric and radio

The red light (R) wavelength is most active at around 660 nm and the far-red (FR) wavelength at around 730 nm. It is thought that the R⟷FR reaction is caused by a phytochrome, a pigment, that exists in two forms. One form absorbs the red light, which changes the pigment to the second form. Then the second form absorbs the far-red light, which changes it back to the red-absorbing form, and thus renders it biologically inactive.

Quality of light available appears to be more important than light intensity (46). Under field conditions, type and depth of canopy influence both intensity and quality of light at the soil surface. Also, the greater the quantity of light available, up to the highest values in the field, the better the plant growth (10, 11, 23, 37).

Intensity   Under full sunlight in the summer, light intensity reaches levels between 85,000 and 110,000 1x (7900 to 10,200 ft-c) (23, 36). It is very difficult to duplicate in a growth chamber or greenhouse the light intensity and quality that occur during the summer in the field.

Forage plants grown under high light intensity tend to have smaller leaves, shorter internodes, shorter overall height, and larger root systems than do plants grown under low light intensity. As the light intensity increases, transpiration increases, which may limit plant growth or cell division and enlargement.

Most forage species are sun plants rather than shade plants. They require a high light intensity for maximum growth, though some species differ in their response to variations in intensity. Orchardgrass will grow under a lower intensity than will timothy or smooth bromegrass. Of the legumes, red clover will produce more dry matter under lower light intensity than will alfalfa, and alfalfa will produce more dry matter than will birdsfoot trefoil (13, 14). Legumes are generally more sensitive to light intensity than grasses are. Therefore, competition for light is very important when forages are being established along with a small-grain companion crop, or when forage species are grown together.

Lowered light intensity will result in production of less dry matter and a reduction of root growth. When the leaf canopy has developed to completely intercept the light available, and when no other factor limits growth, then production cannot be increased. Therefore, a full leaf canopy is needed to give maximum dry matter production (7).

It has been shown that individual leaves of most forage legumes and cool-season grasses are light-saturated at lower light intensities than are tropical grasses (11). Light saturation of cool-season grasses will occur at around 20,000 to 30,000 1x, while tropical grasses will become light-saturated around 60,000 1x (45). Canopy photosynthesis does not exhibit light saturation. The conversion of light energy is around 5 to 6% for tropical grasses, but less than 3% for the cool-season grasses. Therefore, tropical grasses have a greater potential photosynthetic rate.

Tropical grasses also have no photorespiration and can reduce atmospheric CO_2 to less than 5 ppm in a closed atmosphetre (11). Temperate grasses have photorespiration and can reduce CO_2 only to about 50 to 60 ppm. Another biochemical difference found between tropical and temperate grasses is the carbon pathway in photosynthesis. The tropical grasses appear to fix their carbon initially into C-4 compounds as compared to C-3 compounds for the temperate grasses (11). The C-4 pathway in photosynthesis is a more efficient manner of fixing carbon than is the C-3 pathway.

Recently many researchers have investigated different ways and means for increasing the efficiency of photosynthesis and its overall effectiveness. Plant breeders and plant physiologists are attempting to develop plants that will better utilize greater amounts of solar energy in the photosynthetic process. Genetic engineering is helping develop new lines that may prove successful. The final product will mean a greater supply of food for both humans and the food-producing aninals that serve humanity.

Duration   Duration of the photoperiod, or daylength, influences vegetative growth and flowering (10, 11, 23, 37). It is really the length of the dark period, rather than the light period, that controls the response of a plant (57).

Plants are classified according to their flowering response to various lengths of daylight. Short-day plants flower only within a certain period of short daylengths of light (long nights). Sorghums, sudangrass, and bermudagrass are short-day plants. The reverse is true for long-day plants; they flower only within a certain period of long daylengths (short nights) (3). Examples of long-day plants are alfalfa, red clover, sweet clover, birdsfoot trefoil, orchardgrass, smooth bromegrass, and timothy. There are also plants classified as day-neutral. These plants flower regardless of the length of the photoperiod. Vetch is a day-neutral plant. Plant breeders have essentially made most grain species day-neutral but have not done so with many forage crops. Temperature requirements are closely related to light requirements for flowering.

Daylength varies considerably with the seasons at any latitude except at the

TABLE 4-6
HOURS OF DAYLENGTH FROM SUNRISE TO SUNSET AT VARIOUS LOCATIONS IN THE
NORTHERN HEMISPHERE

Location	Degrees latitude	Hours of daylength			
		Mar. 21	June 21	Sept. 21	Dec. 21
Equator	0°	12.1	12.1	12.1	12.1
Mexico City	20°N	12.1	13.2	12.1	10.9
New Orleans	30°N	12.1	14.1	12.1	10.2
Champaign, Ill.	40°N	12.2	15.0	12.2	9.3
Winnipeg, Man.	50°N	12.2	16.4	12.3	8.1
Seward, Ala.	60°N	12.3	18.9	12.4	5.9
North Greenland	70°N	12.5	24.0	12.5	0.0

equator. As shown in Table 4-6, daylengths vary at different latitudes and at different times of the year.

Temperature

All physiological processes are influenced by temperature (37). Temperature limits the extremes where a given cultivar can be grown. Such limits might be that (1) the winter temperature is too low, causing injury or death of dormant plants; (2) the temperature during the growing season may be too high or too low, which may injure or kill growing plants or cause improper growth and development; (3) there may be inadequate periods of favorable temperature for proper development and maturity; and (4) the temperature may favor the development of a destructive insect or disease.

Most plants grow in a temperature range of 0°C (32°F) to 50°C (122°F), with an optimum temperature for each species. Usually a seedling has an optimum temperature range lower than that in the plant's later stages of growth. Roots have an optimum temperature lower than the aboveground parts do (46).

There are also optimum day and night temperatures for plants. Most of the temperate plants have optimum daytime temperature requirements higher than their nighttime requirements. Thermoperiodicity refers to a plant's different responses to cyclic day and night temperatures. Most temperate grasses have an optimum temperature for growth around 20°C (68°F), but will actually grow below this temperature. Tropical grasses have an optimum temperature for growth around 29 to 32°C (85 to 90°C) but grow very little below 16°C (60°F) (11) (Fig. 4-2).

Increasing temperature hastens maturity in most grasses and legumes. As the temperature rises, digestibility of the forage and the nonstructural carbohydrate percentages decrease, but generally the mineral and protein percentages increase (46). Therefore, temperature or time of the year at which the harvest is made influences the feeding value (16, 27).

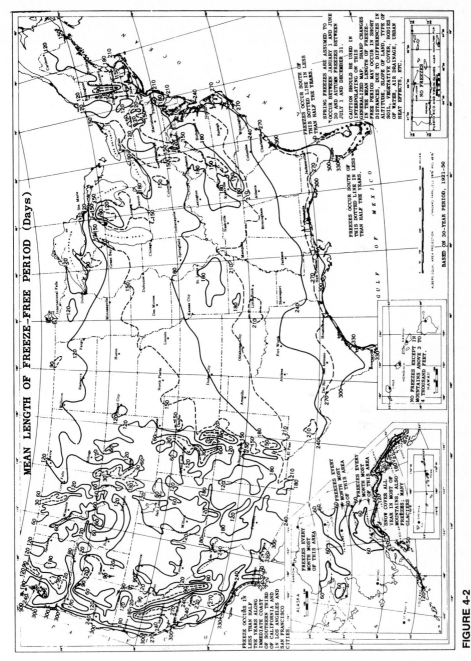

FIGURE 4-2
The mean length of freeze-free period for the United States.

Moisture

Plant growth depends upon adequate soil moisture (24, 36, 53). Several terms refer to the amount of soil moisture. The water held by the soil against the downward pull of gravity is called "field capacity." When plant roots can no longer absorb water from the soil, this is called the "wilting percentage" or "wilting point." The water that remains below the wilting percentage is not available for plant use. The soil moisture between field capacity and wilting point is available for plant use.

Very little of the water that passes through a plant is actually used in the various physiological processes, even though a lot of water passes through a plant in its life cycle. A large amount of water is lost by transpiration to the atmosphere through the stomata (Table 4-7). Various plant species are more efficient than others in water use. The water requirements for a given species may vary considerably under different climatic conditions (Table 4-8) (12, 17a).

Water is very important for normal and productive plant growth. The critical point is the water status within the plant tissue. The amount of soil moisture available for plant growth and the transpiration losses from the leaf surfaces affect the water status within the plant. A loss of water within the plant will affect its physiological processes. Two major physiological processes that could be affected are cell expansion and cell division. Cell expansion is probably the most important: Lack of water causes stomata closure and wilting of plants, which reduces CO_2 entrance and limits photosynthesis.

TABLE 4-7
TRANSPIRATION RATIOS OF VARIOUS CROPS
(Lb Water per Lb Dry Matter)

Crop	5-yr. average
Alfalfa	858
Smooth bromegrass	828
Sweet clover	731
Red clover	698
Crested wheatgrass	678
Soybean	646
Oats	635
Hairy vetch	587
Cowpeas	578
Cotton	562
Barley	521
Wheat	505
Corn	372
Blue grama grass	343
Millet	287
Sorghum	271

Source: H. L. Shantz and L. N. Piemeisel, "The water requirement of plants at Akron, Colo.," *J. Agr. Res.* 34:1093–1190 (1927).

TABLE 4-8
WATER-USE EFFICIENCY OF VARIOUS FORAGE SPECIES
(Dry Matter Produced per Evaporation Transpiration Loss)

Species	Dry matter per evaporation transpiration (kg/ha/cmH₂O)
Blue grama	263 a*
Thickspike wheatgrass	170 b
'Latar' orchardgrass	169 bc
'Fawn' tall fescue	167 bc
Crested wheatgrass	165 bc
Western wheatgrass	156 bcd
Slender wheatgrass	154 bcd
'Travois' alfalfa	147 bcd
'Ladak' alfalfa	141 bcd
Alsike clover	141 bcd
'Vernal' alfalfa	138 bcd
'Remont' sainfoin	137 bcd
'Greenor' intermediate wheatgrass	136 bcd
'Regar' smooth bromegrass	135 bcd
'Freemont' alfalfa	133 cd
'Garrison' creeping foxtail	130 d
'Team' alfalfa	122 d
'Lutana' cicer milkvetch	119 d
'Dawson' alfalfa	115 d
Green needlegrass	76 d

*Column values followed by the same letters do not differ significantly.
Source: M. L. Fairbourn, "Water use by forage species," *Agron. J.* 74:62–66 (1982).

Carbon Dioxide

Carbon dioxide is very important in photosynthesis. It is another environmental factor that cannot be controlled to a great extent. The breakdown of CO_2 into carbohydrates is important in the quality of harvested forage and the over-wintering process due to carbohydrate storage.

As discussed under light intensity, the physiological processes using CO_2 are related to photorespiration and the photosynthetic pathway that fixes carbon into C-3 and C-4 compounds.

FOOD RESERVES

Food reserves result from photosynthesis and the translocation of soluble carbohydrates. Forage crops store available carbohydrates in various plant parts (29). The primary storage organ for alfalfa and red clover is the root; for ladino clover and bahiagrass, the stolon; for orchardgrass and tall fescue, the stem base; for smooth bromegrass and reed canarygrass, the rhizomes; for timothy, the corm. The food reserves are used for regrowth after harvest and for new growth in the spring. Soluble carbohydrates are also used to sustain the plant during winter months or dormancy, and they assist in developing tolerance for cold and heat.

The legumes store their food reserves primarily in the form of starch. Tropical and subtropical grasses also store starch, whereas the temperate grasses accumulate fructosans in their vegetative plant parts (43).

SEASONAL TRENDS OF FOOD RESERVES

Forage plants go through a cyclic pattern where soluble carbohydrates are used for growth and again when they are translocated and stored (Fig. 4-3). This cycle is repeated after each harvest—carbohydrates are used during early regrowth and then stored after some regrowth has occurred (Fig. 4-4). It is essential that a high level of food reserves be maintained to ensure rapid, vigorous, and productive regrowth. A similar pattern is found in both legumes and grasses, but it may be influenced by climatic conditions and by the morphological and growth behavior of the species (Fig. 4-5).

As shown in Figure 4-3, a cyclic food reserve pattern occurs in alfalfa. At a given point during the spring season, the reserves of soluble carbohydrates begin to decrease due to the spring growth (or after harvest) until enough top growth is produced so that an excess of carbohydrates is being manufactured and translocated back into the taproot and crown. In alfalfa the decrease continues

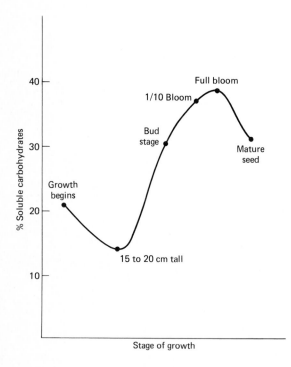

FIGURE 4-3
Production cycle of soluble carbohydrates in alfalfa in the spring to mature seed at Urbana, IL. *(After Graber et al. 1927 45.)*

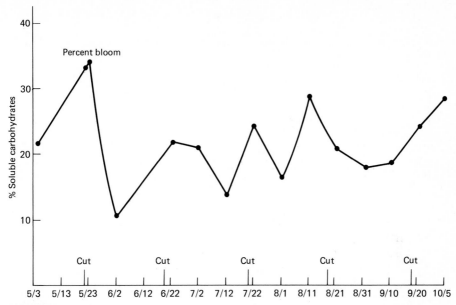

FIGURE 4-4
Production cycle of soluble carbohydrates in 'Saranac' alfalfa roots, harvested every 30 days.
(University of Ill., Urbana, IL)

until around 15 to 20 cm (6 to 8 in) of top growth is produced; then the reserves increase until full bloom and seed set begin. Food reserves are then depleted again for the development of seed. The cycle then restarts, with reserves decreasing until enough regrowth produces an excess of soluble carbohydrates. In grasses the lowest point of the cycle is at about the time of stem elongation. The maximum level of food reserves for most forage crops is usually at maturity or when stem elongation has ceased.

Food reserves are influenced by environmental conditions. Generally speaking, warm weather, limited moisture, and considerable sunlight will hasten development, while cool temperatures, plenty of soil moisture, and cloudy conditions will delay development. In Nevada, it was found that 'Moapa' alfalfa reached 10% bloom in 20 days under a hot 33/27°C (91/80°F) temperature regime but took 42 days under a cool 23/4°C (75/39°F) regime (17). Similarly, in Illinois, it was found that 'WL303' alfalfa grown under a hot 34/27°C (93/80°F) temperature regime bloomed in 22 days, but it took 41 days to bloom when grown in a cool 25/5°C (77/41°F) regime. Therefore, more food reserves may be used under cool conditions than under warmer, more normal temperatures.

Some species, such as birdsfoot trefoil, do not translocate soluble carbohydrates during the warm summer months. Therefore, they do not survive close cuttings as does alfalfa. Once these species' reserves are used for growth in the early spring, little is stored until the advent of cold fall months (21, 47, 48).

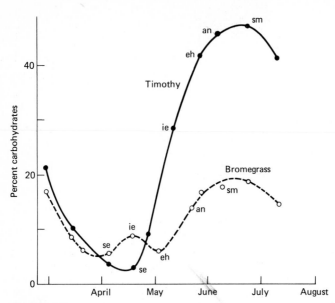

FIGURE 4-5
Production cycles of soluble carbohydrates in various grass species at Madison, WI *(Drawing courtesy of Wis. Agr. Exp. Sta.)*

Therefore, a taller cutting height is required so that these plants have enough leaves remaining to furnish the food and energy needed for regrowth.

INFLUENCE OF HARVESTING

As a result of some species' morphological characteristics and regrowth patterns, some forage species can be harvested frequently, such as ladino clover and bluegrass. Many species need a definite rest period before the second harvest can be taken. The rest periods are tied to the food reserves and the areas of storage. Species needing a certain rest period include alfalfa, red clover, sweet clover, orchardgrass, smooth bromegrass, timothy, tall fescue, and reed canary-grass.

Rate of recovery is directly related to food reserves. If the plant is low-growing, and harvesting does not remove all of the leaves, regrowth may be fairly rapid. Ladino clover has a stoloniferous growth while Kentucky bluegrass has rhizomes. The regrowth of both ladino clover and Kentucky bluegrass is rather rapid. Food reserves are stored in the stolons and rhizomes, and the leaves develop near the soil surface, thereby aiding regrowth and allowing a lower cutting height. By contrast, bunch grasses, such as tall fescue and orchardgrass, store their food reserves in the lower stem and sheath area, and have leaves that develop at a slight height above the soil surface. These species need a definite rest period before leaves can form and begin manufacturing their own energy.

A key adaptation for forage species' survival following defoliation is the location of shoot meristems near the soil surface, as in bunch-type grasses, so that both the tillers and expanding leaf blades continue growth. Bunch grasses, whose meristems are elevated by internode elongation, may be slow to regrow after harvest because new shoot growth depends on initiation of activity in axillary buds whose prior development may have been minimal as a result of apical dominance effects. Grasses that have many internodes near or below the soil surface have many sites for tiller regrowth (6, 16). Grass species that have shoot apices somewhat above the soil surface may be damaged or killed if the mowing height is too low. There is also a critical period when the shoot apices are at the cutting height and regrowth is greatly reduced (6, 38). As an example, the regrowth of smooth bromegrass and timothy may be greatly reduced when harvested too early, or when grown with alfalfa and harvested when the alfalfa is at the early flower stage (33, 44). Selection of earlier-developing cultivars of smooth bromegrass would greatly reduce this danger. When smooth bromegrass and timothy are in the stage of internode elongation, until flower emergence, the shoot apices are above the cutting height and levels of soluble carbohydrates are low (Fig. 4-5). There appears to be an apical dominance present in both timothy and smooth bromegrass. Orchardgrass is a bunch-type grass, but can be harvested at almost any stage of growth without reducing the rate of recovery (40).

CRITICAL FALL HARVEST

Many studies have been conducted to determine the last date to harvest forages in the fall without causing winter injury or winter kill (13, 19, 44, 56). Again, food reserves are the basic consideration. There must be enough time between the last harvest and a killing frost so that sufficient food reserves are stored to bring the plant through the winter and have enough food reserves left for excellent spring growth. Plants need enough top growth before fall dormancy so that enough soluble carbohydrates can be produced. In the northern portion of the United States a general recommendation for cool-season species is that at least 40 to 50 days of fall regrowth at last harvest precede a killing frost (8, 21, 35). This rule does not hold true for the west or southwest. The last harvest occurs when fall dormancy occurs. Generally one should not take a last harvest (after a killing frost) of legumes the year of establishment. An older, well-established stand of legumes may be harvested after a killing frost with little damage, providing there is excellent soil drainage. Regardless, a higher stubble height is recommended for several purposes: (1) it allows some insulation for plants, (2) it helps catch snow, and (3) it reduces ice damage (kill) in the spring when stubble can extend up through the ice sheets (Fig. 4-6).

TEMPERATURE TOLERANCE

There are many differences between forages' abilities to tolerate cold (Table 4-9) (8, 21, 35, 49, 50). Cold tolerance within plants is brought on by a shorter

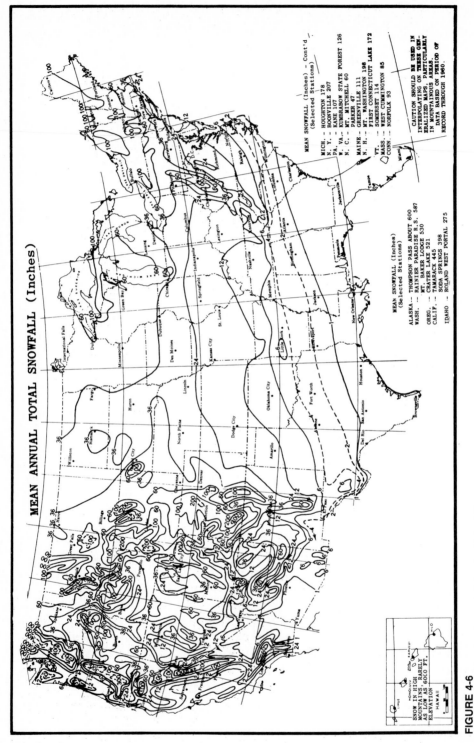

FIGURE 4-6
The mean annual total snowfall for the United States.

TABLE 4-9
ORDER OF RESISTANCE OF SOME FORAGE SPECIES
TO COLD TOLERANCE*
(In Order of Resistance)

Legumes	Grasses
Sweet clover	Crested wheatgrass
Alfalfa	Western wheatgrass
Red clover	Smooth bromegrass
Crownvetch	Reed canarygrass
Birdsfoot trefoil	Tall fescue
Ladino clover	Orchardgrass

*Various cultivars within a species may change the order.
Sources: R. J. Bula and D. Smith, "Cold resistance and chemical composition in overwintering alfalfa, red clover, and sweet clover," *Agron. J.* 46:397–401 (1954); G. A. Jung and D. Smith, "Trends of cold resistance and chemical changes over winter in the roots and crowns of alfalfa and medium red clover," *Agron. J.* 53:359–366 (1961); G. A. Rogler, "Response of geographical strains of grasses to low temperatures," *J. Amer. Soc. Agron.* 35:547–559 (1943); O. C. Ruelke and D. Smith, "Overwintering trends of cold resistance and carbohydrates in medium red, ladino, and common white clover," *Plant Physiol.* 31:364–368 (1956).

photoperiod and cool temperatures in the fall. In the case of alfalfa, cold tolerance is also increased due to alternating day/night 20/5°C (68/41°F) temperatures and high light intensity, which ensures excellent photosynthesis (57).

Fertility status is also important for good cold tolerance. Excess N, along with moisture, increases fall growth and reduces cold tolerance (1). Potassium aids in developing cold tolerance (20).

There are also differences between forages' heat tolerance (18, 25). It has been found that warm-season grasses, such as bermudagrass and buffalograss, have more heat tolerance than do cool-season grasses.

High soil temperatures appear to have a more adverse effect than do high air temperatures on plant growth (24, 36). High air temperatures are more detrimental during anthesis. Air and soil temperatures interact in their effects, so that lower soil temperatures are beneficial at higher air temperatures, probably because they conserve photosynthates. At low air temperatures, slightly higher soil temperatures are an advantage, promoting growth and development of the crown (17, 36).

There are cultivar differences in cold and heat tolerance. Plant breeders are beginning to work more closely with plant physiologists to determine the mechanisms of tolerance.

The adverse effects of high temperatures may be alleviated by selecting better-adapted cultivars, avoiding mismanagement such as overgrazing or cutting too frequently, avoiding soils subject to waterlogging, avoiding excess competition for light from other species, and minimizing root and crown diseases. Keep in mind that various agronomic management practices that maximize plant growth may be contrary to production of quality forage, so a compromise may be necessary.

QUESTIONS

1 What is an ecosystem? Give an example of how an ecosystem functions.
2 Name the four elements of a forage ecosystem. Describe how each element ties together to form a cyclic pattern.
3 What are the major environmental factors influencing a forage group? Which environmental factor limits the extremes where a species may grow?
4 Which wavelengths of light are most important to plant growth?
5 Why is light intensity so important in forage growth and establishment and maintenance of a particular mixture?
6 Name two forage species that require high light intensity. Name two forage species that will tolerate lower light intensity.
7 Why are food reserves so important in forage growth?
8 Where are food reserves stored in the various forage species?
9 Draw the cyclic pattern of the production of soluble carbohydrates in alfalfa and timothy. Label the various morphological points of development on each curve.
10 Why is cutting height so important in forage management?
11 Describe the importance of the late fall harvest of forages.
12 Name the environmental conditions needed for cold tolerance.

REFERENCES

1 Adams, W. E. and M. Twersky. Effect of soil fertility on winter-killing of Coastal bermudagrass. *Agron. J.,* 52:325–326, 1960.
2 Alberda, T. Responses of grasses to temperature and light. In F. L. Milthorpe and J. D. Ivins (eds.), *The Growth of Cereals and Grasses,* Butterworths, London, pp. 200–212, 1966.
3 Allard, H. A. and M. W. Evans. Growth and flowering of some tame and wild grasses in response to different photoperiods. *J. Agr. Res.,* 62:193–228, 1941.
4 Balasko, J. A. and D. Smith. Influence of temperature and nitrogen fertilization on the growth and composition of switchgrass (*Panicum virgatum* L.) and timothy (*Phleum pratense* L.) at anthesis. *Agron. J.,* 63:853–857, 1971.
5 Black, J. N. The influence of varying light intensity in the growth of herbage plants. *Herb. Abstr.,* 27:89–98, 1957.
6 Booysen, P. DeV., N. M. Tainton, and J. D. Scott. Shoot-apex development in grasses and its importance in grassland management. *Herb. Abstr.,* 33:209–213, 1963.
7 Brown, R. H. and R. E. Blaser. Leaf area index in pasture growth. *Herb. Abstr.,* 38:1–9, 1968.
8 Bula, R. J. and D. Smith. Cold resistance and chemical composition in overwintering alfalfa, red clover, and sweet clover. *Agron. J.,* 46:397–401, 1954.
8 a Carter, P. R., C. C. Sheaffer, and W. B. Voorhees. Root growth, herbage yield, and plant water status of alfalfa cultivars. *Crop Sci.,* 22:425–427, 1982.
9 Clark, R. E. and E. A. Paul. The microflora of grassland. *Advan. Agron.,* 22:375–435, 1970.
10 Cooper, J. P. The use of controlled life cycles in the forage grasses and legumes. *Herb. Abstr.,* 30:71–79, 1960.

11 Cooper, J. P. and N. M. Tainton. Light and temperature requirements for the growth of tropical and temperate grasses. *Herb. Abstr.*, 38:167–176, 1968.

12 Fairbourn, M. L. Water use by forage species. *Agron. J.*, 74:62–73, 1982.

12 a Fishbeck, K. A. and D. A. Phillips. Host plant and rhizobium effects on acetylene reduction in alfalfa during regrowth. *Crop Sci.*, 22:251–254, 1982.

13 Folkins, L. P., J. E. R. Greenshields, and F. S. Nowosad, Effect of date and frequency of defoliation on yield and quality of alfalfa. *Can. J. Plant Sci.*, 41:188–194, 1961.

13 a Fribourg, H. A., R. J. Carlisle, and V. H. Reich. Yields of warm- and cool-season forages on adjacent soils. *Agron. J.*, 74:663–667, 1982.

14 Gist, G. R. and G. O. Mott. Growth of alfalfa, red clover, and birdsfoot trefoil seedlings under various quantities of light. *Agron. J.*, 50:583–586, 1958.

15 Gist, G. R. and G. O. Mott. Some effects of light intensity, temperature, and soil moisture on the growth of alfalfa, red clover, and birdsfoot trefoil seedlings. *Agron. J.*, 49:33–36, 1957.

16 Harris, W. Defoliation as a determinant of the growth, persistence and composition of pasture. In J. R. Wilson (ed.), *Plant Relations in Pasture,* Commonwealth Scientific and Industrial Research Organization, pp. 67–85, 1978.

16 a Henderson, M. S. and D. L. Robinson. Environmental influences on fiber component concentrations of warm-season perennial grasses. *Agron. J.*, 74:573–579, 1982.

16 b Henderson, M. S., and D. L. Robinson. Environmental influences on yield and in-vitro true digestibility of warm-season perennial grasses and the relationship to fiber components. *Agron. J.*, 74:943–946, 1982.

17 Jensen, E. H., M. A. Massengale, and D. O. Chilcote. Environmental effects on growth and quality of alfalfa. *Nev. Agr. Exp. Sta. Bull. T–9*, 1967.

17 a Jodari-Karimi, F., V. Watson, H. Hodges, and F. Whisler. Root distribution and water use efficiency of alfalfa as influenced by depth of irrigation. *Agron. J.*, 75:207–211, 1983.

18 Julander, O. Drought resistance in range and pasture grasses. *Plant Physiol.*, 20:573–599, 1945.

19 Jung, G. A., R. L. Reid, and J. A. Balasko. Studies on yield, management, persistence, and nutritive value of alfalfa in West Virginia. *W. Va. Agr. Exp. Sta. Bull. 581T*, 1969.

20 Jung, G. A. and D. Smith. Influence of soil potassium and phosphorus content on the cold resistance of alfalfa. *Agron. J.*, 51:585–587, 1959.

21 Jung, G. A. and D. Smith. Trends of cold resistance and chemical changes over winter in the roots and crowns of alfalfa and medium red clover. I. Changes in certain nitrogen and carbohydrate fractions. II. Changes in certain mineral constituents. *Agron. J.*, 53:359–366, 1961.

22 Knievel, D. P., A. V. A. Jacques, and D. Smith. Influence of growth stage and stubble height on herbage yields and persistence of smooth bromegrass and timothy. *Agron. J.*, 65:430–434, 1971.

23 Larsen, P. Light requirements in plant production and growth regulation. *Acta Agr. Scand. (Suppl.),* 16:161–172, 1966.

24 Larson, K. L. and J. D. Eastin (eds.). Drought injury and resistance in crops. *Crop Sci. Soc. Amer. Spec. Publ. 2*, 1971.

25 Laude, H. M. Plant response to high temperatures. *Amer. Soc. Agron. Spec. Publ. 5*, pp. 15–30, 1964.

26 Levitt, J. Drought. *Amer. Soc. Agron. Spec. Publ. 5,* pp. 57–66, 1964.

27 Marten, G. C. Temperature as a determinant of quality of alfalfa harvested by bloom stage or age criteria. *Proc. 11th Intl. Grassl. Congr.,* pp. 506–509, 1970.

28 McIlroy, R. J. Carbohydrates of grassland herbage. *Herb. Abstr.,* 37:79–87, 1967.

29 Moss, D. N. Some aspects of microclimatology important in forage plant physiology. *Amer. Soc. Agron. Spec. Publ. 5,* pp. 1–14, 1964.

30 Odum, E. P. *Fundamentals of Ecology,* W. B. Saunders, Philadelphia, 3d. ed., pp. 8–39, 1971.

31 Olmsted, C. E. Growth and development in range grasses. I. Early development of *Bouteloua curtipendula* in relation to water supply. *Bot. Gaz.,* 102:499–519, 1941.

31 a O'Neil, K. J. and R. N. Carrow. Perennial ryegrass growth, water use, and soil aeration status under soil compaction. *Agron. J.,* 75:177–180, 1983.

32 Reckenthin, C. A. Elementary morphology of grass growth and how it affects utilization. *J. Range Manage.,* 9:167–170, 1956.

33 Rhykerd, C. L., C. H. Noller, J. E. Dillon, J. B. Ragland, B. W. Crowl, G. C. Naderman, and D. L. Hill. Managing alfalfa-grass mixtures for yield and protein. *Purdue Univ. Agr. Exp. Sta. Res. Bull. 839,* 1967.

34 Rogler, G. A. Response of geographical strains of grasses to low temperatures. *J. Amer. Soc. Agron.,* 35:547–559, 1943.

35 Ruelke, O. C. and D. Smith. Overwintering trends of cold resistance and carbohydrates in medium red, ladino, and common white clover. *Plant Physiol.,* 31:364–368, 1956.

36 Russell, E. W. The soil environment and gramineous crops. In F. L. Milthorpe and J. D. Ivins (eds.), *The Growth of Cereals and Grasses,* Butterworths, London, pp. 138–152, 1966.

37 Salisbury, F. B. and C. Ross. *Plant Physiology,* Wadsworth, Belmont, Calif., 1969.

38 Scott, J. D. The study of primordial buds and the reaction of roots to defoliation as the basis of grassland management. *Proc. 7th Intl. Grassl. Congr.,* pp. 479–487, 1956.

39 Shantz, H. L . and L. N. Piemeisel. The water requirement of plants at Akron, Colo. *J. Agr. Res.,* 34:1093–1190, 1927.

40 Sheard, R. W. and J. E. Winch. The use of light interception, grass morphology and time as criteria for the harvesting of timothy, smooth bromegrass and cocksfoot. *J. Brit. Grassl. Soc.,* 21:231–237, 1966.

41 Shih, S. C., G. A. Jung, and D. C. Shelton. Effects of temperature and photoperiod on metabolic changes in alfalfa in relation to coldhardiness. *Crop Sci.,* 7:385–389, 1967.

42 Smith, D. Carbohydrate root reserves in alfalfa, red clover, and bindsfoot trefoil under several management schedules. *Crop Sci.,* 2:75–78, 1962.

43 Smith, D. Classification of several native North American grasses as starch or fructosan accumulators in relation to taxonomy. *J. Brit. Grassl. Soc.,* 23:306–309, 1968.

44 Smith, D. The establishment and management of alfalfa. *Wis. Agr. Exp. Sta. Bull. 542,* 1968.

45 Smith, D. Influence of cool and warm temperatures and temperature reversal at inflorescence emergence on yield and chemical composition of timothy and bromegrass at anthesis. *Proc. 11th Intl. Grassl. Congr.,* pp. 510–514, 1970.

46 Smith, D. Physiological considerations in forage management. In M. E. Heath, D. S. Metcalfe, and R. E. Barnes, (eds.), *Forages,* Iowa State Univ. Press, Ames, IA, 3d ed., pp. 425–436, 1973.

47 Smith, D. Removing and analyzing total nonstructural carbohydrates from plant tissue. *Wis. Agr. Exp. Sta. Res. Rep. 41,* 1969.

48 Smith, D. The unusual growth responses of birdsfoot trefoil. *Crops and Soils,* 18(7):12, 1966.

49 Smith, D. Varietal-chemical differences associated with freezing resistance in forage plants. *Cryobiology,* 5:148–149, 1968.

50 Smith, D. Winter injury and the survival of forage plants. *Herb. Abstr.,* 34:203–209, 1964.

51 Spedding, C. R. W. *Grassland Ecology,* Oxford Univ. Press, London, pp. 4–5, 109, 1971.

52 Tansley, A. G. The use and abuse of vegetational concepts and terms. *Ecology,* 16:284–307, 1935.

53 Taylor, S. A. Water condition and flow in the soil-plant-atmosphere system. *Amer. Soc. Agron. Spec. Publ. 5,* pp. 81–107, 1964.

54 Taylor, T. H. and W. C. Templeton, Jr. Grassland ecosystem concepts. In M. E. Heath, D. S. Metcalfe, and R. E. Barnes (eds.), *Forages,* Iowa State Univ. Press, Ames, IA, 3d ed., pp. 44–52, 1973.

55 Templeton, W. C., Jr. and T. H. Taylor. Some effects of nitrogen, phosphorus and potassium fertilization on botanical composition of a tall fescue-white clover sward. *Agron. J.,* 48:569–572, 1966.

56 Torrie, J. H. and E. W. Hanson. Effects of cutting first-year red clover on stand and yield in the second year. *Agron. J.,* 47:224–228, 1955.

57 Tysdal, H. M. Influence of light, temperature, and soil moisture on the hardening process in alfalfa. *J. Agr. Res.,* 46:483–515, 1933.

58 Zarrough, K. M., C. J. Nelson, and J. H. Couetts. Relationship between tillering and forage yield of tall fescue. I. Yield. *Crop Sci.,* 23:333–337, 1983.

59 Zarrough, K. M., C. J. Nelson, and J. H. Couetts. Relationship between tillering and forage yield of tall fescue. II. Pattern of tillering. *Crop Sci.,* 23:338–342, 1983.

PART TWO

FORAGE IMPROVEMENT, PRODUCTION, MANAGEMENT, AND UTILIZATION

FORAGE BREEDING

Forage crops are primarily perennial plants. The perennial nature of forage plants dictates that longevity of stand, combined with palatability and digestibility, plus high yields, are important factors in their utilization.

The plant breeder must know why and how a particular forage species will be utilized. For instance, will it be grown for pasture or hay; will it be harvested as hay or silage; how digestible is it; are there any antiquality factors present? These are only some of the questions that the breeder must answer before genetic studies can be made. The breeder also must be very knowledgeable about the species' mode of reproduction and its genetic makeup.

Breeding forage crops is more difficult than is breeding grain crops. Difficulties arise from methods of pollination, irregularities in fertilization and seed set, and determining how to evaluate and maintain new strains. For example, many of the important forage species are cross-pollinated (Table 5-1). The heterozygosity in cross-pollinated species makes it difficult to increase and maintain the purity of the lines. Most forage breeders hope to maintain a wide gene base, which entails the maintenance of heterozygosity. This tends to cover up inherent weaknesses which, in periods of stress, would eliminate homozygous species. The pure line theory is not accepted in forage breeding programs. Some forage species are self-sterile, which makes it very difficult to develop inbred lines. Some grasses reproduce by apomixis—setting seed without fertilization. Flower parts are small, which makes it difficult to hybridize. Many of the forage species produce poor seeds with low viability. Forages are often grown in mixtures rather than as a monoculture. In addition, forage species often perform differently under grazing conditions than when harvested as a hay crop. Seeds of many forages are produced in climatic regions different from where the forage is produced.

TABLE 5-1

MODES OF POLLINATION AND GROWTH HABITS OF VARIOUS FORAGE GRASS AND LEGUME SPECIES

Species		Chromosome number	Growth habit
Self-pollinated forage grasses			
Lovegrass, weeping	*Eragrostis curvula*	40	Perennial
Sudangrass	*Sorghum bicolor* var. *sudan*	20	Annual
Wheatgrass, slender	*Agropyron trachycaulum*	28	Perennial
Self-pollinated forage legumes			
Lespedeza, common	*Lespedeza striata*	22	Annual
Lespedeza, Korean	*L. stipulacea*	20	Annual
Vetch, common	*Vicia sativa*	12	Winter annual
Vetch, hairy	*V. villosa*	14	Winter annual
Cross-pollinated forage grasses			
Bromegrass, smooth	*Bromus inermis*	42, 56	Perennial
Bermudagrass	*Cynodon dactylon*	30, 36	Perennial
Fescue, meadow	*Festuca elatior*	14, 28, 42, 70	Perennial
Fescue, tall	*F. arundinacea*	42	Perennial
Grama, blue	*Bouteloua gracilis*	Varied	Perennial
Grama, sideoats	*B. curtipendula*	Varied	Perennial
Orchardgrass	*Dactylis glomerata*	28	Perennial
Redtop	*Agrostis alba*	28, 42	Perennial
Reed canarygrass	*Phalaris arundinacea*	14, 28	Perennial
Ryegrass, perennial	*Lolium perenne*	14	Perennial
Timothy	*Phleum pratense*	14, 42	Perennial
Wheatgrass, crested	*Agropyron desertorum*	14	Perennial
Wheatgrass, western	*A. smithii*	42, 56	Perennial
Cross-pollinated forage legumes			
Alfalfa	*Medicago sativa*	32	Perennial
Alfalfa	*M. falcata*	16, 32	Perennial
Birdsfoot trefoil	*Lotus corniculatus*	12, 24	Perennial
Clover, alsike	*Trifolium hybridum*	16	Perennial
Clover, crimson	*T. incarnatum*	14, 16	Winter annual
Clover, red	*T. pratense*	14	Perennial
Clover, white	*T. repens*	32	Perennial
Sweet clover, white-flowered	*Melilotus alba*	16	Annual, Biennial
Sweet clover, yellow-flowered	*M. officinalis*	16	Biennial
Apomictic species			
Dallisgrass	*Paspalum dilatatum*	40, 50	Perennial
Kentucky bluegrass	*Poa pratensis*	28, 56, 70	Perennial
Dioecious species			
Buffalograss	*Buchloe dactyloides*	56, 60	Perennial

Many factors must be considered before a breeding program can be started. The mode of reproduction is very important in relation to plant breeding methods. Most forage species are either cross- or self-pollinated; these are forms of sexual reproduction (Table 5-1). A few forage crops are asexually reproduced.

Another factor in breeding forages is the amount of genetic variability present in forage germ plasms. This may or may not limit the improvement that we may make over a period of time. Generally, there is considerable genetic variability in most forage germ plasms. This in turn must be evaluated to see whether one has qualitative characters—simply inherited traits—to work with, or whether one has quantitative traits, which are controlled by many genes.

Another factor to consider is a species' fertility-sterility relationship. Many variations occur in this area. Species in which the chromosome number has been doubled are called autotetraploids. Chromosome pairing within autotetraploids may be fairly normal, but even at best some irregular pairing causes some sterility. Alfalfa and orchardgrass are examples of autotetraploids (25). Alloploids arise by combining two or more species, as in *Triticale* (rye-wheat hybrid); sterility in the F_1 is experienced in most instances. Some species are self-sterile because of irregularities in chromosome behavior and chromosome number (Table 5-1). An example of chromosome pairing irregularity is smooth bromegrass *(Bromus)* (12, 20). Within other species, such as sweetclover *(Melilotus),* the number of chromosomes may be the same but when crossed the embryos abort or the progeny are sterile (33).

Self-sterility may be caused by a multiple series of genes called S-alleles, which are found, for instance, in red clover and alfalfa. These genes are responsible for self-incompatibility due to a failure of the pollen tube to grow down the stylar tissue of a plant having the same S-alleles. Therefore, pollen tubes will not grow down the stylar tissue within the same plant that produced a particular pollen grain (8). Many different S-alleles exist, however, so that intercrossing is not restricted.

A great deal of variation in seed set, due to selfing, may exist in other species such as orchardgrass, smooth bromegrass, and alfalfa. A complex genetic system for self-sterility exists in these species. Within some self-sterile species, such as red clover, it may be possible to obtain selfed seed by exposing plants to a high temperature, thereby breaking the self-sterility mechanism (22). Once self-sterility is broken, then inbred lines may be developed and hybrids eventually may be produced.

There may be another mechanism resulting in low selfed seed production, called somatoplastic sterility. This is a developmental competition for nutrition among the embryo, the endosperm, and the nucleus. More of the ovules collapse following selfing than following cross-fertilization. Embryo development is much slower after self-fertilization than after cross-fertilization. A physiological competition exists for nutrients in the developing seed.

Cross-pollination enables the plant breeder to transfer genetic traits from one population to another. It increases the variability within a species. Genetic variability is essential to species improvement for it not only extends the species' area of adaptation, but the differences also act as a hedge against encroachments or epiphytotics of diseases and/or insects.

SELECTION OF BREEDING MATERIALS

The first thing a breeder must do is identify the problem and then clearly outline his or her objectives. The breeder who haphazardly makes crosses and hopes to find something good is almost certainly doomed to failure.

One of the keys to a successful breeding program is the proper selection of breeding material, or parents, to be used in developing a new cultivar. The breeder can select surviving plants in old, perennial stands that, through natural selection, possess characteristics desired in a new cultivar (Fig. 5-1). Plants selected in this manner that are leafy and insect-resistant will often have the proper winterhardiness and disease resistance. Many old cultivars were developed in this way, a form of mass selection.

Most mass selections have involved cross-pollinated species. Many parents or clones were selected and then allowed to mate randomly. Seed was harvested and bulked to form the breeder seed for a new cultivar. Following a testing program that indicates superior production or resistance to some pest, the cultivar is later increased and released. Many of these selections have created improvements over the average of the parents' characteristics. Maintaining the parents or clones is often done vegetatively.

The selection process can be repeated in successive generations, thus continually improving the strain. Improvements have been made and new

FIGURE 5-1
Alfalfa source nursery used in a breeding program.

cultivars developed using successive cycles of mass selections in many forage species, such as alfalfa, big bluestem, sideoats, grama, tall fescue, orchardgrass, timothy, red clover, and white clover (2, 9, 11, 14, 21).

USE OF CLONES OR SELECTED PLANTS

Asexual Reproduction

Some species are sterile and must be propagated vegetatively. Vegetative propagation is a very effective way to increase a species to the point that it can be utilized extensively. 'Coastal' and 'Tift' bermudagrasses are good examples of species grown by this procedure. Because of their poor seed-setting capability, many turf grasses, especially the bentgrasses, have been propagated vegetatively for golf courses and other turf purposes.

Apomixis occurs in some grasses. This is the formation of viable embryos without actual union of male and female gametes, as in Kentucky bluegrass.

Individual Plant Progeny

Most of the progeny produced from a single plant, if selfed, would be undesirable because of inbreeding. It was found in smooth bromegrass that some individual plant progeny were as high-yielding as some commercial cultivars (24). Two cultivars of orchardgrass, 'Barge' and 'Skandia II,' were developed in Sweden from a single plant. Most of the seed produced is the result of sib cross fertility within the self-incompatible species.

F_1 Crosses

When self-sterility is found in a species, selected clones may be used as parents in controlled F_1 crosses. The selected clones or parents need to be increased vegetatively on a large scale so that sufficient seed will be produced. Hybrid vigor may be observed in such F_1 progeny. Inbred lines are often used as the parents in such cases, as in orchardgrass and smooth bromegrass (12). There have been some very promising F_1 crosses resulting from noninbred parents.

If a species possesses considerable self-fertility, many times the resulting progeny is not as promising as the parents. White clover, pearl millet, and alfalfa are examples of self-fertile species (1). Cytoplasmic male sterility, found in alfalfa and other forage species, allows controlled crossing to occur in normally self-fertile species.

Nonsterile alfalfa double crosses were suggested in 1942 for producing hybrids, by using F_1 crosses as parents (6). Seed produced in this manner was a mixture of selfed and crossed seed from crossing within each of the two F_1 parents, and of hybrid seed from reciprocal cross-pollination between the two F_1 parents. With diploid forage species, four different homozygous lines, possessing self-sterile alleles, could be used, so that a true double-cross hybrid could be produced.

Synthetic Cultivars

By selecting particular clones, one can recombine several to many clones that possess the various characteristics wanted in a superior cultivar. The steps involved in developing a synthetic forage cultivar are outlined in Fig. 5-2. The parents or clones used in a synthetic need to be high-yielding because the end product is the result of blending several crosses. Other factors may affect the final combined average yield of the synthetic: the extent of cross-pollination, the yield of the F_1 crosses from the various crosses, the polyploidy relationship, the number of generations, and the number of parents in the synthetic. Generally most of the seed-yielding ability of a synthetic occurs between the Syn-1 and Syn-2 generations. Seed yields are usually stable by the Syn-3. Seed produced out of its region of adaption may result in a shift in its genetic composition, thus changing its potential yield. This is another reason why a limited generation program was developed in the commercial seed trade. Many cultivars of the various forage species have been developed using the synthetic approach (3, 20, 31).

The yield of a synthetic cultivar can be predicted by using clones with the best general combining ability. General combining ability is more important for a synthetic because many parents may be involved and the progeny are the result of many uncontrolled crosses. Specific combining ability is important for hybrid seed production because the crossing is controlled and the specific crosses will result in more productive progeny. Yields of alfalfa, sweet clover, and smooth bromegrass synthetics have been predicted by using the general combining ability of the clones making up the synthetic (3, 20).

Recurrent Selection

Recurrent selection is a method of plant breeding in which selected plants are intercrossed to produce a new population. This is the first cycle (cycle 1) of the recurrent selection. Then, individual plants are again selected for the various desired characteristics and intercrossed, producing cycle 2 of the recurrent selection. Selection continues until the desired progress is achieved or until progress does not seem justified.

If one selects plants or clones for certain observable characteristics, this is called phenotypic recurrent selection. If one selects plants because the progeny of the inbred or crossed plants perform in a certain manner, this is called genotypic recurrent selection. The performance of the progeny denotes a certain genotype or genetic makeup of the selected parents (15). Many characteristics can be improved by recurrent selection—plant type, yield, and disease or insect resistance (7, 13).

In recent years genetic engineering and gene splicing have been investigated in forages. This area of research includes tissue culturing of callus material for generating new lines resistant to a particular problem. It appears that gene splicing, the placing of a gene segment of a chromosome into another species, is

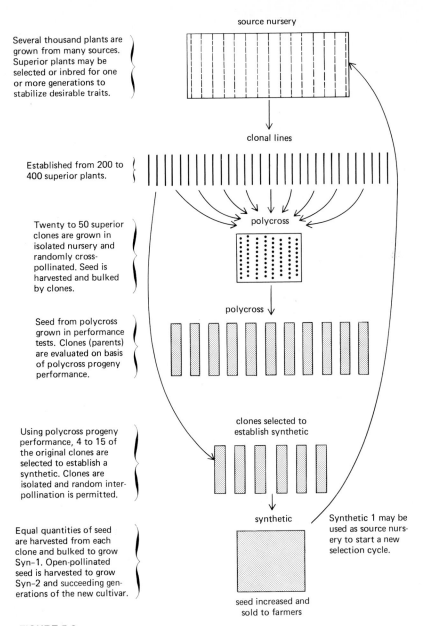

Several thousand plants are grown from many sources. Superior plants may be selected or inbred for one or more generations to stabilize desirable traits.

source nursery

clonal lines

Established from 200 to 400 superior plants.

Twenty to 50 superior clones are grown in isolated nursery and randomly cross-pollinated. Seed is harvested and bulked by clones.

polycross

Seed from polycross grown in performance tests. Clones (parents) are evaluated on basis of polycross progeny performance.

polycross

Using polycross progeny performance, 4 to 15 of the original clones are selected to establish a synthetic. Clones are isolated and random inter-pollination is permitted.

clones selected to establish synthetic

Equal quantities of seed are harvested from each clone and bulked to grow Syn-1. Open-pollinated seed is harvested to grow Syn-2 and succeeding generations of the new cultivar.

synthetic

Synthetic 1 may be used as source nursery to start a new selection cycle.

seed increased and sold to farmers

FIGURE 5-2
An outline for developing a synthetic forage cultivar.

possible. These investigations may lead to a new species or a forage crop that is much more valuable because of its quality, growth, or resistance to some pest.

EVALUATION OF BREEDING MATERIALS

All plant breeders evaluate the crosses or cultivars that they develop before releasing them. In fact, with cultivar review boards and plant variety protection, it is necessary to evaluate all proposed releases. The newly developed cultivar has to be unique or improved in some characteristic before it can be protected or released.

Forage yield trials are the most common method of evaluation. Dry weight yields are taken, along with data on winterhardiness, longevity, disease and insect resistance, drought resistance, leafiness, nutritive value, digestibility values, and antiquality factors. If a new cultivar does not excel in one or several of these traits, it is unlikely that it will succeed in the trade.

Various methods are used in testing a given species or trait. Greenhouse and growth chambers are used to study physiological responses, or insect and disease resistance. Space plants in the field are used to determine winterhardiness, environmental responses, and regrowth following cutting. When inbred lines can be produced, the inbred progeny will assist the breeder in determining the genetic makeup of the parents.

Again, when crossing is done, general combining ability is the tool used to evaluate the performance of top-crossed, polycrossed, diallele-crossed, or open-pollinated progenies in cross-pollinated species. Specific combining ability is used in hybrid seed production or single crosses, when crossing is controlled. Specific combining ability will also assist in developing a very specific recombination for a limited number of parents for a synthetic cultivar.

Laboratory tests are conducted to determine the cultivar's nutritive value or digestibility value. In vitro techniques measure forage quality. Rumen extracts are used in determining digestibility, and detergent techniques make it possible to study individual plants and determine their potential feeding value (27, 29). Plant breeders are developing new, improved, high-quality cultivars and cultivars with low antiquality factors. Such developments take the cooperation of the breeder, the physiologist, and the animal nutritionist.

The animal nutritionist could greatly assist the plant breeder when developing a new cultivar or germplasm. Besides the actual composition of the forage species, quality should be measured. A forage's nutritive value is often designated TDN (total digestible nutrient), or digestible energy. Chemical analysis of the forage and feces determines the content of crude protein, crude fiber, nitrogen-free extract, and ether extract, yielding a percentage value for TDN (32).

Digestibility can be estimated using the forage fed and measuring the feces. Another pertinent measurement is the rate of consumption (23).

Eventually the new cultivar needs to be evaluated using tester animals in the field under grazing, silage, or hay conditions. Persistence under grazing pressure is essential to the evaluation of a new cultivar. These tests will measure gain per day, production per unit area grazed, and economic return. Farmers need to evaluate silage or hay in relation to total production, ease of harvesting, and cash value of the end product.

Current research is under way at several universities to develop salt-tolerant forages that could facilitate the use of seawater to irrigate forage crops in certain regions of the world. The population growth of developing countries currently exceeds the growth of food production. Moreover, for improved human diets, these countries need additional high-quality animal proteins and minerals. The availability of salt-tolerant forages irrigated with seawater would make this possible by providing feed for milk- and meat-producing animals.

QUESTIONS

1 What are some of the characteristics that plant breeders look for when developing a new cultivar? What characteristics do they wish to avoid?

2 What is the difference between qualitative and quantitative inherited characteristics?

3 What is meant by the S-alleles self-sterility complex? In which species is it found?

4 Describe mass selection.

5 How is a synthetic cultivar developed?

6 Name several ways in which a plant breeder evaluates his or her breeding material.

7 Why is it important that the breeder and foundation seed classes be maintained in the area of their adaptation?

REFERENCES

1 Atwood, S. S. Genetics of cross-incompatibility among self-incompatible plants of *Trifolium repens. J. Amer. Soc. Agron.,* 32:955–968, 1940.

2 Barnes, D. K., C. H. Hanson, F. I. Frosheiser, and L. J. Elling. Recurrent selection for bacterial wilt resistance in alfalfa. *Crop Sci.,* 11:545–546, 1971.

3 Bolton, J. L. A study of combining ability of alfalfa in relation to certain methods of selection. *Sci. Agr.,* 28:97–126, 1948.

4 Brink, R. A. and D. C. Cooper. Double fertilization and development of seed in angiosperms. *Bot. Gaz.,* 102:1–25, 1940.

5 Davis, R. L. An evaluation of S_1 and polycross progeny testing in alfalfa. *Agron. J.,* 47:572–576, 1955.

6 Davis, W. H. and I. M. Greenblatt. Cytoplasmic male sterility in alfalfa. *J. Hered.,* 58:301–305, 1967.

7 Devine, T. E., C. H. Hanson, S. A. Ostazeski, and T. A. Campbell. Selection for resistance to anthracnose *(Colletotrichum Trifolii)* in four alfalfa populations. *Crop Sci.,* 11:854–855, 1971.

8 East, E. M. and A. J. Mangelsdorf. A new interpretation of the hereditary behavior of self-sterile plants. *Proc. Natl. Acad. Sci.,* 11:166–171, 1925.

9 Fryer, J. R. The maternal-line selection method of breeding for increased seed setting in alfalfa. *Sci. Agr.,* 20:131–139, 1939.

10 Getty, R. E. and I. J. Johnson. The nature and inheritance of sterility in sweetclover, *Melilotus officinalis* Lam. *J. Amer. Soc. Agron.,* 36:228–237, 1944.

11 Harlan, J. R. The breeding behavior of sideoats grama in partially isolated populations. *Agron. J.,* 42:20–24, 1950.

12 Hayes, H. K. and A. R. Schmid. Selection in self-pollinated lines of *Bromus inermis* Leyys., *Festuca elatior* L., and *Dactylis glomerata* L. *J. Amer. Soc. Agron.,* 35:934–943, 1943.

13 Hill, R. R., Jr., C. H. Hanson, and T. H. Busbice. Effect of four recurrent selection programs in two alfalfa populations. *Crop Sci.,* 9:363–365, 1969.

14 Jenkins, T. J. The method and technique of selection, breeding, and strain-building in grasses. *Imp. Bur. Plant Genet. Herbage Plants,* 3:5–34, 1931.

15 Johnson, I. J. and A. S. El Banna. Effectiveness of successive cycles of phenotype recurrent selection in sweetclover. *Agron. J.,* 49:120–125, 1957.

16 Johnson, I. J. and J. E. Sass. Self- and cross-fertility relationships and cytology of autotetraploid sweet clover, *Melilotus alba. J. Amer. Soc. Agron.,* 36:214–227, 1944.

17 Johnson, I. J. and A. M. Hoover, Jr. Comparative performance of actual and predicted varieties in sweetclover. *Agron. J.,* 45:494–498, 1953.

18 Jones, J. and F. C. Elliott. Two rapid assays for saponin in individual alfalfa plants. *Crop Sci.,* 9:688–691, 1969.

19 Kalton, R. R. and R. C. Leffel. Evaluation of combining ability in *Dactylis glomerata* L. III. General and specific effects. *Agron. J.,* 47:370–373, 1955.

20 Knowles, R. P. Studies of combining ability in bromegrass and crested wheatgrass. *Sci. Agr.,* 30:275–302, 1950.

21 Law, A. G. and K. L. Anderson. The effect of selection and inbreeding on the growth of big bluestem (*Andropon furcatus* Muhl.). *J. Amer. Soc. Agron.,* 32:931–943, 1940.

22 Leffel, R. C. Pseudo self-compatibility and segregation of gametophytic self-incompatibility alleles in red clover (*Trifolium pratense* L.). *Crop Sci.,* 3:377–380, 1963.

23 Mott, G. O. Evaluating forage production. In *Forages,* Iowa State Univ. Press, Ames, IA, 3d ed., pp. 126–135, 1973.

24 Murphy, R. P. and S. S. Atwood. The use of I_1 families in breeding smooth bromegrass. *Agron. J.,* 45:23–28, 1953.

25 Myers, W. M. and H. D. Hill. Increased meiotic irregularity accompanying inbreeding in *Dactylis glomerata* L. *Genetics,* 28:383–397, 1943.

26 Pederson, M. W. and Li-Chun Wang. Modification of saponin content of alfalfa through selection. *Crop Sci., 11:833–835, 1971.*

27 Skenk, J. S. and F. C. Elliott. Plant compositional changes resulting from two cycles of directional selection for nutritive value of alfalfa. *Crop Sci.,* 11:521–524, 1971.

28 Smith, D. C. Pollination and seed formation in the grass. *J. Agr. Res.,* 68:79–95, 1944.

29 Tilley, J. A. and R. A. Terry. A two-stage technique for the *in vitro* digestion of forage crops. *J. Brit. Grassl. Soc.,* 18:104, 1963.

30 Tysdal, H. M. and B. H. Crandall. The polycross performance as an index of the combining ability of alfalfa clones. *J. Amer. Soc. Agron.,* 40:293–306, 1948.

31 Tysdal, H. M., T. A. Kiesselbach, and H. L. Westoner. Alfalfa breeding. *Nebr. Agr. Exp. Sta. Bull. 124,* 1942.

32 Van Soest, P. J. Development of a comprehensive system of feed analysis and its application to forages. *J. Anim. Sci.,* 26:119–128, 1967.

33 Webster, G. T. Fertility relationships and meiosis of interspecific hybrids in *Melilotus. Agron. J.,* 42:315–322, 1950.

34 Weiss, M. G., L. H. Taylor, and I. J. Johnson. Correlations of breeding behavior with clonal performance of orchardgrass plants. *Agron. J.,* 43:594–602, 1951.

35 Williams, R. D. Self- and cross-sterility in red clover. *Welsh Plant Breeding Sta. Bull. H-12,* pp. 181–208, 1931.

CHAPTER 6

SEED PRODUCTION

Seed production is a specialized type of agriculture. In the past, farmers often grazed or harvested the first legume growth and then harvested a seed crop as a byproduct. This was generally not satisfactory. Seed production involves sufficient vegetative growth, with abundant flower production, and good pollination.

The ability of a given forage species to produce sufficient quantities of seed plays a very important role in the species' desirability and availability. The price of seed is directly related to supply and demand, as well as to the seed-producing ability of a given cultivar. The price of seed often increases when a cultivar or new species is released and seed supply is limited, or when growing conditions are unfavorable for seed production. Unlike the case in grain crops, in some forage species the seed shatters readily when ripe, resulting in an increase in seed costs. Birdsfoot trefoil, for example, is an excellent seed producer, but a great deal of its seed shatters. Furthermore, birdsfoot trefoil is indeterminate in maturation: it sets seed all summer. Thus, it is difficult to determine the best time to harvest it, thereby creating a low supply of seed. The difficulty in producing high seed yields in forages centers, to a large extent, around the failure of flowers to set seed. Disease and insects can be incriminated in some cases, whereas in others lack of pollination, soil fertility, or a caustic shortage may be involved. Numerous problems exist in harvesting and conditioning forage seeds. A seed's viability starts to deteriorate immediately after the seed reaches maturity, and all one can do is retard the change by regulating the time and method of harvest and treatment after harvest. Forage seed production is a very specialized industry.

AREAS OF PRODUCTION

Most seed production of the certified cool-season grasses and legumes is located in the far western portion of the United States. However, some forage species are grown for either certified or noncertified seed in areas other than the west. Most of the native and introduced warm-season grass cultivars are produced for seed in the southern Great Plains. Seed production of most of the improved red clover cultivars now occurs in the west, but a great deal of common red clover seed is produced in Illinois, Indiana, Ohio, and Missouri. Similarly, alfalfa seed production shifted from the Great Plains to the west in the late 1940s, but Nebraska, Kansas, and Oklahoma still maintain considerable acreage devoted to it. Seeds of such species as lespedeza, crimson clover, and timothy are produced in the midwest and southeast (15, 27, 39).

Photoperiod response, climatic conditions, and the presence of a crop's pollinating insects determine the area where seed may be produced. An example of a species whose seed production depends on its photoperiod response is ladino clover. This species is primarily vegetative in the southern United States; therefore, seed production must take place in northern areas such as northern California, Oregon, and Washington, where summer days are adequately long to induce flowering. Virtually all alfalfa seed, too, is produced in these three states. Similarly, seed production for various forage species often takes place in areas of the United States other than where the species are grown for forage (Table 6-1).

CLIMATIC FACTORS

The climatic conditions of the west are best characterized by high air temperatures and low humidity, as well as by dry conditions for harvesting (27). The number of alfalfa florets setting pods decreases when the air temperature exceeds 38°C (100°F). The optimum temperature for alfalfa seed is around 27°C (80°F). As the relative humidity decreases, seed production usually increases. High light intensity is also essential for good seed production. Clear skies and warm temperatures favor the pollination and insect activity that are essential for good seed set.

CULTURAL AND MANAGERIAL FACTORS

Cultural and managerial practices for forage seed production are very important. Cultural practices involve soil tillage, fertilization, cultivation, planting, and so on. Management practices involve the how, when, and why—that is, the time of fertilization, split versus single application of fertilizer, amount of irrigation, and so on. Management practices are often more important to the successful production of a forage seed crop than are strictly cultural ones. The conditions needed for high seed yields involve practices that result in healthy, vigorous plant growth with a high level of carbohydrate food reserves, in addition to proper fertility, water control, and pollinating insects.

Seeding in Rows

Most forages grown for seed production are planted in rows 60 to 122 cm (24 to 48 in) apart and at a very low seeding rate. The advantages of row planting over solid stands are: (1) easier weed control or cultivation; (2) better penetration of insecticides; (3) more erect plant growth, which allows better light penetration; (4) easier control and application of irrigation water; (5) much less seed needed than for solid seedings; (6) facilitation of roguing; (7) reduced foliar disease problems; (8) greater accessibility of flowers to pollinating insects; and (9) higher seed yields (2, 3). Some disadvantages of row planting are: (1) more time and labor are involved in the seeding, and (2) more equipment is needed. Both of these disadvantages involve special row seeders, which make the seeding process slower than broadcast seeding.

Fertilization Requirements

Grasses need N fertilization (5, 14, 31). Some grasses grow primarily during the summer, develop their flower buds during this period, and respond to one heavy N application. Other grasses, which initiate their flower primordia during the fall or winter, respond to split N applications in the early fall and spring. However, dates and rates of N application for cool-season grasses vary from spring or fall to split-fall and spring in different parts of the country. Similarly, warm-season grasses grown for seed respond differently to different N fertilization rates and dates of application. The total amount of N applied depends upon the grass species, soil fertility, and amount of irrigation water needed. As the grass stand becomes older, more N is needed to keep seed production sufficiently high.

Legume seed fields do not need N. If properly nodulated with compatible or host-specific symbiotic N-fixing rhizobia. N will be fixed and obtained from soil air. Legumes need other nutrients: P, K, B, and S. Most soils in the west have a good supply of P and K. Local soil tests dictate the amounts needed of each nutrient.

Water Management

A good flush of growth and excellent flower production are needed for good seed production (17, 28). Water requirements for this vary with species, though one can control the amount and time of water application with irrigation. In addition to the species, irrigation requirements depend upon the soil type, temperature, evaporation, natural precipitation, and the length of the growing season. The shallow-rooted species, such as ladino clover, need more frequent water applications than does alfalfa. In fact, alfalfa, with its taproot, will survive a much longer dry period than will ladino or red clover.

During seed harvest, dry weather is very important (28). Unexpected rainfall when a seed crop is mature results in greatly reduced seed yields and lower seed quality. Some seed may shatter when drying out in the field following a rain.

TABLE 6-1
AREAS WHERE VARIOUS FORAGE CROPS ARE GROWN FOR FORAGE AND SEED PRODUCTION

Kind	Species	Production area	
		Forage	**Seed**
Legumes	Alfalfa	Entire U.S.	U.S., Dakotas S to Oklahoma
	Clover		
	Alsike	N and NW U.S.	W U.S.
	Arrowleaf	S and SE U.S.	Gulf states
	Crimson	SE U.S.	Oregon
	Ladino	E ½ U.S.	NW U.S.
	Red	E ½ and NW U.S.	Midwest and NW U.S.
	Crownvetch	N ½ and NE U.S.	Pennsylvania and NE U.S.
	Lespedeza	SE U.S.	Kentucky, Tennessee, North Carolina Georgia, Missouri
	Sainfoin	W dry regions U.S.	Montana
	Sweet clover	Mid-U.S.	Minnesota and Dakotas S to Texas
	Trefoil, birdsfoot	Central ½ and NE U.S.	Central and NE U.S.
	Vetch		
	Cicer	Great Plains and NW U.S.	Texas and Oklahoma
	Common	S U.S.	Texas and Oklahoma
	Hairy	Most U.S.	Texas and Oklahoma
Cool-season grasses	Bluegrass	N midwest to NE U.S.	Kentucky, N central and NW U.S.
	Bromegrass, smooth	N ½ U.S.	N Central U.S. and Canada
	Fescue, tall	Florida to Canada to Nebraska	Oregon and Kansas

74

Hardinggrass	S ⅓ U.S.	Gulf states and Texas
Orchardgrass	Midwest to NE U.S.	Oregon
Redtop	Central to S U.S.	Central Illinois
Reed canarygrass	N ½ U.S.	NW U.S. and Canada
Rescuegrass	S ⅓ U.S.	S U.S.
Ryegrass	SE U.S.	Oregon
Timothy	Central and NE U.S.	Central and NE U.S.
Wheatgrasses	W U.S.	Great Plains
Warm-season grasses		
Native		
Bluestem, big and little	S Great Plains	Oklahoma, Texas, Kansas
Buffalograss	S Great Plains	Nebraska, New Mexico
Indiangrass	S Great Plains	Colorado, Arizona
Lovegrass	S Great Plains	Colorado, Arizona
Sideoatgrass	S Great Plains	Colorado, Arizona
Switchgrass	S Great Plains	Colorado, Arizona
Bahiagrass	S ⅓ U.S.	S U.S.
Introduced		
Bermudagrass, coastal	S ½ U.S.	None
Bluestem, Plains and Caucasian	S Great Plains	Oklahoma, Texas
Buffalograss	S Great Plains	Oklahoma, Texas
Carpetgrass	S ⅓ U.S.	Gulf states
Centipedegrass	Gulf states	Gulf states
Dallisgrass	S ½ U.S.	Gulf states
Pangolagrass	Gulf states	Gulf states
Rhodesgrass	Gulf states	Gulf states, Texas

Weed Control

Weed control varies for different species and kinds of weeds. Weeds may be a serious problem when stands are thin because of the competitive nature of a weed and the delay in harvest that weeds can cause. Most seed fields are seeded in rows, so cultivation is the most practical method of weed control. Herbicide application such as a preemergence is very important; a postemergence applied during the dormant growth period may also be effective. Dodder is probably the most important weed in legumes in the west. It can be controlled by chlorpropham (CIPC) if properly applied. The seed producer can remove most weed seeds, including dodder and field bindweed, by means of seed cleaners and magnetic separators, but this adds to the cost of the final product (27, 39).

Insect Control

Insect control in legume seed fields is more serious than it is with grass species. Some of the insect pests in alfalfa seed fields are lygus bugs, alfalfa seed chalcid, aphids, mites, weevils, midges, and stink bugs. Lygus bugs are also very serious pests in the various clovers.

DDT, toxaphene, and other persistent insecticides were used in the past to control lygus. These insecticides have now been replaced by systemics and other insecticides of low toxicity to pollinating insects. Insect control must be carefully planned and applied so that pests are controlled without harm to pollinators (9, 27). Burning of stubble fields to control insects and disease is very important for grass seed production in Oregon, Washington, and Minnesota.

Disease Control

Most diseases are best controlled by selecting and seeding resistant cultivars. There appear to be more diseases that greatly reduce seed yields of grasses than of legumes. These diseases primarily attack the seed and inflorescence of grasses. Several of these diseases are ergot, nematode seed gall, dwarf bunt, seed smuts, and blind seed (18, 19, 20).

In addition to growing resistant cultivars, one may help control diseases with crop rotation and field burning. Some fungicides are used on Kentucky bluegrass to control stripe rust, *Puccinia striiformis* West.; leaf rust, *P. poaenemoralis* Otth.; stripe smut, *Ustilago striiformis* Niessl., and flag smut, *Urocystis agropyri* (Preuss.) Schroet. (18, 19, 20).

SEED CERTIFICATION

Certified seed ensures the purchaser of cultivar identity and genetic purity. Seed is certified in each state by a seed certifying agency. The minimum rules and standards for certification were developed by the Association of Official Seed Certifying Agencies (AOSCA) (4). Seed fields are inspected, and the entire processing channel is checked to ensure that no cultivar is contaminated.

There are four recognized classes of seed. Each class is produced to maintain its cultivar identify and genetic purity and handled with a specific color seed tag:

1 Breeders seed is the original seed produced and designated as such by the originating or sponsoring breeder or institution. It carries a white seed tag.

2 Foundation seed is produced from the breeders seed. It also carries a white seed tag.

3 Registered seed is produced from foundation seed. With some species or in some states this seed class may be eliminated in order to reduce the number of generations between breeders and certified seed. It carries a purple seed tag.

4 Certified seed is produced from foundation or registered seed. It carries a blue seed tag.

Each forage seed field is inspected during the production of certified seed for proper isolation, freedom from off types, noxious weeds, and seed-borne diseases. This ensures the purity of the cultivar during seed production and subsequent harvest and conditioning. Certified seed labels will indicate the percentage of pure seed germination, hard seed, other crop seed, inert matter, and weed seed content. When a grower purchases from a farmer seed that is not certified, there is no guarantee of the quality of the seed or that the grower is not merely purchasing a new weed problem.

Recently the Organization of Economic Cooperation and Development (OECD) was developed for the export of improved cultivars and maintenance of high-quality seed. The basis for maintaining quality seed was seed certification. A special tag is issued for the export seed trade certifying that the seed meets quality standards. As an example, Europe demands quality seed of improved cultivars because seed production there is poor, seed conditioning plants do not exist, and the climate is not favorable for seed production.

GENETIC SHIFT

Most of our forage seed is produced and certified in areas outside a cultivar's region of origin or adaptation. Rules have been established to limit the number of years or generations that a cultivar can be increased and the number of years a seed field may be harvested. This is a rule to safeguard a cultivar's genetic identity, winterhardiness, and other characteristics. Environmental selection pressure may first be put upon a new cultivar in the developmental stage. As this cultivar is then increased outside the region of its adaptation, a new environmental pressure is placed upon it. For example, the San Joaquin Valley and southern California as a whole produce 40 to 50% of the alfalfa seed produced nationwide. The climate of this area is very mild, with mild winters and mild to hot summers. But consider the development of a cold-resistant alfalfa cultivar for the midwest or upper Great Lakes area. Alfalfa seed of a northern-adapted cultivar produced from one generation of increase in Arizona produced stands that were taller than the original material, with fewer short plants, more fall growth, and more susceptibility to winter injury (10, 11, 16). As the generations increased, the shift was to more plants similar to a southwestern cultivar. A genetic shift might be in the reverse if southern species were increased for several years in a much colder region. Another type of genetic shift may occur in relation to flowering responses or photoperiod (13, 16). Data have shown that genetic shifts have occurred in red clover, ladino clover, and alfalfa (10, 11, 12, 26). These support the concept of limited generations in the increase of certain forage cultivars, especially those that have considerable genetic variability.

CULTIVAR PROTECTION

A plant cultivar protection act was passed by the Ninety-first Congress in 1970 and signed into law (27). This law was passed to encourage the development of cultivars of sexually propagated species; it provides protection to those who have bred, developed, or discovered them. It is similar to the protection a patent gives an inventor. The developer must present evidence of uniqueness of the cultivar before protection will be granted. This law is administered by the Agricultural Marketing Service of the USDA. Under this law, fees are collected for the protection and registration of a cultivar.

POLLINATING INSECTS

Pollination is the transfer of pollen from the male to the female portion of the flower. Pollination of legumes usually occurs by means of insects, whereas in grasses pollination usually occurs by wind. However, some important warm-season grasses reproduce by apomixis and set seed without fertilization of their ovules.

Some species are self-fertile and self-pollinated. Some of these species are soybeans, common lespedeza, crimson clover, wheat, oats, barley, and millet. Other species, which are cross-pollinated, include alfalfa, red clover, sweet-clover, birdsfoot trefoil, smooth bromegrass, orchardgrass, reed canarygrass, and timothy.

The actual transfer of pollen within the cross-pollinated species involves either the wind or an insect, as well as timeliness of pollen release and receptivity of the stigma. Alfalfa is an example of a legume whose pollination involves an insect and some special occurrences within the flower before fertilization can be accomplished (7, 24, 27). The pistil of the legume flower is held under pressure within the keel. The pressure derives from very turgid cells at the base of the keel. Fertilization cannot occur until the keel is opened, releasing the pistil, or reproductive column, and allowing it to strike the standard. This action is called tripping (Figs. 6-1, 6-2, 6-3) (37). When a honeybee visits a flower to collect nectar or pollen, the bee puts pressure on the keel, and the cells holding the keel together are broken. The reproductive column snaps out in force, hitting the bee. In addition to the tripping action, when the bee visits a flower it also ruptures the stigmatic membrane, which releases a sticky stigmatic fluid. The surface of the stigma then touches the bee and can pick up pollen from previously visited flowers. Pollen on the stigma then germinates, grows down the style, and unites with an ovule, completing fertilization.

FIGURE 6-1
Three alfalfa florets untripped.

FIGURE 6-2
Three alfalfa florets tripped.

FIGURE 6-3
Close-up of a tripped alfalfa floret.

FIGURE 6-4
Honeybee visiting red clover florets.

Tripping of the floret is not necessary for all legumes. Red clover and birdsfoot trefoil florets are manipulated by the bee, forcing the reproductive column out of the keel in a pistonlike, reversible action (27, 29). To be an effective pollinator of alfalfa, though, the bee must collect pollen (38), and since a honeybee soon learns to avoid the tripping action of the alfalfa flower, the degree of pollination is very low when a honeybee has its choice of flowers to visit. Honeybees also learn how to collect nectar without tripping the flower (38).

A high percentage of the pollinating insects in southern California are honeybees (*Apis* spp.). Most honeybees are nectar collectors (Fig. 6-4). A few honeybee strains, however, have been selected to collect pollen, and these make much more effective pollinators. In the San Joaquin Valley of California and southern California, honeybees exclusively are used to pollinate alfalfa seed fields, especially if there are very few other flowers or species blooming in close proximity, or if there is a very high population of bees (Fig. 6-5, 6-6) (36).

Other bee species are highly efficient pollinators, since they are primarily pollen collectors. Two species that have been cultured to pollinate alfalfa are the leaf-cutter bee (*Megachile* spp.) and the alkali bee (*Nomia* spp.) 28.

The leaf-cutter bee nests in grooved, laminated boards composed of wood, particle board, or polystyrene plastic that can be clamped together to form holes or tunnels (Fig. 6-7). The blocks or nesting boards are placed in shelters that can be moved from field to field (Fig. 6-8). The unique characteristic of the leaf-cutter bee is that it collects pollen to be placed around the egg in the nest. This makes the leaf-cutter a very effective pollinator of alfalfa (28).

In contrast, the alkali bee nests in the soil (Fig. 6-9). These bees have also

FIGURE 6-5
Cages used for honeybee pollination of alfalfa in California.

FIGURE 6-6
The moving of honeybee hives from various pollinating cages.

been cultured, using gravel and soil beds lined with plastic so that the moisture level of the beds may be controlled. Such plastic-lined beds can be moved if desired, though this is difficult. Alkali bees also need alkaline conditions. Salt is often added to the surface of the soil to satisfy this requirement (28).

FIGURE 6-7
Leaf-cutter bee placing a leaf segment into a nesting site.

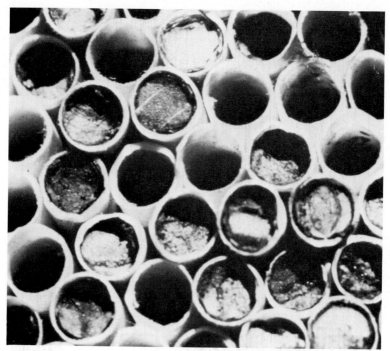

FIGURE 6-8
Close-up of the leaf-cutter bee nest.

The bumblebee (*Bombus* spp.) is also an effective pollinator. Bumblebees are very difficult to culture. They nest primarily in undisturbed soil. Since seed fields are cultivated, bumblebees are found in roadside nests, wooded areas, or undisturbed sod. They are too aggressive to domesticate (27).

HARVESTING

Harvest of forage seed involves special equipment. Some species have a tendency to shatter their seed before all of it is ripe. The seed of reed

FIGURE 6-9
Alkali bee on the ground at a nesting site.

canarygrass and birdsfoot trefoil shatters readily. In the case of birdsfoot trefoil, the pods tend to burst open when ripe, releasing seed in all directions. As the pod matures, it also dries up and shrinks, slowly rupturing the seed pod. Shattering is worse during alternating dry and damp days (21, 28).

Most forage seed is now harvested with a combine, either by direct combining or by windrowing the crop and then threshing with a combine (27, 28, 29) (Fig. 6-10). When direct combining is involved, a desiccant may be used to aid the drying down of the foliage. The windrow method allows one to cut the green foliage and dry it out in a windrow. Later the combine is used to pick it up without cutting or shaking the plant, minimizing seed shatter. Alfalfa seed is harvested when about one-half to three-fourths of the seed pods are yellowish brown or buckskin-colored.

With some grasses, such as Kentucky bluegrass, a stripper may be used to harvest the seed. This machine has a comblike action that strips the seed off the plant (27, 31, 39).

SEED CONDITIONING AND STORAGE

Following the field harvest, seed is dried and later cleaned. Cleaning involves the removal of trash, weed seeds, other crop seeds, and foreign material by size and shape separation (screens) or density separation (air pressure, vacuum, or

FIGURE 6-10
Harvesting alfalfa seed.

FIGURE 6-11
Close-up of high-quality alfalfa
seed.

gravity table). A hammer-mill-type machine or rubber rollers help break the seed away from various appendages attached to it. Fans help blow away lightweight trash when the seed passes over a shaker. There are special felt rollers to remove certain weed seeds, such as dodder in alfalfa. Another separation method involves spraying water and iron filings over alfalfa seed. Since dodder seed has a rough seed coat, iron fillings adhere more readily to the dodder seed, and a magnet is used to help separate these seeds. Again, felt rollers are used to remove seeds with rough coats (23, 27, 30).

Legume seed is also scarified. This is the scratching or polishing of the seed coat so that the hard seeds will absorb water and germinate. Scarifying can be done in several ways—by buffing after treatment with a special oil; treating the seeds with acid or heat; irradiating them electrically; or abrading them mechanically (23, 27, 28, 30).

Some legume seeds are now inoculated before being sold. This step involves adding *Rhizobia* bacteria by means of a slurry and vacuum treatment.

Some legume seeds are also pelleted. A claylike substance and fertilizer are added to the legume seed to increase its size (30). This may help produce a more precise seeding rate. It also adds a small amount of fertilizer to help ensure a quick start, and it may help reduce the amount of seed sown per unit area. Pelleted seed is a reliable method to inoculate and get a stand established under adverse planting conditions. However, pelleted seed needs more moisture for germination than unpelleted seed does. Because of the heavier weight per seed, seeding rates of pelleted seed may have to be adjusted to ensure the proper number of seeds per unit area. Forage producers should be aware of the weight difference between coated and uncoated seed to be certain that they are buying the correct amount of actual seed.

Seed is then packaged and stored in low-humidity and low-temperature control chambers. It is often packaged in moisture-proof bags or containers to ensure that the low humidity is maintained when shipped from the drier western states to more humid areas (6). To maintain maximum viability, most seeds are stored at a fixed moisture content and at a fixed temperature, usually between 0 and 4°C (32 to 39°F). In this temperature range, water in the seeds does not freeze, but the enzyme activities within the seeds, which allow germination and vitality, are retarded. Most storage chambers are held below a relative humidity of 50% (4). A rule of thumb is that relative humidity (RH) + temperature (F°) = <100 for proper seed storage. Seeds samples for purity tests are collected to determine percentages of the seed crop, weed seeds, other crop seed, inert matter, germination, and hard seed in compliance with the Federal Seed Act of 1939 and various state laws (Fig. 6-11). A label is attached with this data, along with the date of the germination test and the name and address of the firm labeling the seed (4, 6).

QUESTIONS

1 Why are most of the forage (legume and cool-season grass) seed-producing areas located in the far west?
2 Why are most forage species grown in rows for seed production?
3 Discuss water management in relation to seed production.
4 Which element is most important for grass seed production?
5 Define a cultural practice and a managerial practice.
6 Which weed is probably the greatest problem for legume seed production?
7 What problem must one be cautious of when applying insecticides to legume crops? To grasses (both warm- and cool-season)?
8 Describe the steps involved in pollination of alfalfa.
9 Name two effective pollen collectors. Why are they good pollinators?
10 How is dodder seed removed from alfalfa seed?
11 Describe the practice of limited generations and its role.
12 What is meant by genetic shift?

REFERENCES

1 Ahlgren, G. H. *Forage Crops,* McGraw-Hill, New York, 2d ed., pp. 455–473, 1956.
2 Ahring, R. M., C. M. Talliaferro, and L. Q. Morrill. Effects of cultural and management practices on seed production of Plaines bluestem. *J. Range Manage.,* 26:413–446, 1973.
3 Ahring, R. M., C. M. Talliaferro, and C. C. Russell. Establishment of Old World bluestem grasses for seed. *Okla. Agr. Exp. Sta. Tech. Bull. T-149,* p. 24, 1978.
4 Association of Official Seed Certifying Agencies. *AOSCA Certification Handbook, Publ. 23,* 1971.
5 Atkins, M. D. and J. E. Smith, Jr. Grass seed production and harvest in the Great Plains. *USDA Farmers' Bull. 2226,* 1967.
6 Bass, L. N., Te May Ching, and F. L. Winter. Packages that protect seeds. In *Seed, USDA Yearbook Agr.,* pp. 330–338, 1961.

7 Bohart, G. E. Insect pollination of forage legumes. *Bee World,* 41:64–97, 1960.

8 Bohart, G. E. Pollination of alfalfa and red clover. *Ann. Rev. Entomol.,* 2:1–28, 1957.

9 Bolton, J. L. *Morphology and Seed Setting: Alfalfa,* Interscience, New York, pp. 97–114, 1962.

10 Bula, R. J. and C. S. Garrison. Fall regrowth response to Ranger and Vernal alfalfa as related to generation of seed increase and area of seed production. *Crop Sci.,* 2:156–159, 1962.

11 Bula, R. J., R. G. May, C. S. Garrison, C. M. Rincker, and J. G. Dean. Comparisons of floral response of seed lots of Dollard red clover. *Crop Sci.,* 5:425–428, 1965.

12 Bula, R. J., R. G. May, C. S. Garrison, C. M. Rincker, and J. G. Dean. Floral response, winter survival, and leaf mark frequency of advanced generation seed increases of Dollard red clover. *Crop Sci.,* 9:181–184, 1969.

13 Bula, R. J., R. G. May, C. S. Garrison, C. M. Rincker, and D. R. McAllister. Growth responses of white clover progenies from five diverse geographic locations. *Crop Sci.,* 4:295–297, 1964.

14 Cooper, H. W., J. E. Smith, and M. D. Atkins. Producing and harvesting grass seed in the Great Plains. *USDA Farmers' Bull. 2112,* 1957.

15 Crop Reporting Board. Seed crops annual summary 1977. *USDA Stat. Rep. Serv.,* 1978.

16 Garrison, C. S. and R. J. Bula. Growing seeds of forages outside their regions of use In *Seed, USDA Yearbook Agr.,* pp. 401–406, 1961.

17 Grandfield, C. O. Alfalfa seed production as affected by organic reserves, air temperature, humidity, and soil moisture, *J. Agr. Res.,* 70:123–132, 1945.

18 Hardison, J. R. Chemotherapeutic eradications of *Ustilago striiformis* and *Urocystis agropyri* in *Poa pratensis* 'Merion' by root uptake of α(2,4-dichlorophenyl)-α phenyl-s-pyrimidinemethanol (EL-273). *Crop Sci.,* 11:345–347, 1971.

19 Hardison, J. R. Chemotherapy of smut and rust pathogens in *Poa pratensis* by thiazole compounds. *Phytopathology,* 61:1396–1399, 1971.

20 Hardison, J. R. Commercial control of *Puccinia striiformis* and other rusts in seed crops of *Poa pratensis* by nickel fungicides. *Phytopathology,* 53:209–216, 1963.

21 Harmond, J. E., J. E. Smith, Jr., and J. K. Park. Harvesting the seeds of grasses and legumes. In *Seed, USDA Yearbook Agr.,* pp. 181–188, 1961.

22 Harrison, J. R. Disease problems in forage seed production and distribution. In *Grassland Seeds,* Van Nostrand, Princeton, NJ, 1957.

23 Klein, L. M., J. Henderson, and A. D. Stoesz. Equipment for cleaning seeds. In *Seed, USDA Yearbook Agr.,* pp. 307–321, 1961.

24 Larkin, R. A. and H. O. Graumann. Anatomical structure of the alfalfa flower and an explanation of the tripping mechanism. *Bot. Gaz.,* 116:40–52, 1954.

25 Laude, H. M., E. H. Stanford, and J. A. Enloe. Photoperiod, temperature, and competitive ability as factors affecting the seed production of selected clones of ladino clover. *Agron. J.,* 50:223–225, 1958.

26 McLennan, J. E., R. Greenshields, and R. M. MacVicar. A genetic analysis of population shifts in pedigree generations of Lasalle red clover. *Can. J. Plant Sci.,* 40:509–515, 1960.

27 Pedersen, M. W. and C. S. Garrison. Legume and grass seed production. In *Forages,* Iowa State Univ. Press, Ames, IA, 3d ed., pp. 105–113, 1973.

28 Pedersen, M. W., G. E. Bohart, V. L. Marble, and E. C. Klostermeyer. Seed production practices. In C. H. Hanson (ed.), *Alfalfa Monograph,* Amer. Soc. Agron., Madison, WI, 1972.

29 Pederson, M. W., L. G. Jones, and T. H. Rogers. Producing seeds of the legumes. In *Seed, USDA Yearbook Agr.,* pp. 171–181. 1961.

30 Purdy, L. H., J. E. Harmond, and G. B. Welch. Special processing and treatment of seeds. In *Seed, USDA Yearbook Agr.,* pp. 322–330, 1961.
Rogler, G. A., H. H. Rampton, and M. D. Atkins. Production of grass seeds. In *Seed, USDA Yearbook Agr.,* pp. 163–171, 1961.

31 Rollin, S. F. and F. A. Johnston. Our laws that pertain to seeds. In *Seed, USDA Yearbook Agr.,* pp. 482–492, 1961.

32 Smith, Dale. Influence of area of seed production on the performance of Ranger alfalfa. *Agron. J.,* 47:201–205, 1955.

33 Smith, Dale. Performance of Narragansett and Vernal alfalfa from seed produced at diverse latitudes. *Agron. J.,* 50:226–229, 1958.

34 Tysdal, H. M. Influence of tripping, soil moisture, plant spacing, and lodging on alfalfa seed production. *J. Amer. Soc. Agron.,* 38:515–535, 1946.

35 Tysdal, H. M. Is tripping necessary for seed setting in alfalfa? *J. Amer. Soc. Agron.,* 32:570–585, 1940.

36 Vansell, G. H. and F. E. Todd. Alfalfa tripping by insects. *J. Amer. Soc. Agron.,* 38:470–488, 1946.

37 Vansell, G. H. Use of honeybees in alfalfa seed production. *USDA Cir. 876,* 1951.

38 Wheeler, W. A. and D. D. Hill. *Seed Production of Grassland Seeds,* Van Nostrand, Princeton, NJ, 1957.

CHAPTER 7

NUTRITIVE EVALUATION OF FORAGES

Forages contain not only a high percentage of fibrous carbohydrates, but also contain protein, minerals, and vitamins, to name a few of their constituents. Forages are generally divided into the readily digestible portion and the less-digestible, fibrous components. The highly digestible portion includes protein, sugars, starch, and organic acids. Total digestible nutrients (TDN) vary with stage of maturity within and between species.

NUTRITIVE VALUE OF FORAGES

The nutritive availability of the various components of forages vary, as shown in Table 7-1. The cellular portion represents the available protein, carbohydrates, starch, pectin, and acids, whereas the cell wall is composed of lignin, cellulose, and hemicellulose. The cell wall, or structural, portion of the plant represents the total fibrous fraction and contributes 98% of the nondigestible fraction.

Forage species vary in their composition and nutritive value, as shown in Table 7-2. Relative values of energy and protein content of common feedstuffs can be compared to those of corn and soybean meal (Table 7-2a). Digestibility may be expressed as dry matter, energy, organic matter, or TDN. One can express digestibility as apparent digestibility: a percentage of the balance of forage ingested, less the dry matter lost in the feces. Digestibility also may be expressed in terms of digestible energy and TDN; this method tries to equate the various forages or feeds on an energy basis. Another method is true digestibility: the actual digestibility of a forage, measured as the difference between intake and fecal loss of undigested material. This value is always higher than apparent digestibility because part of the feces is a metabolic loss. The fibrous portion of a

TABLE 7-1
NUTRITIVE COMPONENTS OF FORAGES

Component	Availability	Factors limiting animal utilization
Cellular contents		
Soluble		
carbohydrate	100%	Intake
Starch	90+	Passage and intake
Organic acids	100	Intake and toxicity
Protein	90+	Fermentation
Pectin	98	Fermentation
Plant cell wall		
Cellulose	Variable	Lignification, cutinization, and silicification
Hemicellulose	Variable	Lignification, cutinization, and silicification
Lignin, cutin, and silica	Indigestible	Limit use of cell wall
Tannins and polyphenols	Limited (?)	Inhibit proteases and cellulases

TABLE 7-2
NUTRITIVE VALUE OF COMPOSITIONS OF VARIOUS FEEDS

Feed	Crude protein	TDN	Cell wall	Lignin
Corn grain	11%	89%	10%	1%
Alfalfa hay (early)	22	67	41	6
Alfalfa (all analyses)	17	57	52	8
Birdsfoot trefoil	16	61	44	10
Corn silage	8	70	45	3
Orchardgrass (very early)	15	71	46	4
Sorghum silage	8	58	59	8
Smooth bromegrass before bloom	17	62	62	5
Tall fescue ('Ky 31')	10	59	74	5
Timothy (all analyses)	8	54	68	7
'Coastal' bermuda-grass	8	57	75	9
Barley straw	5	47	80	8
Wheat straw	4	46	85	12

TABLE 7-2a

FEED EVALUATION FACTORS FOR ESTIMATING RELATIVE VALUES OF THE
ENERGY AND PROTEIN CONTENT OF COMMON FEEDSTUFFS COMPARED
TO CORN AND SOYBEAN MEAL
(As is or wet basis)

Feed	% Fiber	Feed evaluation factors	
		Corn	Soybean meal
Dry forages			
Alfalfa hay, low quality	Over 36	0.263	0.153
Alfalfa hay, average	30–36	0.296	0.212
Alfalfa hay, high quality	Below 30	0.296	0.259
Bromegrass hay	(Average)	0.415	0.060
Clover, red, hay		0.412	0.106
Marsh or swamp hay		0.383	0.037
Mixed hay, good, less than 30 percent legume		0.427	0.039
Oat hay		0.423	0.049
Oat straw		0.326	−0.035
Orchardgrass hay		0.450	0.028
Sorghum-sudangrass hay		0.421	0.035
Soybean hay, average		0.236	0.196
Timothy hay		0.469	−0.003
Wheat hay		0.427	0.011
Silages			
Corn, dent, well-matured, well-eared		0.265	−0.011
Oats (headed out)		0.163	0.017
Peas and oats		0.165	0.021
Sunflower		0.109	0.007
Grains			
Barley		0.908	0.093
Beet pulp		0.931	−0.051
Brewers' grains		0.374	0.464
Corn, dent		1.000	0.000
Corn and cob meal		0.918	−0.018
Distillers' dried corn grain with solubles		0.710	0.350
Linseed meal		0.201	0.699
Milo grain		0.916	0.056
Molasses, cane		1.058	−0.169
Oats		0.806	0.095
Rye		0.765	0.116
Screenings (low fiber)		0.534	0.134
Screenings (high fiber)		0.432	0.134
Soybean seed		0.352	0.746
Soybean meal		0.000	1.000
Sunflower meal		−0.267	1.114
Wheat		0.875	0.125
Wheat bran		0.619	0.218
Wheat standard middlings		0.743	0.222
Whey		0.839	0.140
Yeast, brewers'		−0.113	0.937

TABLE 7-3
CHEMICAL ANALYSIS OF PLANT TISSUE

Fraction	Reagent	Treatment	Yield
Neutral-detergent fiber (NDF)	Na lauryl sulfate EDTA, ph 7.0	Boil 1 hr	Total plant cell wall
Acid-detergent fiber (ADF)	Acetyl trimethylammonium bromide in 1 normal H_2SO_4	Boil 1 hr	Lignocellulose + SiO_2
Lignin	$KMnO_4$, pH 3.0	Treat 1½ hr at 20°C	Lignin as loss in weight by oxidation
Cellulose	None	Ash residue from lignin step	Loss in weight
Silica (SiO_2)	Concentrated HBr (48%)	Treat ash dropwise, 1 hr at 25°C	Residue is SiO_2, and soil silicates
Hemicellulose	None	Calculate as NDF − ADF	Difference

forage is very important to ruminants. A minimal amount is needed in the diet for adequate rumen function. The fibrous portion is also very important for milk production. Forages that are finely ground or pelleted will not maintain the normal ratio of volatile fatty acids in the rumen, nor will the milk fat level be high enough. Therefore, fiber needs to be maintained, not destroyed.

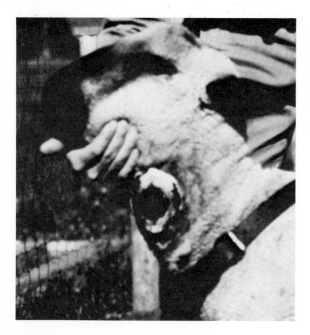

FIGURE 7-1
Sheep fitted with an esophageal fistula used to sample the diet during grazing. A plastic lined canvas bag is attached to collect the consumed forage. *(Courtesy USDA-ARS, Agron, and Plant Genetics Dept., University of Minnesota)*

METHODS FOR DETERMINING NUTRITIVE VALUE

There are various chemical tests used to determine the composition of the various fractions of the cell (Table 7-3). Physical methods of determining nutritive value of forages have included measurements of solubility, resistance to grinding, density, and porosity. A more recent method of determining the nutritive value is the use of infrared reflectance at selected wavelengths (8a, 10). This method is rather versatile in that many components of quality can be ascertained and compared to results of conventional methods. A correlation can be run with values derived from conventional means. Crude fiber, protein, TDN, and dry matter are some of the values easily obtained in this way. The infrared reflectance method can be run rapidly. Its results are easy to obtain, no large supply of chemicals is needed, and it is environmentally clean.

Biological methods are also available for evaluating forages. These methods include: (1) production of milk; (2) weight gains of cattle, sheep, goats, rabbits, meadow voles, or crickets; and (3) digestion trials by cattle, sheep, goats, and rabbits (Figs. 7-1, 7-2). In vitro dry matter disappearance has been used a great deal; this method involves water, rumen fluid, and physiological buffers (5, 6) (Figs. 7-3, 7-4). Another method of determining digestibility uses a nylon bag placed inside the rumen of live animals (23).

Crude or total protein is measured by obtaining the N content and multiplying it by 6.25. This figure includes the nonprotein N (NPN). NPN includes glutamine, glutamic acid, asparagine, aspartic acid, gamma-aminobutyric acid, and nitrates. In silages, the NPN includes ammonia, amines, and amine salts.

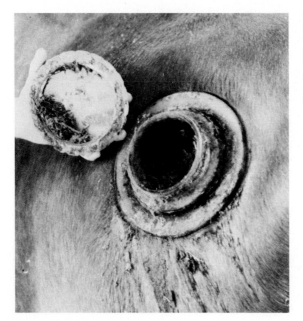

FIGURE 7-2
Cow fitted with a rumen cannula. When the cap is removed, rumen contents can be removed to use in in vitro rumen fermentation analysis. *(Courtesy USDA-ARS, Agron, and Plant Genetics Dept., University of Minnesota)*

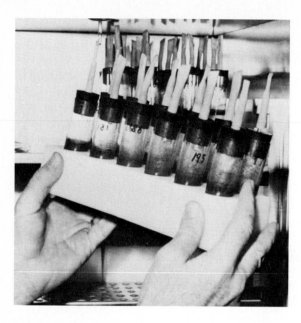

FIGURE 7-3
Series of corked centrifuge tubes
fitted with bunsen valves to
enable escape of gasses during
in vitro rumen fermentation in the
incubator. *(Courtesy USDA-ARS,
Agron, and Plant Genetics
Dept., University of Minnesota)*

There are many digestible organic acids in forages. Soluble carbohydrates include glucose, sucrose, fructose, fructosan, and amylose starches. Starch is highly digestible and is the main storage carbohydrate in many plants.

Cellulose is insoluble, and digestible only by microbial action. It is broken down to volatile fatty acids, which in turn are absorbed by the animal through the rumen wall. Pectin is found in the middle lamella of the plant cell wall and is very digestible. Hemicellulose makes up part of the plant cell wall. It is not highly digestible because it is linked to lignin. Lignin is found in the plant cell

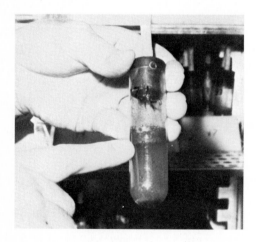

FIGURE 7-4
Floating forage particles in a centrifuge
during in vitro rumen fermentation.
*(Courtesy USDA-ARS, Agron, and Plant
Genetics Dept., University of Minnesota)*

wall, and is not digestible. Cutin is a waxlike covering on the leaf surface. It is linked to lignin and lowers the digestibility of cellulose and hemicellulose. When forages are heated to about 60°C (140°F), the protein becomes tied up with lignin and is not digestible. This is called the maillard reaction (18, 26). Tannin is often associated with lignin when it is present. Tannin creates a bitter taste, reduces palatability, and precipitates proteins.

Plant lipids or ether extracts may influence palatability. Glycosides may also be present in forages (23). Vitamins A and D are present in high-quality forages. Vitamin B is snythesized in the rumen. Carotene, a yellow pigment, is a source of vitamin A. Sunlight destroys carotene rapidly. Therefore, silages and freshly cut forage are higher in vitamin A than hay is.

Mineral composition of forages is usually fairly high if the forage is provided with adequate soil fertilization. Forage mineral concentration varies greatly with the soil fertility level (Table 7-4) (21).

FACTORS AFFECTING NUTRITIVE VALUE

Environmental factors, along with various methods of harvesting, have a great influence on the nutritive value of a forage. Climatic factors, such as light, temperature, and moisture, greatly influence the final quality value of forages. Light is essential to photosynthesis, which produces sugars and organic acids. Light alone increases digestibility of plant tissue.

Increasing temperature increases the metabolic processes within a plant, decreases the production of sugars, increases lignification, and lowers the plant's digestibility (18). Moisture may be excessive or lacking. If moisture is limited,

TABLE 7-4
MINERAL COMPOSITION LEVELS OF FIRST-CUT ALFALFA UNDER
DIFFERENT LEVELS OF FERTILIZATION

Element	Low (less than)	Sufficient	High (more than)
N	2.51%	2.51–3.70%	3.70%
P	0.26	0.26–0.70	0.70
K	2.41	2.41–3.80	3.80
Ca	0.50	0.50–3.00	3.00
Mg	0.31	0.31–1.00	1.00
S	0.31	0.31–0.50	0.50
Zn	20.0 ppm	20.0–71.0 ppm	71.0 ppm
B	30.1	30.1–80.0	80.0
Mn	21.0	21.0–200.0	200.0
Fe	30.0	30.0–250.0	250.0
Cu	3.0	3.0–30.0	30.0

Source: D. Smith, "Chemical composition of herbage with advance in maturity of alfalfa, medium red clover, ladino clover, and birdsfoot trefoil," *Wisc. Agr. Exp. Sta. Res. Report 16* (1964).

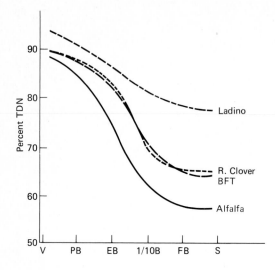

FIGURE 7-5
Percentage of TDN of four legume
species with advance in maturity.
(Courtesy Wisc. Ag. Exp. Sta.) (21)

maturity may be hindered or the plant may go into a state of dormancy. In most species this lowers digestibility.

Nutritive quality is reduced with physiological maturity. Figure 7-5 shows that the percent TDN lowers with maturity. This one factor probably plays the greatest role in forage quality.

Nitrogen fertility greatly influences the crude protein content, especially with grasses. Nitrogen also depresses soluble carbohydrates in stems and leaves of grasses, and increases lignification. Nitrogen increases alkaloid levels, which influence digestibility.

GRAZING TRIALS

Animal performance is the final aspect in determining forage quality. Many forages are grazed, so grazing experiments are needed to determine animal response or animal product per unit area of land. Grazing trials should measure quality and quantity of the forage produced (8).

The area selected for a grazing trial should be representative of the area where the forage will be grown. There should be sufficient replication of the site and forage grown. Factors to be considered by the experimenter are site, uniformity of soil type, topography, and previous treatments. The design should be simple. Randomized block designs will usually control soil variability.

Animal variability creates the most error when studying forage quality. It has been suggested that two to three pastures per treatment, with 5 to 11 animals per pasture, are adequate (12). When studying the yield of animal products produced per area, three or more replications are needed, with at least two animals per pasture. Pasture size dictates the number of animals needed.

Tester Animals

Selecting the proper animal for a grazing trial is very important. One should select the type or class of animal with which the forage is to be utilized. For example, one should select beef animals in lieu of dairy animals for studying the rate of gain per day or per unit area; conversely, one should select dairy animals to study the influence of forage on milk production. Several researchers have reviewed the experimental design, blocking effect of animals, and so on, that are recommended when using animals to evaluate forages (14).

Forage production and animal production are generally not synonymous (Fig. 7-6). Forage growth may exceed animal demands at various times of the year, while at other times animal demands will exceed forage production. Therefore, the researcher and the farmer experience similar problems in relation to management and testing procedures. Animal scientists have developed a technique known as tester animals and the put-and-take method (14). In this technique, the number of tester animals is selected on the basis of the minimum number of animals or the carrying capacity at the low-production cycle of the forage. The put-and-take animals are a reserve group of animals used for grazing the excess forage beyond that which the tester animals could remove. The total number of animals, both tester and the put-and-take group, vary throughout the forage production cycle. The tester animals are used for obtaining the actual data related to gain or production per day or per hectare (17), which also

FIGURE 7-6
High-quality forages are leafy, fine-stemmed, palatable, and highly digestible.

indicate quality. The put-and-take group, along with the tester animals, will give the carrying capacity, or animal grazing days per hectare. Water is usually withheld from the animals for one day prior to going on test and for one day prior to coming off test. This is called the shrink test: getting rid of excessive body water in order to obtain a more accurate body weight (13).

There are many factors to consider when evaluating the nutritive value of a forage. Time, money, labor, laboratory facilities, availability of animals, the use of the forage, and many other factors must be considered by the researcher.

QUESTIONS

1 What portion of plant tissue is most digestible? What portion is not digestible or is very low in digestibility?
2 What are the various methods of describing digestibility?
3 Describe the various components that comprise the nutritive value of a forage.
4 What are the factors that affect the nutritive value of forages?
5 Describe the use of animals in grazing trials.

REFERENCES

1 Campbell, A. G., D. S. M. Phillips, and E. D. O'Reilly. An electronic instrument for pasture yield estimation. *J. Brit. Grassl. Soc.,* 17:89–101, 1962.
2 Crampton, E. W. Interrelationship between digestible nutrient and energy content, voluntary dry matter intake, and overall feeding value of forages. *J. Anim. Sci.,* 16:546–552, 1957.
3 Donefer, E. Forage solubility measurements in relation to nutritive value. *Proc. Natl. Conf. Forage Qual. Eval. Util.,* Nebr. Center Cont. Ed., Lincoln, pp. Q1–Q6, 1970.
4 Evans, R. A. and M. B. Jones. Plant height times total ground cover versus clipped samples for estimating forage production. *Agron. J.,* 50:504–506, 1958.
4 a Fales, S. L., D. A. Holt, V. L. Lechtenberg, K. Johnson, M. R. Lodisch, and A. Anderson. Fractionation of forage grass carbohydrates using liquid (water) chromatography. *Agron. J.,* 74:1074–1077, 1982.
4 b Henderson, M. S. and D. L. Robinson. Environmental influences on fiber component concentrations of warm-season perennial grasses. *Agron. J.,* 74:573–579, 1982.
4 c Henderson, M. S., and D. L. Robinson. Environmental influences on yield and in-vitro true digestibility of warm-season perennial grasses and the relationships to fiber components. *Agron. J.,* 74:943–946, 1982.
5 Johnson, R. R. Techniques and procedures for *in vitro* and *in vivo* rumen studies. *J. Anim. Sci.,* 25:855, 1966.
6 Johnson, R. R. The development and application of *in vitro* rumen fermentation methods for forage evaluation. *Proc. Natl. Conf. Forage Qual. Eval. Util.,* Nebr. Center Cont. Ed., Lincoln, pp. M1–M18, 1970.
7 Klingman, D. L., S. R. Miles, and G. O. Mott. The cage method of determining consumption and yield of pasture herbage. *Agron. J.,* 35:739, 1943.
8 Marten, G. C. Measurement and significance of forage palatibility. *Proc. Natl. Conf. Forage Qual. Eval. Util.,* Nebr. Center Cont. Ed., Lincoln, pp. D1–D55, 1970.

8 a Marten, G. C., J. L. Halgerson, and J. H. Cherney. Quality prediction of small grain forages by near infrared reflectance spectroscopy. *Crop Sci.*, 23:94–96, 1983.

9 Matches, A. G. Pasture research methods. *Proc. Natl. Conf. Forage Qual. Eval. Util.*, Nebr. Center Cont. Ed., Lincoln, pp. I1–I32, 1970.

10 Morris, K. H., R. F. Barnes, J. E. Moore, and J. S. Shenk. Predicting forage quality by infrared reflectance spectroscopy. *J. Anim. Sci.*, 43:889, 1976.

11 Mott, G. O. Animal variation and measurement of forage quality. Symposium on forage evaluation. IV. *Agron. J.*, 51:223–226, 1959.

12 Mott, G. O. Evaluating forage production. In *Forages,* Iowa State Univ. Press, Ames, IA, 3d ed., pp. 126–135, 1973.

13 Mott, G. O. Grazing pressure and the measurement of pasture production. *Proc. 8th Intl. Grassl. Congr.*, pp. 606–611, 1960.

14 Mott, G. O. and H. L. Lucas. The design, conduct, and interpretation of grazing trials on cultivated and improved pastures. *Proc. 6th Intl. Grassl. Cong.*, pp. 1380–1385, 1952.

14 a Munoz, A. E., E. C. Holt, and R. W. Weaver. Yield and quality of soybean hay as influenced by stage of growth and plant density. *Agron. J.*, 75:147–149, 1983.

15 Owen, J. B. and W. J. Ridgman. The design and interpretation of experiments to study animal production from grazed pasture. *J. Agr. Sci.*, 71:327–335, 1968.

16 Peterson, R. G. and H. L. Lucas. Experimental errors in grazing trials. *Proc. 8th Intl. Grassl. Congr.*, pp. 747–750, 1960.

17 Peterson, R. G., H. L. Lucas, and G. O. Mott. Relationship between rate of stocking and per animal and per acre performance on pasture. *Agron. J.*, 57:27–30, 1965.

17 a Pitman, W. D. and E. C. Holt. Environmental relationships with forage quality of warm-season perennial grasses. *Crop Sci.*, 22:1012–1016, 1982.

18 Raymond, W. R. The nutritive value of forage crops. *Advan. Agron.*, 21:1–108, 1969.

19 Shenk, J. S., F. C. Elliott, and J. W. Thomas. Meadow vole nutrition studies with alfalfa diets. *J. Nutr.*, 101:1367, 1971.

20 Shrivastava, J. P., D. A. Miller, and J. A. Jackobs. Estimating alfalfa yields. *Agron. J.*, 61:649–651, 1969.

21 Smith, D. Chemical composition of herbage with advance in maturity of alfalfa, medium red clover, ladino clover, and birdsfoot trefoil. *Wisc. Agr. Exp. Sta. Res. Report 16,* 1964.

22 Teare. I. D., G. O. Mott, and J. R. Eaton. Beta attenuation—a technique for estimating forage yield *in situ. Rad. Bot.*, 6:7–11, 1966.

23 Tinnimit, P. and J. W. Thomas. Forage evaluation using various laboratory techniques. *J. Anim. Sci.*, 43:1058, 1976.

24 Tyree, W. A., W. H. Pfander, and P. C. Stone. Response of crickets to amounts of forage in the diet. *J. Dairy Sci.*, 59:164, 1976.

25 Van Soest, P. J. Estimations of nutritive value from laboratory analysis. *Proc. Cornell Nutr. Conf.*, pp. 106–117, 1971.

26 Van Soest, P. J. The chemical basis for the nutritive evaluation of forages. *Proc. Natl. Conf. Forage Qual. Eval. Util.*, Nebr. Center Cont. Ed., Lincoln, pp. U1–U19, 1970.

27 Van Soest, P. J. Symposium of factors influencing the voluntary intake of herbage by ruminants: Voluntary intake in relation to chemical composition and digestibility. *J. Anim. Sci.*, 24:834, 1965.

28 Waldo, D. R. Factors influencing the voluntary intake of forages. *Proc. Natl. Conf. Forage Qual. Eval. Util.*, Nebr. Center Cont. Ed., Lincoln, pp. E1–E22, 1970.

CHAPTER 8

ANIMAL DISORDERS RELATED TO FORAGES

Since forages play such an important role in overall animal production, it is crucial not only to properly manage the crop but at the same time to be alert to any potential problems that may arise. Most forage crops are more palatable and nutritious during the plant's young, immature stage. Most producers take advantage of this quality factor, but this is also when the crops are most likely to cause animal disorders. In each chapter dealing with a specific forage species, most of the potential disorders that may arise from it are discussed. It is important to understand the basic nutritional disorders in relation to forage management and how they may be controlled or treated. Forage disorders are real dangers, but even potentially dangerous forages can be utilized safely by knowing the danger and by taking reasonable precautions against it.

BLOAT

Bloat is usually observed in ruminant animals grazing legumes. Bloat seldom occurs in the feed lot, but it has been found. Feed lot bloat has been generally associated with the consumption of some succulent legume (7). Bloat is the result of the formation of a stable foam in the rumen of an animal. Gases normally lost by belching are retained, and pressure is increased in the left chamber of the rumen; finally, the eructation mechanism is inhibited. When this occurs, the animal lacks oxygen and eventually will choke to death (2).

It has been reported that bloat may occur in only 1 to 2% of dairy animals, which is not an extremely high percentage (6). But when a very high producing animal dies, it is a great loss to the individual producer concerned.

There are two kinds of bloat: (1) gaseous and (2) foaming. Bloat is the result of a complex interaction of plant, animal, and microbial factors. There appears to be some hereditary difference in cattle that have a tendency to bloat more easily than do others (7).

A great deal of research has been conducted in an effort to determine the cause of bloat. In the past, many researchers thought that plant saponins were the cause (26). Others have indicated that pectins and particulate matter may cause bloat (21, 30). There is a great deal of support for the theory that bloat is caused by Fraction 1 of the plant protein material (23). The Fraction 1 protein is an 18-S protein, believed by Canadian workers to be a primary cause of bloat (23). It is a high-molecular-weight protein and is water-soluble. The 18-S refers to 18 Svedberg units of relative density (23).

One can reduce the incidence of bloat by not turning hungry animals directly into a lush, green legume pasture. It is suggested that some dry hay or silage should be fed before allowing the animals to graze such a pasture. One could cut two swatches around the pasture, allowing this material time to dry, so that when the animals go to the pasture they have available some dry roughage before they consume the fresh legumes. Cattle should be put on pasture gradually instead of turning them in suddenly. Birdsfoot trefoil, crownvetch, and the lespedezas are legumes that do not cause bloat. Any of these legumes could be put into a mixture to help reduce the incidence of bloat. Birdsfoot trefoil is the most widely used of the three (17, 30). Bloat is also greatly reduced when at least 50% of the pasture is grass. Ready access to salt and water when grazing a legume pasture will also help reduce the incidence of bloat (30). Another good bloat control is not grazing pastures during heavy dews (30).

In recent years, poloxalene has proven to be a good bloat deterrent (2, 27). Cattle should be put on poloxalene several days before turning them into a legume pasture that may cause bloat. Be sure that all the cattle have eaten enough of the compound to control bloat. This takes approximately 1 to 2 g of poloxalene per 45 kg (100 lb) of body weight per day (27). Cattle should be fed poloxalene with hay or some roughage before they are turned out on legume pasture for the first day. Poloxalene may also be put in salt blocks or added to molasses liquid-feeders near the cattle's water or rest areas. One lick wheel should be provided for every 25 cattle. Salt and minerals should be available for free choice at all times. For the first few days of grazing, one should check the cattle at regular intervals for any signs of bloat. If bloat develops in an animal it should be drenched with a concentrated poloxalene material for quick relief.

GRASS TETANY

Grass tetany is a disorder that is associated with Mg deficiency in the blood of animals, especially ruminants. Cattle that have just calved or that are lactating heavily may have a tendency to show grass tetany, especially when grazing lush spring grass (20, 30). High nitrate content of forage and a low energy intake will also increase the incidence of grass tetany (15, 20). One of the best ways to reduce this disorder's incidence among animals grazing a grass pasture is to include a legume in the mixture, because of the legume's higher Mg content (15, 20).

Grass tetany occurs most often during the early spring, but it may also occur during the fall, especially when the mean temperature is between 8 and 14°C (46 and 58°F) and the season has been preceded by an exceptionally wet period, which tends to create lush, rapid growth of grass. In addition, many cases of tetany have been associated with grazing a small-grain pasture during the early spring or late winter months (18, 30).

Soil fertility plays an important role in grass tetany. Most soils sre sufficiently high in Mg. Fertilizing grasses with N will increase the Mg concentration of the grass, but applying high rates of K to forage grasses will decrease the Mg content (18, 19, 20, 30). Therefore, when high rates of N and K are applied in combination to a grass, the result may be a decrease of magnesium in the animals' blood and an increase in the incidence of grass tetany.

There are several preventions or treatments for grass tetany. One has already been mentioned—that legumes should be added to the grass. Other methods are free choice hay feeding and deferring grazing until plant growth is 12.5 to 17.5 cm (5 to 7 in) high. Another method of prevention is to feed a 15% mineral mixture to cattle for about 1 month prior to turning them in to a pasture of a rapidly growing perennial grass in the early spring (16, 24). In the case of acute grass tetany, one should give an intravenous injection of a solution containing d-sacclonate and gluconate salts of Ca with dextrose as well as P and Mg to the cattle that have been affected, before they lose consciousness (15, 30). Subsequent treatment involves daily administration of MgO. An MgO drench also may be given to acutely affected cattle (16).

FESCUE FOOT

Tall fescue has been severely criticized because of the chances of cattle getting fescue foot poisoning when grazed on the plant for extended periods of time. The criticism is largely unjustified, however, because the disorder is usually associated with poor management. Fescue foot is a soreness, not a foot rot. If cattle develop fescue foot they can be cured by removing them from tall fescue for several weeks. Feeding dry hay or grain while animals are on tall fescue pasture will help reduce the incidence of fescue foot.

The first stages of fescue foot in cattle are soreness and stiffness, rapid breathing, an increased body temperature, loss of weight, and a rough haircoat (Fig. 8-1) (37). Later, a slight limp may develop in the hind quarters, usually in the left hind leg. As a result of poor blood circulation, a dry gangrene develops, which may affect the hooves or tails and cause them to eventually drop off (30, 37). In contrast to these symptoms, cattle foot rot produces a swelling between the animal's hooves. Fescue foot is thought to be caused by an alkaloid, possibly perloline, that has been found in tall fescue and linked to this particular disease (8, 10, 37). Producers have also indicated that fescue foot may occur more rapidly when cattle have grazed fescue over long periods of time; in low wet areas, where there are consistently heavy dews; or in swampy areas (8, 10, 37).

FIGURE 8-1
Early stages of fescue foot poisoning of cattle. Note rough haircoat, swollen left hoof, and arched back.

Adequate fertilization and proper management are the two best preventive methods for reducing the incidence of fescue foot. When excess growth is left and grazed the following year, the incidence of fescue foot is much greater. Therefore, complete removal of the roughage late in the winter each year is highly advisable to prevent buildup of any potential disease organisms in fescue forage (9, 10, 22).

PLANT ALKALOIDS

Alkaloid- or nitrogen-based chemicals occur naturally in plants and are associated with some fungal growth. High N applications are associated with alkaloid problems.

Some of the diseases or animal disorders that are associated with alkaloids are ryegrass staggers, which is found in sheep, and *Phalaris* staggers, which is found in sheep, dairy, and beef animals (1, 32). These diseases are often found when new growth of perennial grasses occurs in the fall, when the alkaloid content may be very high (32). They are most common when grazing new growth following heavy rains, during especially cool climate conditions, and when high levels of nitrogen are present (1, 32). There are different types of symptoms associated with staggers in animals. The most common symptom is sudden collapse and death of an animal that has been grazing on forage. The disease affects the central nervous system (30).

One may prevent staggers by placing cobalt pellets in the rumen. One can also select a low-alkaloid line of reed canarygrass as a preventive measure. When an acute case occurs, there is no responsive treatment available (30).

NITRATE POISONING

Nitrate poisoning results when cattle consume too much forage that is very high in nitrates, or drinking water that is high in nitrates. Nitrate poisoning will kill livestock and cause the abortion of the fetus in a pregnant animal (30, 36). The level of nitrates may increase in weeds, grasses, and legumes because of drought or cloudy conditions, cold weather, or excessive nitrogen in the soil (34, 35, 36). When plants that are high in nitrogen are eaten by cattle, the oxygen-carrying capacity of the blood is affected (36). Nitrate poisoning is found most often in the northeastern part of the United States; there is lower incidence in the southwest due to higher light conditions there. Soils that are high in nitrogen and low in phosphorus and potassium often favor the accumulation of nitrates in plants (25, 36). All unusual growing conditions in usually "safe" forage should alert one for possible nitrate poisoning, such as drought followed by added N fertilization or insect invasion, or low but adequate rainfall, or any combination causing unusual growth.

There are numerous weed species that are high in nitrate content, such as Canada thistle, stinging needle, redberried elder, boneset, and (under drought) redroot pigweed (36). A good weed control program is very effective in helping reduce the incidence of nitrate poisoning in pastures.

Research conducted at Missouri compared various forage species in relationship to their nitrate accumulation. It was found that with high fertilization, sudangrass, tall fescue, and orchardgrass may accumulate high levels of nitrates. In the same study, it was found that timothy, smooth bromegrass, and ladino clover are intermediate accumulators of nitrates, while Kentucky bluegrass, wheat, and alfalfa are low accumulators of nitrates (25). Pearl millet will often contain higher levels of nitrate than will sudangrass (30). For emergency treatment of nitrate poisoning, one should inject Mg or a 2% solution of methylene blue into the blood stream. As preventive measures, nitrogen applications should be limited to pure grass stands. If silage is potentially high in nitrates, test it for nitrate content. Many of the lower leaves of corn have concentrations of nitrates higher than in the higher leaves, so when grazing a droughty cornfield with livestock, one should check to see if only the lower leaves are eaten, or avoid grazing it entirely (36).

PRUSSIC ACID POISONING

Prussic acid—HCN or hydrocyanic acid—is a potential problem in grazing sorghums. Sorghums contain the glucoside dhurrin, and prussic acid is produced upon the hydrolysis of this material in the rumen (30). Prussic acid is also a problem in many weeds and in grasses such as johnsongrass and sudangrass. Pearlmillet does not produce prussic acid (30). There are different cultivars of sudangrass, such as common sudangrass or 'Piper,' that have a low level of dhurrin, while sudangrass-sorghum hybrids often contain a high level of dhurrin (2, 30). As the nitrogen level increases in the soil, so does the dhurrin content in the plant (12). Also, where there are low levels of phosphorus or where the N and P are not in proper balance, there is a greater incidence of prussic acid poisoning (12).

Losses from prussic acid poisoning can be reduced by postponed grazing of sorghums, sudangrass, or sudangrass-sorghum hybirds until they are at least 45 to 50 cm (19 to 23 in) tall (12, 14). These grasses should be grazed down completely, then cattle should be moved to another field so that they will not be forced to graze the dangerous young regrowth. Another potential danger of prussic acid poisoning in sudangrass or johnsongrass is grazing following drought, frost, or freeze. The frost causes a rapid release of prussic acid from the glucosides when the plant cells break down in the rumen (12, 14) Forage may be dangerous from 1 to 6 days following a frost. To reduce the danger of prussic acid poisoning, one can cut and dry the grass or wait until the forage dries standing in the field. Therefore, for several days following a killing frost, cattle should be kept off of sudangrass and johnsongrass. It would probably be fairly safe to graze after 5 or 6 days following a killing freeze. There is no danger in hay made from this material. A further precaution in reducing prussic acid poisoning would be to keep a proper balance of soil fertility, especially of nitrogen and phosphorus.

BLEEDING DISEASE

Sweet clover contains a chemical called coumarin (4), an aromatic compound which gives the plant its characteristic odor when cut. During heating or spoilage of sweet clover hay or silage, coumarin is converted to a toxic chemical called dicoumarol (4). Dicoumarol prevents blood clotting and is used in commerical rat poisons. Affected animals will bleed to death from slight wounds or internal hemorrhaging. This disease can be cured by giving cattle injections of vitamin K and prevented by not feeding moldy hay or silage made from sweet clover (4).

ERGOT

Ergot poisoning in cattle is caused by grazing diseased grasses such as wildrye, small grains, johnsongrass, dallisgrass, and some cultivars of bahiagrass. Ergot is a purple or white fungus growth in the seedheads. It causes abortions and even death in cattle that consume it. The infected seedheads should be removed from the pasture (8).

SLOBBERS

Slobbers is a result of excessive salivation that is associated with the feeding of red clover hay. It is thought to be the result of an alkaloid produced by the fungus *Rhizoctonia leguminicola* Gough and Elliot (28, 30). Red clover has an excessive amount of pubescence present, which may lend itself to the buildup of this particular fungus. Slobbers is associated mostly with horses but also to some extent with cattle (28, 30).

Moldy hays, which may cause a respiratory disease in cattle that is sometimes associated with the feeding of alfalfa hay, will sometimes also create a slobber condition (22, 28, 30).

ESTROGENIC COMPOUNDS

Breeding failures have been associated with estrogenic compounds found in many legumes, particularly ladino clover, alfalfa, subterranean clover, and the annual medics (31). Estrogenic compounds have also been found in some grasses. These estrogenic compounds are known as isoflavones and coumestan compounds. Coumestarol is the main estrogenic compound found in alfalfa and ladino clover (3). It is thought that estrogenic compounds create an over-stimulation of the reproductive organs, resulting in irregular estrus, cystic ovaries, and decreased fertility in dairy animals (3).

An imbalance of plant nutrients, especially deficiencies of P, N, and S, has been found to increase the concentration of estrogenic compounds (30). It has also been found that when alfalfa is relatively disease-free, it will contain lower levels of estrogenic compounds than when it is considerably diseased (3).

Plant breeders are developing cultivars that are lower in estrogenic compounds and that are also resistant to most diseases. This, along with good fertility programs, will help reduce the incidence of irregularities in breeding resulting from these compounds.

MINERAL DEFICIENCIES

Excellent management includes a good fertility program. Proper concentration of the nutrients needed in forage production is also necessary for good animal health. The micronutrients, such as copper, cobalt, iron, and manganese, which are needed in small amounts, may also cause toxicity symptoms if they occur at extremely high levels (22, 29, 30). However, deficiency symptoms will be exhibited under extremely low mineral levels. Forage grasses have a tendency to be lower in calcium, phosphorus, magnesium, and sulfur than is legume forage. In summary, an excellent fertility program will eliminate any potential mineral deficiencies both in forage growth and in forage that is being consumed.

QUESTIONS

1 Name three potential animal disorders that one will likely find when feeding legumes.
2 Name three animal disorders commonly found when feeding grasses.
3 How may bloat be prevented in cattle?
4 One of the greatest problems in utilizing tall fescue is fescue foot poisoning. Describe how this disease attacks an animal and how one may prevent its incidence.
5 What are plant alkaloids and what types of diseases do they cause?
6 Where and when might one find an increased tendency for nitrate poisoning in cattle?
7 What species is associated with prussic acid poisoning? How may this problem be controlled or eliminated?

REFERENCES

1 Aasen, A. J., C. C. J. Culvenor, E. P. Finney, A. W. Kellock, and L. W. Smith. Alkaloids as a possible cause of rye grass staggers in grazing livestock. *Aust. J. Agr. Res.,* 20:71–86, 1969.

2 Bartley, E. E., G. W. Barr, and R. Mickelson. Bloat in cattle. XVII. Wheat pasture bloat and its prevention with poloxalene. *J. Anim. Sci.,* 41:752, 1975.

3 Bickoff, E. M., R. R. Spencer, S. C. Witt, and B. E. Knuckles. Studies on the chemical and biochemical properties of coumestrol and related compounds. *USDA Tech. Bull. 1408,* pp. 1–95, 1969.

4 Campbell, H. A. and K. P. Link. Studies on the hemorrhagic sweet clover disease. IV. The isolation and crystallization of the hemorrhagic agent. *J. Biol. Chem.,* 138:21–33, 1941.

5 Chessmor, R. A. Grazing pure legume pastures. *Proc. Forage Legume Conf.,* Noble Found., Ardmore, OK, 1977.

6 Clarke, R. T. J., and C. S. W. Reed. Legume bloat. In A. T. Phillipson (ed.), *Physiology of Digestion and Metabolism in the Ruminant,* Oriel Press, New Castle, England, pp. 599–606, 1970.

7 Dougherty, R. W. Bloat in ruminants. In *Forages,* Iowa State Univ. Press, Ames, IA, 2d ed., pp. 607–616, 1962.

8 Ellis, J. J. and S. G. Yates. Mycotoxins of fungi from fescue. *Econ. Bot.,* 25:1–5, 1971.

8 a Fales, S. L. and K. Ohki. Manganese deficiency and toxicity in wheat: Influences on growth and forage quality of herbage. *Agron. J.,* 74:1070–1073, 1982.

9 Fribourg, H. A. and R. W. Loveland. Production, digestibility, and perloline content of fescue stockpiled and harvested at different seasons. *Agron. J.,* 70:745–747, 1978.

10 Fribourg, H. A. and R. W. Loveland. Seasonal production, perloline content, and quality of fescue after N-fertilization. *Agron. J.,* 70:741–745, 1978.

11 George, J. R. and J. L. Phull. Cation concentration of N- and K-fertilized smooth bromegrass during the spring grass tetany season. *Agron. J.,* 71:431–436, 1979.

12 Gillingham, J. T., M. M. Shirer, J. J. Starnes, N. R. Page, and E. F. McClain. Relative occurrences of toxic concentrations of cyanide and nitrate in varieties of sudangrass and sorghum-sudangrass hybrids. *Agron. J.* 61:727–730, 1969.

13 Gopeland, B. P., A. C. Fesser, and S. A. Grandt. A possible role for leaf cell rupture in legume pasture bloat. *Crop. Sci.,* 18:129–133, 1978.

14 Gorashi, A. M., P. N. Drolson, and J. M. Scholl. Effect of stage of growth, temperature, and N and P levels on the hydrocyanic acid potential of sorghums in the field and growth room. *Crop Sci.,* 20:45–47, 1980.

15 Grunes, D. L., P. R. Stout, and J. R. Brownell. Grass tetany in ruminants. *Advan. Agron.,* 22:331–374, 1970.

16 Horvath, D. J. and J. R. Todd. Magnesium supplement for cattle. *Proc. 23d Ann. Tex. Nutr. Conf.,* pp. 96–104, 1968.

16 a Howarth, R. E., B. P. Goplen, S. A. Brandt, and K. J. Cheng. Disruption of leaf tissues by rumen microorganisms: An approach to breeding bloat-safe forage legumes. *Crop Sci.,* 22:564–568, 1982.

17 Jones, W. T. and J. W. Lyttleton. Bloat in cattle. XXXIV. A survey of legume forages that do and do not produce bloat. *N. Z. J. Agr. Res.,* 14:101–107, 1971.

18 Karlen, D. L., R. Ellis, Jr., D. A. Whitney, and D. L. Grunes. Influence of soil moisture on soil solution cation concentrations and the tetany potential of wheat forage. *Agron. J.,* 72:73–78, 1980.

19 Karlen, D. L., R. Ellis, Jr., D. A. Whitney, and D. L. Grunes. Soil and plant parameters associated with grass tetany of cattle in Kansas. *Agron. J.,* 72:61–64, 1980.

20 Kemp, A. and M. L. 'Hart. Grass tetany in grazing milking cows. *Neth. J. Agr. Sci.,* 5:4–17, 1957.

21 Mangan, J. L. Bloat in cattle. XI. Foaming properties of proteins, saponins, and rumen liquor. *N. Z. J. Agr. Res.,* 2:47–61, 1959.

22 Matches, A. G. Anti-quality components of forages. *Crop Sci. Soc. Amer. Spec. Publ. 4,* Madison, WI, 1973.

23 Miltimore, J. E., J. M. MacArthur, J. L. Mason, and D. L. Ashby. Bloat investigations. The threshold fraction 1 (18-S) protein concentration for bloat and relationships between bloat and lipid, tannin, Ca, Mg, Ni and Zn concentrations in alfalfa. *Can. J. Anim. Sci.,* 50:61–68, 1970.

24 Murdock, L. W., C. W. Absher, N. Gay, and J. A. Bolling. Grass tetany in beef cattle. *Ky. Agr. Ext. Serv. Cir. ASC–16,* 1975.

25 Murphy, L. S. and G. E. Smith. Nitrate accumulation in forage crops. *Agron. J.,* 59:171–174, 1967.

26 Pedersen, M. W., D. E. Zimmer, D. R. McAllister, J. O. Anderson, M. D. Wilding, G. A. Taylor, and C. F. McGuire. Comparative studies of saponin in several alfalfa varieties using chemical and biochemical assays. *Crop Sci.,* 7:349–352, 1967.

27 Phillips, D. S. M. The use of "Pluronics" administered in the drinking water as a means of bloat control in cattle. *N. Z. J. Agr. Res.,* 11:85–100, 1968.

28 Rainey, D. P., E. B. Smalley, M. H. Crum, and F. M. Strong. Isolation of salivation factor from *Rhizoctonia leguminicola* on red clover hay. *Nature,* 205:203–204, 1965.

29 Reid, R. L. and G. A. Jung. Effects of elements other than nitrogen on the nutritive value of forage. In *Forage Fertilization,* Amer. Soc. Agron., Madison, WI, pp. 395–435, 1974.

30 Reid, R. L. and G. A. Jung. Forage-animal stresses. In *Forages,* Iowa State Univ. Press, Ames, IA, 3d ed., pp. 639–653, 1973.

31 Rossiter, R. C. Factors affecting the oestrogen content of subterranean clover pastures. *Aust. Vet. J.,* 46:141–144, 1970.

32 Simons, A. B. and G. C. Marten. Relationship of indole alkaloids to palatability of *Phalaris arundinacea* L. *Agron. J.,* 63:915–919, 1971.

33 Sullivan, J. T. and R. J .Garber. Chemical composition of pasture plants, with some reference to the dietary needs of grazing animals. *Pa. Agr. Exp. Sta. Bull. 489,* 1947.

34 Sund, J. M. and M. J. Wright. Control weeds to prevent lowland abortion in cattle. *Down to Earth,* 15(1):10–13, 1959.

35 Sund, J. M., M. J. Wright, and J. Simon. Weeds containing nitrates cause abortion in cattle. *Agron. J.,* 49:278–279, 1957.

36 Wright, M. J. and K. L. Davidson. Nitrate accumulation in crops and nitrate poisoning in animals. *Advan. Agron.,* 16:197–247, 1964.

37 Yates, S. G. Toxicity of tall fescue forage: A review. *Econ. Bot.,* 16:295–303, 1962.

CHAPTER 9

PASTURES AND THEIR IMPROVEMENT

Pastures are usually considered to be those areas used primarily for grazing that are seeded or vegetatively propagated with improved or introduced forage species, in contrast to native rangelands. The improved pastureland of the United States and Canada is found primarily in the central and eastern parts of these countries, although irrigated pastureland, and irrigated cropland in rotation with pastureland, are found in the western regions. One may classify pastures into three categories: permanent pastures, rotation pastures, and supplementary or emergency pastures.

This chapter will cover primarily the improvement of permanent pastures. Pasture management will be discussed in a later chapter.

ROTATION PASTURE

Pastures that are grown on a more-or-less regular interval with other crops are referred to as rotation pastures. A rotation pasture may be a mixture of grasses and legumes, but if grasses are grown in a monoculture, nitrogen fertilizer will be necessary for good production. The species grown in rotation pastures depends in part on the number of years that a forage wll be utilized in a rotation. Pastures that will be grown for only 1 year will usually include annual species such as sudangrass, pearlmillet, ryegrass, small grains, or crimson clover. If the pasture will be grown for 2 or more consecutive years, then biennials and perennials will be grown, such as alfalfa, red clover, ladino clover, smooth bromegrass, orchardgrass, timothy, tall fescue, and other cool-season grasses. In the southern states, bermudagrass and dallisgrass, as well as ladino clover and crimson clover, are commonly grown.

Most forages used in rotation pastures are deep-rooted, tall-growing, rapid in establishment, and highly productive. Most rotation pastures are more productive than are permanent pastures and rangelands largely because they are on good land suitable for row crop production, and because they may be managed more intensively. Species selection will vary according to the region and soil type.

Rotation pastures must be reseeded periodically; therefore, they are rather expensive to maintain. Forage species, especially grasses in a cropland rotation, improve soil structure, help prevent soil erosion, reduce leaching, and, if legumes are grown in the mixture and plowed down, increase the nitrogen content of the soil. Yields of row crops are generally increased when seeded on terminated rotational pastureland. Forage crops grown in a rotation also help reduce weeds and diseases that may affect row crops.

SUPPLEMENTARY PASTURES

In contrast to rotation pastures, supplementary, or emergency, pastures are generally grown for 1 year and must be highly productive within a short period after seeding. Most of the species used in this way are annuals.

An unexpected shortage of pasture, caused by loss of stand as a result of disease, flooding, drought, or winter killing, can be corrected economically by replacing the old stand with small grains, such as oats or winter rye, or a warm-season species, such as sudangrass or pearlmillet. Increasing the herd size may require a supplementary pasture until a productive, permanent pasture can be established. If properly established and fertilized, supplementary pastures can supply enough forage in a short time when other pastures are unreliable, unavailable, or unproductive.

PERMANENT PASTURES

Permanent pastures are those maintained indefinitely for grazing purposes. Most of the species that constitute permanent pastures are perennials or self-reseeding annuals. Permanent pastures are generally established by seeding or sprigging or have become established by means of aggressively spreading species that will compete with and crowd out other species that are present. A permanent pasture is very seldom plowed or brought into a rotation with other crops. Many permanent pastures are in areas too wet, steep, or rocky for cultivated crops.

Most permanent pastures are less productive than other pastures are because most permanent ones lack some of the basic nutrients required for good plant growth. Although good managers will apply these nutrients in the form of fertilizer, many of these pastures consist of unproductive or undesirable species. If this is the case, the pasture should be reseeded. These pastures produce good growth primarily in the early spring or late fall, so that only early spring grazing and limited fall grazing are realized. It has been estimated that approximately

70% of permanent pastures in the United States need improvement by introduction of new forage species and proper fertilization.

PERMANENT PASTURE IMPROVEMENT

The simple addition of needed fertilizer and lime in many cases improves the quality of plant growth, and extends the growing period further into the season. The major element needed in permanent grass pastures is nitrogen, although other elements may be limiting. A soil test will indicate which other fertilizers are needed for any given pasture. In high-rainfall areas, soils may be acidic. In these areas the addition of lime to such soils will reduce the acidity and often increase the forage productivity. Added plant nutrients and lime not only increase forage yields, but also the proportion of desired species and the stand density.

RESEEDING

One method of improving permanent pastures has been by complete seedbed preparation and the reseeding of desirable, productive perennial species. Tillage is needed when one reseeds a new permanent pasture rather than renovating an old one. Plowing will produce a better seedbed and provide more help in the establishment of legumes than will using a field cultivator or disk (7, 25). In a study in Wisconsin, it was found that plowing may require 20 to 60% less energy in establishing a legume than does surface cultivating equipment (25). Plowing must be restricted to slopes of less than 5%. If sloping land is involved, plowing should be done on the contour and as late as possible in the fall to help reduce erosion. Fall plowing will help break up the grass sod and allow one to work up a firmer seedbed in the spring. In the south, where many species are planted in the fall, one needs to plow in the summer.

If the new pasture seeding is not managed properly, the tall-growing species may die out and unproductive native grasses and weeds will predominate. Another factor that many people overlook is the lack of proper annual fertilization or top dressing of permanent pasture. Do not plan on the feces droppings of grazing animals providing adequate fertilizer return. Without proper fertilization, legumes soon die out because of low nutrient levels, especially K (18, 25).

PASTURE RENOVATION

As mentioned earlier, old permanent pastures are generally low in fertility, rather weedy, not very productive during the droughty summer months, and overgrazed. The most effective way to improving poor-quality permanent pastures has been found to be complete renovation, rather than just fertilization or reseeding. Renovation is the improvement of a pasture by partial or complete destruction of the sod, plus fertilizing, controlling weeds, and seeding desirable forage species (29). There are a number of steps involved in renovation, and it requires some expense and time. Production may be increased two to five times provided there is good establishment of the new species (18, 19, 24, 27). Additional benefits from renovation are: (1) carrying capacity is increased; (2) good growth is provided during the summer slump period; (3) weeds are controlled; and (4) results are long-lasting, if managed properly. The deep-rooted species overcome any moisture deficiencies.

The steps involved in a successful renovation program are as follows:

1 Test the soil to determine the level of soil nutrients and apply corrective fertilizer. Lime is often the first requirement in high-rainfall areas; it should be applied approximately 6 months prior to establishment. It is an advantage to disk in the lime before the main tillage operation is done because tillage helps in incorporating the lime throughout the soil. Apply P and K prior to tillage so that these nutrients can be mixed throughout the seedbed.

2 Kill the competitive species by tearing up the old sod. Plowing is considered the best method of killing the competition, mixing in the fertilizer, and preparing the seedbed. Much of the value of renovation may be lost because of competition from unkilled plants. Prior to the tillage, one should graze the existing grass as closely as possible. This allows tillage tools to tear into the old sod more easily and gets the legumes off to a better start with early, vigorous growth. A late-summer and fall tillage in the north, or fall and winter tillage in the south, provides a good seedbed for establishment of seedlings the following spring. (18).

If one cannot use a plow, a field tiller or heavy disk can be used. Since it is very important to kill the old sod, repeated use of a disk, field cultivator, or chisel plow is needed to work up the seedbed. For best results, it is suggested that at least 40 to 60% of the old grass sod be disturbed when establishing some clovers, and 80 to 100% for alfalfa, crownvetch, and birdsfoot trefoil (19). Herbicides are sometimes used to help kill or retard growth of the old sod when one uses disking or the newer method of minimum tillage (35). The seedbed should be made level before seeding, followed by some method of compaction.

3 Prepare a firm seedbed. If the seedbed was plowed, repeated disking and harrowing should be done to prepare a firm, level seedbed. The major consideration is that the forage seed be placed in close contact with moist soil. The typical forage seed is about one-fourth to one-twentieth the size of soybeans or corn, respectively. Once the forage seed absorbs moisture, it needs to continue to do so to ensure the development of a good root system. Drying out leads to death of the root system.

To ensure close contact of the seed and soil, compaction is very important. Merely pulling a corrugated field roller or a cultipacker over the new seeding will aid in compaction.

When should one seed? The answer is simple—when moisture is available and one can be assured of obtaining a stand. A late-winter or early-spring establishment is better in the upper midwest, while a late summer seeding is better in the lower midwest to the southern United States. Spring seedings have more time to become established and to develop an excellent root system going into the winter. In most regions, there is usually sufficient moisture for spring seeding. In the south, the soils hold more moisture (except for some piedmont and coastal plain soils that are extremely droughty), and the region receives more rainfall than do other areas. But the spring rains and soil temperatures sometimes work against each other. As the soil temperatures rise, grasses and weed grasses germinate and develop more readily. Therefore, greater competition arises during the establishment period, limiting the success of legume establishment. In general during the fall, sufficient rainfall is present, and there is a long enough time before a killing frost to allow proper establishment of both grasses and legumes. Inadequate fall moisture is the most serious problem for fall planting in the south.

4 Seed a productive pasture mixture. Alfalfa is usually the first choice among the legumes, provided that the soil will grow alfalfa. If the pH is over 6.5 and drainage is reasonable, one should seed alfalfa. Red clover will help in the first year because it has good seedling vigor. Ladino clover is fine if the field is usually well-supplied with moisture or if there is poor drainage. Ladino clover is the most reliable legume for permanent pastures in the south. Birdsfoot trefoil also has some merit in a permanent pasture. Although it is rather slow in establishment, once it is established it reseeds itself and spreads rather quickly. It is adapted to a wide range of soils, and to soils of varying depths. It is long-lived, winter-hardy, resistant to grazing, and yields well. Birdsfoot trefoil does not cause bloat. All legumes should be inoculated with the proper N-fixing *Rhizobia*.

Of the grasses, several species may be chosen in the north—orchardgrass, tall fescue, and smooth bromegrass. In the south and southwest, tall fescue, tall wheatgrass, and even bermudagrass are fairly productive if properly managed and irrigated. Timothy will not produce as much if grazed continually, but if grazed rotationally, it may persist a long time and produce fairly well. Orchardgrass and tall fescue are the best competitors but can present some management problems. Both are high yielders and fit well with the new, rapid-growing cultivars of alfalfa. Of all the grasses, tall fescue is the best for winter grazing or for deferred grazing and is probably used most in the south. Smooth bromegrass is a sod-forming grass and may take over the sward if not managed properly to maintain the legume portion. Bermudagrass and bahiagrass are good choices for the humid south (5, 7).

5 Control grazing. One should not overgraze the first year. If the old sod was not entirely killed, the renovated field should be kept grazed closely until the livestock begin eating the top of the newly established legumes. At that time, remove the livestock to allow the legumes to become well-established; this takes approximately 4 to 6 weeks. Then graze or harvest the pasture to best suit your needs and to favor the legumes. Do not allow the livestock to graze after September 1 from Minnesota to New York; or September 25 for central Missouri to southern Ohio, so that the forage plants may go into the winter months both healthy and with high levels of food reserves. Further south and in the southwest, the last harvest should be taken about 30 days before fall dormancy. Most seedings are made in the late summer to early fall in the south and southwest, so no harvests are taken until the following spring. Many good seedings are killed because of overgrazing in the fall of the first year. The pasture may be grazed rotationally in later years in order to maintain the legumes.

6 Clip or control weeds. If weeds become established along with the forage species, generally one close clipping 60 days after seeding will remove the weed competition and will give a competitive advantage to the newly seeded forage or grasses and legumes. Making silage out of the first growth will aid in killing weed seeds that have developed. Many different herbicides are effective in controlling different weeds in new or established stands.

7 Maintain fertility. This step is often neglected in a pasture renovation program. Legumes and grasses continue to need phosphorus and potash. When grasses grow in a monoculture, N is needed in addition to P and K. Rate and frequency of P and K application will vary with the soil. It is considered that for every metric ton of forage removed from the field, there are 27 kg of K_2O and 5.4 kg of P_2O_5 removed (60 lb K_2O per ton and 12 lb P_2O_5 per ton, respectively) (10, 18). If the stand reverts to more than 70% grass, one should disregard the legume component and fertilize the stand as if it were a solid grass sward. This means an annual application or split applications of nitrogen, depending upon the grass species and the percent stand present.

OTHER METHODS OF RESEEDING A PERMANENT PASTURE

Besides the conventional method of seedbed preparation, one may use other methods of reseeding.

Tread-in Method

If the land is too hilly or rocky, one can use a large number of livestock on a small area and allow the animals to till the soil with their hooves as they closely graze the growth. With this method the seeding rate should be increased by 20% over a conventional seeding in a well-prepared seedbed.

Disking

A once-over renovator or heavy disk may be used to eliminate the need for a tillage that disturbs the soil and many times leaves the soil too rough to go over later with mowing equipment. With this method, special attention must be given to the management of grass regrowth. After planting the legume, the grass should be grazed or mowed closely to control the grass competition. When the livestock begin grazing the tops of the newly established legumes, remove the animals until the legumes are established.

Approved herbicides may be applied to help suppress the grass (Fig. 9-1).

FIGURE 9-1
Pasture renovation involving the
application of roundup in fall and
spring seeding of alfalfa in an old
grass pasture.

These are generally applied with a disking or once-over renovator at the time of seeding. When using herbicides, be sure to read the label carefully and follow the manufacturer's recommendations closely.

Once-Over Seeder

Recently the use of a once-over seeder or no-till drill has become popular in renovation programs. This method is generally used to introduce or reintroduce a legume in a desirable grass (Fig. 9-2) (25a). This method is quite different than conventional tillage systems in that the forage is seeded in rows and no sod is disturbed between the rows. The steps involved are as follows: (1) graze closely to reduce the above-ground grass competition; (2) fertilize with lime and corrective amounts of P and K; (3) apply 2,4-D about 2 weeks before seeding to kill broadleaf weeds; (4) apply paraquat or a good contact killer immediately preceding the seeding step (Fig. 9-3) (this step may not be necessary if the pasture is grazed heavily and seeded on time); (5) seed legumes into the grass sod with a once-over seeder or no-till drill, about 0.32 to 0.64 cm (1/8 to 1/4 in) deep and at the rates of 13 to 20 kg/ha (12 to 18 lb/A) for alfalfa or red clover or 7 to 9 kg/ha (6 to 8 lb/A) for birdsfoot trefoil (17a); (6) control competition by grazing when legumes are 8 to 15 cm (3 to 6 in) tall (Fig. 9-4); (7) graze the pasture rotationally through the remainder of the season if renovated in spring or, if renovated in the fall, graze rotationally the next year with 7 to 10 days grazing and 30 to 35 days rest; and (8) maintain fertility.

BENEFITS FROM PASTURE RENOVATION

Besides increased dry matter production (Table 9-1), there are additional benefits from pasture renovation. An Illinois test showed that when beef cattle

FIGURE 9-2
John Deere power-till seeder is designed for performance in a variety of soil conditions.
(Courtesy John Deere Co.)

FIGURE 9-3
Close-up showing the rowing effects of alfalfa-red clover using a zip seeder, paraquat, and 2,4-D.

FIGURE 9-4
Pasture renovation involving alfalfa, red clover on left and alfalfa on right six weeks after seeding using paraquat, 2,4-D, and a zip seeder.

grazed renovated clover-grass pasture, the birth rate increased from 75 to 89%. In this test, weaning calves averaged 22.7 kg (50 lb) more per head on a renovated clover-grass pasture than did those on grass with nitrogen added (11).

Tests in southern Indiana indicated that a tall fescue–red clover–ladino clover renovated pasture allowed beef calves running in the herd with their mothers to gain 0.23 kg (0.5 lb) more per day than did calves running on straight nitrogen-fertilized fescue pasture. The conception rate in these same cows increased from 72 to 92% when renovated fescue pastures were seeded to red and ladino clover (17). With the increased rate of gain and a price of $1.00 or $1.20 per kg of beef, over a grazing season this would mean an increase of $7,420 to $8,904, respectively, in favor of the clover-renovated pasture for 100 cows. Another test at Tennessee, a 3-year study, showed that calves on a fescue–red clover–ladino clover renovated pasture averaged 48 kg (105 lb) more than did calves grazed on straight grass (7).

Milk production is also increased when grass pastures are renovated. Tennessee tests showed renovated grass-clover forage produced 2.7 kg (6 lb) of milk per cow per day more than did cows grazing on fescue alone (7), which amounts to 830 kg (1800 lb) more milk over a 305-day lactation period. At current milk prices, that is a significant bonus per cow per year.

In Wisconsin it was estimated that it takes about 269 kg of N/ha (240 lb/A) on grasses to produce as much dry matter as produced on an alfalfa-grass mixture without N (14).

Most southern pastures can be improved by renovation and sod-seeding with

TABLE 9-1
PRODUCTION LEVELS FROM UNIMPROVED, FERTILIZED, AND RENOVATED
PERMANENT PASTURES AT VARIOUS LOCATIONS
(kg/ha; lb/A in parentheses)

Production	Unimproved	Fertilized	Renovated
Beef			
Alabama	115 (103)	330 (294)	— —
Iowa	120 (107)	— —	375 (334)
Indiana	160 (143)	350 (312)	390 (348)
Wisconsin	245 (219)	391 (349)	430 (384)
Forage			
Wisconsin			
26–35% slope	1365 (1218)	— —	3420 (3051)
15–25% slope	1990 (1775)	— —	3770 (3363)
Wisconsin	2195 (1958)	5875 (5240)	— —
Wisconsin	3180 (2837)	— —	5155 (4598)
Alabama	— —	4075 (3635)	3360 (2997)

Sources: H. L. Ahlgren, M. L. Wall, R. J. Muckenhien, and F. V. Burcalow, "Effectiveness of renovation in increasing yields of permanent pastures in southern Wisconsin," *J. Amer. Soc. Agron.* 36:121–131 (1944); O. N. Andrews, "Bahiagrass," *Ala. Polytech. Inst. Ext. Serv. Circ. 548* (1963); C. R. Krueger and J. M. Scholl, "Performance of bromegrass, orchardgrass, and reed canarygrass grown at five nitrogen levels and with alfalfa," *Wis. Agr. and Life Sci. Res. Rept. 69* (1970); J. M. Scholl, M. F. Finner, A. E. Peterson, and J. M. Sund, "Pasture renovation in southwestern Wisconsin." *Wis. Agr. and Life Sci. Res. Rept. 65* (1970); and J. M. Scholl, W. H. Paulson, V. H. Brungardt, and M. W. Ream, "A comparison of production from nitrogen-fertilized and legume-grass pastures for beef cattle in southwestern Wisconsin," *Wis. Agr. and Life Sci. Res. Rept. R2680* (1974).

legumes or annual winter grasses. Tall fescue and bermudagrass respond to complete renovation, fertilization, and seeding legumes. Summer grass pastures, such as bermudagrass or dallisgrass, are especially well adapted for sod-seeding with small grains, ryegrass, annual legumes, or perennial legumes. This practice will increase forage quality, lengthen the grazing season, and reduce winter feed bills.

Perennial winter grass pasture, such as tall fescue and orchardgrass, should be renovated as in the upper midwest, by interseeding alfalfa, ladino clover, or red clover.

QUESTIONS

1 What are the differences between permanent, rotation, and supplemental pastures?
2 Describe the type of forage species grown in rotation pastures.
3 Why are permanent pastures lower in production than are rotation or supplemental pastures?
4 What are various methods of improving pastures?
5 Describe the steps involved in a complete pasture renovation.
6 What percentage of the old sod should be killed when renovating an old pasture? Why?
7 Why is compaction of the seedbed before seeding so important?

8 Discuss the fertility maintenance step in renovation.
9 Discuss other methods of pasture renovation.
10 Why is the once-over seeder a good method of improving an old pasture?
11 What are some of the advantages of pasture improvement?

REFERENCES

1 Abbring, F. T., C. L. Rhykerd, C. H. Noller, S. J. Donahue, K. L. Washburn, Jr., and V. L. Letchtenberg. Nitrogen fertilization of orchardgrass and soil nitrate content. *Ind. Agr. Exp. Sta. Res. Progr. Rept. 364,* 1969.

2 Ahlgren, H. L., M. L. Wall, R. J. Muckenhien, and F. V. Burcalow. Effectiveness of renovation in increasing yields of permanent pastures in southern Wisconsin. *J. Amer. Soc. Agron.,* 36:121–131, 1944.

3 Ahlgren, H. L., M. L. Wall, R. J. Muckenhien, and J. M. Sund. Yields of forage from woodland pasture on sloping land in southern Wisconsin. *J. Forest.,* 44:709–711, 1946.

4 Alexander, C. W. and D. E. McCloud. Influence of time and rate of nitrogen application on production and botanical composition of forage. *Agron. J.,* 54:521–522, 1962.

5 Andrews, O. N. Bahiagrass. *Ala. Polytech. Inst. Ext. Serv. Circ. 548,* 1963.

6 Burger, A. W., T. S. Ronningen, and A. O. Kuhn. Pasture renovation. *Maryland Agr. Exp. Sta. Bull. 449,* 1954.

7 Burns, J. D. Renovating grass pastures. *Tenn. Agr. Ext. Serv. Circ. 714,* 1976.

8 Carter, L. P. and J. M. Scholl. Effectiveness of inorganic nitrogen as a replacement for legumes grown in association with forage grasses. I. Dry matter production and botanical composition. *Agron. J.,* 54:161–163, 1962.

9 Coats, R. E. Sod-seeding—Brown loam tests. *Miss. Agr. Exp. Sta. Bull. 554,* 1957.

10 Decker, A. M., H. J. Retzer, M. L. Sorna, and H. D. Kerr. Permanent pastures improved with sod seeding and fertilization. *Agron. J.,* 61:243–247, 1969.

11 Hinds, F. C., G. F. Cmark, and G. E. McKibben. Fescue for the cow herd in southern Illinois. *Ill. Res., Univ. of Ill. Agr. Exp. Sta.,* 16:6–7, 1974.

12 Hoveland, C. S. and E. M. Evans. Cool season perennial grass and grass-clover management. *Ala. Univ. Agr. Exp. Sta. Circ. 175,* 1970.

13 Jung, G. A., J. A. Balasko, and G. E. Toben. The response of hillside pastures to fertilizer applied by airplane. *W. Va. Agr. Exp. Sta. Bull. 545,* 1967.

14 Krueger, C. R. and J. M. Scholl. Performance of bromegrass, orchardgrass, and reed canarygrass grown at five nitrogen levels and with alfalfa. *Wis. Agr. and Life Sci. Res. Rep. 69,* 1970.

15 Leonard, W. H., R. M. Love, and M. E. Heath. Crop terminology today. *Crop Sci.,* 8:257–261, 1968.

16 Mott, G. O., R. E. Smith, W. M. McVey, and W. M. Beeson. Grazing trials with beef cattle on permanent pastures at Miller-Purdue Memorial Farm. *Ind. Agr. Exp. Sta. Bull. 581,* 1953.

17 Rohweder, D. A. and W. C. Thompson. Permanent pastures. In *Forages,* Iowa State Univ. Press, Ames, IA, 3d ed., pp. 596–606, 1973.

17 a Schaeffer, C. C. and D. R. Swanson. Seeding rates and grass suppression for sod-seeded red clover and alfalfa. *Agron. J.* 74:355–358, 1982.

18 Scholl, J. M., M. F. Finner, A. E. Peterson, and J. M. Sund. Pasture renovation in southwestern Wisconsin. *Wis. Agr. and Life Sci. Res. Rept. 65,* 1970.

19 Scholl, J. M., and C. Llambias. Response of a permanent grass sward to rate of nitrogen and to nitrogen combined with phosphorus and potassium fertilizer. *Proc. 9th Int. Grassl. Congr.*, pp. 1335–1337, 1966.

20 Scholl, J. M., W. H. Paulson, and V. H. Brungardt. A pasture program for beef cattle in Wisconsin. *Wis. Lancaster Cow-Calf Day Proc.*, 1970.

21 Scholl, J. M., W. H. Paulson, V. H. Brungardt, and H. W. Ream. A comparison of production from nitrogen-fertilized and legume-grass pastures for beef cattle in southwestern Wisconsin. *Wis. Agr. and Life Sci. Res. Rep. R2680*, 1974.

22 Smith, D. Winter injury and the survival of forage plants. *Herb. Abstr.*, 34:203–209, 1964.

23 Sprague, M. A. Seedbed preparation and improvement of unplowable pastures using herbicides. *Proc. 8th Intl. Grassl. Congr.*, pp. 264–266, 1961.

24 Sprague, V. G., R. R. Robinson, and A. W. Clyde. Pasture renovation. I. Seedbed preparation, seedling establishment, and subsequent yields. *J. Amer. Soc. Agron.*, 39:12–25, 1947.

25 Sund, J. M., M. J. Wright, E. H. Zehner, and H. L. Ahlgren. Comparisons of the productivity of permanent and rotation pastures on plowable cropland. II. Second six-year cycle and summary. *Agron. J.*, 50:637–640, 1958.

25 a Taylor, R. W. and D. W. Allinson. Legume establishment in grass sods using minimum-tillage seeding techniques without herbicide application: Forage yield and quality. *Agron. J.*, 75:167–172, 1982.

26 Taylor, T. H., J. M. England, R. E. Powell, J. F. Freeman, C. K. Kline, and W. C. Templeton, Jr. Establishment of legumes in old *Poa pratensis* L. sod by use of paraquat and strip tillage for seedbed preparation. *7th Brit. Weed Control Conf.*, 64:3–65, 1964.

27 Taylor, T. H., E. M. Smith, and W. C. Templeton, Jr. Use of minimum tillage and herbicides for establishing legumes in Kentucky bluegrass swards. *Agron. J.*, 61:761–766, 1969.

28 Templeton, W. C., Jr., C. F. Buck, and N. W. Bradley. Renovated Kentucky bluegrass and supplementary pastures for steers. *Ky. Agr. Exp. Sta. Res. Bull. 709*, 1970.

29 Thompson, W. C., T. H. Taylor, and W. C. Templeton, Jr. Renovating grass fields. *Ky. Coop. Ext. Serv. Leaflet 277*, 1965.

30 VanKeuren, R. W. and G. B. Triplett. Seeding legumes into established grass swards. *Proc. 11th Intl. Grassl. Congr.*, pp. 131–134, 1970.

31 Wedin, W. F. and D. A. Rohweder. Pasture trials at Albia, 1940–1960. *Iowa Coop. Ext. Serv. Pam. OEF* 61–23, pp. 1–5, 1962.

32 Wedin, W. F. and R. L. Vetter. Pasture for beef production in western Corn Belt, U.S.A. *Proc. 11th Intl. Grassl. Congr.*, pp. 842–845, 1970.

33 Wedin, W. F., R. L. Vetter, J. M. Scholl, and W. R. Woods. An evaluation of birdsfoot trefoil in pasture improvement. *Agron. J.*, 59:525–528, 1967.

34 Winch, J. E., G. W. Anderson, and T. L. Collins. Chemical renovation of roughland pasture. *Proc. 10th Intl. Grassl. Congr.*, pp. 982–987, 1966.

CHAPTER 10

FORAGE FERTILIZATION

One of the most important components of forage production is proper soil fertility. Successful forage producers understand the importance of soil in a forage program, as well as some of the basic properties and functions of soils. It has often been said that the missing link in forage production is proper soil management and soil fertility.

Soil is very important to plants, animals, and humans. It provides mechanical support for plants, and most of the nutrients that are needed for plant growth. Sixteen essential elements are known to be necessary for plant growth. Some of these elements—carbon, hydrogen, and oxygen—come from the air. Nitrogen comes from both air and soil. But the majority—phosphorus, potassium, calcium, sulfur, iron, magnesium, manganese, zinc, copper, boron, molybdenum, and chlorine—come only from the soil.

Soil is the residue of the weathering of parent material, and of the decomposition of plant and animal materials that have become associated with it. The characteristics of a particular soil at a given site are the result of parent material, climate, living organisms, topography, and time. Soil formation is a very complex mechanical, physical, chemical, and biological process.

One may describe many physical properties of the soil. Soil texture refers to the size of the soil particles. Soil structure refers to the arrangement of these particles. Using a mechanical analysis one can determine the percentage of the various size particles or soil separates. The main soil separates are very coarse sand, coarse sand, medium sand, fine sand, very fine sand, silt, and clay. The texture of the soil is determined predominantly by the mixture of sand, silt, and clay. In addition to these components, organic matter may or may not be present in the soil. Organic substances such as animal manure, crop residue, and leaf mulches greatly benefit the soil structure. Many producers will grow green manure crops purely for the purpose of adding organic matter to improve the soil structure.

In addition to proper soil structure and texture, temperature of the soil plays a very important role in plant growth and soil characteristics. Proper soil temperature must be present before germination can occur, and this temperature varies with different species. When soil temperatures are below 10°C (50°F), nitrification, or the breakdown of ammonia to nitrate, is reduced (2, 8, 23, 54). Alternating freezing and thawing improves the structure of cloddy soils. Cold temperatures also reduce plants' absorption of some elements, particularly phosphorus.

SOIL TESTING

Soil testing is the most important single step in determining the rate of application of fertilizer and lime. One must know the level of fertility and especially the pH before one can profitably apply fertilizer. Soil test results, combined with the knowledge of what each nutrient does for plant growth, comprise the basis for planning a producer's fertility program for each field tested.

It is very easy to collect a soil sample for every 1 to 2 ha (2.5 to 5 A) in areas having similar soil type, topography, cropping history, and so on. One should develop a regular pattern in a field and make a very accurate map so that the results can be interpreted correctly and the proper amount of fertilizer can be applied to each area. The most common mistake in this process is taking too few samples.

One should soil test every 3 to 4 years. Late summer and fall are the best seasons for collecting soil samples. Potassium results are most reliable at these times. Since this nutrient is rather water-soluble and readily absorbed by growing plants, it will be at the lower level of its availability in the late summer or early fall as compared with its availability in the spring. Following freezing and thawing during the winter, if there is sufficient soil moisture, a higher test value will be obtained in the spring, which will be false or inflated.

Most soil tests report soil acidity, the P_1 (available phosphorus) level, and K, the potassium level. Some states report total phosphorus, or P_2, which includes both the P_1 and P_2 in the soil. P_1 is the water-soluble portion; P_2 is the total or acid-soluble portion. Most soil tests will not indicate the nitrogen present because the particular test used is rather unreliable. Some soil tests may provide the availability of secondary and micronutrients, if this information is requested. Tests may be made for the secondary and micronutrients, but their interpretation is less reliable than that for the tests of pH, P, and K.

Organic matter is sometimes reported in soil tests. Color is related to the organic matter content of the soil. Light-colored soils usually have less than 2½% organic matter, whereas medium-colored soils have from 2½ to 4½%

organic matter. Dark-colored soils may have more than 4½% organic matter. Sands are usually excluded from the determination of organic matter content.

LIME

One of the most serious limitations to soil productivity is soil acidity. On most cropland areas in the United States, additional nitrogen has been used without a corresponding addition of limestone, and since nitrogen fertilizers are acid-forming, most U.S. soils are consequently increasing in soil acidity. A soil test every 4 years is the best way to keep a check on soil acidity levels. It requires about 4 units of lime to reduce the acidity resulting from 1 unit of nitrogen applied as ammonia or urea, and as much as 9 units of lime to neutralize the acidity resulting from 1 unit of nitrogen applied as ammonium sulfate.

There are several ways in which soil acidity affects plant growth. As the soil pH becomes lower and acidity increases, several situations may exist: (1) the concentration of soluble metals may become toxic—an excess of solubility of aluminum and manganese has been established experimentally; (2) populations and activity of organisms responding to transformations involving nitrogen, sulfur, and phosphorus may be altered; (3) calcium may become deficient as the pH decreases—this usually occurs when the soil's cation exchange capacity is extremely low; (4) symbiotic nitrogen fixation is reduced greatly on acid soils—the symbiotic relationship requires a very narrow range of soil acidity compared to that necessary for the growth of plants that do not rely on nitrogen fixation; (5) acid soils are poorly aggregated and have poor tilth—this is particularly true for soils low in organic matter; and (6) other minerals' availability to plants may be greatly reduced, as shown in Fig. 10-1, which indicates the relationship between the soil pH and nutrient availability. The wider the white bar in the figure, the greater the nutrient availability. For example, phosphorus availability is greatest when the pH is between 6.5 and 7.5. Potassium availability drops off very rapidly below a pH of 6.0. Molybdenum availability increases greatly as the soil acidity decreases, so molybdenum deficiencies are usually corrected by proper liming. When soil pH exceeds 7.0, iron, manganese, boron, copper, and zinc decrease in availability. When pH exceeds 8.5, one should apply leaching treatments of gypsum and sulfur.

Various forage crops may be classified according to their sensitivity to pH, as shown in Table 10-1. For example, alfalfa and smooth bromegrass are very sensitive to pH levels, whereas other plants, including alsike clover, lespedezas, tall fescue, and reed canarygrass, may tolerate moderate acidity. It is highly recommended that when alfalfa and clover, or forage crops in general, are grown, the pH be at least 6.5 to 7.0. When a monoculture of alfalfa is grown, the pH should be 6.8 to 7.

Growers often ask how much lime should be put on the soil to bring the pH up to at least 6.5. Fig. 10-2 illustrates the number of tons needed to bring the pH up to the proper productivity. There are several assumptions that must be

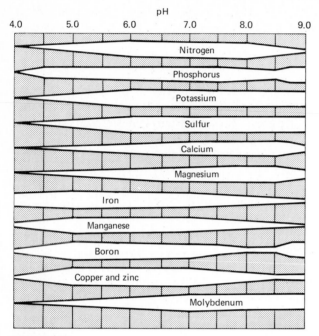

FIGURE 10-1
Available nutrients in relation to soil pH.

TABLE 10-1
FORAGE CROPS CLASSIFIED ACCORDING TO SENSITIVITY TO
SOIL pH

Very sensitive to acidity (optimum pH 6.5–7.0)	Slight tolerance to acidity (optimum pH 6.0–6.5)	Moderate tolerance to acidity (optimum pH 5.5–6.0)
Alfalfa	Corn	Alsike clover
Barley	Crimson clover	Bahiagrass
Smooth bromegrass	Dallisgrass	Bermudagrass
Sweet clover	Kentucky bluegrass	Birdsfoot trefoil
	Ladino clover	Lespedeza
	Orchardgrass	Meadow fescue
	Red clover	Oats
	Ryegrass	Pearlmillet
	Timothy	Redtop
	Wheat	Reed canarygrass
		Rye
		Soybeans
		Sudangrass
		Tall fescue
		Vetch

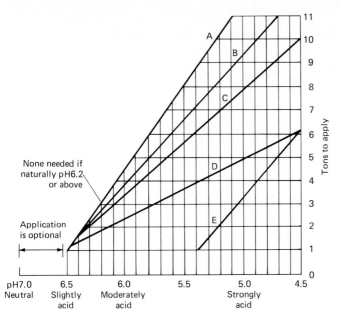

FIGURE 10-2
Suggested limestone application rates based on soil type, pH, and cropping
systems with alfalfa, clover, or lespedeza.

remembered when using this figure: (1) a plowing depth of 23 cm (9 in) is
considered proper. If the plowing depth is less than this, then one should reduce
the amount of limestone; if it is more, one should increase the lime application
rate proportionally. (2) The measure of the fineness of limestone is determined
by the following figures: 90, 60, and 30% of the limestone that will pass through
an 8-, 30-, or 60-mesh screen, respectively. If the limestone is not as fine as
indicated, or if a quick reduction of the pH is desired, then one should apply
more limestone than is indicated by the chart. (3) The calcium carbonate
equivalent, total neutralizing power, is 90%. The rate of application may be
adjusted according to the deviation from 90. When using this figure, one should
determine the class into which the soil fits (Fig. 10-2). The A line within the
figure includes dark-colored silty clays and silty clay loams. The B line includes
light- and medium-colored silty clay loams and dark-colored silts and clay loams.
The C line refers to light- and medium-colored silty clay loams, dark- and
medium-colored loams, and dark-colored sandy loams. The D line includes
light-colored loams, light- and medium-colored sandy loams, and sands. And the
E line refers to muck and peat soils.

One should purchase limestone on the basis of quality; premium limestone
should sell for a premium price. Limestone quality is measured by the fineness
and the neutralizing power. The value of lime in correcting soil acidity problems
can be easily calculated using the various efficiency factors given in Table 10-2.

TABLE 10-2
COMPARATIVE VALUE OF LIMESTONE PARTICLE SIZES

Size fraction	Efficiency factor
Through 60 mesh	100
30 to 60 mesh	50
8 to 30 mesh	20
Over 8 mesh	5

Depending upon the initial cost, the entire amount of limestone may be applied at one time. If cost is a factor, as when 6 tons or more per acre are needed, one should add the limestone in split applications—about two-thirds in the first application, and the remaining one-third 3 to 4 years later. But if alfalfa, other legumes, or especially smooth bromegrass are to be grown, one should apply all of the lime at once.

Limestone does not react with the acid in soil very far from the soil particle. Special tillage operations should be used to properly mix the lime with the soil; it is suggested that if the limestone is applied to the top of the soil, then one should disk in the limetone, plow down the material, and work the soil (Figs. 10–3, 10–4). This distributes the limestone throughout the plow depth much more evenly than if it is just plowed directly under. If lime is plowed down without disking in first, the resulting plant growth may appear as light and dark green stripes corresponding to the lime distribution.

Growers often ask what is the difference between calcitic limestone and dolomitic limestone. Dolomitic limestone is a mixture of $CaCO_3$ and $MgCO_3$,

FIGURE 10-3
Proper forage establishment begins with a proper soil pH level. One should apply limestone about 6 months before seeding and work it well into the seedbed.

FIGURE 10-4
Various effects of fertilizer on top growth and root development.

whereas calcitic limestone includes only $CaCO_3$. As shown in Table 10-3, these two different types of limestone, in varying finenesses, were applied to alfalfa and grown under greenhouse conditions. The soil pH at the onset was 5.0. Generally, the maximum percent alfalfa yield is reached when calcitic limestone passes through a 60-mesh screen; the dolomitic limestone yield is slightly lower with a 60-mesh screen (33).

Alfalfa responds to the proper pH (Table 10-4). Comparing 5-year averages,

TABLE 10-3
EFFECT OF DIFFERENT LIMESTONES AND FINENESSES
ON ALFALFA YIELDS UNDER GREENHOUSE CONDITIONS

Particle size (mesh)	Alfalfa yield, % maximum	
	Calcitic limestone	Dolomitic limestone
No lime*	48	53
5–8	55	56
8–20	59	57
20–30	79	74
30–40	84	75
40–50	86	91
50–60	93	91
60–80	100	89
80–100	100	100
>100	100	95

*pH of control was 5.0.
Source: T. A. Meyer and G. W. Volk, "Effect of particle size of limestone on soil reaction, exchangeable cations and plant growth," *Soil Sci.* 73:37–52 (1952).

TABLE 10-4
RESPONSE OF ALFALFA TO pH AND
ANNUAL FERTILIZER TREATMENT

Annual fertilizer treatment	5-year average t/ha	
	pH 5.0–5.5	pH≥6.5
0—0—0	7.21	11.22
0—150—300 (3)*	14.47	16.04

*(3) plus boron treatment.
Source: D. Smith, "Effects of potassium top dressing a low fertility silt loam soil on alfalfa herbage yields and composition and on soil K values," *Agron. J.* 67:60–64 (1975).

more than 11 t/ha was obtained with no fertilizer and a pH of 6.5 or more, as compared with 14.5 t/ha at a pH of 5.0 to 5.5. Fertilizer plus liming yielded 16 t. Consequently, as one applies fertilizer of P and K plus boron, corresponding yield increases result (33). One should keep in mind that if legumes are grown in combination with grasses, it is advisable to apply enough lime to increase the pH for several reasons: (1) to maintain the legume in the stand; (2) to meet the bacteria's needs and so maintain the legume's proper nitrogen-fixing ability; and (3) to better balance the availability of other nutrients, which occurs when the pH is increased up to 7.0 (Fig. 10-5).

FIGURE 10-5
Second-growth alfalfa showing the effect of proper soil pH. Plot on left had a pH of 5.5, while the plot on the right had a pH of 6.8. All other nutrients were equal on both plots.

NITROGEN

The use of nitrogen fertilizer is essential in present-day agriculture if people are to continue to provide adequate crop production to meet the ever-increasing world demand for food. Nitrogen is probably the most important element needed for forage grasses. With the world's shortage of nitrogen fertilizer and energy, nitrogen fertilizer should be used in the most efficient manner possible. Nitrogen is essential for the formation of protein and must be obtained from either commercial fertilizer, nitrogen-fixing bacteria, organic matter, or rainfall. Organic matter is provided primarily by plant residues, such as leaves, stems, and roots, and animal residues, such as manure and urine. The organic matter not only contributes nitrogen, but also provides for the improvement of soil structure, water-holding capacity, and nutrient-exchange capacity.

Most plants absorb nitrogen in the form of nitrate (NO_3), although some is absorbed in the form of ammonia (NH_3). Some plants will use NH_4^+ if NO_3^- is not present. When temperatures are above 10°C (50°F), the ammonia in fertilizer must be broken down to the nitrate form by bacteria before it becomes available for plants. Therefore, fertilizers that are in the form of ammonium nitrate contain some of the readily available nitrate nitrogen. Under cooler soil conditions, this form of N is much more available than the ammonia form, which makes it especially good for early application to cool-season grasses.

The main sources of nitrogen fertilizers and their percentages of nitrogen are as follows: anhydrous ammonia, 82%; urea, 46%; ammonium nitrate, 33.5%; ammonium sulfate, 21%; and nitrate solutions, 28 to 32%. Heavy applications of nitrogen fertilizer will increase the soil acidity and will require an increase in limestone applications, provided the pH is below 7.0.

Another source of nitrogen is plowing down a stand of alfalfa. Table 10-5 illustrates the amount of nitrogen that a first-year stand of alfalfa would contribute to a following crop of corn or wheat. This is also compared to the contribution of plowing down soybeans. Alfalfa adds much more nitrogen to these crops than soybeans do. From a complete stand of alfalfa, there is some

TABLE 10-5
CONTRIBUTION OF NITROGEN FOLLOWING VARIOUS STAND DENSITIES OF ALFALFA OR SOYBEANS IN RELATION TO CORN AND WHEAT PRODUCTION
(lb/A)

Crop to be grown	After soybeans	1st yr after alfalfa or clover plants/sq. ft				2d yr after alfalfa or clover plants/sq. ft		
		>5	5	2–4	<2	>5	5	<5
Corn	40	140	100	50	0	40	30	0
Wheat	10	40	30	10	0	—	0	0

carryover of nitrogen into the second year. The amount of nitrogen removed is also related to the amount of residue plowed under. Most of the N is removed with the forage removal. In addition to releasing nitrogen, alfalfa also improves the tilth of the soil and adds organic matter.

Many researchers have studied the effect of nitrogen on forage grass yields and have found significant results related to the amount of nitrogen added (2, 3a, 8, 14, 21, 27, 35, 36, 38, 40, 42, 44, 48, 53, 58). A general rule is that in temperate regions, forage grass yields will increase proportionally to the amount of nitrogen applied, up to the rates of 336 to 404 kg/ha (300 to 450 lb/A) annually. It has been shown that when one exceeds 500 kg/ha, yields may decrease because of the thinning of the stand and increased competition between plants (58).

Most nitrogen fertilizers are top-dressed in split applications during the growing season. Grasses will consume nitrogen luxuriously if it is applied in excess, resulting in the presence of nonprotein nitrogen in the plant. As the amount of nitrogen given to forage grasses is increased, this will also increase the amount of other elements absorbed from the soil, especially K (48). Therefore, when one is fertilizing grasses with higher rates of nitrogen, P and K should be applied properly or they will become limiting factors (Fig. 10-6).

The forage grass species to be grown, the period of use, and the yield goal determine the optimum level of nitrogen fertilization (Table 10-6). A lower rate of nitrogen is recommended for fields where there is an inadequate stand, or where moisture may limit production. Kentucky bluegrass, for example, is a shallow-rooted species that is also susceptible to drought. Consequently, the

FIGURE 10-6
Highly fertilized grass species require better management, such as increased carrying capacity, earlier harvest, and more frequent harvest.

TABLE 10-6
NITROGEN FERTILIZATION OF HAY AND PASTURE GRASSES
(kg/ha; lb/A in parentheses)

Species	Time of application			
	Early spring	After 1st harvest	After 2d harvest	Early Sept.
Kentucky bluegrass	67–90 (60–80)	—	—	
Orchardgrass	84–140 (75–125)	84–140	—	
Smooth bromegrass	84–140 (75–125)	84–140	—	56 (50)*
Reed canarygrass	84–140 (75–125)	84–140	—	56 (50)*
Tall fescue—deferred use	—	112–140 (100–125)	112–140	56 (50)*

*Optional if extra fall growth is needed.

most efficient use of nitrogen by bluegrass is from an early spring application. September applications may be a second choice if there is sufficient moisture available and fall pasture is needed. Timothy should receive nitrogen applications at approximately the same rate and time as Kentucky bluegrass.

Orchardgrass, smooth bromegrass, tall fescue, and reed canarygrass are more drought-tolerant than are bluegrass and timothy, and can use higher rates of nitrogen more efficiently than the shallow-rooted species can. In these species, more uniform pasture production is obtained by splitting high rates of nitrogen into two or more applications.

The early spring application of nitrogen depends on the geographic location where the forage species is grown. In northern Illinois, for example, the application of nitrogen should be in mid-April, whereas in central Illinois, it should be in early April, and in the southern part of the state, in mid-March. This corresponds to other states with similar locations. The further south in the United States, the earlier the nitrogen application in the spring. Depending on organic matter and other fertility levels, spring growth may be adequate without extra nitrogen. Therefore, the first application may be delayed until after the first harvest or grazing period, so one could distribute production more uniformly throughout the summer. But total production would likely be reduced if nitrogen is applied after the first harvest rather than in the early spring. The second application is generally made immediately after the first harvest or the first grazing period. When deferring summer grazing of tall fescue, one should make a late summer application or apply additional N after the second harvest.

Legume-grass mixtures should not receive nitrogen if the legume content makes up 30% or more of the mixture. The main objective is to maintain the legume, so the emphasis should be on proper P and K fertilization rather than on nitrogen. It is considered that if the legume is at least 30% of the sward, sufficient nitrogen may be fixed by the legume for the growth of the grass component of the mixture (23, 36, 42, 56, 58). After the legume has declined to less than 30% of the mixture, the objective of the fertility program is to increase the yield of grass. The suggested rate of nitrogen application is about 56 kg/ha (50 lb/A) when the legume makes up 20 to 30% of the mixture, and 112 kg/ha (100 lb/A) or more when the legume is less than 20% of the mixture (58).

Manures generally are considered to contain 10 lb of nitrogen, 5 lb of P_2O_5, and 10 lb of K_2O per ton, or 4 kg of nitrogen, 2 kg of P_2O_5, and 4 kg of K_2O per metric ton. There is some variation in the content depending upon the source and the method of handling (Table 10-7). Regardless of the source, only 50% of the total nitrogen will be available to the crop during the first year after application because of organic N and volatilization (Fig. 10-7).

Research has indicated that when good stands of alfalfa or clovers are grown with grasses, they will fix approximately 90 to 112 kg/ha (80 to 100 lb/A) of nitrogen (56). It was found in Maryland that on pure stands of orchardgrass and tall fescue, an excess of 180 kg/ha (160 lb/A) of nitrogen must be applied to equal the contribution of ladino clover grown with either of the grasses (56). Individuals in Wisconsin have found that it takes more than 270 kg/ha (270 lb/A) of nitrogen per acre applied to grass stands to equal the contribution of alfalfa to the mixture (48).

There has been considerable research conducted regarding nitrogen application on good stands of legumes and grasses. This has shown a slight increase in dry matter production with such combinations (26, 36, 48, 59). However, the response is due primarily to the increase in the grass mixture and not from the legume. It is generally considered that nitrogen fertilization of a good grass-legume mixture is not recommended and should not be added until the legume content in the stand is below the 30% level. Research in Illinois indicated that when NH_4 and N-Serve, a nitrification inhibitor, were applied to an alfalfa-

TABLE 10-7
AVERAGE COMPOSITION OF VARIOUS MANURES IN DRY AND LIQUID FORM

	Dry form					
	Nitrogen (N)		Phosphorus (P_2O_5)		Potassium (K_2O)	
	kg/t	lb/T	kg/t	lb/T	kg/t	lb/T
Dairy	5	11	2	5	5	11
Beef	6	14	4	9	5	11
Hogs	4	10	3	7	3	8
Chickens	8	20	7	16	3	8

	Liquid form (per 1000 gal)					
	kg	lb	kg	lb	kg	lb
Dairy	12	26	5	11	10	23
Beef	10	21	3	7	8	18
Hogs	25	56	14	30	10	22
Chickens	34	74	31	68	12	27

FIGURE 10-7
Animal waste is an excellent source of plant nutrients.

orchardgrass mixed stand, there was a significant increase in dry matter, due primarily to the grass content in the forage mixture (Table 10-8). But when a pure stand of alfalfa received NH_4 plus N-Serve at varying rates, there was no significant increase in yield; in fact, there was a decrease because of the knife applicator used for the anhydrous ammonia.

TABLE 10-8
RESPONSE OF ALFALFA-ORCHARDGRASS AND PURE ALFALFA
STANDS TO NH_4 PLUS N-SERVE AT ILLINOIS

Application rate (kg/ha)		Yield (t/ha)	
NH_4	N-Serve	Alfalfa-orchardgrass	Alfalfa
0	0	9.33	—
270	0.55	11.57*	—
0	0	—	15.87
0	0 (knife only)	—	14.37†
270	0	—	14.50†
540	0	—	14.44†
270	0.55	—	14.42†
270	1.12	—	14.64†
540	0.55	—	15.09
540	1.12	—	14.37†

*Significantly different than control at the .01 level.
†Significantly different than control at the .05 level.

Effects of Nitrogen on Forage

Nitrogen fertilization will increase the concentration of P and K in the forage, provided their availability is sufficiently high in the soil. NH_4 will reduce the concentration of Ca, K, and Mg in the forage. Therefore, if NH_4 is used on grasses, one should be alert to the possibility of grass tetany, if the level of Mg is already on the low side (16, 42).

When the amount of nitrogen applied to forage grasses is increased, the nitrate nitrogen level will increase correspondingly. Therefore, one should be alert for the possibility of nitrate toxicity if excessive amounts of nitrogen are applied (42, 51a, 58). If the harvest is delayed or the grasses are allowed to become more mature, the nitrate nitrogen level will decrease in the plant tissue. A further concern is that nitrogen application to the legume stand will reduce the rhizobia activity and nitrogen fixation within the nodules.

Nitrogen fertilization will decrease the plants' level of water-soluble carbohydrates, primarily polysaccharides (42). Consequently, this may have a slight effect upon silage fermentation, since carbohydrates are required to make good-quality silage. Nitrogen application will also increase the level of total organic acids and alkaloids in grasses. Carotene levels are also increased as nitrogen is applied (42).

Since nitrogen fertilization tends to decrease the available carbohydrates in plants, it is more difficult to make high-quality silage from heavily N-fertilized grasses. This is especially true for cool-season grasses, as they already tend to be very low in available carbohydrates. But nitrogen application to corn does not appear to have any great effect on silage quality since there is generally a high level of carbohydrates readily available (42). In the western corn belt and areas where drought damages the growth, high levels of NO_3 may be found in the stalk, so 30 to 45 cm (12 to 18 in) stubble heights are recommended.

There are conflicting reports concerning the effect of nitrogen fertilization on the palatability and intake of forages. There are probably equal positive and negative results (42). Generally, comparing no added nitrogen and moderate levels of nitrogen applied to forage grasses, the added nitrogen will consistently increase animal preference. However, there is no such difference between moderate levels of nitrogen and high levels of nitrogen. One must also keep in mind that as the nitrogen level is increased, there is also a greater demand for water uptake, so moisture may become deficient a little sooner with higher nitrogen fertilization (3b). Some researchers have suggested that when reed canarygrass and tall fescue are fertilized with nitrogen, the alkaloid concentration will increase, thus reducing the forage's intake and palatability (37, 58). There appears to be no consistent effect of nitrogen fertilization on the digestibility of grasses (42).

Many studies have indicated that nitrogen fertilization of forages grasses will increase the beef gains in kg/ha (58). A lot of this work has been done with warm-season grasses, such as 'Coastal' bermudagrass, dallisgrass, and bahiagrass. Most of the warm-season grasses are grown where sufficient amounts of moisture are present. Thus, a direct correlation has been found between increased rates of nitrogen and beef gains in kg/ha (14, 38, 41, 53). Increasing nitrogen fertilization on most grasses will increase the carrying capacity, thereby yielding a greater production of red meat per hectare than from unfertilized plots (58). Cool-season grasses also respond to nitrogen fertilization, but probably not to the same extent as the warm-season grasses do.

PHOSPHORUS

Phosphorus (P) is one of the major elements plants need to live. It plays a very important role in the growth of both plants and animals. Many of the soils of the southeast are deficient in phosphorus, whereas those in the midwest and far west vary in their levels of phosphorus (59). Phosphorus is especially needed for good grass and legume growth and for nitrogen fixation by bacteria. It is crucial in the establishment of legumes and grass seedlings because it is utilized very early in the plant's life cycle and because it is very important for good root growth. Without an adequate supply of P, the plant will not reach its maximum growth or yield potential, nor will it reproduce normally. The functions of phosphorus range from the primary mechanism of energy transfer to the coding of genes, which cannot be performed by any other nutrient (59). In the photosynthesis process, phosphorus is essential for the transfer of energy and synthesis within the plant and for the eventual breakdown of carbohydrates. Phosphorus is considered the essential nutrient in the plant's electron-transport system. It is probably the main energy source within the plant because it is tied in with the phosphorylation that is required in many reactions. It provides the energy used in the synthesis of sucrose, starch, and proteins (7).

Adequate phosphorus fertilization will improve root exploration in the soil since this element enhances root development. When root development is increased, added phosphorus decreases the amount of water that a plant needs (12). When adequate phosphorus is applied, the foliage increases, thereby shading the soil and so reducing the water loss. Within grain crops, P is essential in speeding up the maturity or development of the crop, especially during the pollination and seed set stages.

Phosphorus does not readily leach out of the soil and, with continued annual applications, will gradually accumulate. Many soils fix or tightly hold this element. When P is fixed or tightly held in the soil, the amount of leaching and availability to plants is minimized. This varies with soil type. If the soil pH exceeds 7.5, P becomes less available. Usually no more than 10 to 20% of the P that is applied is recovered in the first year of growth. Therefore, it is very important to retest the soil every 3 to 4 years. Plant growth is highly dependent upon the amount of P that has been applied or built up over the years from previous applications.

Soil tests will generally report the available phosphorus (P_1) or the total phosphorus (P_2). High-phosphorus-supplying soils are generally very favorable for good root penetration and branching throughout the soil profile. Some soils supply low amounts of P for several reasons: (1) there may be a low supply of available phosphorus in the soil profile because the parent material was low in P, or phosphorus was lost in the soil-forming process, or phosphorus is made unavailable due to high pH or calcareous material; (2) poor internal drainage, which restricts plant growth and so creates a low-phosphorus-supplying soil; (3) a dense, compact layer that inhibits root penetration or branching; (4) only a shallow layer above the bedrock, sand, or gravel; and (5) droughtiness, strong acidity, or other conditions that may restrict crop growth or reduce root depth.

Within various areas of the United States, or in a particular state, one can determine the soil's phosphorus-supplying power by determining the primary parent material. Figure 10-8 shows that areas with high P availability may have very high phosphorus content, or the parent material may have been very high in phosphorus. As the depth of these soils diminishes, or the parent material changes in phosphorus content, the classification of the soil changes to a medium supplying power. Areas with low P availability may have soils that are very weathered or have a highly restricted rooting zone. Furthermore, the parent material, native vegetation, and natural drainage may vary within a region, causing a variation in the phosphorus-supplying power in that area.

Since phosphorus will be lost from the soil system not only through crop

FIGURE 10-8
Phosphorus-supplying power in Illinois.

removal or soil erosion, and since there are minimum values required for optimum forage yields, it is recommended that soil test levels be built up to 40, 45, and 50 lb/A in high-, medium-, and low-phosphorus-supplying regions, respectively. These values will ensure that the soil's phosphorus availability will not limit crop yield.

Research has shown that on average, it takes about 9 units of P_2O_5 to increase the P_1 soil test by 1 unit. Therefore, the recommended rate of buildup of phosphorus is equal to nine times the difference between the soil test goal and the actual soil test value. The amount of phosphorus recommended over a 4-year period for various soil test levels is shown in Table 10-9. Some soils will fail to reach the desired goal in 4 years with P_2O_5 applied at the suggested rate, whereas

TABLE 10-9
AMOUNT OF PHOSPHORUS (P_2O_5) REQUIRED
ANNUALLY PER ACRE TO BUILD UP THE SOIL
(Based on Buildup Occurring Over a 4-year Period; 9 lb/A
of P_2O_5 Required to Change P_1 Test 1 lb)

P_1 test (lb/A)	Soil supplying power		
	Low	Medium	High
4	103	92	81
6	99	88	76
8	94	83	72
10	90	79	68
12	86	74	63
14	81	70	58
16	76	65	54
18	72	61	50
20	68	56	45
22	63	52	40
24	58	47	36
26	54	43	32
28	50	38	27
30	45	34	22
32	40	29	18
34	36	25	14
36	32	20	9
38	27	16	4
40	22	11	0
42	18	7	0
44	14	2	0
45	11	0	0
46	9	0	0
48	4	0	0
50	0	0	0

others may exceed this goal. Therefore, it is recommended that every field be retested every 4 years.

Most plants use phosphorus in the H_2PO_4 form, but the phosphorus content of commercial fertilizer is usually measured in the P_2O_5 form. Several sources of phosphorus in commercial fertilizers and their normal percentages of P_2O_5 are: superphosphate, 20%; triple superphosphate, 45%; and rock phosphate, 41%. There are some additional sources of phosphorus that may also contain nitrogen, such as diammonium phosphate, 53% P_2O_5 and 21% N; monoammonium phosphate, 48% P_2O_5 and 11% N; and ammonium phosphate sulfate, 20% P_2O_5 and 16% N.

Prior to the seeding of perennial forage crops, one should broadcast and incorporate all of the buildup phosphorus needed, plus all the maintenance phosphorus that is economically feasible, prior to seeding. On low-fertility soils, one should apply approximately 34 kg/ha (30 lb/A) of P_2O_5, using a band seeder. If a band seeder is used, one may safely apply a maximum of 34 to 45 kg (30 to 40 lb) of potash (K_2O) in the band, along with the phosphorus. Usually, up to 336 kg/ha (300 lb/A) of K_2O can be safely broadcast in the seedbed without damaging the seedlings (12).

Top-dress applications of phosphate and potassium or perennial forage crops can be applied at any convenient time, usually after the last harvest or in early September. It is preferred that all of the phosphorus be applied after the last harvest in September so that the plant has sufficient phosphorus going into the winter to maintain good root structure and early spring vigor. Potash is usually applied in split applications—half of the required amount, or maintenance fertilizer, at the same time in September, after the last harvest, and the remaining half after the first harvest the following spring (5, 6).

As long as the recommended rates of application and broadcast placements are used, the water solubility of P_2O_5 listed on a fertilizer label is of little importance under most field conditions on soils that have medium to high levels of available phosphorus. There are two exceptions, however, when water solubility is important: (1) When band-placing small amounts of fertilizer to stimulate early growth, at least 40% of the applied P_2O_5 should be water-soluble for acid soils, and preferably 80% for calcareous soils, as shown in Table 10-10. Most of the phosphorus in commercial fertilizers is highly water-soluble. Fertilizers with a phosphate water solubility greater than 75 to 80% will not increase dry-matter yields more than will those with water solubility levels of 50 to 80%. (2) For calcareous soils, a high degree of water solubility is desirable, especially for soils that tests have shown to be low in available phosphorus.

Numerous researchers have reported the effect of P on various forage crops (1, 3, 7, 8, 10, 11, 12, 24, 25, 27, 28, 30, 32, 35, 36, 37, 40, 43, 45, 53, 54, 55, 57). In Virginia, based on a 3-year average, alfalfa yields were found to increase from 7.2 to 11.7 t/ha when 100 kg (202 lb/A) of P were applied per hectare (Table 10-11). At the same location, results indicated a significant increase in orchard-grass yields with phosphorus applied at rates of 25 and 100 kg/ha (50 and 202

TABLE 10-10
CHARACTERISTICS OF SOME COMMON PROCESSED PHOSPHATE MATERIALS

Material	Percent			
	P_2O_5 content	Water solubility	Citrate solubility	Total available phosphate
Superphosphate 0—20—0	16–22	78	18	96
Triple superphosphate 0—45—0	44–47	84	13	97
Monoammonium phosphate 11—48—0	46–48	100	—	100
Diammonium phosphate 18—46—0	46	100	—	100
Ammonium polyphosphate 10—34—0 11—37—0	34–37	100	—	100

lb/A) (28). In another trial, using a 4-year average, yields increased significantly when P was applied to an irrigated mixture of smooth bromegrass, timothy, orchardgrass, and bluegrass, with occasional red clover plants (Table 10-12) (40). Native warm-season grasses and various other grasses also respond to P application (Table 10-13) (53).

As indicated earlier, phosphorus increases the water efficiency of alfalfa. Phosphorus rate studies over a 3-year period showed that increasing rates of P application decreased water use per metric ton by 15 cm, as yields increased 6.5 t/ha (Table 10-14) (51).

Recent studies, using regression analysis, showed that when smooth brome-grass and alfalfa were grown together and the yields were expressed as a

TABLE 10-11
THREE-YEAR AVERAGE RESPONSE OF ALFALFA AND ORCHARDGRASS TO P APPLICATIONS

P applied, kg/ha (lb/A)	D. M. yield, t/ha	
	Alfalfa	Orchardgrass
0 (0)	7.2	5.7
25 (22)	—	8.7
100 (89)	11.7	9.8

Source: J. A. Lutz Jr., "Effect of partial acidulated rock phosphate and concentrated superphosphate on yield and chemical composition of alfalfa and orchardgrass," *Agron. J.* 65:286–289 (1973).

TABLE 10-12
RESPONSE TO APPLIED P OF SMOOTH
BROMEGRASS, TIMOTHY, ORCHARDGRASS,
BLUEGRASS, AND OCCASIONAL RED CLOVER
PLANTS UNDER IRRIGATION

Applied P, kg/ha (lb/A)	D. M. yield, t/ha
0 (0)	12.2
20 (18)	15.8
40 (36)	16.7
60 (54)	19.1

Source: G. W. Rehm, J. T. Nichols, R. C. Sorensen, and W. J. Moline, "Yield and botanical composition of an irrigated grass-legume pasture as influenced by fertilization," *Agron. J.* 67:64–68 (1975).

percentage of the control, and related to the application of N and P fertilizers and the available soil P, yields were increased by 251% over the controls. It was also noted that the proportion of alfalfa in the sward was significantly reduced—by 29%—when N and P fertilizer was supplied. When N was applied at 90 kg/ha (80 lb/A) and P at 20 kg/ha (18 lb/A), the average increase in forage yield was 74%. These rates gave the most economical return for the fertilizer invested and did not significantly increase the nitrate N in the soil profile (35, 36).

It is generally concluded that phosphorus increases the nodulation on legumes primarily because of increased root and nodule mass. Since a larger root system is developed, the potential for nodulation is further increased. The root mass develops earlier in the life cycle of plants that are fertilized with P, so that nodules can develop earlier, resulting in earlier nitrogen fixation and faster growth (59).

TABLE 10-13
RESPONSE OF VARIOUS GRASSES TO 90 KG/HA
OF P

Grass	Increase over check (%)
'Midland' bermudagrass	31
'Morpa' weeping lovegrass	12
'Plains' bluestem	16
Native range grass	20

Source: C. M. Taliaferro, F. P. Horn, B. B. Tucker, R. Totusek, and R. D. Morrison, "Performance of three warm-season perennial grasses and a native range mixture as influenced by N and P fertilization," *Agron. J.* 67:289–292 (1975).

TABLE 10-14
EFFECT OF PHOSPHORUS ON WATER-USE EFFICIENCY OF ALFALFA

Initial applied P, kg/ha (lb/A)	Alfalfa	
	D. M. yield, t/ha	Water use per ton, cm
49 (44)	16.4	36.1
98 (87)	18.2	29.0
196 (175)	21.7	22.9
294 (262)	22.9	21.1

Source: C. O. Stansberry, C. D. Converse, M. R. Maise, and O. J. Kelley, "Effect of moisture and P variables on alfalfa hay production on the Yuma Mesa," *Soil Sci. Amer. Proc.* 19:303–310 (1955).

Effects of P on Animal Health

Phosphorus is present in every living cell, in the nucleic acid fraction and in membranes as a part of phospholipids. It is very important in all phases of reproduction. In ruminants, it is very important for the proper metabolism and health of the rumen microorganisms. Phosphorus is essential in proper utilization of energy and in the process of building muscle tissue in animals. In combination with calcium, phosphorus may comprise more than 70% of the ash in an animal's body. Thus, these two elements are closely related to animal health and metabolism. It is very important to keep a proper balance of calcium and phosphorus in relation to vitamin D. A desirable ratio of Ca:P is between 2:1 and 1:1. It is very important to keep this balance; even though one element may be at the minimum, the other element may be in excess of the balance, consequently creating an imbalance within the animal's body. When vitamin D is insufficiently provided in the ration, the ratio becomes less important and more efficient utilization is made of these elements (59).

About 99% of the Ca and 80% of the P in the body are present in bones and teeth. Calcium and phosphorus are also found in the soft tissues of the body and in the blood. In dairy animals, calcium and phosphorus make up 50% of the ash of milk. Therefore, a very liberal supply of these two elements is required. The calcium and phosphorus requirements for a dairy animal, especially one that is lactating, are very important. Ca:P ratios are also important for the animal's future health and longevity, and the health of the calf.

All livestock need phosphorus. A P deficiency is often indicated by lack of appetite or a deprived appetite, which is referred to as "pica." This is indicated by the animal's eating bones, wood, clothing, and other material that might be available (54).

POTASSIUM

Potassium is another of the major essential elements for plant growth. It is used in larger amounts than is phosphorus. Within live plant tissue, the average percentage of potash is approximately 8 to 10 times that of phosphorus; hay or dry matter contains up to 4 times as much potash as phosphorus. Potassium aids plants in resisting disease, insects, cold weather, and drought, and it functions in stomatal opening and closure and sugar transport. Since plants take up large amounts of potassium, the soil must be replenished with potash or deficiencies will show up readily, particularly when the forage is harvested as hay or silage rather than grazed.

Both legumes and grasses are very heavy users of potassium, especially grasses because of their extensive fibrous root system (Fig. 10-9). Generally, the first species within a grass-legume mixture to show a deficiency or die out will be the legume because of its lack of an efficient uptake of K when in competition with a grass (Fig. 10-10). Most soils in the United States vary in K content from area to area, particularly the sandy soils, which are low in potash and must be fertilized frequently with high rates of potassium.

There are three forms of potassium in the soil. (1) The soluble K is the portion that is water-soluble; this constitutes only a small portion of the total potassium in the soil. (2) There is also an exchangeable K, which is held on the soil colloids and is easily available. This also comprises a small percentage of the total potassium in the soil. (3) In addition, the clay fraction of the soil contains a nonexchangeable K, which is neither soluble nor readily available to plants. The nonexchangeable K makes up the largest portion of the total K in the soil, except in highly acid, sandy soils or soils that are high in organic matter, where the nonexchangeable K is relatively low.

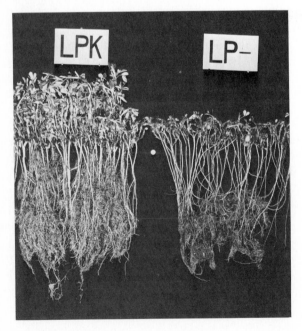

FIGURE 10-9
The effect of K on top growth and root development.

FIGURE 10-10
The effect of annual application of K_2O to alfalfa for 4 years compared to that of no K_2O applications.

The soluble portion of the K in the soil is subject to leaching precisely because of its solubility. It is important to maintain a relatively high level of soluble K for the entire life cycle of a plant, and within a growing season. Following droughty periods, in which the soluble K becomes less available on sandy soils or highly organic soils, a potassium deficiency may show up late in the growing season. Therefore, it is important to apply potassium to the crop annually. Some of the nonexchangeable K is made available each year from the soil, but only in limited amounts.

It has been shown that it takes approximately four units of K_2O to increase a soil test by one unit. As a result the recommended rate for applying potassium to increase the soil test value to a desired goal is equal to four times the difference between the soil test goal and the actual soil test value (Table 10-15).

A particular state or region may be subdivided to indicate areas with low or high potassium-supplying power (Fig. 10-11). Since the only significant loss of potassium applied to the soil is through crop removal or soil erosion, it is recommended that the soil test level of potassium be built up to a minimum of 300 to 400 lb of exchangeable potassium for soils in regions with low and high cation exchange capacity, respectively (Table 10-15). These values are higher than required for maximum yield but, as in the recommendations for phosphorus, this will ensure that potassium availability will not limit the crop or the forage yield.

One should apply one-half of the recommended amount of potassium to a forage crop after the last harvest in the fall, and the remaining amount after the first harvest the following spring. Top dressing in this manner will ensure the

TABLE 10-15
AMOUNT OF POTASSIUM (K_2O) REQUIRED
ANNUALLY PER ACRE TO BUILD UP THE SOIL
(Based on Buildup Occurring over a 4-year Period; 4
lb/A K_2O Required to Change the K Test 1 lb)

| K test* (lb/A) | Soil cation exchange capacity | |
	Low*	High†
50	210	250
60	200	240
70	190	230
80	180	220
90	170	210
100	160	200
110	150	190
120	140	180
130	130	170
140	120	160
150	110	150
160	100	140
170	90	130
180	80	120
190	70	110
200	60	100
210	50	90
220	40	80
230	30	70
240	20	60
250	10	50
260	0	40
270	0	30
280	0	20
290	0	10
300	0	0

*Soil tests taken before May or after Sept. 30 should be adjusted downward as follows: subtract 30 lb for dark-colored soils in central Illinois; 45 lb for light-colored and fine-textured lowland soils in central and northern Illinois; and 60 lb for medium- and light-colored soils in southern Illinois.

†Low cation-exchange-capacity soils are those with CEC less than 12 meq/100 g soil; high are those with CEC equal to or greater than 12 meq/100 g soil.

availability of sufficient amounts of potassium throughout the growing season, and will also provide sufficient potassium in the plant tissue for winter survival (Fig. 10-12).

Research has indicated that for every ton of dry matter removed from the

FIGURE 10-11
Potassium-supplying power in Illinois. The
shaded areas are sands with low
potassium-supplying power.

field, approximately 13 kg (12 lb) of P_2O_5 and 55 kg (60 lb) of K_2O are removed. The conversions of P to P_2O_5 and K to K_2O are shown in Table 10-16. To maintain an adequate fertility level, the amount of nutrient removed by the previous forage crop should be applied to the next as a top dressing. Table 10-17 indicates the number of pounds per acre that one should supply to maintain a potential yield goal for a forage crop (Fig. 10-13). The main sources of potash fertilizers and their percentages of K_2O are: muriate of potash, 60%; sulfate of potash, 51%; and sulfate of potash (magnesia), 22% K_2O and 18% MgO.

Workers in Wisconsin found that most of the applied potassium was absorbed by roots near the soil surface when K_2SO_4 was placed at varying depths below a 2-year stand of alfalfa (39a). It was also found that when the next three harvests were taken, 41% and 29% of the potassium was removed from the upper 7.6 cm (3 in) and 22.9 cm (9 in) from the surface, respectively. In the same study, there were no differences observed in the plant tissue's sulfur concentration from soil that contained sufficient amounts of sulfur. Therefore, it appears that sufficient amounts of K and P can be removed from the upper portion of the soil, and that top dressing of these elements would suffice (10).

The main source of K used as a fertilizer in agriculture throughout the world is potassium chloride (KCl). It has been shown that grasses are less sensitive to chloride than are legumes (4). There appears to be a chloride toxicity level at

FIGURE 10-12
Annual fertilization of K_2O is needed for winterhardiness.

which legumes are very sensitive. From numerous research studies, investigators have set the upper limit as 336 kg/ha (300 lb/A) of KCl. This should be the maximum level at the time of seeding, due to the so-called "salt effect." This writer believes that this effect results from chloride toxicity rather than from the potassium salt. For instance, research indicates that chlorine will depress the yield of red clover (25). In a study using KCl and K_2SO_4, it was found that the highest yield was obtained at 200 kg/ha (178 lb/A) of K using K_2SO_4, whereas at that same level, there was severe yield depression when KCl was used (47). There is also a possibility that some response to S was shown. In Wisconsin, it was found that when KCl was used, the level of Cl found in young alfalfa tillers was as high as 12 to 14% in the plant tissue at the time of death (47).

K has a reputation for being consumed luxuriously by plants, especially grasses. Plants have a tendency to absorb amounts of K far in excess of what is actually required for proper growth and development (29). As shown in Table 10-18, the chemical composition for the critical level of K is between 2.3 and

TABLE 10-16
PLANT NUTRITION CONVERSION

P (phosphorus) $\times$ 2.29 = P_2O_5
P_2O_5 $\times$ .44 = P
K (potassium) $\times$ 1.2 = K_2O
K_2O $\times$ .83 = K
ppm $\times$ 2 = lb/A [assuming that an acre plow depth of 6⅔ in (17 cm) weighs 2000 lb (908,000 kg)]

TABLE 10-17
MAINTENANCE FERTILIZER REQUIRED FOR
VARIOUS FORAGE YIELDS OF ALFALFA, GRASS,
OR ALFALFA-GRASS MIXTURES

Forage yield (T)	P_2O_5 (lb/A)	K_2O (lb/A)
2	24	100
3	36	150
4	48	200
5	60	250
6	72	300
7	84	350
8	96	400
9	108	450
10	120	500

2.5%, which is adequate for growth (32, 59) (Fig. 10-14). It is not unusual to find plants that are grown under high levels of K to range from 3.5 to 4.5% (59). This excessive percentage has not been shown to offer any advantage for either plant life or the animals consuming the forage.

The proper balance of P and K is important in maintaining productive stands of alfalfa. Table 10-19 shows the effect of proper balance of P and K on alfalfa stands in New Jersey (3). At the end of the fifth year, there was almost twice the production in metric tons per hectare when P and K were applied, compared

FIGURE 10-13
The effect of P and K on alfalfa top growth.

TABLE 10-18
CHEMICAL COMPOSITION AND NUTRIENT
REMOVAL BY ALFALFA

Nutrient	Critical level chemical composition of tissue	Removal per Ton
N	2.7–3.3%	*
P_2O_5	0.26–0.32	14 lb
K_2O	2.3–2.5	60
Ca	0.9–1.2	30
Mg	0.25–0.31	6
S	0.15–0.31	6
B	24–36 ppm	0.05
Fe	100–136	0.33
Mn	28–38	0.12
Cu	7–9	0.01
Zn	19–25	0.05
Mo	3.3–3.9	Trace

*Nitrogen fixation by *Rhizobia.*

with the control. In addition, the number of years that a fertilized stand maintained at least 50% alfalfa was at least twice that for the control.

It has also been shown that K has a direct effect on winter survival of alfalfa. In Canada, it was found that a significant increase in the number of plants that survived was observed when potassium was applied beyond the control level, at a rate of 112 to 224 kg/ha (100 to 200 lb/A), whether the temperature was −4°C (25°F) or −9°C (15°F). The number of stems per plant increased in a linear manner as the rate of K was increased (Table 10-20).

FIGURE 10-14
Potash-deficiency symptoms on alfalfa leaflets.

TABLE 10-19
EFFECT OF P AND K FERTILIZER ON MAINTAINING ALFALFA YIELDS IN NEW JERSEY

Applied fertilizer, kg/ha (lb/A)		Yield each year, t/ha						Yrs 50% or more alfalfa
P_2O_5	K_2O	1	2	3	4	5	Avg.	
0 (0)	0 (0)	12.1	14.6	8.7	7.6	9.0	10.3	2
84 (75)	336 (300)	13.7	18.4	13.9	16.8	17.9	16.1	5

Source: F. E. Bear and A. Wallace, "Alfalfa, its mineral requirements and chemical composition," *New Jersey Agr. Exp. Sta. Bull. 748,* 1950.

Temperature has an effect upon the uptake and translocation of nutrients. It has been found that the percent of K in plants increases with temperature and the amount of K applied (Fig. 10-15). It is suggested, when forages are grown in cooler environments, that greater levels of K are needed for maximum growth than is the case in warmer climates (49).

Many researchers have studied the amount of K removed in various forages. It is generally concluded that as the yield of hay increases, the amount of K removal increases proportionally (Table 10-21). Yields above 10,000 kg/ha (8920 lb/A) contain from 2.5 to 3.5% K; thus it is very important to maintain a high level of K when expecting high yields. Research at Illinois also indicates that sufficient levels of K should be above 2.4 to 2.5%; the minimum critical level has been established at approximately this range (32).

Many investigators have studied the effect of K on nitrogen fixation and nodule formation (Fig. 10-16) (2a). A recent study has shown that high rates of K fertilization increase the number of nodules (Table 10-22) (13). It was shown that per plant, the increased nitrogen fixed was a result of the nodule number and not of the acetylene reduction rates; there was a linear correlation between the number of nodules and the acetylene reduction rate. Therefore, it appears that K fertilization will increase alfalfa yields by increasing the numbers of nodules and so the extent of nitrogen fixation.

TABLE 10-20
EFFECT OF K ON WINTER SURVIVAL OF ALFALFA AND LIVING STEMS PER PLANT, ONTARIO

K_2O rate, kg/ha (lb/A)	Winter survival, %		Living stems/plant	
	−4°C (25°F)	−9°C (15°F)	−4°C (25°F)	−9°C (15°F)
0 (0)	73	56	2.8	1.9
112 (100)	97	60	3.4	2.6
224 (200)	90	80	3.8	3.0
336 (300)	97	80	4.3	3.8

Source: R. E. Blaser and E. L. Kimbrough, "Potassium nutrition of forage crops with perennials," in *The Role of Potassium in Agriculture*, Amer. Soc. Agron., Madison, WI, pp. 423–445, 1968.

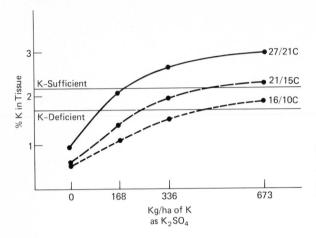

FIGURE 10-15
Percent K in 'Vernal' alfalfa forage at first-flower stage grown at three different day-night temperatures, assuming 2.2% as sufficient and 1.75% as deficient (47).

CALCIUM

Calcium is another major element needed for top forage production. Calcium, along with phosphorus, comprises about 75% of the mineral matter in the bodies of livestock and about 90% of their skeletons. The main source of calcium is limestone. It helps reduce soil acidity, it adds strength to plants, and it is needed in the formation of the bones and teeth of animals. Calcium is also needed in particularly large amounts by growing animals, pregnant aninls, and those that are producing milk. Most of the calcium needed by livestock can be furnished by forages, but limestone must be added to acid soils to supply the plants with amounts of calcium adequate for proper growth.

In chemical composition, green tissues of forages range from 0.9 to 1.2% calcium. Researchers have shown that very young alfalfa plants prior to bloom, or at approximately 7.5 cm (3 in) in height, will range from 1.76 to 3.00% calcium, whereas red clover at that stage will be in the range of 1.2 to 2.0% (7).

MAGNESIUM

Magnesium is another important element needed in forage production and is supplied by dolomitic limestone. Magnesium is a necessary component of chlorophyll, which plays an important role in photosynthesis and carbohydrate

TABLE 10-21
SUMMARY OF K REMOVAL RELATED TO ALFALFA YIELD EXPECTATIONS

Alfalfa hay yield, kg/ha (lb/A)	K concentration, %	K_2O_3 removal, kg/t
2200–5600 (2000–5000)	1.5–2.2	14.8–21.8
5700–10,000 (5010–9000)	1.9–2.8	18.5–27.6
10,000 (9000) or more	2.5–3.5	24.7–34.6

FIGURE 10-16
As one increases the N application to forage grasses, one must maintain a proper balance of K_2O.

production. Most legumes contain amounts of magnesium sufficient to meet the needs of animals. Soils that are high in potassium and low in magnesium may produce an increased incidence of grass tetany (16, 30). This disease of cattle can be controlled by feeding the animals extra magnesium or by growing legumes in the pasture mixture (16).

The chemical composition of alfalfa tissue ranges from 0.25 to 0.31% magnesium. Some workers have also shown that the percentage of magnesium in the bud stage may exceed this amount, ranging from 0.31 to 1.0% in alfalfa and 0.20% in red clover (7, 25, 32, 54).

Generally, researchers refer to both Ca and Mg levels in legumes in recommending application strategies for dolomitic limestone fertilizer. Legumes

TABLE 10-22
EFFECT OF K ON NODULATION OF ALFALFA

K application, kg/ha (lb/A)	Nodules/plant	Increase over control, %
0	42	—
448 (400) as K_2SO_4	69	166
673 (600) as K_2SO_4	111	271
673 (600) as KCl	93	223

Source: S. H. Duke, M. Collins, and R. M. Soberalske, "The effects of potassium fertilization on nitrogen fixation and nodule enzymes on nitrogen metabolism in alfalfa," *Crop Sci.,* 20:213–218 (1980).

are considered very efficient users of soil magnesium. They contain more magnesium than do grasses growing in the same environment; for instance, red clover dry matter contains 0.27% magnesium, whereas grasses have 0.19% when grown under the same conditions (3). Mg deficiencies are generally associated with acid or sandy soils. Suggested soil test magnesium levels are 60 to 70 lb/A on sandy soils and 120 to 150 lb/A on silty loams or fine-textured soils.

Research has shown that Ca and Mg levels in legumes may be depressed by K fertilization (4, 31, 54). The concentrations of both Ca and Mg may be reduced below the sufficient range, but there is very little data available to show yield responses when Ca or Mg have been applied.

SULFUR

Recently, much more attention has been given to sulfur as a fertilizer element for forage legumes. Sulfur increases root growth and helps maintain a dark green color. It is needed for protein formation and nitrogen fixation by legumes (11). Most of the sulfur used by plants comes from the air or from organic matter. Sulfur is also present in many low-analysis fertilizers. There are a few soils in the southeastern United States that need sulfur fertilization. Because of the Clean Air Act, sulfur is being removed from burning coal, so there is now an increased concern for adding sulfur to some fertilizer recommendations. Sulfur is removed from the soil at approximately the same rate as is Mg (Table 10-18).

Sulfur deficiencies are found generally on highly leached or sandy soils and soils that are low in organic matter. Sulfur may also become deficient under continuous alfalfa production. Most sulfur soil test readings are reported as parts per million (ppm), and it is suggested that for alfalfa production, if the soil test level falls below 7 ppm, one should apply sulfur. For readings between 7 to 12 ppm of S, it is suggested that sulfur be applied on a trial basis, but when sulfur readings are above 10 to 12 ppm, there is very little chance for a positive response. Limited data are available to show the response of alfalfa to sulfur application. In Minnesota, on soils that were considered sufficient in sulfur, the yield was doubled when elemental S was applied as gypsum (Table 10-23) (18). Sulfur will also significantly reduce the death rate of young alfalfa seedlings. It has also been shown that where sulfur is deficient, the alfalfa stand is much less

TABLE 10-23
RESPONSE OF ALFALFA TO S APPLICATIONS

Treatment	S application rate, kg/ha (lb/A)	D. M. yield, kg/ha (lb/A)
Control	0 (0)	4020 (3586)
Sulfur (elemental)	56 (50)	9410 (8394)
Gypsum	56 (50)	9690 (8643)

Source: W. K. Griffith, "Satisfying the nutritional requirements of established legumes," in *Forage Fertilization,* Amer. Soc. Agron., Madison, WI, pp. 147–169, 1974.

TABLE 10-24
EFFECT OF S ON ALFALFA YIELD, CONCENTRATION OF S IN TISSUE, AND AMOUNT
REMOVED WHEN HARVESTED AT EARLY BLOOM

S application rate, kg/ha (lb/A)	D. M. yield, kg/ha (lb/A)	S concentration, %	S removed, kg/ha (lb/A)
0 (0)	7409 (6609)	0.16	12 (11)
40 (35)	15,114 (13,482)	0.25	38 (34)
80 (71)	17,287 (15,420)	0.33	55 (49)

Source: W. K. Griffith, "Satisfying the nutritional requirements of established legumes," in *Forage Fertilization,*
Amer. Soc. Agron., Madison, WI, pp. 147–169, 1974.

winter-hardy (18). Other researchers have found that alfalfa yields will continue
to increase up to a maximum of 80 kg/ha (72 lb/A) when S is applied (18). It was
shown that alfalfa yields over 17,000 kg/ha (15,000 lb/A) dry matter contain
0.33% of S and remove 55 kg/ha (49 lb/A) of S (Table 10-24) (Fig. 10-17) (18).

It is important to keep the proper balance between nitrogen and sulfur. It is
suggested that a ratio of 10:1 N:S be considered optimum for maximum forage
production and animal utilization. If the ratio exceeds 15:1, it is considered that
yield and protein production may be depressed.

MICRONUTRIENTS

The micronutrients needed for forage production are B, Fe, Mn, Cu, Zn, and
Mo. All of these nutrients are essential for plant growth and are just as

FIGURE 10-17
The effect of sulfur applied to alfalfa.

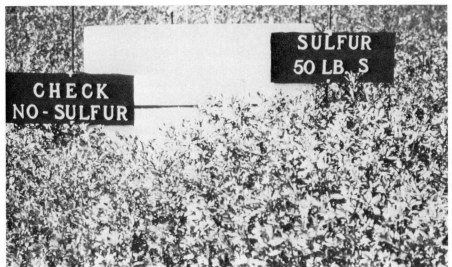

important for the proper life cycle of a forage crop as are any of the major elements. They are required in very small amounts, usually reported in ppm (Table 10-18).

Boron

Legumes, especially alfalfa, are sensitive to low boron levels. Adding boron to alfalfa fertilizer is an accepted practice where deficiencies may occur. Boron deficiency is associated with low moisture levels or droughty conditions, sandy soils, and light-colored soils. Boron is relatively immobile in the plant and the youngest tissue may show boron-deficiency symptoms first (18). When there is a boron deficiency, alfalfa may appear stunted, with the internodes shortened, and the plant may take on a general yellowish character (Fig. 10-18). This condition is sometimes confused with drought problems.

Boron is held slightly to the soil's clay fraction and is easily leached unless it is utilized by the legume or retained in the organic-matter fraction. Boron availability is associated with the level and the decomposition of organic matter, as well as with the soil texture and the soil pH (18). When the organic matter decomposes, boron is released for plant use. The soil pH has a direct effect on the availability of boron. A low pH inhibits the activity of microorganisms and

FIGURE 10-18
Boron is needed for top alfalfa production.

FIGURE 10-19
The effect of boron on alfalfa weevil larvae.

the eventual release of boron. Consequently, it is very important to have a soil pH of 6.5 to 7 to reduce a potential boron deficiency. Excessive pH will also reduce boron availability (Fig. 10-1).

Properly fertilized and healthy legumes will usually contain 24 to 35 ppm of boron (Table 10-18). If the boron level falls below 20 ppm, an application may be advisable. For legumes, boron application is recommended at the rate of 3 to 5 kg/ha (2 to 4 lb/A) of actual boron. It is recommended that light-textured or permeable, well-drained soils receive annual applications of boron to avoid deficiencies. Boron can be applied with blended fertilizers or by itself, as a top-dressed material. Soils that are very heavy or dark-colored, and high in organic matter, may need only one application every 2 or 3 years. Boron is available in several different types of carriers. The main form of boron fertilizer is borax, which ranges from 11 to more than 21% boron. The label indicates the percent boron. If boron is applied annually, it is advisable that if one plans to grow corn the year after an alfalfa stand's last year in production, no boron be applied to the alfalfa because corn is rather sensitive to high boron concentrations.

Boron is needed for both plant growth and the control of alfalfa weevil. Data from the University of Illinois indicate that high levels of boron in alfalfa plant tissue may reduce the level of hatching of alfalfa weevil eggs, thus providing a potential biological control (Fig. 10-19). Under extremely high levels of boron, the alfalfa weevil may become completely sterilized or incapable of complete metamorphosis (50).

In addition to providing a healthy plant stand, boron is essential for proper legume nodulation (18). Other workers have shown that boron is necessary for legume seed production.

Molybdenum

Molybdenum is needed in both nitrogen assimilation, or protein production, and nitrogen fixation. A deficiency of molybdenum appears to resemble a nitrogen deficiency in legumes. When molybdenum is deficient, nitrogen metabolism is reduced, so the problem appears as a nitrogen deficiency. Research has shown that protein content may be increased by addition of molybdenum, which is directly to the nitrogen concentration of the plant tissue (18).

As the pH is reduced, so is the availability of molybdenum for plant growth. Only trace amounts of Mo are required. The chemical composition of an alfalfa plant ranges from 3.3 to 3.9 ppm of Mo; deficient plants contain less than 0.5 ppm (Table 10-18). It is sometimes falsely claimed that a small amount of Mo will substitute for lime application. Caution should be exercised in adding Mo to inoculated legume seed as a solution, since it can kill the bacteria.

Other Micronutrients

Alfalfa can absorb Zn more efficiently from soils than can many other crops (6). The chemical composition of alfalfa ranges from 19 to 25 ppm of Zn (Table 10-18). Zinc is important in the enzyme system of plants. Generally, the Zn level within the plant tissue must be below 6 ppm before a response will be shown to Zn application. Zinc deficiency is indicated by white or striped leaves (18). It is usually found on acid, sandy soils that have recently been limed.

Copper is important in the enzyme system of plants and in the formulation of hemoglobin in animals. Copper deficiency may appear on peat or muck-type soils and heavily weathered, sandy soils. The availability of copper is decreased on high-pH soils, and the element may show up as a plant deficiency on extremely low-pH soils. On mineral soils, application rates of 11 to 17 kg/ha (10 to 15 lb/A) of copper sulfate are usually high enough for good legume growth. For legumes grown on muck soils, this rate should be doubled (18).

Iron is essential in the formation of chlorophyll and also is in the blood hemoglobin of animals. Most soils contain sufficient amounts of iron, except where it is tied up by very high pH. Addition of sulfur will increase the availability of iron in soil (18). Excessive P fertilizer may induce deficiency of iron and zinc.

Manganese is required in small amounts by both plants and animals. Manganese is important in the oxidative enzyme system of plants. Soils that have an extremely high pH may be deficient in available manganese. Manganese also reduces the availability of iron for plant growth (18) and may produce Mn toxicity in plants grown on very acid soils.

QUESTIONS

1 Why is pH so important for the overall fertility program of forages?
2 Name at least two forage crops that fall into each of the following classes: those that are very sensitive to acidity; those that can tolerate moderate acidity.
3 Assuming that one had a light- to medium-colored silty clay loam and the pH was 5.5, how many tons of limestone would be recommended to bring the pH to the proper level for alfalfa growth?

4 If one has a choice between calcitic limestone and dolomitic limestone, which should be used and why?

5 Assuming that one were to plow down an old alfalfa stand that had more than five plants per square foot, what would be the contribution of nitrogen to the succeeding corn crop and the succeeding wheat crop? What would be the contribution of nitrogen for the second year of corn following the plow down of that same stand of alfalfa?

6 Describe the role of nitrogen in forage production in relation to both grasses and legumes.

7 What is meant by the phosphorus-supplying power and the potassium-supplying power of the soil?

8 Why do various soils vary in their supplying power of various nutrients?

9 How much P_2O_5 and K_2O are required to build up the soil by one unit in a soil test?

10 What is the role of phosphorus in forage production? What is the role of potassium?

11 Why is phosphorus important in forages' utilization of water?

12 How much P_2O_5 and K_2O are removed in an 8-ton yield of alfalfa? How much fertilizer must one add to maintain the proper level?

13 Why is it important to maintain the proper balance of P and K fertilizer when applying them to forages?

14 Describe the role of K in forage fertilization.

15 In which area of the United States would the rate of forage fertilization be greater, the upper midwest or the middle midwest?

16 Discuss the roles of boron, sulfur, magnesium, and manganese in forage fertilizer.

REFERENCES

1 Ahlgren, G. H. *Forage Crops,* McGraw-Hill, New York, pp. 371–384, 1956.

2 Allen, S. E., G. L. Terman, and H. G. Kennedy. Nutrient uptake by grass and leaching losses from soluble and S-coated urea and KCl. *Agron. J.,* 70:264–273, 1978.

2 a Barta, A. L. Response of symbiotic N_2 fixation and assimilate partitioning to K supply in alfalfa. *Crop Sci.,* 22:89–92, 1982.

3 Bear, F. E. and A. Wallace. Alfalfa, its mineral requirements and chemical composition. *New Jersey Agr. Exp. Sta. Bull. 748,* 1950.

3 a Belesky, D. P. and S. R. Wilkinson. Response of 'Tifton 44' and 'Coastal' bermudagrass to soil pH, K, and N source. *Agron. J.,* 75:1–4, 1983.

4 Blaser, R. E. and E. L. Kimbrough. Potassium nutrition of forage crops with perennials. In *The Role of Potassium in Agriculture,* Amer. Soc. Agron., Madison, WI, pp. 423–445, 1968.

5 Brown, B. A. Potassium fertilization of ladino clover. *Agron. J.,* 49:477–480, 1957. *Agron. J.,* 49:477–480, 1957.

6 Brown, B. A., R. I. Munsell, and A. V. King. Potassium and boron fertilization of alfalfa on a few Connecticut soils. *Soil Sci. Soc. Amer. Proc.,* 10:134–140, 1945.

7 Brown, J. C. and J. H. Graham. Requirements and tolerance to elements by alfalfa. *Agron. J.,* 70:367–373, 1978.

8 Christians, N. E., D. P. Marten, and J. F. Wilkinson. Nitrogen, phosphorus, and potassium effects on quality and growth of Kentucky bluegrass and creeping bent-grass. *Agron. J.,* 71–564–567, 1979.

9 Collins, K. L., G. C. Nedermann, C. L. Rhykerd, and C. H. Noller. How do we apply 600 lbs. potash (K_2O) on alfalfa? *Better Crops with Plant Food,* 56(3):18–19, 1972.

10 Doll, E. C., A. L. Hatfield, and J. R. Todd. Vertical distribution of top-dressing fertilizer phosphorus and potassium in relation to yield and composition of pasture herbage. *Agron. J.,* 51:645–648, 1959.

11 Drlica, D. M. and T. L. Jackson. Effects of stage of maturity on P and S critical levels in subterranean clover. *Agron. J.,* 71:824–828, 1979.

12 Duell, R. W. Fertilizing forage for establishment. In *Forage Fertilization,* Amer. Soc. Agron., Madison, WI, pp. 67–93, 1974.

13 Duke, S. H., M. Collins, and R. M. Soberalske. The effects of potassium fertilization on nitrogen fixation and nodule enzymes on nitrogen metabolism in alfalfa. *Crop Sci.,* 20:213–218, 1980.

13 a Fales, S. L., and K. Ohki. Manganese deficiency and toxicity in wheat; influences on growth and forage quality of herbage. *Agron. J.,* 74:1070–1073, 1982.

14 Fribourg, H. A., K. M. Barth, J. M. McLaren, L. A. Carver, J. T. Connell, and J. M. Bryan. Season trends in *in vitro* dry matter digestibility of N-fertilizer bermudagrass and of orchardgrass-ladino pastures. *Agron. J.,* 71:117–120, 1979.

15 George, J. R., M. E. Pinheirro, and T. B. Bailey, Jr. Long-term potassium requirements of nitrogen-fertilized smooth bromegrass. *Agron. J.,* 586–591, 1979.

16 George, J. R. and J. L. Thill. Cation concentration of N- and K-fertilized smooth bromegrass during the spring grass tetany season. *Agron. J.,* 71:431–436, 1979.

17 Graven, E. H., O. J. Attoe, and D. Smith. Effect of liming and flooding on manganese toxicity in alfalfa. *Soil Sci. Soc. Amer. Proc.,* 29:702–706, 1965.

18 Griffith, W. K. Satisfying the nutritional requirements of established legumes. In *Forage Fertilization,* Amer. Soc. Agron., Madison, WI, pp. 147–169, 1974.

19 Hall, R. L. Some implications of the use of potassium chloride in nutrition studies. *J. Aust. Inst. Agr. Sci.,* 37:249–252, 1971.

20 Hannaway, A. B. and J. H. Reynolds. Seasonal changes in organic acids, water-soluble carbohydrates, and neutral detergent fiber in tall fescue as influenced by N and K fertilization. *Agron. J.,* 71:493–496, 1979.

21 Hanson, C. L., J. F. Power, and C. J. Erickson. Forage yield and fertilizer recovery by three irrigated perennial grasses as affected by N-fertilization. *Agron. J.,* 70:373–375, 1978.

22 Heichel, G. H. and C. P. Vance. Nitrate-N and rhizobium strain roles in alfalfa seedling nodulation and growth. *Crop. Sci.,* 19:512–518, 1979.

23 Hojjati, S. M., W. E. Templeton, Jr., and T. H. Taylor. Nitrogen fertilization in establishing forage legumes. *Agron. J.,* 70:429–433, 1978.

24 Jackobs, J. A., T. R. Peck, and W. M: Walker. Efficiency of fertilizer top dressing on alfalfa. *Ill. Agr. Exp. Sta. Bull. 738,* 1970.

25 Laughlin, W. M., M. Blom, and P. F. Martin. Red clover yield and composition as influenced by phosphorus, potassium rate and source, and chloride. *Commun. Soil Sci. Plant Analysis* 2(1):1–10, 1971.

26 Lee, C. T., and D. Smith. Influence of nitrogen fertilizer on stands, yields of herbage and protein, and nitrogenous fractions of field-grown alfalfa. *Agron. J.,* 64:527–530, 1972.

26 a Leyshon, A. J. Deleterious effects on yields of drilling fertilizer into established alfalfa stands. *Agron. J.,* 74:741–743, 1982.

27 Ludwick, A. E. and C. B. Rumberg. Grass hay production as influenced by N-P top dressing and by residual P. *Agron. J.,* 68:933–937, 1976.

28 Lutz, J. A., Jr. Effect of partial acidulated rock phosphate and concentrated superphosphate on yield and chemical composition of alfalfa and orchardgrass. *Agron. J.,* 65:286–289, 1973.

29 Lutz, J. A., Jr. Effects of potassium fertilization on yield and K content of alfalfa and on available subsoil K. *Commun. Soil Sci. Plant Analysis,* 4(1):57–65, 1973.

29 a Mallarino, A. P., W. F. Wedin, R. D. Voss, and C. P. West. Phosphorus requirements of alfalfa, smooth bromegrass, orchardgrass, and reed canarygrass pastures under two grazing pressures. *Agron. J.,* 75:291–294, 1983.

30 Markus, D. K. and W. R. Battle. Soil and plant responses to long-term fertilization of alfalfa. *Agron. J.,* 57:613–616, 1965.

31 Martel, Y. A. and J. Zizka. Yield and quality of alfalfa as influenced by additions of S to P and K fertilization under greenhouse conditions. *Agron. J.,* 69:531–535, 1977.

32 Melsted, S. W., H. L. Motto, and T. R. Peck. Critical plant nutrient composition values useful in interpreting plant analysis data. *Agron. J.,* 61:17–20, 1969.

33 Meyer, T. A. and G. W. Volk. Effect of particle size of limestone on soil reaction, exchangeable cations and plant growth. *Soil Sci.,* 73:37–52, 1952.

34 Miller, D. A. and R. K. Smith. Influence of boron on other chemical elements in alfalfa. *Commun. Soil Sci. and Plant Analysis,* 8(1):465–478, 1979.

34 a Morris, R. J., R. H. Fox, and G. A. Jung. Growth, P uptake, and quality of warm and cool-season grasses on a low available P soil. *Agron. J.,* 74:125–129, 1982.

35 Nuttall, W. F. Effect of nitrogen and phosphorus fertilizers on a bromegrass and alfalfa mixture grown under two systems of pasture management. II. Nitrogen and phosphorus uptake and concentration in herbage. *Agron. J.,* 72:295–298, 1980.

36 Nuttall, W. F., D. A. Cooke, J. Waddington, and J. A. Robertson. Effect of nitrogen and phosphorus fertilizers on a bromegrass and alfalfa mixture grown under two systems of pasture management. I. Yield, percentage legume in sward, and soil test. *Agron. J.,* 72:289–294, 1980.

37 Odom, O. J., R. L. Haaland, C. S. Hoveland, and W. B. Anthony. Forage quality response of tall fescue, orchardgrass, and phalaris to soil fertility level. *Agron. J.,* 72:401–402, 1980.

38 Perry, L. J., Jr., and D. D. Baltensperger. Leaf and stem yields and forage quality of three N-fertilized warm-season grasses. *Agron. J.,* 71:355–358, 1979.

39 Peterson, L. A. and D. Smith. Recovery of K_2SO_4 by alfalfa after placement at different depths in a low fertility soil. *Agron. J.,* 65:769–772, 1973.

39 a Peterson, L. A., S. Smith, and A. Krueger. Quantitative recovery by alfalfa with time of K placed at different soil depths for two soil types. *Agron. J.,* 75:25–30, 1983.

40 Rehm, G. W., J. T. Nichols, R. C. Sorensen, and W. J. Moline. Yield and botanical composition of an irrigated grass-legume pasture as influenced by fertilization. *Agron. J.,* 67:64–68, 1975.

41 Rehm, G. W., R. C. Sorensen, and W. J. Moline. Time and rate of fertilization on seeded warm-season and bluegrass pastures. II. Quality and nutrition content. *Agron. J.,* 69:955–961, 1977.

41 a Reneau, R. B., Jr., G. D. Jones, and J. B. Friedericks. Effect of P and K on yield and chemical composition of forage sorghum. *Agron. J.,* 75:5–8, 1983.

41 b Reynolds, J. H., and W. H. Wall, III. Concentrations of Mg, Ca, P, K, and crude protein in fertilized tall fescue. *Agron. J.,* 74:950–954, 1982.

42 Rhykerd, C. L. and C. H. Noller, The role of nitrogen in forage production. In *Forages,* Iowa State Univ. Press, Ames, IA, 3d ed., pp. 416–424, 1973.

43 Robinson, R. R., V. G. Sprague, and C. F. Gross. The relation of temperature and phosphate placement to growth of clover. *Soil Sci. Amer. Proc.*, 22:225–228, 1959.

44 Saibro, J. C., C. S. Hoveland, and J. C. Williams. Forage yield and quality of phalaris as affected by N fertilization and defoliation regimes., *Agron. J.*, 70:497–500, 1978.

45 Sheard, R. W. Nitrogen in the P band for forage establishment. *Agron. J.*, 72:89–97, 1980.

46 Smith, A. E. and G. V. Calvert. Fescue forage production and quality response to sequential nitrogen applications. *Agron. J.*, 71:647–649, 1979.

47 Smith, D. Effects of potassium top dressing a low fertility silt loam soil on alfalfa herbage yields and composition and on soil K values. *Agron. J.*, 67:60–64, 1975.

48 Smith, D. Influence of nitrogen fertilization on the performance of alfalfa–bromegrass mixture and bromegrass grown alone. *Wis. Agr. Exp. Sta. Res. Rep. R2384*, 1972.

49 Smith, D. Influence of temperature on the yield and chemical composition of five forage legume species. *Agron. J.*, 62:520–523, 1970.

50 Smith, R. K., D. A. Miller, and E. J. Armburst. Effect of boron on alfalfa weevil oviposition. *J. Econ. Entom.*, 67:130, 1974.

51 Stansberry, C. O., C. D. Converse, H. R. Haise, and O. J. Kelley. Effect of moisture and P variables on alfalfa hay production on the Yuma Mesa. *Soil Sci. Amer. Proc.*, 19:303d–310, 1955.

51 a Stritzke, J. F. and W. E. McMurphy. Shade and N effects on tall fescue production and quality. *Agron. J.*, 74:5–8, 1982.

52 Sullivan, E. F. and L. F. Marriott. Effect of soil acidity on establishment and growth of orchardgrass seedlings in pot cultures. *Agron. J.*, 52:147–148, 1960.

53 Taliaferro, C. M., F. P. Horn, B. B. Tucker, R. Totusek, and R. D. Morrison. Performance of three warm-season perennial grasses and a native range mixture as influenced by N and P fertilization. *Agron. J.*, 67:289–292, 1975.

54 Tisdale, S. L. and W. L. Nelson. *Soil Fertility and Fertilizers,* Macmillan, New York, 2d. ed., 1966.

55 Vickers, J. C. and J. M. Zak. Effects of pH, P and Al on the growth and chemical composition of crownvetch. *Agron. J.*, 70:748–751, 1978.

56 Wagner, R. E. Influence of legume and fertilizer nitrogen on forage production and botanical composition. *Agron. J.*, 46:167–171, 1954.

57 Watschke, T. L., D. V. Waddington, D. J. Wehner, and C. L. Forth. Effect of P, K and lime on growth, composition, and P absorption by merion Kentucky bluegrass. *Agron. J.*, 69:825–829, 1977.

58 Wedin, W. F. Fertilization of cool-season grasses. In *Forage Fertilization,* Amer. Soc. Agron., Madison, WI, pp. 95–118, 1974.

59 Woodhouse, W. W., Jr., and W. K. Griffith. Soil fertility and fertilization of forages. In *Forages,* Iowa State Univ. Press, Ames, IA, 3d ed., pp. 403–416, 1973.

CHAPTER 11

FORAGE ESTABLISHMENT AND WEED CONTROL

A very important step in a good, efficient forage program is establishment of an excellent stand. This cannot be overemphasized. It is most important to obtain a proper plant population during the seeding year.

Many factors affect forage establishment. Most of the general guidelines for proper establishment of each forage species are discussed in the appropriate chapters beginning with Chapter 16, on alfalfa, through Chapter 31.

Thick, vigorous stands of grasses and legumes are needed for high yields. A thick stand of grass covers nearly all of the soil surface. A thick stand of alfalfa is considered 295 plants/sq.m (30 plants/sq.ft) for the seedling year, 109 plants/sq.m (10 plants/sq.ft) the second year, and 52 plants/sq.m (5 plants/sq.ft) for succeeding years (13, 29).

Production of high-yielding forages begins by selection of disease- and insect-resistant cultivars that will quickly establish themselves and recover rapidly after each harvest. Such cultivars should respond to fertilizer, and if harvested at the optimum time and protected from insects, they produce high-quality forage.

ESTABLISHMENT

Basically, the establishment of forage crops begins with the proper soil environment. Soil tests indicate whether the pH of the soil is suitable for maximum legume production. Under most situations, the optimum pH value should fall between 6.5 and 7.5. The optimum value of alfalfa production may vary somewhat depending upon soil characteristics, such as texture, the content of organic matter, or the lime level in the subsoil.

The available phosphorus level should be at least 50, and the exchangeable K should be at least 300 to 400. In addition to phosphorus and potassium, elements such as magnesium, calcium, sulfur, boron, molybdenum, manganese, and other trace elements are needed for proper growth.

To ensure a highly productive stand of forages, in addition to providing adequate fertility, one must also select the best-adapted, high-yielding, persistent cultivar for a specific soil, climate, and intended utilization. There are many cultivars available on the market, and it is important to purchase high-quality seed, which will ensure rapid establishment. Quality is best ensured by using certified seed or brand seed from a reputable seed company.

Legume seeds must be properly inoculated with appropriate, effective strains of nitrogen-fixing bacteria *Rhizobium*. These bacteria, which live in a symbiotic relationship with the legume, can take nitrogen from the soil air and fix it in a form that plants can use. It is generally concluded that once a legume has been grown in a particular soil, the native population of *Rhizobium* is adequate to inoculate that legume. However, because of differences in the effectiveness of different strains of *Rhizobium* in the soil, it is generally a good practice to inoculate the seed so that one will have the most effective strain available for that particular legume.

Some of the alfalfa seed currently on the market is preinoculated by the selling company, so that no further inoculation is needed before planting. In the case of other legumes, though, there has been very little preinoculation of the seed prior to sale. Specific inoculum is available and should be applied prior to seeding. Since there is no commercial test for *Rhizobium* presence in the soil, one should inoculate seed to be sure of nodulation. It is important that the inoculum be maintained in a refrigerator or under cool conditions before using it because the bacteria are very sensitive to light and heat. Inoculated seed should be planted as soon as possible.

Seeding rates vary with different species and also with different areas of the United States. For example, seeding rates for alfalfa range from 11 to 13 kg/ha (10 to 12 lb/A) in the high plains to more than 33 kg/ha (30 lb/A) in some areas of California (21). There are approximately 200,000 to 220,000 seeds per pound of alfalfa seed; therefore, each pound of seed planted per acre provides about 5 seeds per square foot. It is common to find that under field conditions, approximately 30 to 40% of the seed sown provides viable seedlings during the seeding year (1). Approximately 50% of those seedlings will survive into the second year. Thus, the seeding rate for alfalfa should be in the general range of 18 to 20 kg/ha (16 to 18 lb/A) to provide the proper seedling survival per 0.3 sq.m (1 sq.ft) that is favorable for high-yielding stands. This seeding rate results in 12 to 18 plants per 0.3 sq.m (1 sq.ft) 1 year after seeding (Table 11-1).

There are several reasons for low seedling survival of small-seeded legumes and grasses (Fig. 11-1): (1) the seeds have a very small food reserve to feed the germinating seedling; (2) the extremely small seeds make it difficult to maintain close contact with the soil; (3) many producers plant the seed too deeply in the soil; (4) the seedlings are very weak competitors against weeds, insects, and

TABLE 11-1
SEEDS SOWN AND SEEDLING NUMBERS OF ALFALFA FOLLOWING
ESTABLISHMENT

Seeding rate, kg/ha (lb/A)	Seeds sown per 0.3 m²	Seedling number, per 0.3 m²	
		2 mo after seeding	1 yr later
18–20 (16–18)	80–90	24–36	12–18

Source: M.B. Tesar and J.A. Jackobs, "Establishing the stand," in *Alfalfa Science and Technology*, Amer. Soc. Agron., Madison, WI, pp. 415–435, 1972.

disease; (5) the little seedlings are very susceptible to droughty conditions following seeding because their root systems are extremely small compared with those of many of the more aggressive weeds and not nearly as deep in their penetration; and (6) allelopathic effects on alfalfa germination reduce the stand (see Chapter 16) (7a, 7b).

It is essential that legume seeds be planted at a very shallow depth. One of the most common problems associated with forage establishment is planting the forage seed too deeply. Normally, one should seed legumes no deeper than 0.6 to 1.2 cm (1/4 to 1/2 in) in heavy soils that have good water-holding capacity (21, 25). To ensure a good establishment in lighter soils, and in the more arid areas of

FIGURE 11-1
Seedling vigor is very important in establishment.

TABLE 11-2
EMERGENCE OF ALFALFA SOWN AT VARYING DEPTHS

Planting depth		Avg. number of seedlings	
cm	in	emerged per ft of row	Emergence, %
1.25	0.5	14.14	60.9
2.50	1.0	10.61	45.7
3.75	1.5	6.16	26.5

˙Based on assumption of 24 viable seeds sown per ft of row.
Source: D. Smith, "Seeding an establishment of legumes and grasses," in *Forage Management,* Kendall/Hunt, Dubuque, IA, 3d ed., pp. 15–29, 1975.

the country, a slightly greater seeding depth may be required so that seed and developing roots are in contact with moisture. It has been shown that the percent of established seedlings is greatly reduced when planting depths are increased (Table 11-2).

Spring is the generally recommended time of year for seeding forage crops in the midwest, whereas in the southern half of the United States, many forage crops are seeded from late summer to early fall. In the south, summer forages such as johnsongrass, dallisgrass, bahiagrass, and sericea lespedeza are planted in the spring. The best time for spring seeding, for both north and south, is essentially as early as one can get to the field. Seeding time varies from the latter part of February in Kentucky and Missouri, to the latter part of March and early April in the upper midwest (progressing north into Iowa and northern Illinois), and perhaps to early May in Wisconsin and Minnesota. Spring seeding with a companion small grain, such as spring oats, is generally done when the oats are seeded.

Spring seedings in the northern quarter of Illinois are usually more successful than are late summer seedings in the southern quarter. The frequency of success in the southern one-quarter to one-third of Illinois indicates that late summer seeding may be more desirable than is spring seeding.

Seedings in mid-summer to late summer or early fall are very popular in several areas of the United States primarily south of the 40th to 41st parallel. This method ensures a good establishment of forages without the use of a companion crop or herbicide, since a killing frost eliminates the annual weeds in the stand. During late summer, moisture may be somewhat limited, but generally there is also less competition from weeds. In this area of the United States, there is also often an early fall rain that ensures good germination and proper establishment. In other areas of the country, normal seeding dates range from early to mid-August in the north to mid-September in the central portion.

Late summer seeding dates are governed by the average number of days before a killing frost, allowing at least 35 to 45 days before the average killing frost. For example, the suggested seeding date for the northern one-quarter of Illinois is from August 10 to 15, whereas in the central part of the state it is

FIGURE 11-2
Close-up of an alfalfa plant heaved from the soil, a result of
alternating freezing and thawing, plus a poor root system, low
fertility, and low food reserves.

August 30 to September 5, and in the southern quarter approximately September 15 to 20. Seeding should be done as close to these dates as possible to ensure that the plants become well-established before winter. It is important that the plants develop an adequate root system and sufficient food reserves to carry them through the winter (Fig. 11-2). The dates suggested for Illinois correspond to those for other states at approximately the same latitude. In the west and southwest, seedings can be made up to October 15, allowing enough time for an excellent establishment. These dates vary somewhat with species; for instance, birdsfoot trefoil and reed canarygrass, because of their slower seedling establishment, should not be seeded as late as alfalfa or orchardgrass. The seeding rates for each of the various forage species are presented in their respective chapters.

Seeding Methods

There are two basic methods of seeding forage crops, band seeding and broadcast seeding. Band seeding involves placing a band of phosphate fertilizer (0-45-0) at about 5 cm (2 in) deep in the soil with a grain drill, then placing the forage seed on the soil surface directly above the fertilizer band (8). The fertilizer should be covered with soil before the forage seeds are dropped. This process occurs naturally when the soil is in good working condition. Press wheels should be rolled over the forage seed to firm it into the soil surface. Many seeds are placed 1.25 to 2.5 cm (1/2 to 1 in) deep with this seeding method (Fig. 11-3) (8).

Band seeding, if done correctly, may make it more reliable than other seeding methods—on a variety of soil types—for several reasons: (1) the readily available phosphorus promotes rapid development of large, healthy seedlings with excellent root systems (Fig. 11-4); (2) fewer weeds are fertilized than there would be if the phosphorus were broadcast; and (3) because of the rapid early development, seedlings can survive more adverse soil and climatic conditions than they would otherwise. Researchers have found varying results when comparing the seeding of forage crops using a band method and a broadcast method. The general conclusion is that if the soil is low in available phosphorus, it is advantageous to band-seed (5, 20, 26). Another situation where band seeding is suggested is early seeding in a cold, wet soil. But if seeding is done later in the spring, under more ideal conditions—such as warmer weather and high phosphorus levels—there is no advantage in seeding forages with a band

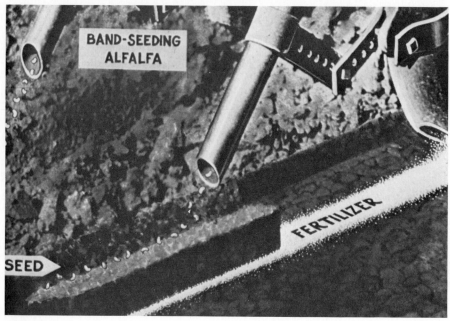

FIGURE 11-3
Band seeding of alfalfa showing the placement of fertilizer and seed.

method over a broadcast method, provided one uses a culti-packer or corrugated roller (Fig. 11-5) (21).

Broadcast seeding involves spreading seed uniformly over a properly pre-pared soil surface, then pressing the seed into the soil surface with a corrugated roller (Fig. 11-6). The fertilizer is applied earlier, during the seedbed prepara-tion. The seedbed is usually disked and smoothed with a harrow. Most soil conditions are too loose after these tillage operations alone, and should be allowed to settle by rain or firmed with a corrugated roller before seeding. The best seeding tool for broadcasted seed is the double-corrugated roller-seeder (8, 21) (Fig. 11-7). It is very important to obtain good seed-to-soil contact when seeding forage crops; this is one reason that broadcasting is done with a culti-packer or corrugated roller. The culti-packer or first corrugated roller firms the seedbed and soil surface, so that the seed will not be buried when the second roller rolls over it. The seed is then placed on top of the soil and pressed into it by the second culti-packer, which forms ridges over the previous furrows and furrows over the previous ridges. This type of seeding places the seed at a uniform shallow depth (21, 27).

Companion Crops vs. Clear Seeding

Until recently, most forage crops were seeded in the spring with a companion crop, such as spring oats or, in some areas, winter wheat. However, clear

FIGURE 11-4
Rapid root development is important in a successful establishment, resulting from planting seed at the proper depth.

seeding, without a companion crop (formerly called a nurse crop), has become an accepted practice in many areas (25). The companion crop provides a quick ground cover in the spring, thus controlling soil erosion by wind or rain on sloping soils. The quick ground cover also helps reduce the invasion of weeds into the young forage seedlings. However, the companion crop itself competes strongly with the young forage seedlings for nutrients, light, and moisture, and may be as much if not more competition than the natural weed would be.

The companion crop should be removed as early as possible to eliminate any undue competition (Fig. 11-8). It is suggested that the companion crop be harvested as a forage crop instead of as a grain crop. A companion crop is used when seeding forages mainly so that a harvest of some forage can be taken during the year of establishment. Oats could be removed as a silage crop from the headed stage to the milk stage for dairy animals, or up to the soft dough stage for beef animals, thus providing quality forage for both species. The same would hold true for wheat or barley companion crops. In most cases, wheat and barley are harvested as grain crops. Leaving these companion crops in the field until maturity provides further competition for the young forage seedlings, thus

FIGURE 11-5
An excellent seedbed involves compaction of seed with a corrugated field roller into a
well-prepared soil.

increasing the risk of lodging and consequent smothering of the young seedlings.

Under optimum moisture conditions, seeding a legume forage crop without a
companion crop generally produces 50 to 60% of a normal yield during the
seedling year, whereas a grass forage may yield from 10 to 60%. The number of

FIGURE 11-6
Close-up of alfalfa emergence resulting from excellent compaction.

FIGURE 11-7
A brillion-type seeder utilizes two sets of corrugated rollers when broadcasting the seed of forage crops.

harvests taken during the seedling year is generally limited by the time of seeding in the early spring and subsequent moisture or environmental conditions. The number of harvests taken during the seeding year is normally one less than that taken from an established stand in the same area. From a new seeding,

FIGURE 11-8
Spring oats is an excellent nurse crop for establishing forage crops.

the crowns are generally smaller and the yields per cutting somewhat less than those of an established stand. These differences can be decreased by having the same number of stems per unit area in a new stand as in an old stand. In the year following seeding, the yield of a clear stand of either a legume or a grass is comparable to that of a stand seeded with a companion crop.

Management for clear seeding is more demanding than that for seeding with a companion crop. It is usually essential to use a herbicide to assist the establishment of a pure-seeded legume in the spring. However, a herbicide is seldom used in a late summer-early fall seeding. Proper herbicide selection helps achieve maximum stands and, therefore, excellent yields.

WEED CONTROL

Another problem in establishing and growing quality forage is the control of weeds. Legumes and grasses are often interplanted, thus making it more difficult to control broadleaf or grassy weeds in the mixture than would be the case otherwise. It is much more difficult to control weeds in forages than in row crops. For instance, since corn and soybeans are grown in rows, it is much easier in these crops than in forages to maximize the effectiveness of weed control by supplementing chemical control with row cultivation (Fig. 11-9). Good weed control in forage crops involves a number of cultural practices. Most weeds in small-seeded legumes and grasses can be controlled or avoided by one or more of the following methods: (1) using clean seed; (2) controlling weeds before seeding; (3) seeding at the proper date; (4) maintaining the competitive nature of the crop; (5) mowing; (6) flaming; (7) cultivation; and (8) chemicals.

Clean Seed

One of the best cultural practices for eliminating weed competition in forage crops is the use of certified seed. This ensures the seeding of an improved cultivar, thus avoiding the planting of any noxious or secondary weed seed. The seed of many serious problem weeds are nearly the same size, shape, and weight as is legume seed. Once the seed is contaminated, it is very difficult to eliminate the weed. However, there are several different mechanical seed-cleaning devices in use at present that eliminate most noxious weed seeds from forage crops. These devices are quite ingenious and take advantage of very small physical differences between the seeds to be separated, such as size, shape, weight, coat hairiness, and appendages. All these differences can prove useful in eliminating weed seed from forage seed. One of the best examples of planting clean seed and thus avoiding the problem of planting weed seeds is that of elimination of dodder. Dodder is very difficult to remove from legume seed, and the process takes advantage of differences in seed characteristics. The seedcoats of dodder are rough and pitted and stick to felt cloth. But the seedcoats of legumes, which are smooth and waxy, do not stick to such material. When seeds of both types pass between felt-covered rollers rotating in opposite directions, the dodder

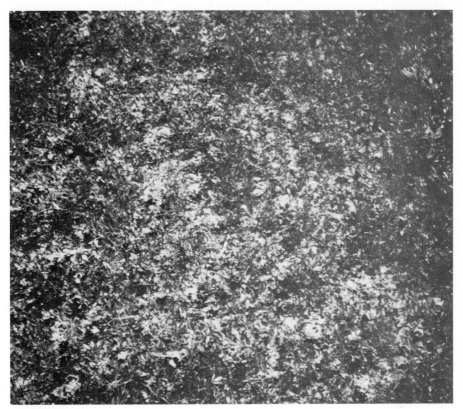

FIGURE 11-9
Excellent weed control using Eptam, preplant, and 2,4-DB postemergence.

seeds are carried to the outside, whereas the legume seeds remain between the rollers and pass on out into the clean-seed troughs. These rollers are slanted so the legume seeds slide out the lower end.

Weed Control before Seeding

Most forage species are not vigorous growers and offer very little competition to aggressive weeds that may be growing in the sward. When a serious perennial weed may persist or exist where the legume or forage crop is to be seeded, control of that weed before seeding is necessary. Such methods as spot treatments or growing crops that will competitively remove or eliminate the weed species from the area can be utilized for this.

There are two methods of controlling annual weeds prior to the seeding of a crop. The first such method is rotation of row crops and a small grain in which weeds are to be effectively controlled for 2 years or more, which usually reduces the weed seed population in the soil. The second method is the proper preparation of the seedbed before seeding by repeating the process a couple of times. The timing permits the killing of one or more crops of the germinating weed seeds. When preparing the seedbed, it is preferred that the equipment used should not bring deeply buried weed seed to the surface. Deeply buried seed will usually remain dormant for several years, as long as the surrounding soil is not disturbed.

Proper Date of Seeding

The number of weeds in a forage crop may be controlled by the date of seeding, since forages may be planted during either the spring or fall. The seriousness of the problem depends upon whether the weed is a winter or summer annual; the more serious problem is that of summer-annual weeds. If a legume seeding can be done early enough in the spring, the problem will be eliminated because the legume will have a competitive advantage over these weeds, which germinate later in the spring. As the seeding date is delayed, and the soil warms, the greater is the possibility that summer weed seeds will germinate. Late summer or early fall seedings generally do not have the summer-annual weed problem encountered with spring seeding. However, in some areas of the United States, problems that may arise with a fall or late summer seeding are fall drought, winter injury, and excessive rainfall, which result in poor establishment. If the forage crop can be seeded early enough and become well-established by the following spring, the forage seedlings will competitively eliminate any summer-annual weed problem.

If winter-annual weeds are the major problem, then a spring seeding is preferred over a late summer–early fall seeding. Spring-seeded forage crops will be well-established by early fall, and will competitively reduce the invasion of any winter-annual weed. With present-day herbicides, more growers are establishing forage crops in the early spring, taking advantage of ideal seeding conditions with the potential of early spring rains. If forage crops are seeded when the soil temperatures and moisture are especially favorable, they will germinate readily and grow fast enough to crowd out most weeds (Fig. 11-10).

Competitive Nature of the Crop

Probably the most effective control of weeds is to maintain a thick, productive stand of the forage crop. Most legumes are competitive or vigorous enough that they usually reduce weed growth, and a decrease in the stand of forage crops greatly enhances the possibility of weed establishment in the sward. It is very important to maintain proper fertility, needed drainage, and proper moisture conservation when moisture may be limiting, as well as to control disease and insects, harvest at the proper time, use adapted cultivars, and employ other cultural practices that help maintain a fixed stand of the forage crop. Provided that these steps are taken, and the proper seeding rate is used, weeds can often be controlled in forage crops.

FIGURE 11-10
The result of an effective weed control program when establishing forage legumes.

Mowing

Most weeds will not survive in a vigorously growing stand of alfalfa that is harvested two or more times during a growing season. Repeated mowing weakens perennial weeds and most likely kills annual weeds. Most weeds that are growing in forage crops do not form crowns like those of alfalfa. By cutting very close to the soil surface, one removes the nodal areas of the weed where potential regrowth may occur, thus killing the weed. The only weeds that can survive or prosper under such close, repeated mowings would be those with a prostrate growth habit. Herbicides or some other means must be used to eliminate these weeds.

Flaming

Prior to recent increases in the cost of petroleum products, propane and diesel burners were used to control many weeds in established alfalfa stands. One effective way of controlling a winter-annual broadleaf weed is the use of flaming just before the resumption of spring growth. This treatment not only retards or kills the weeds but also burns the residue that is left, thus removing sites that may harbor insects or disease. When dodder is a problem, flaming immediately after the harvest has been an effective control. This will usually result in at least 1 day's suppression of the alfalfa's growth. Most of these treatments are used only as spot treatments or in dodder-infested areas.

Cultivation

Since it is essential in the production of alfalfa seed or any other legume seed to eliminate any potential weeds, cultivation may be used as a means of control.

Cultivation is used primarily in legume seed production when the legume is seeded in rows. Row planting permits weed control by cultivation. It also permits a directed and shielded application of herbicides. Cultivation is used for summer annuals seeded in rows, such as pearlmillet and sorghum.

Cultivation has also been used in establishing legume seedings, but there is very little experimental evidence to support such a procedure. Again, the best weed control in a solid seeding of forage crops is a highly competitive, healthy stand. The main tillage tool used in helping suppress or control weeds is a spike-toothed harrow. This tool kills many annual weeds without seriously injuring the crowns of legumes. If a disk-harrow is used, considerable damage occurs to alfalfa by cutting the crown and so providing an area of entry for diseases. However, the disk-harrow has been used to some extent, along with fertilizer, in solid grass seedings of forages. The disk-harrow helps eliminate or reduce the competitiveness of aggressive weeds, while the fertilizer encourages the forage grass to become better established and eliminate or competitively reduce the weeds that were formerly in the field.

Chemical Weed Control

Small-seeded legumes vary widely in their tolerance to different herbicides. Therefore, the tolerance of the legume, as well as the susceptibility of the weeds, is very important in choosing the appropriate herbicide to apply to a forage crop (Fig. 11-11).

Herbicides can be applied to the forage at various times. They can be applied

FIGURE 11-11
Close-up illustrating the effect of Eptam on grass and broadleaf weed control in alfalfa.

prior to planting—as preplant treatments—to eliminate troublesome weed species. A herbicide can also be applied as a preemergence treatment, after seeding and prior to the emergence of either weed seedlings or the forage crop itself. Another treatment method is postemergence, or the application of a herbicide to an established stand of forage and/or the weedy plant (12, 14, 15, 18, 19, 24).

Preplant Herbicides   Weed controls in forage legumes differ according to the forage species, type of seeding, and whether the legume is sown alone or in combination with a grass. Alfalfa and most clovers can be established without a companion crop. When alfalfa, birdsfoot trefoil, and several of the clovers are seeded alone, there are several herbidices that may be used as preplant soil-incorporated treatments to control most annual grasses and many annual broadleaf weeds (Fig. 11-12):

1 Eptam 7D (EPTC) may be applied at a rate of 0.17 to 0.21 l/ha (3.5 to 4.5 pt/A or 3 to 4 lb/A) and incorporated immediately before planting. It is most effective on grasses and also gives some control of nutsedge, especially at the higher rates (15, 19, 24).

2 Balan (benefin) can be preplanted and incorporated at the rate of 1.14 to 1.52 l/ha (3 to 4 qt/A or 1 1/8 to 1 1/2 lb/A). The rate varies with the soil type. Planting may be delayed up to 10 weeks after the incorporation of Balan, but planting immediately after incorporation is preferred. Balan will control many

FIGURE 11-12
Check plot in center showing numerous broadleaf weeds; Eptam on left; Balan on right. Both herbicides resulted in excellent weed control.

annual grasses and some broadleaf weeds. Eptam persists in the soil from 1 to 4 months, while Balan will last longer—3 to 8 months (15, 19, 24).

3 Tolban (profluralin) can be used as a preplant treatment at the rate of 0.05 to 0.1 l/ha (1 to 2 pt/A or 1/2 to 1 lb/A). It controls annual grasses and some broadleaf weeds. Some injury to alfalfa may occur on soils with a calcareous surface layer or soils high in salt content (24).

Preemergence Herbicides   Chem-Hoe (propham) can be used as a surface treatment. (It should not be incorporated.) Propham can be applied as a preemergence treatment when rainfall is expected or when irrigation is used. This herbicide controls most annual grasses and some annual broadleaf weeds in alfalfa, birdsfoot trefoil, and clovers. It should be applied at the rate of 4.4 to 5.6 kg/ha (4 to 5 lb/A). Moisture is required for propham to be effective as a surface treatment. It is degraded in less than 1 month in certain soils.

Postemergence Herbicides   After legumes have emerged and broadleaf weeds have begun to emerge, there are several herbicides that one may apply as postemergence treatments. Butoxone and Butyrac (2,4-DB) may be used as postemergence controls of broadleaf weeds in alfalfa and most clovers after a companion crop is removed, or in a pure legume stand or in a legume-grass establishment without a companion crop. Rates vary according to whether the herbicide has an ester or an amine formulation. Suggested rates are 0.38 to 0.71 l/ha (1 to 2 qt/A or 1/2 to 3/4 lb/A) for an ester formulation and 0.38 to 1.14 l/ha (1 to 3 qt/A or 1/2 to 1 1/2 lb/A) for an amine formulation. An ester formulation may be used on alfalfa and birdsfoot trefoil, but not on most clovers. One should not graze livestock or cut hay from treated fields within 30 days after application. It is suggested that the weed be in the 1- to 3-leaf stage before treatment. It is essential that broadleaf weeds be quite small when herbicides are applied. Grassy weeds are not controlled with 2,4-DB (14, 15, 19, 24).

Dinoseb, or Premerge-3, can be applied when the crop has two trifoliate leaves and broadleaf weeds are very small. Premerge is used in pure stands of alfalfa or clovers for chickweed control. Application rates are 0.57 to 0.76 l/ha (1 1/2 to 2 qt/A); the ammonium salt of this herbicide is applied at 0.8 to 1.1 kg/ha (3/4 to 1 lb/A) and the amine salt at 1.1 to 1.7 kg/ha (1 to 1 1/2 lb/A). Premerge can also be applied on an established stand (see below). Repeated applications may be needed for weeds that emerge after treatment, or for very heavy infestations (14, 15, 19). One should not graze or feed treated forages to livestock for 6 weeks after application. Dinoseb is quite toxic to humans and animals.

Established Stands   Broadleaf weeds and grasses often invade an established stand after it has been in production for several years. This not only reduces the yields of total dry matter and protein per hectare but also competitively eliminates the legumes and grasses that one is trying to maintain in the forage stand.

Dinoseb has been used widely to kill all emerged annual weed seedlings in established alfalfa. It is used primarily in the midwest as a chickweed control. It can be applied during the dormant season to control annual winter weeds, or immediately after the first cutting, before regrowth starts, to control annual weeds and dodder. Dinoseb is usually applied at 1.4 to 2.2 kg/ha (1 1/4 to 2 lb of active ingredient in 20 to 50 gal of diesel oil, with water added to make 100 gal of the mixture per acre). The resulting emulsion is sprayed to thoroughly wet all weed foliage. This treatment also suppresses the alfalfa weevil larvae population in forage regrowth (14, 15, 19).

Princep (simazine) can be applied to pure alfalfa stands that have been established for more than 12 months. It is applied to dormant alfalfa or after the last fall cutting but before the ground is permanently frozen. Princep controls many weedy grasses and broadleaf weeds by prohibiting germination of weed seeds, but it does not control most well-established perennial or biennial weeds. It controls fall or early spring weeds, such as chickweed, henbit, shepherd's purse, peppergrass, and downy brome. However, it may injure certain grasses, such as timothy or smooth bromegrass (8b, 12a, 17a). Application rates are 1.1 to 1.7 kg/ha (1 to 1 1/2 lb/A) of Princep 80W, depending on soil type. It should not be used on sandy soils, nor should it be grazed for 30 days after spraying or cut for hay for 60 days after spraying. If excessive rates are used, or applications are made at the wrong time, or unfavorable conditions exist, the treated legume may be injured somewhat (19, 28).

Kerb (pronamide) can be applied on newly established stands, or preferably on old, well-established stands, of alfalfa, clover, birdsfoot trefoil, or crown-vetch. Fall application is considered best for most of the weeds. Kerb controls most annual grasses and broadleaf weeds and helps suppress quackgrass. One should not apply Kerb to a mixture of legumes or forage grasses since this herbicide will injure most grasses. Application rates vary from 2.2 to 4.4 kg/ha (2 to 4 lb/A) of Kerb 50W per 1.1 to 2.2 kg/ha (1 to 2 lb of active ingredients per acre). One should not graze or harvest the forage for 120 days after treatment (19, 24, 28).

Furloe, or Chloro-IPC (chlorpropham), is a very popular herbicide because it can be applied to active, semidormant, or dormant alfalfa that is well established, or to newly seeded alfalfa with three or four trifoliate leaves. The popularity of this herbicide results from its excellent control of chickweed and downy brome. To control these weeds one should make application from October through January using 0.38 to 0.76 l/ha (1 to 2 qt/A) of 4EC. After the beginning of February, 0.76 to 1.14 l/ha (2 to 3 qt/A) should be applied, or if one were using granules, it is suggested that 44 kg/ha (40 lb/A) of the 10 g form be applied. One should not apply Furloe within 40 days of harvest, nor to an alfalfa-grass mixture at any time (28).

Butoxone or Butyrac can be applied to established stands of legumes at the rates suggested above, for their use as postemergence herbicides.

Sinbar (terbacil) can be applied to pure stands of alfalfa that are at least 1 year old. This herbicide controls annual grasses and broadleaf weeds. One should apply 0.5 to 1.7 kg/ha (1/2 to 1 1/2 lb/A) when the alfalfa is dormant during the fall or spring, but not when the ground is frozen. Injury may result when Sinbar is used on certain types of soil or under unfavorable growing conditions.

Sencor, or Lexone (metribuzin), can be used in established stands of alfalfa or alfalfa-forage grass mixtures (7c, 8a). However, high rates of application may result in reducing the forage grass in mixtures. This herbicide should be applied during the dormant season or early in the spring before the alfalfa begins to grow, or in the fall after growth has ceased. Application rates vary from 0.8 to 2.2 kg/ha (3/4 to 2 lb/A) depending upon the weed species that one wishes to control. For the control of chickweed, henbit, and downy brome, it is suggested that 0.8 to 1.1 kg/ha (3/4 to 1 lb/A) of the wettable powder be applied. Higher rates—1.7 kg/ha (1 1/2 lb/A)—control the mustards, white cockle, dandelion, and sand bur. At the maximum of 2.2 kg/ha (2 lb/A), quackgrass may be suppressed. One should not graze or harvest for 28 days after application (19).

Paraquat CL has recently been cleared to be used east of the Rockies and north of the southern borders of Colorado, Kansas, Missouri, Kentucky, and Virginia. It is suggested that one should use 0.1 to 0.15 l/ha (2 to 3 pt in 20 to 60 gal of water/A) as a broadcast application for the control of bluegrass, chickweed, henbit, and downy brome, and the suppression of perennial grasses, including orchardgrass, timothy, and smooth bromegrass. Using 0.15 l/ha (3 pt/A) may kill or suppress the harder-to-control weeds and grasses, as well as the perennial species. One should not apply this herbicide if the fall regrowth following the last cutting or harvest is more than 15 cm (6 in) tall (12, 23, 24, 28). The foliage present at the time of application is burned, which may reduce the yield of the first harvest. Weeds and grasses should be succulent and growing at the time of application. Weeds that germinate after the application will not be controlled. One should not graze, cut, or harvest for 60 days after application of this herbicide, nor apply it more than once per season.

Roundup (glyphosate) can be used in a renovation program for no-till incorporation of alfalfa into a grass sod. The application rates are 0.76 to 1.14 l/ha (2 to 3 qt/A) on foliage that may be 15 to 20 cm (6 to 8 in) or taller, and that is actively growing. There should be a 3- to 7-day delay after application before the incorporation of the alfalfa or legume into a grass sod, or glyphosate will eliminate the stand (Fig. 11-13).

Karmex (diuron) can be used to control many annual weeds in established alfalfa that is at least 1 year old and dormant or semidormant. It is applied at rates of 1.7 to 3.4 kg/ha (1 1/2 to 3 lb/A). Diuron is relatively persistent and will give a season-long weed control. It is not suitable for use on light, sandy soils. An area that has been treated should not be replanted with any crop for 2 years after application because of the broad spectrum of plants that is controlled by diuron (19).

Many other herbicides have been used experimentally on forage crops and result in satisfactory control of various weeds. However, because of the various federal regulations and the amount of research data that must be submitted to

FIGURE 11-13
Zero-till establishment of alfalfa involving the application of roundup in the fall and interseeding alfalfa in the spring.

justify the registration of a herbicide, new herbicides of potential worth on forage crops may be effectively (and perhaps justifiably) eliminated. Nevertheless, the hectarage affected in this way is relatively low compared with that for other row crops or specialty crops, so that the justification for registering a new herbicide for forage crops is rather limited.

Harbicides for Grass Pastures   For a complete weed program we include good management, proper grazing, excellent fertilization, and the proper reseeding necessary to maintain excellent grass pastures, as well as the use of various herbicides:

2,4-D   2,4-D can be applied to control most broadleaf weeds; the application rates range from 0.5 to 1.1 kg/ha (1/2 to 1 lb/A). Higher rates are needed for more resistent perennial weeds. Dairy cattle should not graze treated fields for at least 7 days after treatment (19, 28).

2,4,5-T   2,4,5-T is more effective than 2,4-D on woody or shrubby plants. It may be used in combination with 2,4-D or alone. One should not graze dairy animals on treated forage for at least 6 weeks. Beef animals should not graze a pasture that has been treated with 2,4,5-T within 2 weeks of slaughter. It is illegal to spray 2,4,5-T on ditch banks, lakes, ponds, or around homes. Kuron, or 2,4,5-TCPPA (silvex), will control woody plants as effectively as 2,4,5-T.

Silvex can be used around waterways, for water impoundments, and recreational areas. The use of silvex involves the same grazing restrictions as those for 2,4,5-T (19).

Banvel Banvel (dicamba) is very effective on many broadleaf weeds when used at a rate 1 to 4 pt/A. Grazing should be delayed 7 to 40 days and harvest for 37 to 70 days, depending on the amount that has been applied. Banvel is a very effective control for legumes, so it should not be used where desirable legumes are present (19). In fact, legumes are easily killed at the recommended rates. When applying this herbicide, one should, therefore, take precautions to prevent injury to nearby plants that are susceptible to Banvel.

SUMMARY

High-yielding forage crops can be achieved only from well-established, vigorous stands. Vigorous stands will eliminate weed invasion and help reduce soil erosion. Therefore, when establishing a high-yielding cultivar, one must make sure that proper seeding rates are used with optimum procedures. The management practices involved in excellent establishment are proper fertility, seeding rate, seeding depth, seeding date, and method of establishment.

Since legumes and grasses are usually grown together as a forage crop, there is no simple recommendation for weed control. Herbicide choices are limited because of the susceptibility of most legumes to herbicides. Conversely, if one wishes to eliminate a specific grass in a legume stand, it is very difficult to use a selective herbicide to remove that grass and still maintain the desired perennial forage grass in the stand. Most researchers conclude that a good weed-control program for forage crops depends on the elimination of weeds in the rotational crops that precede the planting of a forage. Good cultural weed-control programs involve proper crop rotation, crop competition, the use of excellent clean seed and feed, and the elimination of weeds around fencerows, ditch banks, and roadways leading to the field.

QUESTIONS

1 Outline the proper planting method for establishing a forage legume in your area. Indicate the fertility level, seeding rate, and seeding method, and why you have chosen this method.
2 Why is inoculation so important in crops of forage legumes?
3 Why is it so important to prepare an excellent seedbed when establishing a forage legume?
4 Why can we not seed less than 1.1 kg/ha (1 lb/A) and get a sufficient stand?
5 Compare the broadcast and band seeding methods of establishing forages.
6 Why are weeds difficult to eliminate from forage crops?
7 What are the various methods of weed control in forages?
8 Why is it difficult to use a field cultivator to control weeds in an established stand of forages?

9 What are meant by preplant, preemergence, and postemergence treatments?
10 Outline an excellent preplant treatment for forage crops and a postemergence spray program for an established stand of legumes with a specific weed problem.
11 List two herbicides that may control a specific weed problem in grass pastures and indicate the various precautions one must implement when using each.

REFERENCES

1 Ahlgren, G. H. *Forage Crops,* McGraw-Hill, New York, pp. 303–367, 1956.
2 Beveridge, J. L. and C. P. Wilsie. Influence of depth of planting, seed size and variety on emergence and seedling vigor in alfalfa. *Agron. J.,* 51:731–734, 1959.
3 Blaser, R. E., W. L. Griffeth, and T. H. Taylor. Seedling competition in compounding forage seed mixtures. *Agron. J.,* 48:118–123, 1956.
4 Blaser, R. E., T. Taylor. Griffeth, and W. Skrdla. Seeding competition in establishing forage plants. *Agron. J.,* 48:1–6, 1956.
5 Brown, B. A. Band versus broadcast fertilization in alfalfa. *Agron. J.,* 51:708–710, 1959.
6 Burton, J. C. Nodulation and symbiotic nitrogen fixation. In *Alfalfa Science and Technology,* Amer. Soc. Agron., Madison, WI. pp. 229–246, 1972.
7 Buxton, D. R. and W. F. Wedin. Establishment of perennial forages. I. Subsequent yields. II. Subsequent root development. *Agron. J.,* 62:93–100, 1970.
7 a Cope, W. A. Inhibition of germination and seedling growth of eight forage species by leachates from seeds. *Crop Sci.,* 22:1109–1111, 1982.
7 b Dutt, T. E., R. G. Harvey, and R. S. Fawcett. Feed quality of hay containing perennial broadleaf weeds. *Agron. J.,* 74:673–676, 1982.
7 c Dutt, T. E., R. G. Harvey, and R. S. Fawcett. Influence of herbicides on yield and botanical composition of alfalfa hay. *Agron. J.,* 75:229–233, 1983.
8 Decker, A. M., T. H. Taylor, and C. J. Willard. Establishment of new seedings. In *Forages,* Iowa State Univ. Press, Ames, IA, 3d. ed., pp. 384–395, 1973.
8 a Evers, G. W. Seedling growth and nodulation of arrowleaf, crimson, and subterranean clovers. *Agron. J.,* 74:629–632, 1982.
8 b Evers, G. W. Weed control on warm season perennial grass pastures with clovers. *Crop Sci.,* 23:170–171, 1983.
9 Klebesadel, L. J. and D. Smith. Effects of harvesting an oat companion crop at four stages of maturity on the yield of oats, on light near the soil surface, and on soil moisture, and on the establishment of alfalfa. *Agron. J.* 52:627–730, 1960.
10 Klebesadel, L. J. and D. Smith. Light and soil moisture beneath several companion crops as related to the establishment of alfalfa and red clover. *Bot. Gaz.,* 121:39–46, 1959.
11 Kust, C. A. Herbicides or oat companion crops for alfalfa establishment and forage yields. *Agron. J.,* 60:151–154, 1968.
12 Linscott, D. L., A. A. Akhavein, and R. D. Hagin. Paraquat for weed control prior to establishing legumes. *Weed Sci.,* 17:428–431, 1969.
12 a Luu, K. T., A. G. Matches, and E. J. Peters. Allelopathic effects of tall fescue on birdsfoot trefoil as influenced by N fertilization and seasonal changes. *Agron. J.,* 74:805–808, 1982.
13 Marten, G. C., W. F. Wedin, and E. F. Hueg, Jr. Density of alfalfa plants as a

criterion for estimating productivity of an alfalfa-bromegrass mixture on fertile soil. *Agron. J.,* 55:343–344, 1963.

14 Moline, W. J. and L. R. Robison. Effects of herbicides and seed-rates on the production of alfalfa. *Agron. J.,* 63:614–616, 1971.

15 Peters, E. J. and R. A. Peters. Weeds and weed control. In *Alfalfa Science and Technology,* Amer. Soc. Agron. Madison, WI, pp. 555–573, 1972.

16 Rolston, M. P., W. O. Lee, and A. P. Appleby, Volunteer legume control in legume seed crops with carbon bands and herbicides. I. White clover. *Agron. J.,* 71:665–670, 1979.

17 Rolston, M. P., W. O. Lee, and A. P. Appleby. Volunteer legume control in legume seed crops with carbon bands and herbicides. II. Red clover and alfalfa. *Agron. J.,* 71:671–675, 1979.

17 a Scheaffer, C. C., D. L. Rabas, F. I. Frosheiser, and D. L. Nelson. Nematicides and fungicides improve legume establishment. *Agron. J.,* 74:536–538, 1982.

18 Schmid, A. R. and R. Behrens. Herbicides vs. oat companion crops for alfalfa establishment. *Agron. J.,* 64:157–159, 1972.

19 Schreiber, M. M. Weed control in forages. In *Forages,* Iowa State Univ. Press, Ames, IA, 3d ed., pp. 396–402, 1973.

20 Sheard, R. W., G. J. Bradshaw, and D. L. Massey. Phosphorus placement for the establishment of alfalfa and bromegrass. *Agron. J.,* 63:22–27, 1971.

21 Smith, D. Seeding an establishment of legumes and grasses. In *Forage Management,* Kendall/Hunt, Dubuque, IA, 3d ed., pp. 15–29, 1975.

22 Sund, J. M., G. P. Barrington, and J. M. Scholl. Methods and depths of sowing forage grasses and legumes. *Proc. 10th Intl. Grassl. Congr.* (Finland), Sec. 1:319–322, 1966.

23 Taylor, T. H., E. M. Smith, and W. C. Templeton, Jr. Use of minimum tillage and herbicide for establishing legumes in Kentucky bluegrass (*Poa pratensis* L.) swards. *Agron. J.,* 61:761–766, 1969.

24 Temme, D. G., R. G. Harvey, R. S. Fawcett, and A. W. Young. Effects of annual weed control on alfalfa forage quality. *Agron. J.,* 71:51–54, 1979.

25 Tesar, M. B. J. A. Jackobs. Establishing the stand. In *Alfalfa Science and Technology,* Amer. Soc. Agron., Madison, WI, pp. 415–435, 1972.

26 Tesar, M. B., K. Lawton, and B. Kawin. Comparison of band seeding and other methods of seeding legumes. *Agron. J.,* 46:189–194, 1954.

27 Triplett, G. B. and M. B. Tesar. Effects of compaction, depth of planting, and soil moisture tension on seeding emergence of alfalfa. *Agron. J.,* 52:681–684, 1960.

28 Triplett, G. B., Jr., R. W. Van Keuren, and J. D. Walker. Influence of 2,4-D, pronamide and simazine on dry matter production and botanical composition of an alfalfa-grass sward. *Crop Sci.,* 17:61–65, 1977.

29 Van Keuren, R. W. Alfalfa establishment and seeding rate studies. *Ohio Rep.,* 58(2):52–54, 1973.

CHAPTER 12

FORAGE MIXTURES

Mixtures of legumes and grasses are usually desired when forages are used for grazing. Yields tend to be greater with mixtures than with either a legume or a grass alone (Table 12-1). Mixtures give more uniform forage production over the season and can extend the grazing season. As shown in Fig. 12-1, in the south, forage mixtures or certain species grown in a monoculture can provide forage for the entire year. There are some exceptions to this general rule, as when climatic and soil conditions favor a particular species over a mixture. Compared with grasses, legumes reduce the danger of tetany in the spring, give better nutrient balance, and the deep-rooted species provide more production during the droughty period of the year. But grasses are a desirable addition to legume seedings because they tend to fill in when legumes die out, help reduce bloat hazard in ruminant animals, reduce late winter-heaving damage, increase the drying rate for hay, and may improve animal acceptance. Mixtures of two or three well-chosen species are usually higher-yielding than are mixtures of five or six species in which some of the components are not particularly well-suited to the soil, climate, or use.

There are numerous forage mixtures suited for specific regions of the United States and also for a particular state. Table 12-1a gives an example of proper mixtures to use with a companion crop in Illinois (Fig. 12-2). Table 12-2 shows that in Minnesota, seeding rates and mixtures may change slightly from those used in Illinois. Forage mixtures in Wisconsin (Table 12-3) differ from those in either Illinois or Minnesota because of different utilization necessitated by variations in soil types (Fig. 12-3). These areas range from a moderately well drained soil to a poorly drained or droughty-type soil (Fig. 12-4).

Periodically one may encounter a situation where the perennial forage crop is

TABLE 12-1
DRY MATTER YIELDS OF ALFALFA ALONE AND WITH THREE DIFFERENT GRASSES OVER A 3-YEAR PERIOD, UNIVERSITY OF ILLINOIS

| Alfalfa | | Grass seeding rate, kg/ha* (yield, t/ha) | | | | | | | | | |
| Seeding rate*, kg/ha | D.M. yield, t/ha | Orchardgrass | | | Smooth bromegrass | | | Reed canarygrass | | | Average |
		3.4	6.7	10	4.5	9.0	13.5	4.5	9.0	13.5	
6.7	42.2	44.1	42.7	42.1	41.8	42.9	43.0	43.2	42.5	42.7	—
13.5	42.9	43.2	44.3	44.1	44.4	43.4	43.3	43.9	42.0	44.3	—
20.2	43.9	44.5	43.6	43.5	43.8	44.8	43.9	42.4	44.8	42.8	43.3

*Seeding rates in lb/A: alfalfa, 6, 12, 18; orchardgrass, 3, 6, 9; smooth bromegrass and reed canarygrass, 4, 8, 12.

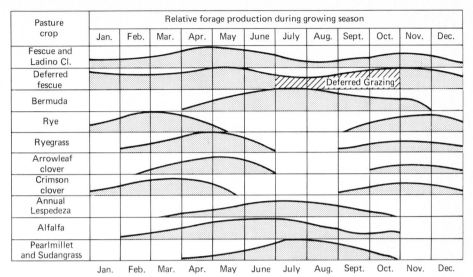

Pasture crop	Relative forage production during growing season											
	Jan.	Feb.	Mar.	Apr.	May	June	July	Aug.	Sept.	Oct.	Nov.	Dec.
Fescue and Ladino Cl.												
Deferred fescue												
Bermuda												
Rye												
Ryegrass												
Arrowleaf clover												
Crimson clover												
Annual Lespedeza												
Alfalfa												
Pearlmillet and Sudangrass												

Jan.   Feb.   Mar.   Apr.   May   June   July   Aug.   Sept.   Oct.   Nov.   Dec.

FIGURE 12-1
The comparative growth of various summer and winter forage species in the south.

killed as a result of winter injury or some disease or insect problem. Consequently, a temporary forage may be needed for pasture, hay, or silage. There
are numerous species that may be used as supplementary forage crops (Table
12-4). All the species listed in this table are annual grasses and are very
productive in the late summer or early winter. These grasses need to be seeded
each year on a well-prepared seedbed. With equal fertility and management, the
total season's production of these grasses may be less than that from a perennial
grass. The sudangrasses and sorghums, however, will fill a need for quick
supplementary pasture, green chop, or silage crop. The sudangrasses and
sorghums are tall, juicy grasses that are rather difficult to make into high-quality
hay. Compared with the sorghum-sudan hybrids, sudangrass and sudan × sudan
hybrids have finer stems, and so will dry more rapidly. Crushing the stems with a
hay conditioner will help speed drying.

The millets are warm-season annual grasses that are drought-tolerant.
Pearlmillet has been evaluated in grazing trials and is a suitable alternative for
summer-annual pastures (4). Pearlmillet requires a warmer soil for rapid
establishment than does sudangrass. Seeding of pearlmillet should be delayed
until the soil temperature of the seedbed is 21°C (70°F).

A comparison between pearlmillets and sudangrasses reveals: (1) there is no
prussic acid problem with pearlmillet (4); (2) pearlmillet is not as susceptible to
leaf disease as is sudangrass; and (3) pearlmillet is more drought-tolerant than is
sudangrass, and so produces more pasture during the hot, dry period in late
summer.

TABLE 12-1a

SUGGESTED FORAGE MIXTURES SEEDED WITH A COMPANION CROP IN
ILLINOIS

(Seeding Rates in kg/ha, Left, and lb/A, Right)

For hay crops					
Central, northern Illinois			**Southern Illinois**		
Moderate to well-drained soils					
Alfalfa	13	12	Alfalfa	9	8
			Orchardgrass	7	6
Alfalfa	9	8			
Bromegrass	7	6	Alfalfa	9	8
			Tall fescue	7	6
Alfalfa	9	8			
Bromegrass	5	4			
Timothy	2	2			
Poorly drained soils					
Alsike clover	6	5	Reed canarygrass	9	8
Timothy	5	4	Alsike clover	5	4
Reed canarygrass	9	8	Tall fescue	7	6
Alsike clover	4	3	Alsike clover	5	4
Birdsfoot trefoil	6	5	Redtop	5	4
Timothy	2	2	Alsike clover	5	4
Droughty soils					
Alfalfa	9	8	Alfalfa	9	8
Bromegrass	7	6	Orchardgrass	5	4
Alfalfa	9	8	Alfalfa	9	8
Tall fescue	7	6	Tall fescue	7	6
			Alfalfa	9	8
			Bromegrass	7	6
For horse pastures					
Central, northern Illinois			**Southern Illinois**		
Moderate to well-drained soils					
Alfalfa	7	6	Alfalfa	9	8
Smooth bromegrass	7	6	Orchardgrass	4	3
Kentucky bluegrass	2	2	Kentucky bluegrass	6	5
Poorly drained soils					
Smooth bromegrass	7	6	Orchardgrass	7	6
Kentucky bluegrass	2	2	Kentucky bluegrass	6	5
Timothy	2	2	Ladino clover	0.6	0.5
Ladino clover	0.6	0.5			

TABLE 12-1a (continued)

For horse pastures (continued)			
Central Illinois			
Moderate to well-drained soils		**Poorly drained soils**	
Alfalfa	7	6	Orchardgrass	7	6
Orchardgrass	4	3	Kentucky bluegrass	2	2
Kentucky bluegrass	2	2	Ladino clover	0.6	0.5

For pasture renovation					
Central, northern Illinois			**Southern Illinois**		
Moderate to well-drained soils					
Alfalfa	9	8	Alfalfa	9	8
Red clover	5	4	Red clover	5	4
Poorly drained soils					
Birdsfoot trefoil	5	4	Alsike	2	2
Red clover	5	4	Ladino clover	0.6	0.5
			Red clover	5	4

For rotation and permanent pastures					
Central, northern Illinois			**Southern Illinois**		
Moderate to well-drained soils					
Alfalfa	7	6	Alfalfa	9	8
Bromegrass	6	5	Orchardgrass	5	4
Timothy	2	2			
			Alfalfa	9	8
Alfalfa	7	6	Tall fescue	7	6
Orchardgrass*	5	4			
			Tall fescue	9	8
Alfalfa	7	6	Ladino clover	0.6	0.5
Orchardgrass*	5	4			
Timothy	2	2	Alfalfa	9	8
			Bromegrass	7	6
Orchardgrass*	7	6	Timothy	2	2
Ladino clover	0.6	0.5			
			Orchardgrass	7	6
Red clover	9	8	Ladino clover	0.6	0.5
Ladino clover	0.6	0.5			
Orchardgrass*	7	6	Tall fescue	11	10
Red clover	9	8	Orchardgrass	9	8
Ladino clover	0.6	0.5			
Tall fescue	7–9	6–8	Red clover	9	8
			Ladino clover	0.6	0.5
Birdsfoot trefoil	6	5	Orchardgrass	7	6
Timothy	2	2			

TABLE 12-1a (continued)

<table>
<tr><td colspan="5" align="center">For rotation and permanent pastures</td></tr>
<tr><td colspan="2" align="center">Central, northern Illinois</td><td colspan="3" align="center">Southern Illinois</td></tr>
<tr><td colspan="5" align="center">Moderate to well-drained soils</td></tr>
<tr><td>Bromegrass</td><td>9</td><td>8</td><td>Red clover</td><td>9   8</td></tr>
<tr><td>Ladino clover</td><td>0.6</td><td>0.5</td><td>Ladino clover</td><td>0.6 0.5</td></tr>
<tr><td></td><td></td><td></td><td>Tall fescue</td><td>7–9 6–8</td></tr>
<tr><td>Tall fescue</td><td>11</td><td>10</td><td></td><td></td></tr>
<tr><td>Orchardgrass*</td><td>9</td><td>8</td><td></td><td></td></tr>
</table>

Poorly drained soils					
Alsike clover	4	3	Alsike clover	2	2
Ladino clover	0.5	0.25	Tall fescue	7	8
Timothy	5	4	Ladino clover	0.6	0.5
Birdsfoot trefoil	6	5	Reed canarygrass	9	8
Timothy	2	2	Alsike clover	4	3
			Ladino clover	0.6	0.5
Reed canarygrass	9	8			
Alsike clover	4	3			
Ladino clover	0.5–0.6	0.25–0.5			
Alsike clover	2	2			
Tall fescue	9	8			
Ladino clover	0.6	0.5			

Droughty soils					
Alfalfa	7	6	Alfalfa	9	8
Bromegrass	6	5	Orchardgrass	5	4
Alfalfa	7–9	6–8	Alfalfa	9	8
Orchardgrass*	5	4	Tall fescue	7	6
Alfalfa	7–9	6–8	Red clover	9	8
Tall fescue	7	6	Ladino clover	0.6	0.5
			Orchardgrass	7	6
Red clover	9	8			
Ladino clover	0.6	0.5	Red clover	9	8
Orchardgrass*	7	6	Ladino clover	0.6	0.5
			Tall fescue	7–9	6–8
Red clover	9	8			
Ladino clover	0.6	0.5			
Tall fescue	7–9	6–8			

For hog pastures		
All soil types, anywhere in Illinois		
Alfalfa	7	6
Ladino clover	2	2

*Central Illinois only.
Source: D.W. Graffis, personal communication, 1980.

FIGURE 12-2
Orchardgrass seeded with alfalfa results in a high-quality forage. *(Courtesy University of Wisconsin)*

FIGURE 12-3
Smooth bromegrass alone (right), compared with alfalfa-smooth bromegrass (left). *(Courtesy University of Wisconsin)*

TABLE 12-2
SUGGESTED FORAGE MIXTURES IN
MINNESOTA

Species	Seeding rate	
	kg/ha	lb/A
Legumes		
Alfalfa		
Alone	11–13	10–12
With grasses	6–9	5–8
Alsike clover in mixtures	2	2
Birdsfoot trefoil	6–7	5–6
Ladino clover in mixtures	0.6–1.1	0.5–1
Red clover		
Alone	10–12	9–11
With grasses	5–9	4–8
Sweet clover	11–13	10–12
Grasses		
Bromegrass in mixtures	9–13	8–12
Orchardgrass in mixtures	1.7–4	1.5–3
Reed canarygrass		
Alone	7–9	6–8
In mixtures	5–7	4–6
Tall fescue		
Alone	11–13	10–12
In mixtures	4–6	3–5
Timothy in mixtures	2–7	2–6
Sudangrass		
18–40 in (70–158 cm)	11–22	10–20
6–14 in (24–55 cm)	28–33	25–30
With 1.5 bu soybeans	11	10
Millet	22–45	20–40
Field pea with 1.5–2 bu oats	50–100	45–90

FIGURE 12-4
Alfalfa does not tolerate poor drainage. *(Courtesy PPT.)*

TABLE 12-3
SUGGESTED FORAGE MIXTURES IN WISCONSIN

Mixtures	Seeding rates	
	kg/ha	lb/A
Well-drained, medium to fine-textured soils that are limed or nonacid		
Alfalfa alone	11–17	10–15
Alfalfa,	9–13	8–12
smooth bromegrass or	3–7	3–6
orchardgrass*	2–5	2–4
Alfalfa,	5	4
smooth bromegrass,	11	10
orchardgrass	2	2
Alfalfa,	9–13	8–12
timothy,	3–5	3–4
orchardgrass*	2–5	2–4
Red clover,	7–9	6–8
timothy	3–6	3–5
Sand soils		
Alfalfa,	9–11	8–10
smooth bromegrass	7–9	6–8

TABLE 12-3 (continued)

Mixtures	Seeding rates	
	kg/ha	lb/A
Poorly drained soils		
Red clover,	6	5
alsike clover,	3	3
ladino clover,	0.3–0.6	0.25–0.5
smooth bromegrass or	9	8
timothy	5	4
Alsike clover,	2	2
ladino clover,	1–2	1–2
timothy or	5	4
smooth bromegrass	9	8
Alsike clover	6	5
timothy or	5	4
smooth bromegrass	9	8
Very wet lowlands		
Reed canarygrass	7–9	6–8
Reed canarygrass,	3–6	3–5
timothy	2–3	2–3
Imperfectly drained and rolling soils intended for permanent pasture		
Birdsfoot trefoil,	7	6
Kentucky bluegrass or	2	2
timothy	3–5	3–4
Well-drained soils		
Smooth bromegrass	17–22	15–20
Reed canarygrass	9–13	8–12
Sudangrass	34–39	30–35
Oats	108	96
Winter rye or winter wheat	135	120

*0.25 to 0.51 kg ladino clover may be added when used for pasture. One to 2 kg timothy may be substituted for orchardgrass or smooth bromegrass on heavier soils.

†For soils not adequately drained and/or limed for alfalfa.

Source: D.A. Rohweder, R. F. Johannes, R.L. LaCroix, W.M. Paulson, G.H. Tempas, and G.G. Weis, "Forage crop varieties and seed mixtures for 1972," *Wis. Coop. Ext. Serv. Publ. A1525 (C463),* 1972.

TABLE 12-4
SUGGESTED FORAGES FOR TEMPORARY OR SUPPLEMENTARY PASTURE, HAY, OR SILAGE

	Seeding rate		
Species	kg/ha	lb/A	Use
Sudangrass drilled*	22–28	20–25	Pasture, silage, green chop, hay
Sorghum-sudangrass* hybrid drilled or in 50-cm (20-in) rows	22–34	20–30	Pasture, silage
Forage sorghums 100-cm (40-in) rows 35-cm (14-in)	6–11 8–17	5–10 7–15	Silage
Oats	72–108	64–96	Silage or hay
Winter rye	100	90	Winter or early spring pasture
Millets	22–45	20–40	Pasture, silage, green chop

*To avoid prussic acid poisoning, sudangrass and sudangrass hybrids should be at least 50 to 75 cm (20 to 30 in) tall before grazing.

Caution must be taken when grazing sudangrass or sorghum-sudan hybrids. Before grazing can occur, plant height must be at least 50 cm (20 in) for pure sudangrass, and up to 75 cm (30 in) for sorghum-sudan hybrids. This will avoid any potential prussic acid poisoning (4).

Corn may be considered as a supplementary forage crop for silage to carry animals through the winter. If the summer yield of forages has been deficient, corn is probably one of the best silage crops in the United States.

Farmers have recently used many experimental mixtures as supplementary pastures or silage crops. Soybeans have been added to the grain sorghums, as well as to millets and sudangrass to increase protein content of the final product. In the north, another silage mixture might include a small grain with soybeans to increase the protein level and ensure sufficient dry matter production. Many of these mixtures are used for silage.

QUESTIONS

1 Why might there be different forage mixtures within various areas of a state or region of the United States?
2 For a particular area within a state, with a given soil condition, discuss a suggested forage mixture for a particular use.

3 Supposing your perennial forage stand was killed, outline a supplementary pasture or forage program appropriate for your farm.

4 Under what type of situation, soil, and specific use would you recommend the planting of alfalfa, red clover, smooth bromegrass, orchardgrass, reed canarygrass, oats, and sudangrass?

REFERENCES

1 Ahlgren, G. H. *Forage Crops,* McGraw-Hill, New York, 2d ed., 1956.

2 Blaser, R. E., W. L. Griffeth, and T. H. Taylor. Seedling competition in compounding forage seed mixtures. *Agron. J.,* 48:118–123, 1956.

3 Decker, A. M., T. H. Taylor, and C. J. Willard. Establishment of new seedings. In *Forages,* Iowa State Univ. Press, Ames, IA, 3d ed., pp. 384–395, 1973.

4 Fribourg, H. A. Summer annual grass and cereals for forage. In *Forages,* Iowa State Univ. Press, Ames, IA, 3d ed., pp. 344–357, 1973.

5 Graffis, D. W. Personal communication. 1980.

6 Heath, M. E. Hay and pasture seedings for the central and lake states. In *Forages,* Iowa State Univ. Press, Ames, IA, 3d ed., pp. 448–459, 1973.

7 Jackobs, J. A. A measurement of the contribution of ten species to pasture mixtures. *Agron. J.,* 55:127–131, 1963.

8 Rohweder, D. A., R. F. Johannes, R. L. LaCroix, W. H. Paulson, G. H. Tempas, and G. W. Weis. Forage crop varieties and seed mixtures for 1972. *Wis. Coop. Ext. Serv. Publ. A1525 (C463),* 1972.

9 Schaller, F. W. and H. E. Thompson. Forage crop varieties and seeding mixtures. *Iowa Coop. Ext. Serv. Pam. 223* (rev.), 1968.

9 a Schaeffer, C. C. and D. R. Swanson. Seeding rates and grass suppression for sod-seeded red clover and alfalfa. *Agron. J.,* 74:355–358, 1982.

10 Smith, D. Seeding and establishment of legumes and grasses. In Forage Management, Kendall/Hunt, Dubuque, IA, 3d ed., pp. 15–29, 1975.

11 Van Keuren, R. W. Alfalfa establishment and seeding rate studies. *Ohio Report,* 58(2):52–54, 1973.

CHAPTER 13

MANAGEMENT AND CLIMATIC FACTORS

INFLUENCING FORAGE PRODUCTIVITY

Management of forages involves much more than just fertilizing and harvesting the crop. The productivity of forages depends to a great extent on the use of good management practices. The management of forages includes: field selection, cultivar selection, fertilization, inoculation, seeding rate, seedbed preparation, seeding method, weed and insect control, timely harvest, rest periods, fall pasture management, and proper harvesting methods. Grazing factors include rotations, animal numbers, types of animals, and the length of time grazed (Fig. 13-1). The agronomic practices involved in good forage establishment were covered in an earlier chapter. At present, approximately 95 to 100% of corn producers fertilize their corn crops, whereas 15 to 50% of forage producers fertilize forage crops, and only 5 to 15% fertilize pasture crops. This one practice—adequate fertilization—would greatly increase forage production and maintain much more productive stands for longer periods of time.

MANAGEMENT—SEEDING YEAR

Forage crops and pastures that are spring-seeded in a companion crop benefit from early removal of the companion crop. In most cases, the companion crop is spring oats, wheat, or barley. These crops should be removed when the grain is in the milk stage, or sooner (6, 17, 28, 35). If small-grain companion crops are harvested for grain, it is important to remove the straw and stubble as soon as possible, because as small-grain vegetative yields increase, greater competition is expressed on underseeded forage seedlings, and fewer satisfactory stands will

FIGURE 13-1
Rotational grazing involves increased carrying capacity, productive species, and improved fertility.

be established. When forage crop seedings are established with a companion crop, depending upon climatic conditions following the removal of the companion crop, there is usually only one harvest taken by late Autust, or a short grazing period, in the upper midwest, and occasionally two harvests by the middle of September in the central midwest, or three harvests in the south and southwest.

Spring-seeded forage crops, without a companion crop, should be ready for harvest 65 to 70 days after early spring seeding (30a). Weeds very likely need to be controlled about 30 days after the seeding unless a preemergence herbicide is used. A postemergence herbicide, such as 2,4-DB, is effective against most broadleaf weeds. If the broadleaf weed in a stand is not extremely competitive, one close clipping when the forage crop is in the headed stage or first-flower stage will control it (18, 26, 31). Leafhoppers often become a problem about 30 to 45 days after an early spring seeding, and need to be controlled to obtain a vigorous, high-yielding stand (41).

Second and third harvests follow the first harvest at approximately 35- to 40-day intervals in the midwest, or 28- to 30-day intervals in the west and southwest. In the northern half of the United States, the last harvest of the season should be made approximately 35 to 40 days before the average killing frost.

Cutting Height

Growers sometimes disagree over recommended cutting heights for forages. From a physiological and morphological standpoint, harvest is recommended at a low stubble height, or around 2.75 to 5 cm (1 1/2 to 2 in), for most cool- and warm-season forages, but at a greater height—10 to 12 1/2 cm (4 to 5 in)—for many dry land forage species. The crownal buds, or sites of regrowth, are located at or slightly below the soil surface in alfalfa and in the basal tillers of grasses, so that if the forage is harvested at a height of 10 to 12 1/2 cm (4 to 5 in), regrowth will be much more uneven than if the harvest were taken at a lower height, and will develop from the axillary buds on the remaining stem stubble in alfalfa or from the axillary node in grasses.

Cutting all of the top growth close to the soil surface may remove some diseases and insects. This helps reduce the buildup of such pests. It is good sanitation to remove all of the top growth.

Root Reserves

In the northern half of the United States, the last harvest in the fall should occur approximately 35 to 40 days before the average killing frost. The reason for this is to allow enough regrowth to occur so that soluble carbohydrates will be built up for excellent winter survival (3, 19, 25, 33, 36). If the forage plant has only 1 or 2 weeks of regrowth after the last harvest, the plant is still using its previous root or crown reserves to generate new growth. A killing frost at this stage means no food reserves for winter survival, and death and/or considerable spring heaving may occur in top-rooted species as a result of a poor or weak root system (33, 36).

In the upper midwest, some cultivars continue to grow during the fall after the September harvest; they have very little fall dormancy. Many researchers believe that a plant needs fall dormancy for excellent winter survival, though this may be debatable. In any case, if an excellent, 1-year-old stand of alfalfa is well-fertilized on a well-drained site and it continues to grow until a killing frost, a late fall harvest may be taken after the killing frost. Compared with the earlier harvest, the difference in management here is the cutting height: the late fall harvest should be cut at a greater stubble height, approximately 10 to 12 1/2 cm (4 to 5 in). The purpose for this is threefold: (1) it leaves some winter insulation; (2) it helps collect snow cover; and (3) it prevents icing over in the spring. Ice sheets generally do not "climb" that high over plants, so small projections of stems can penetrate through the ice, allowing for air exchange, which prevents the smothering of plants (9).

A late fall harvest also aids in removing the egg-laying sites of the alfalfa weevil and overwintering protection for other insects. In addition, a late fall harvest is a good sanitary practice in almost all forage species because it lowers the incidence of disease buildup.

Legumes are much more sensitive to fall management practices than are grasses. Grasses help reduce the incidence of spring heaving that results from freezing and thawing. This is primarily because of their fibrous root system, as compared to the taproot system of most legumes, which are very sensitive to freezing and thawing and are therefore susceptible to heaving. Fall growth of tall-growing grasses tends to aid in collecting snow, which gives extra winter cover and reduces winter injury. Grasses also tend to assist in controlling erosion on sloping soil sites more than all-legume stands do.

MANAGEMENT—ESTABLISHED STANDS

The ultimate goal of a forage producer is to produce the maximum amount of dry matter with the greatest possible degree of digestibility. The general rule of thumb is that with advanced maturity, TDN, crude protein, and digestibility decrease proportionally, whereas crude fiber and total dry matter production increase proportionally (4, 22, 23, 38, 42) (Fig. 13-2). One can harvest forage crops two to ten times per year, depending on the latitude and type of forage grown. Although the number of times a forage crop may be harvested varies by area, the crop generally reaches its highest yield and quality at the same relative stage of maturity (22, 23).

Maintaining an annual legume in a grass stand is important. For example, arrowleaf clover produces enough seed to give good volunteer stands in the fall unless summer rains cause the seed to germinate and the seedlings die. One should graze or harvest the grass stand in late summer, and disk lightly or harrow to ensure volunteer stands of legume. Arrowleaf clover will not germinate readily if it is covered with a heavy mulch or excessive grass growth. This practice is very similar with most annual self-seeding legumes.

The highest dry matter yield per cutting is reached sometime around the full-bloom stage. However, quality is highest at more immature stages, such as prebud for alfalfa, or the boot stage for many grasses (22, 23). Therefore, when deciding upon the proper time to harvest, a compromise must be made between the highest dry matter yield and highest quality. Regardless of the time that one chooses, no method of storage will improve forage quality above that found at harvest (4). Therefore, to produce the highest-quality forage for livestock, the harvest must occur when quality is at a high point (Fig. 13-3). On the other hand, continually harvesting a forage crop at an early stage, such as prior to prebud for alfalfa or the boot stage for grasses, results in a thinning of the stand, particularly in northern latitudes (23).

FIGURE 13-2
Highly fertilized grass pastures require better management of cattle carrying capacity to maintain high-quality forage.

FIGURE 13-3
Alfalfa-smooth bromegrass mixtures make an excellent mixture of quality forage. *(Courtesy University of Wisconsin)*

Many studies have been conducted to determine the best time to harvest alfalfa or other forage crops in relation to yield and quality. The best compromise, especially when alfalfa is grown with other forage crops, is to take the first harvest when the alfalfa is at the first-bloom stage, and then determine the date of the last harvest of the season as approximately 35 to 40 days before the average killing frost or onset of dormancy. The remaining harvests should be divided into periods ranging from 30 to 35 days (23, 40). Some researchers prefer to harvest alfalfa according to its maturity rather than by a predetermined calendar schedule (23, 40). A cutting schedule based on plant maturity rather than on a calendar schedule allows the alfalfa or other forage crop to dictate the proper time to harvest and takes into account differences in maturity resulting from different cultivars, years, or locations. For instance, the first-flower stage is easily recognizable and is a satisfactory approximation of the 10% bloom stage. Yields of TDN, protein, and minerals decrease after the early bloom because of the loss of the lower leaves from lodging, aging of the plant, and disease (38).

If one has a large hectarage, harvest of alfalfa should begin by late bud or before the first-bloom stage so that half of the harvest is completed by the optimum first-flower stage. Rotation of the field could be made to ensure that the same field is not harvested too early every time.

Spring harvest management may vary somewhat in northern latitudes or high elevations, where forage stands are susceptible to winter injury. An established stand that has been injured during the winter should be allowed to mature a little longer than usual before the first harvest is taken; in the case of alfalfa, one may wait until 25% bloom has occurred, which assists the plants' recovery. Allowing forage plants to reach later stages of maturity ensures greater carbohydrate reserves in their roots (36). The first harvest may be light and somewhat weedy, but generally the second harvest is back to full productivity.

One cannot overemphasize the importance of fall management in maintaining a forage crop. One should refrain from harvesting or grazing the forage for 35 to 40 days before the first expected killing frost or fall dormancy. This allows the plant to store a high level of food reserves in the roots and crowns. Food reserves are used to develop the hardiness necessary to survive the winter, and to initiate growth during the following spring (37). In the northern United States, the fall growth that occurs in the last few weeks before a killing frost provides a sufficient amount of stubble to ensure some insulation and also provides enough ground cover to help collect a blanket of snow during the winter months.

Many growers are now taking a late fall harvest if the soil is well-drained and extremely high in fertility, and if an excellent, pest-resistant cultivar is grown. Plants should not be close-cut at late fall harvest or after a killing frost (5a). The forage crop does not regrow following a killing frost, so that a late harvest removes the reserves that have been stored during the previous weeks prior to the frost.

An area in the United States in which a great deal of winter injury may occur because of a lack of snow cover is the central midwest. Further north, there is greater assurance of snow cover throughoutt the winter, thus reducing cold injury. But as one progresses further south, snow cover is often completely lacking or variable throughout the year, which allows greater susceptibility to winter injury.

Grasses grown in association with legumes greatly reduce winter losses. The grass provides an additional winter cover, which aids in collection of snow. One should keep in mind that most grasses that are grown in combination with legumes in the northern half of the United States are cool-season, perennial grasses. These grasses provide greater fall growth going into the winter, which results in extra winter cover, and helps reduce winter injury and heaving of legumes the following spring. Some advantages of a grass-legume mixture over a legume or a grass grown alone are: (1) the grass assists in controlling erosion on sloping soils; (2) provided that the grass component is approximately 50% of the sward, there is less chance for bloat while animals are grazing or utilizing these forages as compared with that in a pure stand; (3) generally, the forage cures much more rapidly due to the fluffiness of the grass in the windrow; (4) the invasion of weeds is greatly reduced compared with that in a pure legume stand; and (5) as the stands become older the grasses generally begin to fill in, thereby lengthening the period that a stand will be productive. As indicated in earlier chapters, whenever the legume portion drops below 30% of the stand, it is recommended that nitrogen fertilizer be added to increase the productivity of the forage grass.

Additional benefits from grasses are that they fill in areas where legumes may die out because of high water tables or extremely wet soil conditions, or areas where legumes will not grow. Grasses also tend to increase the organic matter in the soil over that in pure legume stands. When grasses are grown with legumes, it has been shown that the grasses often aid in reducing the lodging of the legume. Lodging, in first harvests of alfalfa or other legumes, may result in lost leaves, which will reduce quality.

Cool-season grasses that are grown with legumes are very productive in early spring, become less productive during the summer months, and then increase in productivity again in early fall. Most cool-season grasses, even though they are harvested in the early spring, produce a leafy aftermath, thus ensuring a fairly high-quality forage.

Legumes help ensure the quality of grass-legume mixtures primarily by increasing the protein concentration of the grass.

CLIMATIC FACTORS

In addition to proper fertility, harvest schedule, grazing management, fall management, food reserves, and spring management, the production and quality of forage may also be influenced by climate. Climatic conditions cannot be controlled as much as the other management factors can, but it should be kept in mind that the climate is a very important factor in the overall production and quality of forage produced. The influence of climatic conditions is thoroughly discussed in Chapter 16 on alfalfa, but it is in order here to summarize some of the climatic conditions that influence quality.

Temperature

Many trials have been conducted under controlled temperature situations, in growth chambers or greenhouses, but when forages are grown outdoors, it is impossible to control temperature regimes. Maturity of forages is greatly influenced by temperature. It has been shown that alfalfa may reach the first-flower stage within 21 days under warm temperatures of 32/24°C (90/75°F) day/night, but this may take approximately 37 days under a cool regime of 18/10°C (65/50°F) (14, 21). For cool-season species, dry matter production is considerably highly under cooler temperature conditions than under warmer regimes. The cooler the temperature, the higher the in vitro digestible dry matter, the percent nonstructural carbohydrate, and the percent Ca and Mg (24, 43). A point of caution: under high K fertilization, Mg may be reduced, causing grass tetany. With warm temperatures, the content of protein and amino acids, as well as total ash, P, K, B, Zn, and Mn, will be higher than under cooler regimes (24). These results are similar in alfalfa, smooth bromegrass, and timothy.

It is thought that the lower concentrations of most minerals under cool temperatures are the result of reduced uptake of the minerals from the soil and, consequently, reduced translocation within the plant tissue itself (24, 36). The cooler temperatures increase dry matter yields, so that there is a dilution of minerals and a higher accumulation of nonstructural carbohydrates within the plant tissue (36).

Research has shown that the prevailing temperature prior to harvest influences chemical composition more than does the temperature following harvest (24). Temperature has the greatest effect upon the concentration of total nonstructural carbohydrates and the digestible dry matter production (5, 21). Therefore, warm temperatures reduce TNC, whereas cool temperatures increase TNC. In summary, forages that are harvested following a cool period should have a much greater feeding value and better silage fermentation than forages harvested after a warm period.

There also has been research conducted to determine the proper time of the day for harvesting a forage crop. It has been shown that total sugars and TNC in forage grasses and legumes usually increase through the morning hours until some time during the afternoon; then the percentage of sugars and TNC decreases until daylight the following morning (36, 43). This is primarily the result of changes in the sucrose concentration. Most of the daily variation of TNC in warm-season grasses and legumes results from changes in both sucrose and starch concentration (36, 38, 43).

Total nitrogen and/or protein is influenced by the amount of nonstructural carbohydrate accumulated during the day. It has been shown that the concentration of nitrogen and other elements is usually lowest sometime in the afternoon, when TNC concentration is the highest (38). Conversely, nitrogen and other elements are generally highest in the morning or at night.

Most of the ecological and physiological factors that determine forage productiveness and the nutritive evaluation of forage have been discussed in Chapters 4 and 7, respectively.

Water

Water stress and excessive water also have a great influence on forage production. When transpirational losses exceed water absorption, metabolism, development, and growth are greatly reduced, thereby reducing overall dry matter production. Most growth occurs during the evening or when transpirational losses are very low (5, 8, 12). Most expansion of leaf and stem cells occurs during the night (27). Water stress has a direct influence on cell enlargement and overall dry matter production. In the field, drought is usually associated with high temperatures. A combination of water stress or drought conditions along with high temperatures depresses overall plant growth (5, 20, 24, 27, 32). Under these conditions, TNC utilization is reduced faster than is the TNC produced by photosynthesis. As a result, under drought conditions, soluble carbohydrates may increase in the plant tissue, and forage quality will not decrease until death or deterioration of the plant actually occurs.

Excessive moisture may depress the growth of some forages, such as alfalfa, but it may encourage the growth of others, such as ladino clover, which has a shallow root system, or some grasses, such as reed canarygrass or tall fescue. Most grasses tolerate wetter soil conditions than do legumes. The increase of disease in plants is often associated with water stress conditions, which in turn reduce the legume stand (10). Short periods of excessive soil moisture do not result in a permanent decrease in stand or overall production. Long periods of excessive soil moisture reduce soil pH and increase toxicity of some elements, such as Al and Mn (7).

It is important, on especially poorly drained soils, to try to increase water removal, whether by tile, internal drainage, or surface drainage. In some poorly drained areas, producers must consider growing a grass with a legume or possibly a monoculture of grass. Plant diseases are highly correlated with excessive moisture and high temperatures. Thus, it is very important to use a species and/or a cultivar of a particular species that is very tolerant to such conditions, if they are prevalent.

Disease and Insect Control

The two greatest factors influencing leaf content and quality are disease and insects. The main control of plant diseases in forage crops is selection of a cultivar that is resistant to a particular disease. Leaves contain most of the plant's protein and minerals, have most of its highly digestible dry matter, and are low in fiber in comparison to the stem portion of the plant, so one should select a cultivar that has leaf-disease resistance. In addition to selecting the proper disease-resistant cultivar, good, clean, timely harvests are important. Close clipping or removing all diseased tissue halts the increase of a particular disease.

Diseases are usually associated with climatic factors such as temperature and rainfall. Compared with other factors, these two have the greatest influence upon the incidence of leaf disease in forages. There is not much that one can do to change the rainfall and temperature, but proper drainage can ensure that a water stress condition does not build up in the soil, which may influence root and crown rots of various forage species. Good internal and surface drainage also help reduce the incidence of disease.

There are numerous insects that consume and reduce forage production. Many insects are leaf-consuming, such as the alfalfa weevil and the potato leafhopper, which sap plants' nutrients and cause leaf drop (Fig. 13-4). There are fewer insects that feed upon grass leaves. But under dry, warm temperatures, grasshopper populations increase to the point that they consume great quantities of forage structures, thereby greatly reducing the yields of forage grasses.

The potato leafhopper is probably one of the most damaging insects that feed on alfalfa. It migrates from the gulf states to the upper midwest from mid- to late May and is active until a killing frost. The female lays 2 to 3 eggs per day on alfalfa stems. The nymphs emerge 7 to 10 days later and become adults within 2 weeks. In Illinois, there are about three to four generations of the leafhopper

FIGURE 13-4
Symptoms of potato leafhopper
feeding on alfalfa.

per year. The 0.3-cm- (1/8-in-) long, lime-green, winged insects cut yields, as well as protein and vitamin A levels, by sucking the sap from plants. Not only does the leafhopper affect the protein and vitamin content, but it also stunts plants by injecting a toxin when it pierces the leaf petiole and feeds upon the plant. Once the plant is attacked in this way, it turns yellow, and a yellow to purplish cast develops, resulting eventually in a red, stunted plant. The very first symptom seen in a field is a wedge-shaped yellow area at the tip of each of the alfalfa leaflets. This is not always easily recognized as leafhopper damage; it can be confused with drainage problems, drought, or boron deficiencies. One should select a leafhopper-tolerant cultivar or utilize an excellent insecticide program to control this pest. Control methods may boost dry matter yields by as much as 25%.

Comparing the potato leafhopper to the alfalfa weevil, the potato leafhopper is present throughout the entire growing season in the midwest, whereas the alfalfa weevil is present during the first or into the second growth period of the season. The alfalfa weevil consumes most of the leaf tissue. Protein, dry matter, and digestibility are greatly reduced when the alfalfa weevil invades an alfalfa stand. The best control method for the alfalfa weevil is to utilize either a tolerant cultivar or a timely harvest, and perhaps a spray program as well. A parasitic wasp and a fungus disease of the weevil offer possible biological controls, and a proper boron application also aids as a control agent for the weevil.

Both the potato leafhopper and the alfalfa weevil cause great yield reduction of forages under low-fertility conditions. It has been found that under high fertility, yields may be reduced by 20%. Under the same environmental conditions, and when alfalfa is grown under a low-fertility range, the yield reduction attributed to the leafhopper may be at least 36% (41). There also have been reported incidences where alfalfa has been completely killed by an extremely high invasion of the alfalfa weevil when no control is practiced. Therefore, it is very important to maintain excellent stands and ensure high soil fertility to help reduce the incidence of buildup of these insects. A good, clean harvest also greatly assists in reducing the buildup of insects within forage stands; close clipping, and complete removal of all infested areas, help reduce the egg-laying sites for insects.

QUESTIONS

1 List as many factors as you can that one must consider in proper management of forage crops.
2 Compare the two different theories of harvest schedules—one by maturity and the other by a harvest date.
3 Why is fall management so important in maintaining a forage stand?
4 What effect does temperature have on forage quality and production?
5 How can one reduce water stress in forages?
6 Since we frequently harvest forage crops, why is it so important to control disease and insects?

REFERENCES

1 Ahlgren, G. H. *Forage Crops*. McGraw-Hill, New York, 2d ed., pp. 65–76, 398–422, 1956.
2 Aldous, D. E. and J. E. Kaufmann. Role of root temperature on shoot growth of two KY bluegrass cultivars. *Agron. J.,* 71:545–547, 1979.
3 Barte, A. L. Effect of root temperature on dry matter distribution, carbohydrates accumulation, and acetylene reduction activity in alfalfa and birdsfoot trefoil. *Crop. Sci.,* 18:637–640, 1978.
4 Baumgardt, B. R. and D. Smith. Changes in the estimated nutritive value of the herbage of alfalfa, medium red clover, ladino clover, and bromegrass due to stage, maturity, and year. *Wis. Agr. Exp. Sta. Res. Rept. 10,* 1962.
5 Brown, R.H., and R. E. Blaser. Soil moisture and temperature effects on growth and soluble carbohydrates of orchardgrass (*Dactylis glomerata*). *Crop. Sci.,* 10:213–216, 1970.
5 a Collins, M. Changes in composition of alfalfa, red clover, and birdsfoot trefoil during autumn. *Agron. J.,* 75:287–291, 1983.
6 Decker, A. M., T. H. Taylor, and C. J. Willard. Establishment of new seedings. In *Forages,* Iowa State Univ. Press, Ames, IA, 3d ed., pp. 384–395, 1973.
7 Erwin, D. C., B. W. Kennedy, and W. F. Lehman. Xylem necrosis and root rot of alfalfa associated with excessive irrigation and high temperatures. *Phytopathology,* 49:572–578, 1959.
8 Evenson, P. D. Optimum crown temperatures for maximum alfalfa growth. *Agron. J.,* 71:798–800, 1979.
9 Freyman, S. Role of stubble in the survival of certain ice-covered forages. *Agron. J.,* 61:105–107, 1969.

10 Frosheiser, F. I. Phytophthora and root rot of alfalfa in Minnesota. *Plant Dis. Rept.,* 51:679–681, 1967.

11 Fulkerson, R. S. Location and fall harvest effects in Ontario on food reserves storage in alfalfa (*Medicago sativa* L.). *Proc. 11th Intl. Grassl. Congr.,* pp. 555–559, 1970.

12 Gavande, S. A. and S. A. Taylor. Influence of soil water potential and atmospheric evaporative demand on transpiration and the energy status of water in plants. *Agron. J.,* 59:4–7, 1967.

13 Garber, L. F. and V. G. Sprague. Cutting treatments of alfalfa in relation to infestations of leafhoppers. *Ecology,* 16:48–59, 1935.

14 Hanson, C. H. and D. K. Barnes. Alfalfa. In *Forages,* Iowa State Univ. Press, Ames, IA, 3d ed., pp. 136–147, 1973.

15 Herst, G. L. and C. J. Nelson. Compensatory growth of tall fescue following drought. *Agron. J.,* 71:559–563, 1979.

16 Jung, G. A., R. L. Reid, and J. A. Balasko. Studies on yield, management, persistence, and nutritive value of alfalfa in West Virginia. *W. Va. Agr. Exp. Sta. Bull. 581T,* 1969.

17 Klebesadel, L. J. and D. Smith. The effects of harvesting an oat companion crop at four stages of maturity on the yield of oats, on light near the surface, on soil moisture, and on the establishment of alfalfa. *Agron. J.,* 52:627–630, 1960.

18 Klingman, D. L. and W. C. Shaw. Using the phenoxy herbicides effectively. *USDA Farmers' Bull. 2183,* 1971.

19 Kust, C. A. and D. Smith. The influence of harvest management on level of carbohydrate reserves, longevity of stands, and yield of hay and protein from Vernal alfalfa. *Crop Sci.,* 1:267–279, 1961.

20 Larson, K. L. and J. D. Eastin. Drought injury and resistance in crops. *Crop. Sci. Soc. Amer. Spec. Publ. 2,* 1971.

21 Laude, H. M. Plant response to high temperatures. *Amer. Soc. Agron. Spec. Publ. 5,* pp. 15–30, 1964.

22 Marten, G. C. and A. W. Hovin. Harvest schedule, persistence, yield, and quality interaction among perennial grasses. *Agron. J.,* 72:378–387, 1980.

23 Matches, A. G., W. F. Wedin, G. C. Marten, D. Smith, and B. R. Baumgardt. Forage quality of Vernal and DuPuits alfalfa harvested by calendar date and plant maturity schedules in Missouri, Iowa, Wisconsin, and Minnesota. *Wis. Agr. Exp. Sta. Res. Rep. 73,* 1970.

24 McCloud, D. E. and R. J. Bula. Climatic factors in forage production. In *Forages,* Iowa State Univ. Press, Ames, IA, 3d ed., pp. 372–383, 1973.

25 Misberg, P., J. B. Washko, and J. D. Harrington. Plant maturity and cutting frequency effect on total nonstructural carbohydrate percentages in the stubble and crown of timothy and orchardgrass. *Agron. J.,* 70:907–912, 1978.

26 Moline, W. J. and L. R. Robinson. Effects of herbicides and seeding rates on the production of alfalfa. *Agron. J.,* 63:614–617, 1971.

27 Nelson, C. J., K. J. Treharne, and J. P. Cooper. Influence of temperature on leaf growth of diverse population of tall fescue. *Crop Sci.,* 18:217–220, 1978.

28 Peters, R. A. Legume establishment as related to the presence of an oat companion crop. *Agron. J.,* 53:195–198, 1961.

29 Portz, H. L. Frost heaving of soil and plants. I. Incidence of frost heaving of forage plants and meterological relationships. *Agron. J.,* 59:341–344, 1967.

30 Russell, W. E., F. J. Olsen, and J. H. Jones. Frost heaving in alfalfa establishment on soils with different drainage characteristics. *Agron. J.,* 70:869–872, 1978.

30 a Schaeffer, C. C. Seeding year harvest management of alfalfa. *Agron. J.,* 75:115–119, 1983.

31 Schmid, A. R. and R. Behrens. Herbicides versus oat companion crops for alfalfa establishment. *Agron. J.,* 64:157–159, 1972.

32 Slayter, R. O. *Plant-Water Relations.* Academic Press, New York, 1967.

33 Smith, D. Carbohydrate root reserve in alfalfa, red clover, and birdsfoot trefoil under several management schedules. *Crop Sci.,* 2:75–78, 1962.

34 Smith, D. Varietal chemical differences associated with freezing resistance in forage plants. *Cryobiology,* 5:148–159, 1968.

35 Smith, D. The establishment and management of alfalfa. *Wis. Agr. Exp. Sta. Bull. 542,* 1968.

36 Smith, D. Physiological considerations in forage management. In *Forages,* Iowa State Univ. Press, Ames, IA, 3d ed., pp. 425–436, 1973.

37 Smith, D. Management, treatments and relation to winter survival. In *Forage Management,* Kendall/Hunt, Dubuque, IA, 3d ed., pp. 59–62, 1975.

38 Smith, D. Management of alfalfa. In *Forage Management,* Kendall/Hunt, Dubuque, IA, 3d ed., pp. 89–99, 1975.

39 Smith, D., M. L. Jones, R. F. Johannes, and B. R. Baumgardt. The performance of Vernal and DuPuits alfalfa harvested at first flower or three times by date. *Wis. Agr. Exp. Sta. Rep. 23,* 1966.

40 Smith, D, G. C. Marten, A. G. Matches, and W. F. Wedin. Dry matter yield of Vernal and DuPuits alfalfa harvested by calendar date and plant maturity schedules in Missouri, Iowa, Wisconsin, and Minnesota. *Wis. Agr. Exp. Sta. Res. Rep. 37,* 1968.

41 Smith, D. and J. T. Medler. Influence of leafhoppers on the yield and chemical composition of alfalfa hay. *Agron. J.,* 51:118–119, 1959.

42 Van Riper, G. E. and D. Smith. Changes in the chemical composition of the herbage of alfalfa, medium red clover, ladino clover, and bromegrass with advanced maturity. *Wis. Agr. Exp. Sta. Res. Rep. 4,* 1959.

43 Vough, L. R. and G. C. Marten. Influence of soil moisture and ambient temperature on yield and quality of alfalfa forage. *Agron. J.,* 63:40–43, 1971.

CHAPTER 14

HARVESTING AND STORAGE OF FORAGES

Profitable livestock production depends upon the feeding of quality forage. This is directly related to the harvesting and storing of the forage itself. There is no method of harvesting or storage, except hay ammoniation, that will increase the nutritive value of a forage crop after harvest. Quality forage is that which is high in energy, protein, digestibility, palatability, and acceptability. Therefore, the starting point of quality forage is the stage of maturity when the forage is harvested and the time that it is left in the field after cutting. Several potential problems may occur while forage is in the field. One is the loss of vitamins from the forage itself during raking or harvesting, especially as a result of the dropping of leaves, which are high in protein and digestibility. Another problem that may be encountered is the advent of rain upon cut forage in the field. For example, during a normal spring in Illinois, the chance of not having rain during a 24-hour period is 65%; for a 30-hour rain-free period the chance drops to 45 to 48%, and to 25% for a 48-hour rain-free period. Therefore, it is very important to try to reduce the chance of rain damage to forage when harvesting by removing the cut hay from the field as rapidly as possible (Table 14-1).

Research shows that under good climatic conditions, 20% of a crop's dry matter can be lost before the crop is put in storage. Under poor conditions, the loss often reaches 30 to 40%. In general, the extent of dry matter losses depends on how long the crop lies in the field. About 4% of the yield is lost each day (7a). One recent advance in harvest technology is the application of a chemical, potassium carbonate, on crops during mowing. It has been shown that the dry-down period may be speeded up by 1 day through the use of such a compound (Fig. 14-1) (34a).

Since there are so many different forage species and mixtures, the question

TABLE 14-1
TIME REQUIRED TO CURE FORAGE, WITH CHANCES OF LIKELY RAIN DAMAGE

Harvest method	Curing time (sunshine hrs)	No. of chances out of 100 for likely rain damage
Field cured; deciding to mow at random	30	34 to 41
Field cured; mowing based on weather forecast	30	15 to 30
Field cured; mowing and conditioning based on weather forecast	15 to 20	12 to 20
Low-moisture silage (45–55% H$_2$O)	18	17 to 19
Silage	12	2 to 6

Source: C. R. Hoglund, "Comparative losses and feeding values of alfalfa and corn silage crops when harvested at different moisture levels and stored in gas-tight and conventional tower silos: An appraisal of research results," *Mich. State Univ. Ag. Econ. Publ. 947* (1964).

often arises as to the best time to take the first harvest and the subsequent regrowth. A summary of the first-harvest schedule that ensures quality forage is presented in Table 14-2. There are many factors that one must consider when beginning to harvest forages: (1) the hectarage that is to be harvested; (2) the various forage species or mixtures that one is harvesting; (3) the type of equipment that is available; (4) the use of the forage; and (5) the weather. The suggested time to harvest each forage species for hay, silage, or pasture is presented in the chapters concerning each species.

There are at least three guidelines developed for growers pertaining to the

FIGURE 14-1
Drying curves for second-cutting alfalfa: solid line, alfalfa treated with potassium carbonate solution; dotted line, untreated forage.

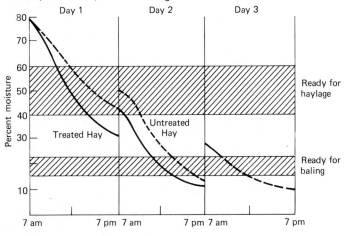

TABLE 14-2
SUMMARY OF WHEN TO HARVEST VARIOUS FORAGE SPECIES AND MIXTURES TO
ENSURE QUALITY FORAGE

Species	First harvest	Regrowth
Alfalfa	First flower	Bud stage of alfalfa
Alfalfa-orchardgrass	When orchardgrass heads	Bud stage of alfalfa
Alfalfa-smooth brome and other grasses	Alfalfa first flower	Bud stage of alfalfa
Red clover	First flower to 25% bloom	First flower of red clover
Red clover-grasses	When grasses head	First flower of red clover
Ladino clover alone or with grasses	10 to 50% bloom of ladino	Every 30 to 35 days
Birdsfoot trefoil alone or with grasses	10 to 50% bloom of birdsfoot trefoil	10 to 50% bloom of birdsfoot trefoil
Smooth brome, orchardgrass, or timothy	When heads emerge	Vegetative
Reed canarygrass or tall fescue	Flag-leaf stage	Every 30 to 40 days
Bermudagrass or bahiagrass	Begin grazing when 20 cm tall or maintain height of 10 cm	Every 2 to 3 wks
	For silage, harvest when 40 cm tall	Every 4 to 5 wks

correct time of forage harvest: (1) stage of maturity or bloom; (2) type of regrowth, whether from the crown or the basal stem area; or (3) calendar date, or the number of days between harvests. Use of the term "first flower" instead of "one-tenth bloom" is suggested because most people underestimate the percent bloom when a crop is growing in the field; "bud stage" or "first flower" will be used here as the suggested first harvest. If a large hectarage is to be harvested, the producer must begin harvest slightly on the early side, such as the bud stage or late-bud stage for alfalfa, and finish the first harvest just beyond the first-flower stage. Another possibility is to select cultivars of a particular species with different periods of maturity, thereby allowing the entire hectarage to be harvested at its optimum growth.

A review of the guidelines as to when to harvest a particular forage mixture of alfalfa grown with various grasses is as follows. One should keep in mind that most orchardgrasses mature and flower earlier than do some other perennial cool-season grasses. Therefore, when alfalfa and orchardgrass are grown together, the orchardgrass is the indicator plant and dictates when the harvest should be taken. When smooth bromegrass or a later-maturing grass species is grown with alfalfa, the alfalfa is the indicator plant for taking the first harvest.

Regardless of the forage species, the suggested time to take a particular harvest is whenever it is at the first flower stage of growth or beginning to flower.

Harvesting by calendar date is preferred by some researchers. A general rule is that this method is satisfactory, provided that there are adequate moisture and fertility. Climatic conditions or rainfall and soil fertility may delay the maturity of a particular species. Therefore, the calendar date may rush or hasten the second or third harvest prior to its optimum stage of growth. Opponents of suggested cutting schedules based on calendar dates say that this method does not allow for the maturity of different cultivars in various years or locations within a particular region; as a result, they prefer that the regrowth be harvested according to the maturity of the plant instead. However, those researchers who support a calendar date of harvest say that as one delays the harvest by waiting strictly until the plant begins to flower, the protein and total digestible nutrient contents of the forage decrease more rapidly.

Most legume species have crowns or crownal buds for regrowth, whereas the grasses generally have regrowth occurring from basal tillers; therefore, cutting height plays a very important role in the regrowth of each of these various species.

ON-THE-SPOT INDICATORS OF QUALITY FORAGES

At present, more hay is being marketed than has been the case previously; as a consequence, one must make some type of visual observation to assess the quality of forage purchased. Some on-the-spot indicators of quality forage are: percent leaves, stage of maturity when harvested, and freedom from mold, weeds, and foreign matter.

Green color in a hay indicates good vitamin content. Green color also indicates that the hay was harvested and stored under ideal conditions, but one should not overestimate color as an indicator of quality. Forage that is cut at early maturity will be lighter green in color, indicating high quality (23).

Leaves are an excellent indicator of quality. The higher the percentage of leaves in the forage, the higher the content of protein and digestible dry matter. A high percentage of leaves also indicates good harvesting methods and storage conditions. In general, forages that have a high percentage of leaves have a higher percentage of minerals and vitamins and a higher energy value than forages that have a low percentage of leaves (23).

It is slightly more difficult to determine quality in grass hay than in legume hay because it is more difficult to determine the grass hay's maturity at harvest time. One method used to determine the quality of grass hay in a bale is to open the bale and look for the elongated stem or flowering head of the grass. The flowering head should pull easily from the base of the node of the stem. When one pulls the inflorescence from the nodal area, one should look at the base of the internode. A black or dark color indicates high quality. If the basal area of the internode is light in color or pale, this indicates that when harvested, the plant was more mature than desired. A pale color also indicates that the plant has a higher fiber content and is lower in digestibility than one with a black area at the nodal base.

When determining the quality of legume hay in bales, one should look for the flowering head and determine the percent bloom. Alfalfa hay should be no more than one-tenth to one-fourth bloom; it should be dark green in color; and it should have fine stems. These three indicators are good visual aids in determining quality alfalfa hay. Seed pods should not be found, nor should a pale green color; these indicate a more mature alfalfa hay that is lower in digestibility and higher in fiber content than desired. Red clover hay can be up to one-half bloom; compared with alfalfa hay, good-quality red clover hay should have a rather deep green to brown color, and its stems should be larger or more medium in size. Compared with alfalfa, red clover hay is normally darker in color when cured.

Some general rules of thumb are that grass hay usually contains 10 to 12% crude protein, and has a 55 to 60% TDN value. Legume hays usually contain 16 to 20% crude protein and are 65 to 70% TDN. Forages grown at higher temperatures contain more stem and cell-wall substances and are less digestible than are forages grown at cooler temperatures. In the northern United States, forage regrowth harvested at the same stage of maturity never contains as high a percent digestible dry matter as does the first-growth forage that is harvested during the early spring. Leaf content is associated with the digestible dry matter. When the forage is not damaged by rain or climatic conditions, digestible dry matter content is not affected appreciably by the curing method. Delaying the harvest of a legume-grass mixture decreases the digestibility of the dry matter and the content of digestible energy (Table 14-3). Protein content of forages

TABLE 14-3
DIGESTIBLE DRY MATTER AND ENERGY VALUE OF AN
ALFALFA-SMOOTH BROMEGRASS MIXTURE HARVESTED AT
DIFFERENT STAGES OF MATURITY

Stage of alfalfa	Digestible D.M., %	Digestible energy, kcal/g
Vegetative	71.4	3.20
First flower	64.6	2.86
Full bloom	58.0	2.54
Seed pod	55.2	2.43

Sources: J. T. Reid, "Quality hay," in *Forages,* Iowa State Univ. Press, Ames, IA, 3d. ed., pp. 532–548, 1973; J. T. Reid, W. K. Kennedy, K. L. Turk, S. T. Slack, G. W. Trimberger, and R. P. Murphy, "Effect of growth stage, chemical composition, and physical properties upon the nutritive value of forages," *J. Dairy Sci.* 42:567–571 (1959); R. E. Roffler, R. P. Niedermeier, and B. R. Baumgardt, "Evaluation of alfalfa-brome forage stored as wilted silage, low-moisture silage, and hay," *J. Dairy Sci.* 50:1805–1813 (1967); J. G. Welch, M. Clancy, and G. W. Vander Noot, "Net energy of alfalfa and orchardgrass hays at varying stages of maturity," *J. Anim. Sci.* 28:263–267 (1969).

TABLE 14-4
PROPOSED MARKET HAY GRADES FOR LEGUMES AND LEGUME MIXTURES

Grade	Stage of maturity (international term)	Definition	Physical description	Typical chemical composition, %			Relative feed value %
				CP	ADF	NDF	
1. Legume hay	Prebloom	Bud to first flower; stage at which stems are beginning to elongate to just before blooming	40 to 50% leaves[†]; green; less than 5% foreign material; free of mold, musty odor, dust, etc.	>19	<31	<40	>140
2. Legume hay	Early bloom	Early to mid-bloom; stage between initiation of bloom and stage in which ½ of the plants are in bloom	35 to 45% leaves[†]; light green to green; less than 10% foreign material; free of mold, musty odor, dust, etc.	17–19	31–35	40–46	124–140
3. Legume hay	Mid-bloom	Mid- to full bloom; stage in which ½ or more of plants are in bloom	25 to 40% leaves[†]; yellow green to green; less than 15% foreign material; free of mold, musty odor, dust, etc.	13–16	36–41	47–51	101–123
4. Legume hay	Full bloom	Full bloom and beyond	Less than 30% leaves[†]; brown to green; less than 20% foreign material; free of musty odor, etc.	<13	>41	>51	<100

6. Sample grade[‡]

Hay that contains more than a trace of injurious foreign material (toxic or noxious weeds and hardware) or that definitely has objectionable odor or is under-cured, heat-damaged, hot, wet, musty, moldy, caked, badly broken, badly weathered or stained, extremely overripe, dusty, that is distinctly low-quality, or contains more than 20% foreign material or more than 20% moisture.

*Chemical analyses expressed on dry matter basis. Chemical concentrations based on research data from north central and northeast states and Florida. Dry matter (moisture) concentration can affect market quality. Suggested moisture levels are: Grades 1 and 2 <14%, Grade 3 <18%, and Grade 4 <20%.
†Proportion by weight.
‡Slight evidence of any factor will lower a lot of hay by one grade.
CP = crude protein; ADF = acid detergent fiber; NDF = neutral detergent fiber; relative feed value is based on digestible dry matter intake.
Source: May Marketing Task Force.

TABLE 14-5
PROPOSED MARKET HAY GRADES FOR GRASSES AND GRASS-LEGUME MIXTURES

Grade	Stage of maturity (international term)	Definition	Physical description	Typical chemical composition, %*			Relative feed value, %
				CP†	ADF	NDF‡	
2. Grass hay	Prehead	Late vegetative to early boot; stage at which stems are beginning to elongate to just before heading; 2 to 3 weeks' growth§	50% or more leaves**; green; less than 5% foreign material; free of mold, musty odor, dust, etc.	>18	<33	<55	124–140
3. Grass hay	Early head	Boot to early head; stage between late boot where inflorescence is just emerging until the stage in which ½ inflorescences are in anthesis; 4 to 6 weeks' growth§	40% or more leaves**; light green to green; less than 10% foreign material; free of mold, musty odor, dust, etc.	13–18	33–38	55–60	101–123
4. Grass hay	Head	Head to milk; stage in which ½ or more of inflorescences are in anthesis and the stage in which seeds are well-formed but soft and immature; 7 to 9 weeks' regrowth§	30% or more leaves**; yellow green to green; less than 15% foreign material; free of mold, musty odor, dust, etc.	8–12	39–41	61–65	83–100

5. Grass hay	Posthead	Dough to seed; stage in which seeds are of doughlike consistency until stage when plants are normally harvested for seed; more than 10 weeks' growth§	20% or more leaves**; brown to green; less than 20% foreign material; slightly musty odor, dust, etc.	<8	>41	>65	<83

6. Sample grade¶

Hay that contains more than a trace of injurious foreign material (toxic or noxious weeds and hardware) or that definitely has objectionable odor or is under-cured, heat-damaged, hot, wet, musty, moldy, caked, badly broken, badly weathered or stained, overripe, dusty, that is distinctly low-quality, or contains more than 20% foreign material or more than 20% moisture.

*Chemical analyses expressed on dry matter basis. Chemical concentrations based on research data from north central and northeast states and Florida. Dry matter (moisture) concentration can affect market quality. Suggested moisture levels are: Grade 2 <14%, Grade 3 <18%, and Grades 4 and 5 <20%.
†Fertilization with nitrogen may increase CP concentration in each grade by up to 40%.
‡Tropical grasses may have higher NDF concentrations than indicated in this table.
**Proportion by weight.
§For grasses that do not flower or for which flowering is indeterminant.
¶Slight evidence of any factor will lower a lot of hay by one grade, except Grade 5.
CP = crude protein; ADF = acid detergent fiber; NDF = neutral detergent fiber; relative feed value is based on digestible dry matter intake.
Source: Hay Marketing Task Force.

depends upon (1) the species grown; (2) the stage of maturity when harvested; (3) the leaf content; and (4) in grasses, the level of nitrogen fertilizer applied.

Recently, the American Forage and Grassland Council proposed various hay grades for legumes and grasses. The proposed hay grades for legumes and legumes mixtures are presented in Table 14-4, and those for grasses and legume-grass mixtures in Table 14-5.

Digestibility and intake data based upon cattle digestion are presented in Table 14-6. Dairy cattle, finishing steers, and beef cows have been used to produce these data. One can also use the same value of the hay for horses, even though they are not ruminants.

MACHINERY EVOLUTION

The greatest challenges to successful forage harvesting are: (1) the amount of labor involved; (2) weather problems, and (3) the quality of the final product.

Prior to World War II, most hay was put up in loose form or stacked, a method involving a mower, a rake, and a great deal of hand labor. Then, also before the war, the stationary baler was developed; soon after that, several significant innovations in machinery brought about a number of changes in the harvesting, storage, and utilization of forage materials. In the 1940s the pickup baler was developed, soon followed by the self-tie pickup baler, which reduced

TABLE 14-6
TYPICAL DIGESTIBLE DRY MATTER (DDM), DRY MATTER INTAKE (DMI), AND DIGESTIBLE DRY MATTER INTAKE (DDMI) VALUES FOR PROPOSED MARKET HAY GRADES DESCRIBED IN TABLES 29-4 AND 29-5*

	Legume hays			Grass hays			
Grade	In vivo DDM, %	DMI, $g/W\ kg^{0.75\dagger}$	DDMI, $g/W\ kg^{0.75}$	In vivo DDM, %	DMI, $g/W\ kg^{0.75}$	DDMI, $g/W\ kg^{0.75}$	Relative feed value$\ddagger$
1	>69	>140	>96	—	—	—	>137
2	66–68	129–139	85–95	>65	>121	>79	120–136
3	59–65	120–128	71–83	62–64	116–121	72–77	102–119
4	<58	<120	<70	58–61	108–115	63–71	90–101
5	—	—	—	<58	<108	<63	<90

*Formulas used to calculate relative feed value: Alfalfa——DDM=34.1080 plus 2.6429 ADF% − .0499 ADF%^{2}; DMI=146.9547 plus 1.0137 NDF% − .0302 NDF%^{2}; grass——DDM=6.1070 plus 3.9963 ADF% − .0663 ADF%^{2}; DMI=9.7914 plus 4.8171 NDF% − .0508 NDF%^{2}; DDMI=DDM × DMI/100; relative feed value=DDMI × 1.4286; where DDM=in vivo dry matter digestibility; DMI=dry matter intake; DDMI=digestible dry matter intake.
$\dagger$g/W kg $^{0.75}$= grams per metabolic weight of an animal.
$\ddagger$Relative feed value is an estimate of overall forage quality. It is calculated from intake and digestibility of dry matter when forages of known composition were fed to cattle. The values are relative; however, they are equally appropriate for all classes of livestock. Relative feed value estimates the intake of digestible energy when the forage is the only source of dietary energy and protein.

the amount of labor involved and created a portable package, the bale. Later, the small round baler was developed; this too reduced labor costs and allowed for a portable package, but added the potential for a package that could be stored in the field.

There was not much done in the way of different methods of packaging or storage until the large round baler was developed in the 1970s. The advantage of the large round baler over the small round baler was the larger package; this also reduced the labor requirement per bale and total field storage losses. The hay cuber was also developed in the 1970s, on a field scale larger than that used previously. This was a fully mechanized, highly portable package, but it was very specific for use in the far west, in low-humidity, low-rainfall areas. The high-density baler, recently developed, also makes a very portable package; it, too, is utilized in low-rainfall areas. A more recent development is the large square baler. More preservatives are also being used in baling higher-moisture hay, along with plastic covers.

In addition to the various balers, another development is the automatic bale pickup wagon, which has improved the pickup and portability of hay packages. Other packaging and storage methods that have been developed since World War II are conditioners and dehydrated forage pellets. The later method provides a highly portable product that facilitates getting the forage material from the producing area to the consuming livestock. Finally, in 1980, the silage bag, which is a large plastic tube, was developed, making it cheaper for growers to start a silage handling program.

Many of the hay harvesting methods developed since World War II have reduced the cost of labor and improved the portability and quality of the final product, but they have increased the capital cost. Most of these innovations have been developed by private individuals instead of by the larger commercial machinery firms. The machinery firms, however, have improved the machines and made them much more accessible to many more growers than before.

TYPES OF FORAGE STORAGE

There are numerous methods of storing forage. Each involves different moisture levels in the forages, different types of final package, and different handling methods (2, 5, 6, 10, 23, 25, 26).

Green Chop

Green chop involves cutting and chopping the forage in the field and hauling it directly to the cattle. No storage is involved other than where the cattle are fed, such as the bunker, or feeding wagon. There is a daily harvest, and no selective grazing is involved. All material that is grown in the field is harvested and taken directly to the animals. The moisture content of green chop is essentially the same as that of the crop in the field, ranging from approximately 70 to 80%, depending upon the forage species.

Silage

Forage harvested for silage is cut and then left in the field for a rather short period of time. Silage is stored in upright cylindrical structures or horizontal bunkers that may be cut into a hillside or into the ground. The moisture content of the final product may range anywhere from 65% for wilted silage, to about 55% for low-moisture silage (29).

Silage is cut and chopped in the field, then hauled to a silo, at which point the high moisture content makes it possible to effectively compact the forage and exclude oxygen. Within a very few hours after the forage is put into the silo, aerobic processes stop; then anaerobic conditions take over and lactic acid is produced. If forage grasses and/or legumes are harvested as direct-cut silage, the moisture content will be too high, preventing proper ensiling. In addition, excessive seepage occurs, resulting in the loss of soluble nutrients. Much of the harvesting and storage of silage is mechanized, from harvest in the field to unloading.

Haylage

Haylage is often called low-moisture silage (5, 10, 25, 29). In fact, moisture content is the only difference between silage and haylage. The moisture content of haylage ranges from around 40 to 55%. Haylage requires a special type of structure that is oxygen-limiting so that one can exclude all oxygen. If oxygen is not excluded and the forage begins to heat up, spontaneous combustion may occur, creating a flame. Haylage is harvested, stored, and handled in a manner almost identical to that for silage, other than in relation to the moisture content.

Hay

Hay is harvested and cured in the field, then stored either in a bale or in a loose or compressed stack. The moisture content of field-cured hay ranges from 12 to 22%, except in high-moisture hay, where the moisture content may be up to 30%, with added preservatives (23). If hay is harvested and stored in bales at too high a moisture content without adding preservatives, in the presence of oxygen, spontaneous combustion may occur. Some hay is chopped and stored in overhead storage.

Harvesting and storing forage in the form of hay involves labor, both mechanical and manual. Labor is involved in removing the hay from the field, storing it, and feeding it to animals. Many types of dry matter loss may occur in field harvesting as a result of poor weather conditions; harvesting at reduced moisture levels, which causes increased leaf shattering; handling the forage in storage; and losses during the feeding process.

Cubes or Pellets

Compared with other storage methods, field cubing involves the lowest moisture level at hay harvest time. Most forage ranges from 8 to 10% moisture or less (2, 6, 7); therefore field curing is involved. Most cubing and pelleting is done in the far west under low-humidity, low-rainfall conditions.

Following cutting, crimping, and windrowing in the field, hay is left until the moisture level is approximately 10% or less; a small amount of moisture is added when the hay is cubed. The process involves mechanization at the point where the hay is pushed through a sieve-like structure, forming a very compact, dense cube. From that step until the final feeding process, there is complete mechanization, using dump trucks, wagons, and conveyors to the feed bunkers.

Dehydration

Dehydration has been associated primarily with alfalfa and bermudagrass. Most of the moisture in forage is removed at a constant rate during dehydration. Dehydrators are rotating drums that have a heat source for removing the moisture rapidly. Following the removal of moisture, the forage goes through a series of hammermills or grinders. The dry material then goes into a pellet mill, where it is forced through various die openings to form a pellet. It is essential to supply some steam or moisture to condition the raw material so that a pellet may be formed. Since pellet formation involves a considerable increase in temperature, the pellets are then cooled, and put into a tank for storage. In the storage tank an inert gas is provided so that the oxygen is removed, ensuring that the products of any combustions that may take place will be nonoxidizing (6, 21).

If pellets are not formed following the grinding step, there may be a separation by means of bulk density so that one can end up with dry ground meal with different protein levels. The final product, alfalfa meal, can then be stored in inert gas (2, 6, 17).

Wet Processing

In recent years, procedures have been developed to remove the juices of alfalfa and to process this juice into a liquid protein component (6, 21). An edible protein that can be used as food for humans has recently been developed by such procedures. The pressed keg, or by-product of the alfalfa, can then be utilized as a high-moisture silage, or it can be dried down further to make a dry meal. The recently developed Pro-Xan Process involves the removal of the alfalfa juices, followed by a dehydration procedure (1, 7). Several steps are involved in this process: (1) removal of the juices; (2) dehydration of the dewatered alfalfa; (3) coagulation and separation of the protein and xanthophyll in the juices; and (4) drying down the coagulum. The product at this point is a juice that contains approximately 8 to 12% solids. It is then heated to 70°C (158°F) or higher to coagulate the protein in the form of a green curd. After pressing, the green material, which contains about 40% solids, is dried and ground into a product called Pro-Xan. A brown-colored material containing amino acids, sugars, minerals, and vitamins is then separated from the green curd. This is an excellent feed supplement (6). In addition, the alfalfa leaf protein itself can be used to feed humans.

FORAGE HARVESTING EQUIPMENT

There are numerous types of equipment involved in forage harvesting and storage. Mowers are used to cut standing forage and are the primary method of harvesting most forages. Most mowers currently used are combined with a crimper or conditioner and a windrower, thereby allowing the forage to be cut, crimped, and placed in a windrow in a single step (Figs. 14-2 and 14-3). Recently, rotary mowers have been developed with fewer moving parts than had been the case previously. Some of these are combined with a crimper and windrower unit.

Rakes are necessary if the mower and crimper do not have a windrow attachment. If the forage is cut, crimped, and left in the field to dry, one must use a side-delivery-type rake or a reel-type rake to make a windrow for pickup purposes (Fig. 14-4). The windrows that are developed from such rakes are continuous and well-adapted for pickup with choppers or balers. "Once-over" equipment for mowing, crimping, and windrowing increases hay quality since fewer leaves are lost than there would be if these processes had to be repeated.

In areas where the forage is sparse and loose hay is put up, there are various types of machines available to pick up the windrow and form a stack. In recent

FIGURE 14-2
Mower-conditioner is an ideal combination.

FIGURE 14-3
Self-propelled windrower, designed for large-scale haying operations that require uniform, fast-curing windrows.

years, automatic haystackers have been developed to pick up the forage from the windrow and blow or compact the loose hay into a form-type stack ranging from 1 ton to 4 tons and larger in size (Fig. 14-5). The forage is blown into a hydraulic system that compacts it into a high-density-type stack.

Balers have been used for many years. Balers pick up forage from a windrow and press it into a rectangular bale, tying it with wire, twine, or plastic-type

FIGURE 14-4
Twin rakes allow one to rake twice as much as with a single rake, as well as making a wide windrow for a large round baler.

FIGURE 14-5
A stack wagon builds 3.3 × 2.3 × 2.6 m (10 × 7 × 8 ft) high hay or stover stacks weighing up to 1½ tons.

FIGURE 14-6
Automatic baler with bale thrower and wagon.

FIGURE 14-7
Balers can be used to make bedding for cattle or low-energy roughage.

material (Fig. 14-6). There is a wide range of bale sizes and densities available (Fig. 14-7). Most of the smaller rectangular bales reach weights of 25 to 35 kg (55 to 77 lb). Larger bales are available at higher densities; these involve three wires, and their weights may reach approximately 60 kg (130 lb) or more. Recently, large square balers have been developed, resulting in bales weighing 455 to 700 kg (1000 to 1500 lb) or more.

FIGURE 14-8
Large round baler with twine tie.

FIGURE 14-9
New Holland round baler featuring autowrap and continuous PTO operation. A full-bale alarm in the tractor cab signals the operator when the bale reaches full size and autowrap engages.

In the mid-1970s, the large round bale was developed (Fig. 14-8). Large round bales are very popular in cow-calf producing areas. Some dairy farmers may use the large round bale to add to their feeding systems. Large round bales range in weight from 400 to 700 kg (900 to 1500 lb), depending upon the crop and baling conditions (Fig. 14-9).

Bale movers and automatic pickups are available for both conventional

FIGURE 14-10
New Holland automatic bale wagon is a 3-bale-wide, 4-ton-pull-type unit. This unit will unload single bales into a conveyor or into feeding barriers along corrals.

FIGURE 14-11
Picking up and moving large round bales.

rectangular bales and large round bales. Automatic bale wagons range from the completely mechanized, self-propelled type to the pull type (Fig. 14-10). Most models that pick up rectangular bales from the field include a stacker or an unloader attachment. Pick rigs for large round bales may involve transportation

FIGURE 14-12
Attachment to move large round bales.

FIGURE 14-13
Moving large round bales.

primarily, picking up the bale from the field and transporting it to a stack or storage area (Figs. 14-11, 14-12, 14-13, and 14-14).

Field choppers, or silage harvesters, have a chopper and blower plus a wagon (Figs. 14-15 and 14-16). Chopped forage is blown into the wagon and transported to a forage blower for the silo-filling process, or dumped directly from the wagon into a bunker-type silo (Fig. 14-17).

FIGURE 14-14
Trailer used to move large round bales.

FIGURE 14-15
Pull-type forage-silage harvester.

Tub grinders, utilizing large round bales or conventional bales, are being used increasingly to aid in the feeding or mechanization process (Fig. 14-18). In addition, there are stack feeders that are used to automatically feed a large round bale or stack and place the hay in a feed bunk (Fig. 14-19).

The cubing process has been popular in the far west. As indicated earlier, the forage must be dried to less than 10% moisture in the field before a field cuber can be used. The cubes are then transported by means of a dump truck or trailer to a storage area, or fed directly to cattle in a bunker-type arrangement. Cubes are convenient for transporting forage within the United States or for overseas

FIGURE 14-16
Self-propelled forage-silage harvester, shown with corn head.

FIGURE 14-17
Forage blower, blowing silage into an airtight silo.

trade. Cubes can be transported easily by truck or rail car to the final area of consumption.

SILAGE

Silage has been gaining in popularity since the 1940s. Silage is the product of controlled anaerobic fermentation of green forage. It is stored in silos or cylindrical containers, bunkers, or "silopress" (a new development using plastic tubes), sometimes in the absence of air and sometimes in the presence of a small concentration of oxygen (Figs. 14-20 and 14-21). The process is called ensiling (3, 19, 29), and it occurs within the container where fermentation takes place. As indicated earlier, there are several types of silage: (1) high-moisture, or direct-cut, silage, in which the forage material going into the silo is 70% or more moisture; (2) wilted silage, in the range of 40 to 70% moisture; and (3) low-moisture silage, or haylage, which is from 40 to 55% moisture.

The ensiling process is controlled by several factors: (1) the makeup of the

FIGURE 14-18
Tub grinder (from top left, clockwise): with conventional bales; with large round bales; large round bale in the field; final product showing the fineness of grind; and being unloaded from a silage wagon.

plant material that goes into the silo; (2) the amount of oxygen or air that is within the silage or allowed to enter the silage material; and (3) the type of bacteria that is on the plant material going into the silo.

The forage material that is chopped and placed in a silo is in small pieces that continue to respire after being placed in the container. While the forage material is respiring and oxidizing, it is using up sugars and oxygen. The oxidization process gives off carbon dioxide, water, and heat (3, 29).

The forage material, along with the bacteria present on it, are placed in the silo. The bacteria may continue to exist for a while and multiply under these conditions while the oxygen is present. The oxygen-preferring bacteria grow and multiply until they use up the oxygen; the anaerobic bacteria then begin to develop and multiply. As this process continues, the bacterial action slows down and silage is eventually developed, forming a fairly stable product in the silo. The completion of the fermentation process takes approximately 21 days (29).

There are several steps involved in the fermentation process: (1) materials are

FIGURE 14-19
A round bale unloader or feeder, feeding bales to cattle on dry lot.

FIGURE 14-20
Silage storage involves concrete stave silos and airtight sealed structures.

FIGURE 14-21
Pressing silage into a plastic tube.

placed into the silo; (2) the plant cells continue to respire; and (3) oxygen is consumed. Carbon dioxide and heat are produced, and the temperature rises from approximately 21°C (69 to 70°F)—the temperature at which plant materials go into the silo—generally to approximately 32°C (90°F), and then decreases over time until the final product, at the end of 20 to 21 days, is approximately 29°C (84°F) (29).

The pH of the plant material going to the silo is around 6.0. Acetic acid is the first acid produced, from day 1 until about 4 days later, when the pH begins to drop, eventually reaching approximately pH 4.2. Acetic acid bacteria decrease at around the end of day 4, then lactic acid begins to form. During day 3 or 4, acetic acid formation essentially stops. Lactic acid formation continues for approximately 2 more weeks (19). Temperatures continue to decline, and bacterial action stops when the pH is lowered to around 4.0 (19, 29, 34). The last phase is the constant product of silage that remains. If no additional air or oxygen is present, the silage will remain constant almost indefinitely, but if oxygen is allowed to seep into the silo, the material continues to break down, or oxidize (29). If insufficient amounts of lactic acid are formed, then butyric acid begins to be produced or protein will be broken down continually, and spoilage will occur (19, 34). Good-quality silage material possesses high concentrations of acetic and lactic acids. Poor-quality silage, with a final pH above 4.2, contains butyric acid as well as high levels of succinic acid and some propionic and formic acids (19).

The maintenance of a low temperature, below 38°C (100°F), is very desirable for making quality silage. At temperatures above 38°C, a sweet, aromatic, tobacco-smelling, dark-brown caramelized silage is formed. The palatability of this material is fairly high, but its protein is less available than is that in quality silage. In addition, if temperatures rise considerably higher than 38°C and if air is allowed to seep into the silo, spontaneous combustion may occur (14).

The moisture content of silage that is to be harvested and stored in most conventional silos should be in the range of 50 to 65%. In airtight or oxygen-limiting structures, the best moisture level for making low-moisture silage, or haylage, is in the 40-to-50% range. When silage is too dry in conventional silos, it is difficult to remove all of the air, so molds and other organisms continue to grow and create spoilage. Conversely, if the material is too wet, creating a lot of seepage, and if air is limited, the plant cells are often smothered before the temperature warms enough for lactic acid to form; as a result, butyric acid begins to form, and a rancid odor develops (11, 19). The nutritive values of various forage silages are presented in Table 14-7).

It is important to get the proper length of cut, compaction, and distribution of the material in the silo, so that the proper amount of air is available to allow acetic and lactic acid formation to begin. It is recommended that the material be both chopped finely and distributed properly throughout the silo. Compaction is probably sufficient in upright silos. Compaction is very important in bunker or horizontal silos, where one must remove as much oxygen as possible.

Many silage additives are available. The two general types of additives for silage are: (1) nutritive additives, such as grains, feeds, and molasses; and (2) chemical, or nonnutritive, additives, such as lactic acid, mineral acid, and formic acid. One would place feed additives with a legume or legume-grass silage primarily for the purpose of adding a carbohydrate for the bacteria in the fermentation process. One would use a feed additive only with a high-moisture legume or grass silage, to ensure a quality product. Some of the grain or feed

TABLE 14-7
NUTRITIVE VALUE OF VARIOUS FORAGE SILAGES

Forage silage	On dry matter basis, %					
	Dry matter	TDN	Protein	Fiber	Ca	P
Corn silage	35	70	8.2	25	0.28	0.20
Alfalfa haylage (early)	50	57	18.5	30	1.25	0.23
Alfalfa haylage (late)	50	57	16.0	34	1.28	0.20
Grass-legume haylage	50	54	15.5	33	1.52	0.37
Oatlage	50	59	9.7	32	0.37	0.30
Forage sorghum	35	56	7.7	35	0.30	0.24
Grain sorghum	35	57	7.3	26	0.25	0.18
Sudangrass	30	59	5.6	35	0.64	0.23

additives that one might add are ground corn or small grains, at the rate of 100 to 300 kg/t (110 to 330 lb/T), depending on the moisture level (19, 29). Molasses can be used at the rate of 20 to 40 kg/t (44 to 88 lb/T) for green forage (19, 29).

Many chemical additives are also available, but results vary widely. Most data indicate that chemical additives do not greatly improve either the preservation of silage or its feeding value (19, 29). The key factors in developing a good-quality silage are storing the material at the proper moisture level and removing oxygen (29).

Total losses with a final product harvested for either hay or silage can be estimated. Earlier work was conducted in Michigan to illustrate the field and storage losses with a legume-grass forage (12). The least amount of total forage losses occurs under the low-moisture, or haylage, condition, as shown in Figure 14-22. There have been numerous trials reported on the percent losses from silage. Using conventional silos, the average monthly loss is approximately 3% (9, 12). The total loss is then determined by the number of months for which the material is stored until feeding. Using limited-air or exclusion-type structures, the total loss from storage until feeding is approximately 4% (12).

Heat affects the feeding value of silage (Table 14-8). Heat and silage color have essentially no effect on the crude protein content, as determined by the Kjeldahl procedure, but digestibility of the protein declines, as measured by the

FIGURE 14-22
Estimated total field and harvest losses and storage losses for grass-legume forages harvested at various moisture levels and with various methods (12).

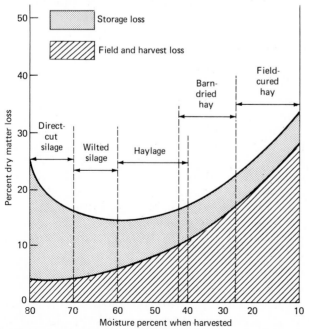

TABLE 14-8
THE EFFECT OF HEAT ON FEED
VALUE OF SILAGE

Color	Crude protein, %	Digestible protein, %
Normal	21.5	67
Brown*	21.0	17
Black*	21.0	3

*Silage heated to at least 60°C (140°F).

detergent analysis method developed by Van Soest (30, 31). Normal silage is approximately 67% digestible protein, whereas black, extremely heated silage is around 3% digestible protein (30).

Caution in Silage Making

Gases that form during silo filling and for a period thereafter can accumulate in sufficient quantity to cause sickness or death. The gases formed in the silage process are carbon dioxide and nitric oxide, which in turn produce nitrogen dioxide (18, 20). Carbon dioxide is a nonpoisonous gas, but it can cause suffocation as a result of lack of oxygen. Carbon dioxide is the most common gas formed at first. It is colorless and odorless. It can be detected by lowering a lighted cigarette lighter, lantern, or candle to the level of the silage. If the flame goes out, the oxygen content is low enough to possibly be dangerous.

Another silage gas formed is nitrogen dioxide (NO_2). This gas is reddish-yellow, with a pungent odor. It is heavier than air and will accumulate during the filling of the silo and up to 6 to 10 days later. One of the first indications of nitrogen dioxide is a laundry-bleach smell (18, 20). Upon further oxidation of nitrogen dioxide, nitric acid (N_2O_5) is formed; in combination with water, this is a highly corrosive acid. When such oxidation occurs within the body, permanent lung damage results (19, 29). Nitrogen dioxide is heavier than air and will remain below the air mass or just above the top of the silage for a number of days.

When the fermentation process begins, oxygen is being used, and nitrates present in the plant tissue are released as nitric oxide (NO). Nitrogen dioxide is produced from fermented plants that contain free nitrate (NO_3) that has not been converted into protein. Some plant species contain more nitrates than do others. When such plants are ensiled, their nitrate may be converted to nitrogen dioxide gas, or it may be lost in the seepage of material. Nitrogen dioxide may escape during fermentation, although nitrate may remain in the silage at a level that may potentially poison livestock (29). Weeds often contain levels of nitrate much higher than those in forage species (29). It has been shown that the maximum percentage of stored nitrates in weeds is 7.5% of the dry matter; in corn and sorghums, approximately 5.0%; in small grains, 4.0%; in perennial grasses, 3.0%; and in legumes, 1.0% (19, 34). Therefore, a grass-legume silage

TABLE 14-9
APPROXIMATE STORAGE CAPACITY FOR CORN SILAGE
IN HORIZONTAL SILOS, WITH SIDES SLOPED OUTWARD
1½ IN PER FT OF DEPTH
(Approximate Tons per Ft of Length)

Bottom width, ft	Depth, ft				
	8	10	12	16	20
20	3.1	4.0	—	—	—
30	4.6	5.9	7.1	9.6	—
40	6.1	7.7	9.3	12.6	16.0
50	7.6	9.6	11.6	15.6	19.8
60	—	11.5	13.8	18.6	23.6
70	—	—	16.1	21.6	27.4
80	—	—	18.3	24.6	31.0
100	—	—	—	30.6	38.6

Source: E. G. Stoneberg, F. W. Schaller, D. O. Hull, V. M. Meyer, N. J. Wardle, N. Gay, and D. E. Voelker, "Silage production and use," *Iowa State Univ. Pam. 417* (1968).

is less likely to produce nitrogen dioxide gas than are other ensiled crops. Weeds in a crop that is being ensiled may produce more nitrogen dioxide than the crop, so caution must be taken when harvesting excessively weedy fields for silage.

Several steps that one might take to minimize the danger of any of these silage gases are: (1) Stay out of the silo immediately after it is filled, at least for several days. If you must enter the silo, first run the silage blower for 15 to 20 minutes to help expel the gases. (2) Be alert for any bleach-like odor or a yellowish-brown gas that might be seen near the silo. In conventional silos, the silo room should be ventilated for at least 2 weeks after filling because the heavier gases may seep out of the silo and into the silo room, where unloading will occur. (3) Make sure that the doors between the silo room and the barn are closed to prevent any of

TABLE 14-10
APPROXIMATE STORAGE CAPACITY FOR CORN SILAGE IN UPRIGHT SILOS AT 70% MOISTURE
(Approximate Tons)

Silo height, ft	Silo diameter, ft								
	12	14	16	18	20	22	24	30	36
30	68	92	121	151	186	225	268	—	—
40	100	135	177	224	276	332	394	617	900
50	133	183	238	302	373	452	538	840	1210
60	—	234	306	387	478	579	689	1076	1560
70	—	—	—	480	592	716	852	1332	1930
80	—	—	—	—	—	—	1030	1600	2300

Source: E. G. Stoneberg, F.W. Schaller, D. O. Hull, V. M. Meyer, N. J. Wardle, N. Gay, and D. E. Voelker, "Silage production and use," *Iowa State Univ. Pam. 417* (1968).

TABLE 14-11
APPROXIMATE STORAGE CAPACITY FOR HAYLAGE IN UPRIGHT SILOS AT 50% MOISTURE
(Approximate Tons)

Silo height, ft	Silo diameter, ft								
	12	14	16	18	20	22	24	30	36
30	41	55	72	90	111	134	160	—	—
40	60	81	106	134	165	198	235	368	540
50	80	109	142	180	223	270	321	500	725
60	—	140	183	231	286	345	410	643	940
70	—	—	—	286	353	426	508	794	1160
80	—	—	—	—	—	—	618	960	1380

Source: E. G. Stoneberg, F. W. Schaller, D. O. Hull, V. M. Meyer, N. J. Wardle, N. Gay, and D. E. Voelker, "Silage production and use," *Iowa State Univ. Pam. 417* (1968).

these gases from entering the barn and killing livestock. Do not allow children or strangers to enter the area where the silage process is occurring—the area should be completely barricaded. If one feels as though he or she might have been exposed to silage gas, that person should see a doctor immediately to prevent any potential lung damage and help reduce the chance that pneumonia will develop.

Estimating Silage Storage

It is somewhat difficult to estimate the amount of silage that can be stored in different types of structures. The amount of silage needed for various types of livestock depends upon the overall feeding program and the type of silage one might store at different levels of moisture. Tables 14-9, 14-10, and 14-11 show the approximate capacity for corn silage in a horizontal structure and an upright structure, and the capacity for haylage in an upright structure, respectively (29).

QUESTIONS

1 What are the various methods of harvesting forage crops?
2 Give the approximate moisture level for each storage condition.
3 What is the advantage of hay harvest and storage over silage? What are the disadvantages of hay harvest and storage compared with silage?
4 Describe the on-the-spot indicators for quality hay.
5 How does maturity affect hay quality?
6 List the various factors that determine protein content of forages.
7 Over a 24-hour period in the upper midwest, what is the chance of harvesting forages without having rain damage?
8 Review the various methods of harvesting forages in relation to the equipment involved.

9 What is oxidization? How can it be prevented or reduced to a minimum?

10 Describe the ensiling process.

11 Are there any dangers in silage making? How would one prevent or detect any of the problems that may be encountered?

12 What are some precautions that one should follow when filling a silo?

13 Which method of harvesting forage would have the greatest amount of storage losses, the smallest amount of storage losses, the smallest amount of field losses, the greatest amount of field losses, and the smallest amount of total losses?

14 What effect does heat have on the feeding value of silage?

REFERENCES

1 Akeson, W. R. and M. A. Stahmann. Leaf protein concentrates: A comparison of protein production per acre of forage with that from seed and animal crops. *Econ. Bot.*, 20:244–250, 1966.

2 Barnes, K. K. Mechanization of storage harvesting and storage. In *Forages,* Iowa State Univ. Press, Ames, IA, 3d ed., pp. 522–531, 1973.

3 Barnett, A. J. G. *Silage Fermentation.* Academic Press, New York, 1954.

4 Bishnoi, U. R., P. Chitapong, J. Hughes, and J. Nishimuta. Quantity and quality of triticale and other small grain silages. *Agron. J.,* 70:439–441, 1978.

5 Byers, J. H. Comparison of feeding value of alfalfa hay, silage and low moisture silage. *J. Dairy Sci.,* 48:206–208, 1965.

6 Chrisman, J., G. O. Kohler, and E. N. Bickoff. Dehydration of forage crops. In *Forages,* Iowa State Univ. Press, Ames, IA, 3d ed., pp. 549–557, 1973.

7 Chrisman, J., G. O. Kohler, A. C. Mottola, and J. W. Nelson. High and low protein fractions by separation milling of alfalfa. *ARS Bull. 74–57,* 1971.

7 a Collins, M. The influence of wetting on the composition of alfalfa, red clover, and birdsfoot trefoil hay. *Agron. J.,* 74:1041–1044, 1982.

8 Goodrich, R. D. and J. C. Meiske. High-energy silage. In *Forages,* Iowa State Univ. Press, Ames, IA, 3d ed., pp. 569–580, 1973.

9 Gordon, C. H. Storage losses in silage as affected by moisture content and structures. *J. Dairy Sci.,* 50:397–403, 1967.

10 Gordon, C. H., J. C. Derbyshire, H. G. Wiseman, E. A. Kane, and C. G. Melin. Preservation and feeding value of alfalfa stored as hay, haylage, and direct-cut silage. *J. Dairy Sci.,* 44:1299–1311, 1961.

11 Henderson, A. R. and B. McDonald. Effect of formic acid on the fermentation of grass of low dry matter content. *J. Sci. Food Agr.,* 22:157–163, 1971.

12 Hoglund, C. R. Comparative losses and feeding values of alfalfa and corn silage crops when harvested at different moisture levels and stored in gas-tight and conventional tower silos: an appraisal of research results. *Mich. State Univ. Ag. Econ. Publ. 947,* 1964.

13 Keith, J. M., W. A. Hardison, J. T. Huber, and G. C. Graf. The effect of physical form on the nutritive value of hay to lactating dairy cows. *J. Dairy Sci.,* 44:1174, 1961.

14 Koegal, R. G. and H. D. Bruhn. Inherent causes of spontaneous ignition in silos. *Ann. Meet. Amer. Soc. Agr. Eng. Paper 69–164,* 1969.

15 Langston, C. W., H. Irvin, C. H. Gordon, C. Bouma, H. G. Wiseman, C. G. Melin, L. A. Moore, and J. R. McCalmont. Microbiology and chemistry of grass silage. *USDA Tech. Bull. 1187,* 1958.

16 Livingston, A. L., M. E. Allis, and G. O. Kohler. Amino acid stability during alfalfa dehydration. *J. Agr. Food Chem.,* 19(5):947–950, 1971.

17 Livingston, A. L., R. E. Knowles, J. W. Nelson, and G. W. Kohler. Xanthophyll and carotene loss during pilot and industrial scale alfalfa processing. *J. Agr. Food Chem.,* 16(1):84–87, 1968.

18 McCalmont, J. R. Farm Silos. *USDA Misc. Publ. 810,* 1960.

19 Noller, C. H. Grass-legume silage. In *Forages,* Iowa State Univ. Press, Ames, IA, 3d ed., pp. 558–568, 1973.

20 Peterson, W. H., R. H. Burris, R. Sant, and H. N. Little. Production of toxic gas (nitrogen oxides) in silage making. *J. Agr. Food Chem.,* 6:121–126, 1958.

21 Pirie, N. W. The production and use of leaf protein. *Proc. Nutr. Soc.,* 28:85–91, 1969.

22 Radolff, H. D. and K. D. Allison. Effect of physical form on alfalfa digestibility and productivity. *J. Dairy Sci.,* 54:773, 1971.

23 Reid, J. T. Quality hay. In *Forages,* Iowa State Univ. Press, Ames, IA. 3d ed., pp. 532–548, 1973.

24 Reid, J. T., W. K. Kennedy, K. L. Turk, S. T. Slack, G. W. Trimberger, and R. P. Murphy. Effect of growth stage, chemical composition, and physical properties upon the nutritive value of forages. *J. Dairy Sci.,* 42:567–571, 1959.

25 Roffler, R. E., R. P. Niedermeier, and B. R. Baumgardt. Evaluation of alfalfa-brome forage stored as wilted silage, low-moisture silage, and hay. *J. Dairy Sci.,* 50:1805–1813, 1967.

26 Slack, S. T., W. K. Kennedy, K. L. Turk, J. T. Reid, and G. W. Trimberger. Effects of curing methods and storage of maturity upon feeding value of roughages. *Cornell Agr. Exp. Sta. Bull. 957,* 1960.

27 Sotola, J. The nutritive value of alfalfa leaves and stems. *J. Agr. Res.,* 47:919–945, 1933.

28 Spencer, R. R., E. M. Bickoff, G. O. Kohler, S. C. Witt, B. E. Knuckles, and A. C. Mottola. Alfalfa products by wet fractionation. *Trans. Amer. Soc. Agr. Engr.,* 13(2):198–200, 1970.

29 Stoneberg, E. G., F. W. Schaller, D. O. Hull, V. M. Meyer, N. J. Wardle, N. Gay, and D. E. Voelker. Silage production and use. *Iowa State Univ. Pam. 417,* 1968.

30 Van Soest, P. J. Structural and chemical characteristics which limit the nutritive value of forages. *Amer. Soc. Agron. Spec. Publ. 13,* pp. 63–76, 1968.

31 Van Soest, P. J. Symposium on factors influencing the voluntary intake of herbage by ruminants: voluntary intake in relation to chemical composition and digestibility. *J. Anim. Sci.,* 24:834, 1965.

32 Waldo, D. R. Factors influencing the voluntary intake of forages. *Proc. Conf. Forage Qual. Eval. Util.,* pp. E-1–22, 1970.

33 Welch, J. G., M. Clancy, and G. W. Vandar Noot. Net energy of alfalfa and orchardgrass hays at varying stages of maturity. *J. Anim. Sci.,* 28:263–267, 1969.

34 Whittenbury, R., P. McDonald, and D. G. Bryant-Jones. A short review of some biochemical and microbial aspects of ensilage. *J. Sci. Food Agr.,* 11:441–446, 1967.

34 a Weighart, M., J. W. Thomas, M. B. Tesar, and C. M. Hansen. Acceleration of alfalfa drying in the field by chemical application at cutting. *Crop Sci.,* 23:225–229, 1983.

35 Woodward, J. E. Making silage by the wilting method. *USDA Leaflet 238,* 1944.

CHAPTER **15**

FORAGES FOR DAIRY, BEEF, HORSES, SHEEP, AND SWINE

Forages play a very important role in animal production in the United States. The forage-rumen, or ruminant, complex is an animal food-production system that involves the transformation of radiant energy through photosynthesis into plant protein, carbohydrates, fats, and other compounds that are found in forage tissue. The forage tissue is then converted by the ruminant animal into the chemical energy found in foods, red meats, and milk, or the energy used for animal power. Forage production, along with its utilization by the ruminant animal, is a highly efficient and economical system of converting energy into human food. Much of the land of the United States that is now well-suited for cropland is used for forages, thus making the overall utilization of land very efficient. In addition, some highly productive soils, where row or grain crops could be grown, have forage crops growing on them. These forages, too, are utilized by ruminant animals and eventually for human food.

The average consumption of food by a U.S. citizen is approximately 645 kg (1420 lb) per year (3). Approximately 42% of this food is of animal origin. About one-third of the average citizen's food supply, or 212 kg (467 lb), is provided by ruminants (3). This is a very important source of protein, energy, nutrients, and minerals for humans. Approximately 23% of our total energy consumption and 45% of our protein are supplied by beef and dairy products (3, 51). At present, there are about four times as many beef animals as dairy animals in the United States. Beef animals provide one-third of our fat supply and are also very important sources of calcium and phosphorus (3, 51).

Forages are very important components of the feed supply for ruminant animals. Approximately 90% of the feed units consumed by ruminants in the United States comes from forages. Concentrates play a significant part in the

239

ration for dry-lot animals and beef cattle, but forages remain very important in the overall feed-supply system. Considering the overall contribution of animals to the energy and protein in our diets, it is calculated that 18% of the energy and 36% of the protein in the American diet originate in forages (3, 51). No other commodity can approach forage in its relationship to the human diet in this country.

DAIRY CATTLE

Dairy animals are very dependent upon high-quality forages for normal maintenance, reproductive efficiency, and production of good quantities of milk. However, it is estimated that a dairy animal fed entirely on high-quality forage cannot maintain a milk output greater than 18 to 20 kg (40 to 44 lb) per day without losing body weight (34). Forages are considered low in energy concentration, and a lactating animal cannot consume enough forage to maintain its body weight and produce high quantities of milk. It has been reported that as little as 1 kg (2 lb) of concentrate per 6 kg (13 lb) of milk produced, fed along with a good-quality forage, may increase the yield of milk by 1.2 to 1.4 times compared with that of cattle fed entirely on forages (28, 34). Other researchers have indicated that by feeding 1 kg (2 lb) of concentrate per 4 kg (9 lb) of milk, milk production may be increased by 1.5 to 1.8 times as much as that of cattle fed entirely on good-quality forage (34).

The daily nutrient requirements of dairy animals are shown in Table 15-1. These values take into account that with the addition of concentrates to an average-quality forage, there will be no deficiencies in the animals' overall requirements, and the animals will provide average rates of milk production and reproduction (28). The average lactating dairy animal receives about 62% of her energy from forages and about 38% from concentrates (34). The average dairy animal, weighing up to 455 kg (1000 lb), consumes 11.5 to 13.5 kg (25 to 35 lb) of dry matter per day (34). Dairy animals weighing up to 560 kg (1300 lb) consume 13.5 to 18 kg (30 to 40 lb) per day, and approximately the same amount of silage or hay containing 35 to 60% dry matter (28). They usually do not eat as much high-moisture silage (20 to 35% dry matter) as low-moisture silage (50 to 60% dry matter) (34). A summary of the effect of forages' dry-matter digestibility on dairy animals' dry matter consumption, digestible energy consumption, and milk production is presented in Table 15-2. Note that at 65% DMD, the energy consumed is nearly 2.4 times the maintenance needed to adequately produce 19.2 kg (42 lb) of milk per day. This is approximately the present average production level of dairy cows in the United States. At 70% DMD, milk production is 29.2 kg (64 lb) per day or about 8200 kg (18,000 lb) per year. These data show the potential for increasing milk production in the United States.

From the standpoint of quality forage fed to dairy animals, alfalfa is the leading species. It is important to examine the potential animal productivity from this species and the impact of its various management factors for attaining the maximum possible dairy production. Forage quality is affected by the plant

TABLE 15-1
DAILY NUTRITIONAL REQUIREMENT FOR DAIRY COWS

Body weight, kg	Feed energy				Total crude protein, g	Calcium, g	Phosphorus, g	Vitamin A, 1000 IU
	NE_1, Mcal	ME, Mcal	DE, Mcal	TDN, kg				
Maintenance of mature lactating cows*								
350	6.47	10.76	12.54	2.85	341	14	11	27
400	7.16	11.90	13.86	3.15	373	15	13	30
450	7.82	12.99	15.14	3.44	403	17	14	34
500	8.46	14.06	16.39	3.72	432	18	15	38
550	9.09	15.11	17.60	4.00	461	20	16	42
600	9.70	16.12	18.79	4.27	489	21	17	46
650	10.30	17.12	19.95	4.53	515	22	18	50
700	10.89	18.10	21.09	4.79	542	24	19	53
750	11.47	19.06	22.21	5.04	567	25	20	57
800	12.03	20.01	23.32	5.29	592	27	21	61
Maintenance plus last 2 months of gestation of mature dry cows								
350	8.42	14.00	16.26	3.71	642	23	16	27
400	9.30	15.47	17.98	4.10	702	26	18	30
450	10.16	16.90	19.64	4.47	763	29	20	34
500	11.00	18.29	21.25	4.84	821	31	22	38
550	11.81	19.65	22.83	5.20	877	34	24	42
600	12.61	20.97	24.37	5.55	931	37	26	46
650	13.39	22.27	25.87	5.90	984	39	28	50
700	14.15	23.54	27.35	6.23	1035	42	30	53
750	14.90	24.79	28.81	6.56	1086	45	32	57
800	15.64	26.02	30.24	6.89	1136	47	34	61
Milk production—nutrients per kg milk of different fat percentages								
% fat								
2.5	0.59	0.99	1.15	0.260	72	2.40	1.65	
3.0	0.64	1.07	1.24	0.282	77	2.50	1.70	
3.5	0.69	1.16	1.34	0.304	82	2.60	1.75	
4.0	0.74	1.24	1.44	0.326	87	2.70	1.80	
4.5	0.78	1.31	1.52	0.344	92	2.80	1.85	
5.0	0.83	1.39	1.61	0.365	98	2.90	1.90	
5.5	0.88	1.48	1.71	0.387	103	3.00	2.00	
6.0	0.93	1.56	1.81	0.410	108	3.10	2.05	
Body weight change during lactation—nutrients per kg weight change								
Weight loss	−4.92	−8.25	−9.55	−2.17	−320			
Weight gain	5.12	8.55	9.96	2.26	500			

*To allow for growth of young lactating cows, increase the maintenance allowances for all nutrients except vitamin A by 20% during the first lactation and 10% during the second lactation.

Source: National Research Council, "Nutrient requirements of dairy cattle," Natl. Acad. Sci., Washington, DC, 1978.

TABLE 15-2

EFFECT OF FORAGE DRY MATTER DIGESTIBILITY ON DRY MATTER
CONSUMPTION, DIGESTIBLE ENERGY CONSUMPTION, AND MILK
PRODUCTION OF A 650-kg DAIRY COW*

DMD, %	DM intake			DE consumed, Mcal/day	Milk produced	
	% body wt.	kg/day	(lb/day)		kg/day	(lb/day)
50	1.54	10.0	(22)	22.0	1.4	(3.1)
55	1.86	12.1	(27)	29.3	6.5	(14.3)
60	2.13	13.8	(30)	36.5	11.5	(25.4)
65	2.56	16.6	(37)	47.6	19.2	(42.3)
70	3.10	20.2	(45)	62.0	29.2	(64.4)

*Based on 1978 National Research Council nutritional requirements for dairy cattle. Assume maintenance 650-kg cow 19.95 Mcal DE/day; 1 kg 4%-fat milk = 1.44 Mcal DE; no body weight changes during lactation.

species and the cultivar's maturity when harvested, and the extent of the losses at harvest and during storage.

The effects of maturity on alfalfa quality in relation to digestible energy consumption and potential milk production are presented in Table 15-3. These data are based on a yield of 8.3 t/ha (3.7 T/A) of alfalfa at bud stage. DMD decreases by 5.8% from the bud stage to the full-bloom stage. The digestible energy consumption of forages at these stages of growth differs by 28.8%; milk production decreases by 45.4%. The potential milk production per hectare decreases by approximately 3340 kg (7350 lb), which, at present-day milk prices of $12 cwt, represents a potential loss of more than $882 per hectare.

Taking the same yield potentials and considering losses accured during storage and harvest, producers may incur additional economic losses. Assuming that there are three different types of losses, the loss can be calculated on a

TABLE 15-3

EFFECTS OF MATURITY ON ALFALFA QUALITY, DIGESTIBLE ENERGY CONSUMPTION,
AND POTENTIAL MILK PRODUCTION*

Stage maturity	DMD, %	DE/AC, Mcal	DE/day consumed, Mcal	Milk produced			
				kg/day	(lb/day)	kg/ha	(lb/A)
Bud	68.0	10,000	56	25.0	(55)	11,053	(9859)
First flower	66.6	9794	51	21.4	(47)	10,131	(9037)
50% bloom	64.4	9470	46	18.2	(40)	9251	(8252)
Full bloom	62.2	9147	40	13.6	(30)	7714	(1881)

*Energy consumption and milk production per day for a 650-kg Holstein cow.

Sources: W. C. Weir, L. G. Jones, and J. H. Meyer, "Effect of cutting interval and stage of maturity on the digestibility and yield of alfalfa," *J. Anim. Sci.* 19:5–19 (1960); and J. G. Welch, C. Martin, and G. W. Vandernoot, "Net energy of alfalfa and orchardgrass hays at varying stages of maturity," *J. Anim. Sci.* 28:263–267 (1969).

TABLE 15-4
EFFECTS OF POSTHARVEST LOSSES ON ALFALFA QUALITY AND MILK
PRODUCTION*

DM loss level, %	DMD, %	DE/AC, Mcal	DE/day consumed, Mcal	Milk production			
				kg/day	(lb/day)	kg/ha	(lb/A)
10	60	7992	36.5	11.5	(25.4)	6219	(5547)
15	57	7171	31.2	7.8	(17.2)	4441	(3961)
25	50	5550	22.0	1.4	(3.1)	877	(782)

*Based on 8.3 t/ha at first flower and a 650-kg Holstein cow.
Source: C. R. Hoglund, "Some economic considerations in selecting forage systems for haylage and silage on dairy farms," *Mich. Agr. Econ. Rep. 14,* 1965.

hectare basis. Three different realistic percentage losses of digestible dry matter are presented in Table 15-4: (1) 10%, the best hay-making procedure with favorable weather during harvest; (2) 15%, forage harvested with at least one rain; and (3) 25%, with more than one wetting rain after harvest. As shown in Table 15-3, delaying the harvest of forage will compound the overall economic losses. Using these data, with good weather during harvest and with a very good hay-making procedure, approximately 56% of the potential milk production can be realized from all the digestible energy of standing forage. With poor weather during harvest, only 40% of the potential can be realized. These data indicate that at the present value of milk, with the losses shown in Table 15-4, $1400 per hectare will be lost under the poorest harvesting method compared with the best.

For top milk production, the average dairy ration should be about 65% digestible and should contain about 16% crude protein (34). High-quality forage contains enough protein to meet this requirement. Digestible dry matter is closely related to plant maturity and decreases by more than 2% each week on late-cut forage (34). A drop in dry matter digestibility decreases milk production and increases the amount of grain needed to maintain top production.

Every good dairy program should have an adequate supply of high-quality forage to feed during lactation. Alfalfa produces excellent hay, haylage, silage, or green chop for dairy cows. A 22.4-t/ha (10 T/A) yield of alfalfa produces as much total digestible nutrient as 350 bushels of corn or 10.5 tons of corn silage do. That yield of alfalfa also produces as much protein as 715 bushels of corn or 150 bushels of soybeans. The key to using any kind of forage and maintaining milk production is quality.

BEEF CATTLE

The beef cow is the basic unit of the beef industry. Probably the greatest cost in beef production is the annual maintenance cost of the beef cow and her calf. Most of her nutritional requirements can be supplied by good-quality pasture or stored forage. Beef gains can be produced at less cost from concentrates; therefore, to make an efficient operation, good pasture grasses and legumes should be used (51). The south, southwest, and western high plains are well-suited for growing cows and calves on a forage program.

A legume should be utilized in the pasture program, if possible, to increase its efficiency. This improves beef grains and production and lowers the cost of production. When utilizing grasses, nitrogen is required for excellent growth and quality forage.

Since grazing is a very important component of the program for beef, stocking rates, grazing intensity, rotational grazing, and limited grazing all affect overall production costs (Fig. 15-1). A light stocking rate increases daily gains, but it reduces the gain per hectare. Heavier stocking rates reduce individual animal gains but increase the overall herd gain per hectare. Pasture rotation permits one to graze younger, more digestible plants, and increases both the animal gains per day and the gain per hectare (4, 5, 6).

There are many managerial factors involved in a good cow-calf progrm for beef production that will not be discussed in this chapter. Many of these factors are influenced by the choice of the forage system. Perhaps the most important factor is the time of calving—whether in the fall, midwinter, spring, or on a noncyclic schedule. Each time period has its advantages and disadvantages, and the choice depends on the climate, soil types, or marketing conditions that best fit into one's farming program. It is much better to fit the calving operation into the forage program in relation to the best forage growth in a particular area than to fit the forage to the time of calving. The normal period of forage production is very important, although the growth period can be changed somewhat by applying fertilizer and, more recently, by various means of stockpiling or delayed grazing.

Many people believe that beef animals can produce excellent gains on lower-quality forage (51). This is not entirely true. To increase production and efficiency in beef cattle, high-quality forage is needed. It is true, though, that

FIGURE 15-1
Well-fertilized, productive, rotational pastures provide excellent grazing for stocker cattle.

much beef in the United States is produced on low-quality forage. Many producers use midwinter to spring calving programs to efficiently utilize low-quality forages or by-products that are available during the winter or fall months. When cool-season forages come into their peak production, the recommended energy intake level for the lactating animal is met; this level is also much better for rebreeding the brood cow.

The nutritional requirements for beef cattle are given in Table 15-5 (27). These recommendations are considered average requirements under average management situations and are not the same as those for animals under stress from disease, parasitism, or extremely cold conditions. It is important that the beef cow, if being fed low-energy residues, be fed high-quality forage, especially three weeks before calving. Bulls can be overfed, resulting in lower breeding efficiency. Beef cows very seldom need any protein supplementation because the ration needs to be only about 9.2 to 9.5% protein and the percent protein of forages very seldom falls below this value, except in very mature grass or low-quality crop residues (51).

For the finishing animal, a higher-quality forage is required. Suggestions are that a grass-legume mixture be used as a pasture, along with either a winter-annual forage species or corn silage. Very little finishing is done entirely on pasture. Many feeder animals are grazed up until approximately the last 90 to 100 days, then finished off in the dry lot. Research has shown that if beef is finished on pasture, it does not pay to feed unlimited grains during periods of high-quality forage (3). It has been shown that an efficient steer-calf program, wintered on accumulated pasture or harvested forage and concentrates, yields daily gains of 0.45 to 0.6 kg (1.0 to 1.3 lb) (3). The average daily concentrate intake should be approximately 1 kg per 100 kg of body weight for the grazing season.

Many beef cattle are finished by feeding corn silage (Fig. 15-2). It is estimated that from 1.5 to 2.5 tons of beef per hectare (0.7 to 1.1 T/A) can be produced from good-quality corn silage (3). Quality beef can be produced on an all-silage ration, but the time on feed is longer than it is when additional grain is fed (3).

In recent years, researchers in both the south and midwest have proposed the utilization of forages deferred from grazing during the summer and grazed or harvested during the late fall or early winter (47). The cool-season grasses lend themselves to an early-spring and summer growth and also to a late-fall growth period. A "three-season grazing system" was proposed at Iowa (Table 15-6) (47). This system increased the calf gains per hectare considerably over the previous method of continuous or rotational grazing during the summer months. In addition, deferred grazing or stockpiling of forage can extend the grazing system in many regions of the United States. This system is applicable in both the temperate and tropical regions. In the south, significant gains can be realized by using warm-season grasses in combination with a long grazing period (Table 15-6a).

The large round bales that have come into existence recently are now being utilized with perennial grasses, primarily tall fescue and bermudagrass. Research

TABLE 15-5
DAILY NUTRIENT REQUIREMENTS FOR BEEF CATTLE BREEDING HERD

Weight*		Daily gain		Minimum dry matter consumption[†]		Roughage,[†] %	Total protein, kg	Digestible protein, kg	NE$_m$, Mcal	NE$_g$, Mcal	ME,[†] Mcal	TDN,[††]		CA, g	P, g	Vitamin A, thousands IU
kg	lb	kg	lb	kg	lb		kg	kg	Mcal	Mcal	Mcal	kg	lb	g	g	IU
						Pregnant yearling heifers—last 3–4 months of pregnancy										
325	716	0.4[‡]	0.9	6.6	14.5	100[§]	0.58	0.34	5.89	0.62	12.6	3.5	7.7	15	15	19
		0.6	1.3	8.5	18.7	100	0.75	0.42	5.89	1.52	16.2	4.5	9.9	18	18	23
		0.8	1.8	9.4	20.7	85–100	0.85	0.50	5.89	2.49	20.1	5.6	12.3	22	20	26
350	772	0.4[‡]	0.9	6.9	15.2	100	0.61	0.35	6.23	0.65	13.2	3.7	8.1	15	15	19
		0.6	1.3	8.9	19.6	100	0.78	0.45	6.23	1.60	16.9	4.7	10.3	19	19	25
		0.8	1.8	10.0	22.0	85–100	0.88	0.51	6.24	2.63	21.1	5.8	12.9	22	21	28
375	827	0.4[‡]	0.9	7.2	15.9	100	0.63	0.36	6.56	0.68	13.7	3.8	8.4	15	15	20
		0.6	1.3	9.3	20.5	100	0.81	0.46	6.56	1.68	17.7	4.9	10.8	19	19	26
		0.8	1.8	11.0	24.2	85–100	0.96	0.55	6.56	2.76	22.1	6.1	13.5	22	22	31
400	882	0.4[‡]	0.9	7.5	16.5	100	0.65	0.38	6.89	0.71	14.2	3.9	8.7	16	16	21
		0.6	1.3	9.7	21.4	100	0.84	0.48	6.89	1.76	18.5	5.1	11.3	19	19	27
		0.8	1.8	11.6	25.6	85–100	1.01	0.57	6.89	2.90	23.0	6.4	14.0	22	22	33
425	937	0.4[‡]	0.9	7.8	17.2	100	0.69	0.40	7.21	0.74	14.8	4.1	9.0	16	16	22
		0.6	1.3	10.1	22.3	100	0.88	0.50	7.21	1.84	19.2	5.3	11.7	19	19	28
		0.8	1.8	12.1	26.7	85–100	1.05	0.60	7.21	3.03	24.0	6.6	14.6	22	22	34

Dry pregnant mature cows—middle third of pregnancy

350	772	5.5	12.2	100§	0.32	0.15	6.23	10.8	3.0	6.6	10	10	15
400	882	6.1	13.4	100	0.36	0.17	6.89	11.9	3.3	7.3	11	11	17
450	992	6.7	14.8	100	0.39	0.19	7.52	13.0	3.6	7.9	12	12	19
500	1102	7.2	15.9	100	0.42	0.20	8.14	14.1	3.9	8.6	13	13	20
550	1213	7.7	17.0	100	0.45	0.22	8.75	15.1	4.2	9.2	14	14	22
600	1323	8.3	18.3	100	0.49	0.23	9.33	16.1	4.4	9.8	15	15	23
650	1433	8.8	19.4	100	0.52	0.25	9.91	17.1	4.7	10.4	16	16	25

Dry pregnant mature cows—last third of pregnancy

350	772	0.4‡	0.9	6.9	13.9	100§	0.41	0.19	7.8	13.2	3.6	8.0	12	12	19
400	882	0.4	0.9	7.5	15.4	100	0.44	0.21	8.4	14.3	4.0	8.7	14	14	21
450	992	0.4	0.9	8.1	16.5	100	0.48	0.23	9.1	15.4	4.2	9.4	15	15	23
500	1102	0.4	0.9	8.6	17.9	100	0.51	0.24	9.7	16.4	4.5	10.0	15	15	24
550	1213	0.4	0.9	9.1	19.0	100	0.54	0.25	10.3	17.5	4.8	10.7	16	16	26
600	1323	0.4	0.9	9.7	20.3	100	0.57	0.27	10.9	18.5	5.1	11.2	17	17	27
650	1433	0.4	0.9	10.2	22.4	100	0.60	0.29	11.5	19.6	5.4	11.9	18	18	29

Cows nursing calves—average milking ability¶—first 3–4 months postpartum

350	772	8.2	18.1	100§	0.75	0.44	9.2	15.9	4.4	9.7	24	24	19
400	882	8.8	19.4	100	0.81	0.48	9.9	17.0	4.7	10.4	25	25	21
450	992	9.3	20.5	100	0.86	0.50	10.5	18.1	5.0	11.0	26	26	23
500	1102	9.8	21.6	100	0.90	0.53	11.1	19.2	5.3	11.7	27	27	24
550	1213	10.5	23.1	100	0.97	0.57	11.9	20.3	5.6	12.3	28	28	26
600	1323	11.0	24.2	100	1.01	0.59	12.3	21.3	5.9	13.0	28	28	27
650	1433	11.4	25.1	100	1.05	0.62	12.9	22.3	6.2	13.7	29	29	29

TABLE 15-5 (continued)
DAILY NUTRIENT REQUIREMENTS FOR BEEF CATTLE BREEDING HERD

Weight* kg	Weight* lb	Daily gain kg	Daily gain lb	Minimum dry matter consumption† kg	Minimum dry matter consumption† lb	Roughage† %	Total protein, kg	Digestible protein, kg	NE_m, Mcal	NE_g, Mcal	ME† Mcal	TDN†‡ kg	TDN†‡ lb	CA, g	P, g	Vitamin A, thousands IU
— Cows nursing calves—superior milking ability[a]—first 3–4 months postpartum																
350	772			10.2	22.4	100[b]	1.11	0.65	12.3		21.0	5.8	12.8	45	40	32
400	882			10.8	23.8	100	1.17	0.69	13.0		22.1	6.1	13.5	45	41	34
450	992			11.3	24.9	100	1.23	0.72	13.6		23.2	6.4	14.1	45	42	36
500	1102			11.8	26.0	100	1.29	0.76	14.2		24.3	6.7	14.8	46	43	38
550	1213			12.4	27.3	100	1.35	0.79	14.9		25.3	7.0	15.4	46	44	41
600	1323			12.9	28.4	100	1.41	0.83	15.5		26.4	7.3	16.1	46	44	43
650	1433			13.4	29.5	100	1.46	0.86	16.2		27.5	7.6	16.8	47	45	45
— Bulls, growth and maintenance (moderate activity)																
300	661	1.00	2.2	8.8	19.4	70–75	0.90	0.55	5.6	3.8	20.4	5.6	12.3	27	23	34
400	882	0.90	2.0	11.0	24.2	70–75	1.03	0.62	6.9	4.1	25.2	7.0	15.4	23	23	43
500	1102	0.70	1.5	12.2	26.9	80–85	1.07	0.62	8.5	3.7	27.0	7.5	16.5	22	22	48
600	1323	0.50	1.1	12.0	26.4	80–85	1.02	0.60	9.8	3.0	26.4	7.3	16.1	22	22	48
700	1543	0.30	0.7	12.9	28.4	90–100[b]	1.08	0.60	11.0	2.0	27.7	7.7	17.0	23	23	50
800	1764	0	0	10.5	23.1	100[b]	0.89	0.50	12.2	0	21.0	5.8	12.8	19	19	41
900	1984	0	0	11.4	25.1	100[b]	0.99	0.55	13.3	0	22.8	6.3	13.9	21	21	44
1000	2205	0	0	12.4	27.3	100[b]	1.05	0.60	14.4	0	24.8	6.9	15.2	22	22	48

*Average weight for a feeding period.
†Dry matter consumption, ME, and TDN requirements are based on the general type of diet indicated in the roughage column.
‡Approximately 0.4 ± 0.1 kg of weight gain/day over the last third of pregnancy is accounted for by the products of conception. These nutrients and energy requirements include the quantities estimated as necessary for conceptus development.
§Average-quality roughage containing about 1.9–2.0 Mcal ME/kg dry matter.
¶5.0 ± 0.5 kg of milk/day. Nutrients and energy for maintenance of the cow and for milk production are included in these requirements.
[a]10 ± 0.5 kg of milk/day. Nutrients and energy for maintenance of the cow and for milk production are included in these requirements.
[b]Good-quality roughage containing at least 2.0 Mcal ME/kg dry matter.
Source: National Research Council, "Nutrient requirements of beef cattle," Natl. Acad. Sci., Washington, DC, 1976.

FIGURE 15-2
Corn silage is an excellent source of energy for beef cattle.

has indicated that the crude protein content of field-stored fescue bales averages around 14.8%, which is very similar to that of the fall regrowth left in the field (22). Generally, the first harvest is put into large round bales; these may be left in the field, but most are stored in other areas. The regrowth of tall fescue following fertilization is either baled or grazed in strips during the winter months. The grazing period is approximately three weeks or until all of the forage standing or remaining in the field is utilized, along with that in the large round bales.

Earlier grazing research compared stands of alfalfa plus grass, and ladino clover plus grass, to a pure stand of grass. It was found that with the added legumes, the number of animal days per hectare was increased by 14%, and the pasture's steer carrying capacity per hectare was increased by 13%. In addition, the average daily gains were increased by 19% and the gain per hectare was increased by 35% when a legume was included with the grass (22).

HORSES

Recently the number of horses in this country has increased; correspondingly, so has the importance of pastures and forages in horse management (Fig. 15-3).

TABLE 15-6
BEEF PRODUCTION AND PERFORMANCE ON VARIOUS COOL-SEASON GRASS PASTURES USING A DEFERRED-GRAZING SYSTEM

Species	Avg. daily gain, kg (lb)		Grazing days		Beef, kg/ha (lb/A)		Total beef, kg/ha (lb/A)
	Spring-summer	Fall	Spring-summer	Fall	Spring-summer	Fall	
Smooth bromegrass	0.73 (1.6)	0.92(2.0)	267	97	457 (407)	222 (198)	679 (605)
Orchardgrass	0.57 (1.2)	0.79(1.7)	301	120	413 (368)	227 (202)	641 (570)
Tall fescue	0.52 (1.1)	0.83(1.8)	316	167	380 (338)	328 (292)	708 (630)
Reed canarygrass	0.65 (1.4)	0.65(1.4)	282	140	388 (346)	223 (198)	611 (544)
Avg.	0.62 (1.3)	0.80 (1.7)	292	131	409 (364)	250 (223)	660 (587)

Source: W. F. Wedin, "Fortifying good management practices with adequate fertilizer programs," in *Forage Highlights*, Amer. Forage Grassland Council Proc., 3d rev., Industry Conf., 1970.

TABLE 15-6a
BEEF PRODUCTION AND PERFORMANCE ON VARIOUS
WARM-SEASON GRASSES

Species	Forage, kg/ha (lb/A)	Grazing days	Beef gains, kg/ha (lb/A)
Coastal bermudagrass	17,120 (15,357)	670	653 (583)
Common bermudagrass	12,313 (10,994)	464	510 (455)
Pensacola bahiagrass	9512 (8493)	451	479 (428)

Source: C. S. Hoveland, W. B. Anthony, J. A. McGuire, and J. G. Starling, "Beef cow-calf performance on coastal bermudagrass overseeded with winter annual clovers and grasses," *Agron J.* 70:418–420 (1978).

There are more than 30 breeds of horses used today in the United States for working cattle, pleasure riding, racing, and for show. Forages play a very important role in this industry in helping reduce overall costs of keeping horses healthy.

Well-managed pastures provide excellent sources of the protein, minerals, vitamins, and energy that are necessary for good nutrition of horses (Table 15-7). A good pasture is probably the very best ration for horses. Mature horses can be maintained on pastures with very little or no additional grain during a normal pasture season. Horses that have regular access to good, green forage have very few nutritional problems in comparison to those that are confined in dry lots (11). Good pasture management is as important for horses as for any livestock species. In general, pastures that are poorly managed and low in fertility do not provide enough feed nutrients to maintain or develop an actively

FIGURE 15-3
Horses need high-quality forage and plenty of pasture exercise.

TABLE 15-7
NUTRIENT REQUIREMENTS OF HORSES, 1100-lb MATURE BODY WEIGHT (b.w.)*

	Dig. energy, Mcal		TDN, lb		Crude protein		Calcium		Phosphorus		Vitamin A, IU per lb	
	Per 100 lb b.w.	Daily	Per 100 lb b.w.	Daily	%	Daily, lb	%	Daily, g	%	Daily, g	Diet	Daily
Mature horses, maintenance	1.49	16.4	0.75	8.2	8.5–10.0	1.4	0.30	23	0.20	14	750	12,500
Mature working horses (per hour above maint.)					8.5–10.0	†	0.20	†	0.20	†	750	†
Light work	0.23	2.5	0.12	1.3								
Moderate work	0.57	6.3	0.29	3.2								
Intense work	1.05	11.6	0.53	5.8								
Gestation (first 8 mo.)	1.49	16.4	0.75	8.2	8.5–10.0	1.4	0.30	23	0.20	14	750	12,500
Gestation (last 3 mo.)	1.67	18.4	0.83	9.2	10.0–11.0	1.6	0.50	34	0.35	23	1500	25,000
Lactation (first 3 mo.)	2.57	28.3	1.28	14.1	13.0–14.0	3.0	0.50	50	0.35	34	1250	27,500
Lactation (after 3 mo.)	2.21	24.3	1.11	12.2	12.0–13.0	2.4	0.45	41	0.30	27	1100	22,500
Foal (2–5 mo.)	3.55	13.7	1.77	6.8	16.0–18.0‡	1.6‡	0.80	33	0.60	20	750	7000
Weanling (5–12 mo.)	3.12	15.6	1.56	7.8	14.0–16.0‡	1.7‡	0.70	34	0.50	25	900	10,000
Yearling (12–24 mo.)	2.15	16.8	1.08	8.4	11.0–13.5	1.7	0.50	31	0.40	22	900	13,000
2 yr. old: show and performance	1.66	16.4	0.83	8.2	10.0–11.0	1.4	0.45	25	0.35	17	900	13,000

*For each 100 lb above or below 1100 lb body weight, add or subtract 7% to the amounts given.
†Daily amount will vary with level of dietary intake.
‡High-quality protein recommended (minimum of 0.65% dietary lysine).
§The following vitamins may be included in certain diets: vitamin D, 125 IU/lb diet; vitamin E, 7 IU/lb diet; vitamin B, complex mixture.

FIGURE 15-4
Productive pastures provide both nutrients and exercise for horses.

growing horse. Furthermore, a poorly managed pasture may be a good source for internal parasites. Pastures also provide exercise for horses (Fig. 15-4). Horses must have good exercise to remain active and healthy.

Since horses are nonruminants, it is important to keep in mind that they do not digest lignin, and the cell wall, which is composed of cellulose and hemicellulose, is only partially digestible; in addition, the digestibility of most forages declines with advanced maturity. However, silage made from either a grain crop or a cereal crop is an exception to this rule. Alfalfa and orchardgrass or corn silage are excellent forages for horses. The data presented in Table 15-8 can be used as a good guide when estimating quality of forage for horses (9, 11, 16, 24). The digestibility of both cellulose and hemicellulose is about 25% lower in mature horses than in cattle (11).

The mineral composition of different forage species varies considerably, and it is very important to maintain a proper mineral balance for horses. The mature horse that is doing medium-type work loses about 60 g (0.14 lb) of sodium chloride daily in sweat and 35 g (0.08 lb) in urine (29). As a consequence, it is important to have an adequate supply of salt available. Even though forages do contain considerable amounts of P and K, these levels decline with advanced maturity of the forage. Legumes contain more calcium than grasses do, and grasses are higher in calcium than most grain feeds are. When legumes are mixed with grains, a good balance of P, Ca, and Mg is present in the diet. There have been some indications that diets with excessive forage legumes contain too much calcium, thereby causing some brittleness of the bones. Most reports suggest that in the proper ratio of calcium to phosphorus, the calcium should be the greater of the two (11, 15). The Ca:P ratio of horse milk is approximately 1.7:1.0, which is generally considered adequate for most horses (11, 15). It is

TABLE 15-8
VARIOUS FORAGES AND THEIR DIGESTIBLE DRY MATTER FOR HORSES

Species and maturity	DDM, %	Dry matter, %			
		Crude protein	Cellulose	Ca	P
Alfalfa					
Bud	76	26.9	20.9	2.2	0.33
10% flower	66	21.3	27.3	1.3	0.24
50% flower	62	19.1	28.5	1.2	0.22
Orchardgrass					
Boot	71	22.8	25.9	0.9	0.43
Headed	63	17.9	28.8	0.8	0.42
Regrowth 42 days	67	20.4	26.5	0.7	0.24
Corn silage					
Milk	54	12.0	28.8	0.31	0.27
Dough-dent	53	10.9	22.9	0.27	0.23

Sources: M. W. Colburn, J. L. Evans, and C. H. Ranage, "Apparent and true digestibility and yield of alfalfa," *J. Anim. Sci.* 51:1450–1457 (1968); J. L. Evans, "Forages for horses," in *Forages,* Iowa State Univ. Press, Ames, IA, 3d. ed., pp. 723–732, 1973; C. H. Gordon, J. C. Derbyshire, H. G. Wiseman, E. A. Kane, and C. G. Melin, "Preservation and feeding value of alfalfa stored as hay, haylage, and direct-cut silage," *J. Dairy Sci.* 44:1299–1311 (1961); and R. R. Johnson and K. E. McClure, "Corn plant maturity. IV. Effects on digestibility of corn silage in sheep," *J. Anim. Sci.* 27:535–540 (1968).

also important to make sure that the Mg concentration is sufficient. Sulfur is usually supplied by the protein in a forage that contains sulfur-bearing amino acids (41). Most of the other trace elements are sufficient in good-quality forage. In addition, green, leafy forages harvested at the early stage of maturity supply a sufficient amount of vitamins and carotene, a precursor of vitamin A (42).

Red clover generally does not make the high-quality forage for horses that alfalfa or other grasses do. Red clover may cause slobbers because of the high percentage of pubescence on the plants (1). When turned on to legume pastures, horses will not bloat, but they can scour.

SHEEP

Sheep are ruminants and are very efficient utilizers of forage. Even though in comparison with beef cattle, sheep production is not extremely high, their numbers are well-distributed in the upper midwest, in the southwestern section of Texas, and in the far west. The demand for lamb has remained relatively constant and prices have remained higher than have cattle prices. Sheep not only provide red meat; their production of wool is also very important (43).

The sheep industry involves two different flock sizes. Range flocks, in the west and the southwest, consist of more than 1000 ewes. Farm flocks dominate the midwest, and usually have fewer than 200 ewes.

Sheep can convert both forage and feed to red meat more efficiently than cattle can (43). It has been reported that it takes approximately 9 kg (19.8 lb) of feed for 1 kg (2.2 lb) of gain in cattle, but only 6 to 7 kg (13 to 15 lb) of feed for 1 kg (2.2 lb) of gain in lambs. By utilizing good pastures, crop residues, and waste land forage, sheep can be raised rather economically in many livestock programs. There has been some mixing of livestock species, such as grazing cattle and sheep together, which increases the pasture gains and carrying capacity (43). Cattle prefer the grass in the pasture mixture, whereas the sheep prefer the legumes and other broadleaf species (36, 46). Sheep are not as selective grazers as are cattle. Sheep are also more flexible in their reproduction than are cattle because of sheep's shorter gestation period, multiple births, and shorter period from birth to market. There is a great deal of interest in the United States in increasing the multiple births in sheep. Use of cross-bred ewes to produce three-way-cross lambs increases the production efficiency and the net dollars returned per investment (43).

A year-round pasture program is essential for sucesssful commercial sheep production. The utilization of a grass-legume combination depends to a large extent upon the time of lambing and upon the nutritional needs of the ewes at different times in their reproductive cycle. Energy is probably the most common deficiency in ewes (43). Although good-quality pastures are high in protein, they may not contain enough energy for a lactating ewe (43). Because of sheep's digestion efficiency, the amount of protein in their ration is more important than is the protein quality, except in young lambs.

Ewes' nutritional needs for maintenance and for the first 15 weeks of the gestation period are relatively small. Most of this can be furnished by medium- to low-quality forage. But nutritional needs increase by about 80% during the last 6 weeks of the gestation period, and then a good-quality forage must be available. Ewes' nutritional needs increase by another 35% during the first 8 weeks of lactation (10), to approximately three times the amount of total digestible nutrients needed for maintenance. The nutrient requirements for sheep are presented in Table 15-9.

Pastures for lambs should be of very high quality. The best pastures for this purpose include small grains for winter and early-spring grazing and perennial grasses and legumes for the rest of the grazing season. The pastures should be grazed closely enough to maintain new regrowth of high quality, but not so close that all of the leaves are removed, as this reduces plant vigor for regrowth. Rotational grazing every 60 to 100 days on three or more pastures allows plants to recover sufficiently and also aids in the control of parasites. Lambs should be moved to a clean, well-drained pasture when they are weaned or have been drenched for internal-parasite control (43).

Many farmers prefer to buy their feeder lambs instead of keeping a commercial ewe flock. There are several advantages to this. Compared with a ewe flock, feeder lambs generally require a lower initial investment and fewer facilities. Their pasture and feed program is shorter and fits better into many farming operations. Feeder lambs can graze corn stubble after harvest. They also respond well to high-quality pasture and supplemental feeding (43).

The average gain for feeder lambs on good pasture should be at least 0.18 kg

TABLE 15-9
DAILY NUTRITIONAL REQUIREMENTS FOR SHEEP

Body weight, kg	Daily dry matter* per animal, kg	Energy		Total protein, %	DP,‡ %	Ca, %	P, %
		TDN, %	DE,† Mcal/kg				
Ewes, maintenance§							
50	1.0	55	2.42	8.9	4.8	0.30	0.28
70	1.2	55	2.42	8.9	4.8	0.27	0.25
80	1.3	55	2.42	8.9	4.8	0.25	0.24
Ewes, nonlactating, first 15 weeks gestation							
50	1.1	55	2.42	9.0	4.9	0.27	0.25
70	1.4	55	2.42	9.0	4.9	0.23	0.21
80	1.5	55	2.42	9.0	4.9	0.22	0.21
Ewes, last 6 weeks gestation or last 8 weeks lactation suckling singles							
50	1.7	58	2.55	9.3	5.2	0.24	0.23
70	2.1	58	2.55	9.3	5.2	0.21	0.20
90	2.3	58	2.55	9.3	5.2	0.21	0.20
Ewes, first 8 weeks lactation suckling singles or last 8 weeks lactation suckling twins							
60	2.3	65	2.86	10.4	6.2	0.50	0.36
80	2.6	65	2.86	10.4	6.2	0.48	0.34
Ewes, first 8 weeks lactation suckling twins							
60	2.6	65	2.86	11.5	7.2	0.44	0.32
80	3.0	65	2.86	11.5	7.2	0.42	0.30
Ewes, replacement lambs and yearlings¶							
30	1.3	62	2.73	10.0	5.8	0.45	0.25
50	1.5	55	2.42	8.9	4.8	0.42	0.23
70	1.4	55	2.42	8.9	4.8	0.46	0.26

(0.4 lb) per day. If lambs are fed 0.68 kg (1 1/2 lb) of grain per day on pasture, daily gains can be increased by 40 to 60%, and the carrying capacity, in kilograms of lamb per hectare, is also increased. When lambs are put on feedlot rations, daily gains should average 0.27 (0.6 lb) (43).

Feeder lambs can be grazed on pure legume pasture, but one should watch for bloat. More poloxalene per unit of liveweight is needed for lambs than for feeder cattle. The recommended rate per day is at least 3 g per 45 kg (100 lb) of liveweight. Poloxalene can be fed with grain, liquid feed, molasses, or in block form, or by whatever method is the most satisfactory. It should be fed to sheep for several days before they are put on a pure legume pasture. It is suggested that some dry hay also be fed to them before they are moved to a legume pasture.

TABLE 15-9 (continued)
DAILY NUTRITIONAL REQUIREMENTS FOR SHEEP

Body weight, kg	Daily dry matter* per animal, kg	Energy		Total protein, %	DP,‡ %	Ca, %	P, %
		TDN, %	DE,† Mcal/kg				
		Rams, replacement lambs and yearlings¶					
60	2.3	60	2.65	9.5	5.3	0.31	0.17
80	2.8	55	2.42	8.9	4.8	0.28	0.16
100	2.8	55	2.42	8.9	4.8	0.30	0.17
		Lambs, fattening*a*					
30	1.3	64	2.81	11.0	6.7	0.37	0.23
40	1.6	70	3.08	11.0	6.7	0.31	0.19
50	1.8	70	3.08	11.0	6.7	0.28	0.17
		Lambs, early-weaned*b*					
10	0.6	73	3.21	16.0	11.5	0.47	0.28
20	1.0	73	3.21	16.0	11.5	0.50	0.30
30	1.4	73	3.21	17.0	9.5	0.51	0.31

*To convert dry matter to an as-fed basis, divide dry matter by percentage dry matter.

†1 kg TDN = 4.4 Mcal DE (digestible energy). DE may be converted to ME (metabolizable energy) by multiplying by 82%.

‡DP = digestible protein.

§Values are for ewes in moderate condition, not excessively fat or thin. For fat ewes feed at next lower weight; for thin ewes feed at next highest weight.

¶Replacement lamb (ewe and ram) requirements start at time they are weaned.

*a*Maximum gains expected: If lambs are held for later market, they should be fed similarly to replacement ewe lambs. Lambs capable of gaining faster than indicated need to be fed at a higher level; self-feeding permits lambs to finish most rapidly.

*b*40-kg early-weaned lambs fed the same as finishing lambs of equal weight.

Source: National Research Council, "Nutrient requirements of sheep," Natl. Acad. Sci., Washington, DC, 1975.

SWINE

Swine are usually grown in confinement, but research and experience have shown that hogs, and especially gestating sows, can use forages in their diet economically (13). By 8 months of age, a hog's digestive system is sufficiently developed to utilize relatively large amounts of high-quality forage (2, 7, 19). A year-round forage program including pasture, hay, silage, and haylage can reduce the feed cost for gestating sows by as much as 50%. Most of the forage that is fed to hogs goes to the sows.

High-quality forage for swine contains the nutrients necessary for successful reproduction. Good-quality forage supplies most of the essential vitamins, minerals, and amino acids. Such forages are high in protein and can furnish most of the 16%-protein ration that is needed by gilts and 14%-protein ration needed by sows (13). The nutrients that are often deficient in a good forage program are phosphorus and salt. These deficiencies can be corrected by feeding sows 0.7 to 0.9 kg (1 1/2 to 2 lb) of grain, plus a mineral mixture of 1 part dicalcium

TABLE 15-10
DAILY NUTRIENT REQUIREMENTS OF BREEDING SWINE*

Air-dry feed intake, g	Bred gilts and sows; young and adult boars	Lactating gilts and sows		
	1800[†]	4000	4750	5500
Digestible energy, kcal	6120[‡]	13580	16130	18670
Metabolizable energy, kcal	5760[‡]	12780	15180	17570
Crude protein, g	216	520	618	715
Indispensable amino acids				
Arginine, g	0	16.0	19.0	22.0
Histidine, g	2.7	10.0	11.9	13.8
Isoleucine, g	6.7	15.6	18.5	21.4
Leucine, g	7.6	28.0	33.2	38.5
Lysine, g	7.7	23.2	27.6	31.9
Methionine + cystine,[§] g	4.1	14.4	17.1	19.8
Phenylalanine + tyrosine,[¶] g	9.4	34.0	40.4	46.8
Threonine, g	6.1	17.2	20.4	23.6
Tryptophan,[a] g	1.6	4.8	5.7	6.6
Valine, g	8.3	22.0	26.1	30.2
Mineral elements				
Calcium, g	13.5	30.0	35.6	41.2
Phosphorus,[b] g	10.8	20.0	23.8	27.5
Sodium, g	2.7	8.0	9.5	11.0
Chlorine, g	4.5	12.0	14.2	16.5
Potassium, g	3.6	8.0	9.5	11.0
Magnesium, g	0.7	1.6	1.9	2.2
Iron, mg	144	320	380	440
Zinc, mg	90	200	238	275
Manganese, mg	18	40	48	55
Copper, mg	9	20	24	28
Iodine, mg	0.25	0.56	0.66	0.77
Selenium, mg	0.27	0.40	0.48	0.55

phosphate and 1 part trace mineral salt, or bonemeal and salt on a free-choice basis (13, 19). Grain also supplies some energy. Forages or pastures can provide a major portion of sows' dietary needs, which results in a significant saving in feed costs, and in less investment per sow for housing facilities and manure disposal.

The most common pastures used for swine in the United States are alfalfa and ladino clover, or a combination of these two. Both are excellent sources of calcium and most of the required vitamins, except vitamin D (31). Ladino clover is less fibrous than is alfalfa, but alfalfa is slightly higher in dry matter, protein, and energy (2, 31).

Another swine forage species is rape, an annual broadleaf species that is neither a legume nor a grass. It is seeded in the early spring in the corn belt and

TABLE 15-10 (continued)
DAILY NUTRIENT REQUIREMENTS OF BREEDING SWINE*

	Bred gilts and sows; young and adult boars	Lactating gilts and sows		
Air-dry feed intake, g	**1800[†]**	**4000**	**4750**	**5500**
Vitamins				
Vitamin A, IU,	7200	8000	9500	11000
or β-carotene, mg	28.8	32.0	38.0	44.0
Vitamin D, IU	360	800	950	1100
Vitamin E, IU	18.0	40.0	47.5	55.0
Vitamin K, mg	3.6	8.0	9.5	11.0
Riboflavin, mg	5.4	12.0	14.2	16.5
Niacin,[c] mg	18.0	40.0	47.5	55.0
Pantothenic acid, mg	21.6	48.0	57.0	66.0
Vitamin B_{12}, μg	27.0	60.0	7.12	82.5
Choline, mg	2250.0	5000.0	5940.0	6875.0
Thiamin, mg	1.8	4.0	4.8	5.5
Vitamin B_6, mg	1.8	4.0	4.8	5.5
Biotin,[d] mg	0.18	0.4	0.48	0.55
Folacin,[d] mg	1.08	2.4	2.8	3.3

*Requirements reflect the estimated levels of each nutrient needed for optimal performance when a fortified grain-soybean meal diet is fed. Concentrations are based upon amounts per unit of air-dry diet (i.e., 90% dry matter).

[†]An additional 25% should be fed to working boars.

[‡]Individual feeding and moderate climatic conditions are assumed. An energy reduction of about 10% is possible when gilts and sows are tethered or individually penned in a stall in environmentally controlled housing. An energy increase of about 25% is suggested for climatic (winter) conditions.

[§]Methionine can fulfill the total requirement; cystine can meet at least 50% of the total requirement.

[¶]Phenylalanine can fulfill the total requirement; tyrosine can meet at least 50% of the total requirement.

[a] It is assumed that usable tryptophan content of corn does not exceed 0.05%.

[b]At least 30% of the phosphorus requirement should be provided by inorganic and/or animal product sources.

[c]It is assumed that most of the niacin present in cereal grains and their by-products is in bound form and thus unavailable to swine. The niacin contributed by these sources is not included in the requirement listed. In excess of its requirement for protein synthesis, tryptophan can be converted to niacin (50 mg tryptophan yields 1 mg niacin).

[d]These levels are suggested. No requirements have been established.

Source: National Research Council, "Nutrient requirements of swine," Natl. Acad. Sci., Washington, DC, 1979.

in the fall in the southern states. It is frequently sown alone or in a mixture with oats, which can be used for early-spring grazing. Its nutritional value is almost equal to that of legumes (13).

Winter rye is another popular species used in swine pastures. It has rapid growth in the fall and early spring and provides good winter grazing. Rye pasture has an excellent carrying capacity, of approximately 12 sows per hectare (13).

Alfalfa in combination with other perennial grasses, such as smooth brome-grass or orchardgrass, provides an excellent pasture for swine. Research has indicated that the combination of smooth bromegrass and alfalfa may equal the

TABLE 15-11
NUTRITIONAL COMPOSITION OF VARIOUS FORAGES AS FED TO SWINE

| Nutrient | Pasture | | | | Silage | | |
	Alfalfa	Ladino clover	Rape	Rye	Corn	Grass-legume	Low-moisture legume
Dry matter, %	24	20	16	17	26	28	55
Digestible energy, kcal/kg	660	525	525	575	800	750	975
Crude protein, %	6.0	5.1	3.0	5.0	2.1	5.0	6.3
Crude fiber, %	5.4	2.9	2.8	2.9	6.4	8.2	11.4

Source: J. R. Foster, "Forages for swine and poultry," in *Forages,* Iowa State Univ. Press, Ames, IA, 3d ed., pp. 715–722, 1973.

value of a pure alfalfa pasture for self-fed pigs (2, 13). Compared with pigs fed on alfalfa pasture, pigs that are fed on smooth bromegrass or on any other perennial grass may consume about a third more protein supplement and 13% more minerals for every unit of gain. (7).

In recent years more silages have been fed to swine. Corn silage is an excellent source of most of the vitamins needed by swine, but it is deficient in protein, energy, and some minerals (19). Research at Illinois has indicated that sows consume only 1.5 kg (3.3 lb) of silage per day (13), though at other institutions, trials have shown that sows consume 4.5 to 6.5 kg (9.9 to 14.3 lb) per day (13). Depending upon the intake and the time during gestation, supplemental feed may be needed when corn silage is fed (13). The best corn silage for swine is that which was harvested in the soft-dough to the hard-dough stage. It is suggested that the silage be cut as finely as possible to eliminate any waste (13).

Good reproductive performance has been found when alfalfa haylage has been fed to sows alone or when mixed with ground corn at the rate of 4 parts haylage to 1 part corn and stored in an airtight silo. This haylage is approximately 45% moisture; its crude protein content is approximately 19 1/2%, and its

TABLE 15-12
FEEDING REQUIREMENTS FOR SWINE ON PASTURE

Swine life-cycle stage	Number per ha	Feed per day, kg	Protein or concentrate, %
Gestation	24	1	13
Lactating litters	12	4–5	13
Growing	75	Full feed	15
Finishing	50	Full feed	12

Source: J. R. Foster, "Forages for swine and poultry," in *Forages,* Iowa State Univ. Press, Ames, IA, 3d ed., pp. 715–722, 1973.

crude fiber content approximately 20% (13, 19). Digestion trials in sows indicate that the crude protein of a mixture of alfalfa haylage and corn is approximately 55 to 60% digestible. Average daily consumption of a haylage-corn mixture has been reported at 3.8 kg (8.4 lb) for gilts and 5.3 kg (11.7 lb) for sows (19). This mixture can be fed to gilts and sows starting about 2 weeks prior to the start of the breeding season and continuing throughout the gestation period (19).

The daily nutritional requirements for swine are shown in Table 15-10. The nutrient composition of forages fed to swine are shown in Table 15-1. And the feed requirements for swine on pasture are summarized in Table 15-12.

QUESTIONS

1 Why and how are forages such an important component in the human food program?
2 In what way do forages contribute to the food supply?
3 Compare the forage quality requirement for beef and dairy.
4 How does quality affect overall milk production and return per hectare?
5 Compare various methods of harvesting losses and how they affect the return per hectare in milk products.
6 Compare the utilization of pasture by beef cattle with that by dairy cattle.
7 How do forages fit into an overall horse program?
8 Why are sheep so important in the utilization of low-quality forage?
9 How do forages fit into a swine program?

REFERENCES

1 Albert, W. W. Utilizing cornfield wastes. *Ill. Agr. Exp. Sta. Cow-calf Info. Round-up,* 1971.
2 Barnhart, C. E. Protein and vitamin supplements for growing-fattening pigs on ladino and on alfalfa. *J. Anim. Sci.,* 11:233–237, 1952.
3 Barrick, E. R. and S. H. Dobson. Forage-land use efficiencies with commercial cattle. In *Forages,* Iowa State Univ. Press, 3d ed., pp. 690–702, 1973.
4 Blaser, R. E., H. T. Bryant, R. C.Hammes, Jr., R. Boman, J. P. Fontenot, and E. C. Polan. Managing forages for animal production. *Va. Poly. Inst. and State Univ. Res. Bull. 45, 1969.*
5 Blaser, R. E., H. T. Bryant, C. Y. Ward, R. C. Hammes, Jr., R. C. Carter, and N. H. MacCold. Symposium on Forage Evaluation: VII. Animal performance and yields with methods of utilizing pastureage. *Agron. J.,* 51:238–242, 1959.
6 Blaser, R. E., D. D. Wolf, and H. T. Bryant. Systems of pasture management. In *Forages,* Iowa State Univ. Press, Ames, IA, 3d ed., pp. 581–595, 1973.
7 Bowden, D. M. and M. F. Clarke. Grass-legume forage fed fresh and as silage for market hogs. *J. Anim. Sci.,* 32:934–939, 1963.
8 Burns, J. C., D. H. Timothy, R. D. Mochrie, D. S. Chamblee, and L. A. Nelson. Animal preference, nutritive attributes, and yield of *Pennisetum flaccidum* and *P. orientale. Agron. J.,* 70:451–456, 1978.
9 Colburn, M. W., J. L. Evans, and C. H. Ranage. Apparent and true digestibility and yield of alfalfa. *J. Anim. Sci.,* 51:1450–1457, 1968.
10 Crampton, E. W. and L. E. Harris. Applied animal nutrition. W. H. Freeman, San Francisco, 2d ed., 1969.

11 Evans, J. L. Forages for horses. In *Forages,* Iowa State Univ. Press, Ames, IA, 3d ed., pp. 723–732, 1973.

12 Fonnesbeck, P. V. and L. D. Symons. Utilization of the carotene of hay by horses. *J. Anim. Sci.,* 26:1030–1038, 1967.

13 Foster, J. R. Forages for swine and poultry. In *Forages,* Iowa State Univ. Press, Ames, IA, 3d ed., pp. 715–722, 1973.

14 Fribourg, H. A., K. M. Barth, J. B. McLaren, L. A. Carver, J. T. Connell, and J. M. Bryan. Seasonal trends of *in vitro* dry matter digestibility of N-fertilized bermudagrass and of orchardgrass-ladino pastures. *Agron. J.,* 71:117–120, 1979.

15 Garces, M. A. and J. L. Evans. Calcium and magnesium absorption in growing cattle as influenced by the age of animal and source of dietary nitrogen. *J. Anim. Sci.,* 32:789–793, 1971.

16 Gordon, C. H., J. C. Derbyshire, H. G. Wiseman, E. A. Kane, and C. G. Melin. Preservation and feeding value of alfalfa stored as hay, haylage, and direct-cut silage. *J. Dairy Sci.,* 44:1299–1311, 1961.

17 Hamilton, R. I., J. M. Scholl, and A. L. Pope. Performance of three grass species grown alone and with alfalfa under intensive pasture management: Animal and plant response. *Agron. J.,* 61:357–361, 1969.

18 Heinemann, W. W. Continuous and rotation grazing of beef steers on irrigated pastures. *Wash. Agr. Exp. Sta. Bull. 724,* 1970.

19 Hoagland, J. M., H. W. Jones, and R. A. Pickett. Haylage as a gestation ration for sows and gilts. *Prudue Univ. Res. Progress Rep. 75,* 1963.

20 Hoglund, C. R. Comparative storage losses and feeding values of alfalfa and corn silage crops when harvested at different moisture levels and stored in gas-tight and conventional tower silos: An appraisal of research results. *Mich. Agr. Econ. Mimeo. 947,* 1964.

21 Hoglund, C. R. Some economic considerations in selecting forage systems for haylage and silage on dairy farms. *Mich. Agr. Econ. Rep. 14,* 1965.

22 Hoveland, C. S., W. B. Anthony, J. A. McGuire, and J. G. Starling. Beef cow-calf performance on coastal bermudagrass overseeded with winter annual clovers and grasses. *Agron. J.,* 70:418–420, 1978.

23 Hoveland, C. S., R. L. Haaland, C. C. King, J r., W. B. Anthony, J. A. McGuire, L. S. Amith, H. W. Grimes, and J. L. Holliman. Steer performance on Ap-2 phalaris and 'Kentucky 31' tall fescue pasture. *Agron. J.,* 72:375–377, 1970.

24 Johnson, R. R. and K. E. McClure. Corn plant maturity. IV. Effects on digestibility of corn silage in sheep. *J. Anim. Sci.,* 27:535–540, 1968.

24 a Jung, G. A., L. L. Wilson, P. J. LeVan, R. E. Kocher, and R. F. Todd. Herbage and beef production from ryegrass-alfalfa and orchardgrass-alfalfa pastures. *Agron. J.,* 74:937–942, 1982.

25 Martin, G. C. and R. M. Jordan. Substitution value of birdsfoot trefoil for alfalfa-grass in pasture systems. *Agron. J.,* 71:55–59, 1979.

26 McCartor, M. M. and F. M. Rouquette, Jr. Grazing pressure and animal perform-ance from pearl millet. *Agron. J.,* 69:983–987, 1977.

27 National Research Council. Nutrient requirements of beef cattle. Natl. Acad. Sci., Washington, DC, 1976.

28 National Research Council. Nutrient requiremments of dairy cattle. Natl. Acad. Sci., Washington, DC, 1978.

29 National Research Council. Nutrient requirements of the horse. Natl. Acad. Sci., Washington, DC, 1978.

30 National Research Council. Nutrient requirements of sheep. Natl. Acad. Sci., Washington, DC, 1975.

31 National Research Council. Nutrient requirements of swine. Natl. Acad. Sci., Washington, DC, 1979.

32 Olson, N. O. The nutrition of the horse. In *Nutrition of Animals of Agricultural Importance,* Pergamon, New York, 17(2):921–960, 1969.

33 Rehm, G. M., R. C. Sorensen, and W. J. Moline. Time and rate of fertilization on seeded as warm-season and bluegrass pastures. II. Quality and nutrient content. *Agron. J.,* 69:955–961, 1977.

34 Reid, J. T. Forages for dairy cattle. In *Forages,* Iowa State Univ. Press, Ames, IA, 3d ed., pp. 664–676, 1973.

35 Reynolds, J. H. C. R. Lewis, and K. F. Laaker. Chemical composition and yield of orchardgrass forage grown under high rates of nitrogen fertilization and several cutting managements. *Tenn. Agr. Exp. Sta. Bull. 479,* 1971.

36 Reynolds, P. J., J. Bond, G. E. Carlson, C. Jackson, R. H. Hart, and I. L. Lindahl. Co-grazing sheep and cattle on orchardgrass sward. *Agron. J.,* 63:533–536, 1971.

37 Rohweder, D. A. and W. C. Thompson. Permanent pastures. In *Forages,* Iowa State Univ. Press, Ames, IA, 3d ed., pp. 596–606, 1973.

38 Rook, J. A. F. and J. E. Storry. Magnesium in the nutrition of farm animals. *Nutr. Abstr. Rev.,* 32:1055–1077, 1962.

39 Rumburg, C. B. Grazing of grass and cicer milkvetch-grass pastures with yearling calves. *Agron. J.,* 70:850–852, 1978.

40 Schertz, K. F., J. A. Viera, and J .W. Johnson. Sorghum stover digestibility as affected by juiciness. *Crop Sci.,* 18:456–458, 1978.

41 Stillions, M. C., S. M. Teeter, and W. E. Nelson. Ascorbic acid treatment of mature horses. *J. Anim. Sci.,* 32:249–251, 1971.

42 Stillions, M. C., S. M. Teeter, and W. E. Nelson. Utilization of dietary vitamin B_{12} and cobalt by mature horses. *J. Anim. Sci.,* 32:252–255, 1971.

43 Terrill, C. E., and I. L. Lindahl, and D. A. Price. Sheep: Efficient uses of forages. In *Forages,* Iowa State Univ. Press, Ames, IA, 3d. ed., pp. 703–714, 1973.

44 Tyrell, H. F. and J. T. Reid. Effect of method of forage preservation on the feeding value of perennial forages. *Proc. Cornell Nutr. Conf.,* pp. 137–145, 1967.

45 Van Keuren, R. W., R. R. Davis, D. S. Bell, and E. W. Klosterman. Effect of grazing management on the animal production from birdsfoot trefoil pastures. *Agron. J.,* 61:422–425, 1969.

46 Van Keuren, R. W. and C. F. Parker. Better pasture utilization—grazing cattle and sheep together. *Ohio Rep.,* 57:12–13, 1967.

47 Wedin, W. F. Fortifying good management practices with adequate fertilizer programs. In *Forage Highlights, Amer. Forage Grassland Council Proc.,* 3d rev. 1970.

48 Wedin, W. F. and A. G. Matches. Cropland pasture. In *Forages,* Iowa State Univ. Press, Ames, IA, 3d ed., pp. 607–616, 1973.

49 Weir, W. C., L. G. Jones, and J. H. Meyer. Effect of cutting interval and stage of maturity on the digestibility and yield of alfalfa. *J. Anim. Sci.,* 19:5–19, 1960.

50 Welch, J. G., C. Martin, and G. W. Vandernoot. Net energy of alfalfa and orchardgrass hays at varying stages of maturity. *J. Anim. Sci.,* 28:263–267, 1969.

51 Wilson, L. L. and J. C. Burns. Utilization of forages with beef cattle and calves. In *Forages,* Iowa State Univ. Press, Ames, IA, 3d. ed., pp. 677–689, 1973.

THREE

FORAGE LEGUMES

CHAPTER 16

ALFALFA

Alfalfa, *Medicago Sativa* L., is the most productive perennial legume grown in the United States. Since alfalfa was the first forage crop to be domesticated, early humans obviously recognized its value as a crop plant. It is the oldest recorded crop grown for forage; records indicate its use as a forage over 3300 years ago. Due to its deep taproot, it is tolerant to drought and very tolerant to heat. Alfalfa produces large quantities of seed under hot, dry climatic conditions.

HISTORICAL BACKGROUND

It is generally agreed that alfalfa originated in Vavilov's "Near Eastern Center"—Asia Minor, Transcaucasia, Iran, and the highlands of Turkmenistan. Iran is often mentioned as the geographic center of alfalfa (6). From extensive phylogenetic studies of the species, Sinskaya (29) inferred that alfalfa had two specific centers of origin. Based on plant degeneration and fungal diseases, Sinskaya concluded that the ancient, mountainous nations of Anterior Asia, including Transcaucasia, were the center of origin of European and Anterior alfalfa. This area is characterized by a continental climate, severe winters, and well-drained soils that have a near-neutral pH (29). The second and independent center of origin is central Asia. The climate of this area is characterized by low humidity, hot, dry summers, and moderately cold winters. Both centers are valuable sources of germ plasm for the plant breeder (18).

The Hittites practiced land cultivation, gardening, and irrigation during the second millenium B.C. (18). Bolton (5) states that during archeological excavations in the Corum/Alacahöyük regions of Turkey, Hittite (1400–1200 B.C.)

brick tablets were discovered that suggested that animals were fed on alfalfa all through the winter season and that alfalfa was regarded as a highly nutritious animal feed.

Historical evidence testifies convincingly to the wide distribution of alfalfa in Media (the mountains of northwestern Iran) in the first millenium B.C. (29). Hence, the Romans referred to alfalfa as the "Median Herb" or medica. Media alfalfa then penetrated into Assyria, probably via the Tigris and Euphrates rivers.

At the end of the fifth century B.C., alfalfa found its way from Media to Greece (15). Theophrastus (6), writing in the fourth century B.C., chronicled the introduction of the plant into Greece. Judging by the frequent allusions to alfalfa by writers of that day, alfalfa must have assumed some importance in Greek agriculture (18). In 126 B.C. alfalfa was taken into China (6).

The Romans acquired alfalfa from the Greek civilization in the second century B.C.; it thrived and quickly spread throughout Italy (18). In 77 A.D., Pliny (6) described the management of alfalfa: "It should be cut when it begins to flower, and again as often as flowers appear; this will happen six times a year, at least four." Also during the first century, alfalfa was cultured around Lucerne Lake in Switzerland. To this day, alfalfa is referred to as lucerne in many European countries.

From Italy, alfalfa appears to have moved into Spain, where the Moors regarded it as the best fodder plant and called it "Al'fol'fa," which means "best forage" (6). Very little is known about alfalfa during the Dark Ages. However, it was introduced rather early from Italy into southern France (18). Alfalfa found its way into Germany and England in about 1578 (18, 29).

Alfalfa spread to North America via two routes: directly from Europe (twice) and from South America. The Portuguese and Spanish took alfalfa into Mexico and Peru (6). At the time of the American gold rush, between 1847 and 1850, alfalfa was brought into California by some trading vessel (6). It soon spread eastward to Utah and Kansas, and by late in the nineteenth century, to Missouri and Ohio. But in 1736, more than 100 years before its introduction into the United States via California, alfalfa was introduced to Georgia, North Carolina, and other eastern states from Europe (5). The typically acid soils and humid climate in the eastern United States, as well as the lack of a proper inoculation, may account for the plant's lack of success in that area (6).

A third important introduction of alfalfa occurred when Wendelin Grimm brought 15 to 20 lb of seed of a variegated strain from Germany when he immigrated to Minnesota in 1857. Over the years and over many generations of plants, Grimm collected seed from those plants that persisted. Thus, through natural and artificial selection, he developed a very winter-hardy strain, called 'Grimm' alfalfa (6, 15).

The history and movement of alfalfa stretches back into the pre-Christian era and follows the path of historic civilization from east to west. From its first center of origin in Iran, alfalfa culture has expanded such that it grows wild in many temperate regions of the world and has become widely acclimated in South

Africa, Australia, New Zealand, and North and South America (6, 15). Thus, alfalfa truly deserves to be crowned as the "Queen of the Forages."

WHERE ALFALFA GROWS

Alfalfa is widely grown in the United States and Canada. It is grown in and is best-adapted to well-drained, highly fertile soils. If drainage is poor and cannot be corrected, other forage species that are tolerant to such conditions must be grown, such as ladino clover or tall fescue.

There are many different strains of *Medicago;* as a result, it lends itself to a wide range of growing environments. For example, *M. falcata* L., a yellow-flowered diploid, grows in areas where the temperature may reach −27°C (−18°F), such as the U.S.S.R. and Alaska, whereas other alfalfas, which are not winter-hardy, have survived temperatures of 60°C (140°F). The amount of available water often sets the production limits of alfalfa. It is adapted to humid areas and also to arid areas if irrigation is available.

In addition to the climatic requirements, alfalfa grows best with a soil pH between 6.5 and 7.5, as well as large amounts of available P, K, and B. As the pH exceeds 7.5, the availability of P, Fe, Mn, B, Cu, Zn, and Mg decreases in both mineral and organic soils. Most of the soils east of the Mississippi River need limestone (Ca) added for productive stands of alfalfa. Of course, the proper *Rhizobium* is needed for adequate growth (26).

HOW ALFALFA GROWS

Alfalfa has a very extensive taproot system. Root growth may penetrate to depths of 2 to 3 m (6 to 9 ft), but under favorable soil conditions it may exceed 6 to 9 cm (20 to 30 ft). This deep taproot system enables the plant to use water held at great depths and so to tolerate droughty conditions (12).

Alfalfa has an upright, perennial growth habit (Fig. 16-1). Flowers are formed in a raceme, with petal colors ranging from white or yellow to various shades of purple. Seed pods form a spiral in the shape of a ram's horn, with 5 to 15 kidney-shaped seeds. Leaves are pinnately trifoliolate, arranged alternately on the stem (5) (Fig. 16-2). One distinctive morphological characteristic of the leaf margin is that the upper one-half to one-third is serrated; in sweet clover, the entire leaf margin is serrated.

A distinctive crown develops at or slightly below the soil surface (Fig. 16-3). Crownal buds are generated in the crown (5). This serves as regrowth and maintains the perennial nature of the species. Alfalfa plant tissue has the ability to form a callous layer (5), which serves the purpose of healing various injuries. In contrast, red clover cannot form a callous layer over an injury; thus it is a shorter-lived species because of the entry of various disease organisms.

During the early fall months, many alfalfas go into fall dormancy. There appears to be an association between fall dormancy and winterhardiness (30). During the shorter days and cooler nights of the fall, dormant alfalfa tends to

FIGURE 16-1
Alfalfa at the flowering stage. *(Courtesy University of Illinois)*

FIGURE 16-2
Close up of alfalfa seed pods and leaves. *(Courtesy University of Illinois)*

FIGURE 16-3
Close-up of an alfalfa crown. *(Courtesy University of Illinois)*

stop growing upright and may seem to grow more prostrate. Plants that lack fall dormancy continue growing erect until a killing frost. It is postulated that fall dormancy and winterhardiness need not be associated. One could select within the alfalfa germ plasm those plants that continue to grow in the early fall and still have sufficient winterhardiness. This alone could give one an additional 30 to 35 days of growth and higher dry matter yields when the crop is harvested after a killing frost.

TYPES OF ALFALFA

Within the *Medicago* genus, there are 60 or more different species. Generally, the species grown to the greatest extent for forage in the United States and Canada are *M. sativa* and *M. falcata* and the crosses between these two species, which are called variegated alfalfas. Most cultivars are of the *M. sativa* type, a tetraploid with 32 chromosomes; *M. falcata* is usually a diploid with 16 chromosomes (5).

Medicago sativa is thought to be native to southwestern Asia (6). It has primarily purple flowers, an erect growth habit with average crown width, and a distinctive taproot. Varying amounts of winterhardiness are found within this species.

Medicago falcata, native to Siberia, has excellent winterhardiness (6). It has primarily yellow flowers, a decumbent growth habit, narrow, deep-set crown, and branching roots. This species has been used in many crosses with *M. sativa* for its winterhardiness trait.

Variegated alfalfas' flowers range in color from white or yellow to blue or purple. Most of the commercial cultivars are of the variegated type. They are intermediate in most characteristics between *M. falcata* and *M. sativa*. They are primarily taprooted and erect in growth, and have sufficient winterhardiness to grow anywhere in the United States and Canada (12).

Regional strains or ecotypes are often called common alfalfa. Most of the present common alfalfas refer to an area of seed production and are not certified. As a result of natural selection, they are adapted to a certain region of the United States. They are primarily erect-growing with purple flowers, and they have varying winterhardiness.

Flemish alfalfas are from northern France. They are purple-flowered, erect-growing with rapid regrowth, and have moderate winterhardiness. The original flemish strains were susceptible to bacterial wilt; however, most of the present flemish alfalfas are resistant to this disease.

Turkestan alfalfas were imported from the U.S.S.R. and Turkey. These alfalfas are used as sources of wilt resistance in various breeding programs, but they are susceptible to leaf diseases, slow in regrowth, and poor at producing seeds.

Strains of alfalfas that are not winter-hardy perform as annuals in the northern United States because they are very susceptible to cold temperatures. These strains are rapid growers and quick in regrowth. Most of these alfalfas are grown in the far southwestern United States. There are also spreading-type alfalfas, most cultivars of which have *M. falcata* germplasm in them. They appear to spread by rhizomes.

USES OF ALFALFA

Alfalfa that has been managed properly has the highest feeding value of all forage species (3). It produces more protein per hectare (ha) than any crop grown in the United States (Table 16-1). Cattle farmers often ask, "Why grow alfalfa in the midwest when I can produce 9408 kg/ha (150 bu/A) of corn?" The reason is very simple: alfalfa's superior protein and energy production. In terms of protein, 22.4 t/ha (10 T/A) alfalfa equals 4036 kg/ha (3600 lb/A) protein, or 18,178 kg (715 bu) of corn, or 4050 kg (150 bu) of soybeans. In terms of energy, 22.4 t/ha (10 T/A) of alfalfa equals 15,807 kg/ha (14,100 lb/A) TDN, or 7989 kg (312 bu) of corn, or 11.55 t (10.5 T) of corn silage.

In addition to its protein and energy production, alfalfa has at least 10 different vitamins and is high in mineral content. It is considered an excellent source of vitamin A. With these important characteristics, alfalfa is often made into a dehydrated meal called "dehy" and pelleted. Pelleting eliminates any dust problem. Pellets are easily handled and are stored under inert gas until they are packaged or shipped. A product called "Pro-Xan" is now being marketed for monogastric and ruminant animals; this involves wet-processing or squeezing out the plant's high-protein juices and then drying it down (19). Protein products are also being developed for human consumption (19). Alfalfa protein is being

TABLE 16-1
ESTIMATED YIELD AND PROTEIN PRODUCTION OF VARIOUS CROPS
GROWN IN THE UNITED STATES

Species	Per acre D.M. yields,* English units	Crude protein production	
		kg/ha	lb/A
Alfalfa	6.5 T	2623	2340
Forage sorghum (silage)	7.0 T	1540	1374
Corn (silage)	6.5 T	1297	1157
Grain sorghum (silage)	5.5 T	1048	935
Sorghum-sudangrass (silage)	4.0 T	706	630
Sudangrass (silage)	3.5 T	855	763
Oats (silage)	3.0 T	874	780
Corn (grain)	150 bu	841	750
Soybeans	50 bu	1345	1200
Wheat	60 bu	585	522
Rice	60 bu	235	210

*Multiply dry matter yield measured in tons by 3 to convert values to silage at 67% moisture or by 3.3 to convert to 70% moisture.

added to low-quality meats, finger foods, and dietary foods. Alfalfa sprouts are becoming a favorite additive to tossed salads, as a source of protein.

Alfalfa has been and will continue to be an excellent hay, silage, and pasture crop. It has always been popular with swine producers as a pasture crop. Horse farmers are now looking to alfalfa as an excellent hay and pasture crop, whether grown as a monoculture or in combination with some grass. Alfalfa-grass combinations make excellent silage or haylage. In fact, more horses are now being fed alfalfa haylage than in the past (22, 40). With the horse population on the increase in recent years, loosely cubed alfalfa is being shipped by railcar from the west to certain eastern centers. Texas, Oklahoma, and New Mexico form a center for horses where alfalfa is normally fed as hay.

In terms of soil conservation, alfalfa-grass combinations are excellent in reducing soil erosion and water runoff. It was shown at Illinois that alfalfa can remove large amounts of nitrate N from the soil (28). By seeding alfalfa in fields having high levels of nitrate N, alfalfa utilizes the N before it moves into the tile water or is washed off into drainage ditches or water reservoirs, thereby reducing nitrate contamination.

DESIRABLE CHARACTERISTICS

Producers want certain characteristics in an alfalfa, depending on its different intended uses. A "dehy" processor would want a rapid-growing, leafy plant with high protein content, whereas a beef producer may be more interested in a very winter-hardy, high-producing, long-lived strain.

Characteristics that are desirable in an alfalfa cultivar are as follows:

1 High-yielding
2 High-quality—leafy, fine-stemmed
3 Long-lived
4 Winter-hardy
5 Rapid regrowth following harvest
6 Resistant to diseases, especially bacterial wilt and *Phytophthora* root rot, anthracnose, and others
7 Resistant to insects, especially weevil, leafhopper, pea aphid, spotted alfalfa aphid, blue alfalfa aphid, and so on
8 Seedling vigor for rapid establishment
9 High in protein content, for alfalfa-meal production and protein extraction
10 Able to withstand grazing, without producing a bloat problem
11 Resistant to soil heaving, or having branch rooting
12 Broad, deep crowns
13 Good seed production
14 Tolerant to acid soils

Many cultivars possess many of the above traits, but no one cultivar possesses all of them. Therefore, the grower must choose a cultivar to be grown in a particular area for a particular use.

MANAGEMENT OF ALFALFA

Establishment

Proper management of alfalfa begins with proper establishment. The selection of a suitable soil is essential for successful alfalfa production. Well-drained, fertile soil is desired. The proper soil pH should be at least 6.5 to 7.5. If Ca is needed to bring up the pH, it is suggested that dolomitic limestone be used since this also contains magnesium. One should apply enough P to obtain an availability-test value of 55, and enough K for a soil-test valve of 400. Annual fertilization must take into account the high removal of nutrients by alfalfa (26). Boron is often needed in alfalfa stands in the eastern half of the United States, and sulfur in those grown on some sandy soils in the southeastern and north central United States.

Seeding

One should select a field that has not been in alfalfa production the previous year, since alfalfa possesses an allelopathic chemical that inhibits alfalfa germination. This chemical is found in extracts of seeds, roots, and top growth of alfalfa (Table 16-2). (Figs. 16-4 and 16-5).

TABLE 16-2
ALLELOPATHIC EFFECTS ON ALFALFA YIELD AND
SEEDLING COUNTS AFTER 6 YEARS OF VARIOUS
CROPPING SEQUENCES AT URBANA, IL

	D.M. yield		Plants per	
Cropping sequence	**t/ha**	**T/A**	**m²**	**ft²**
Corn—alfalfa	8.47	3.78	123	15.0
Corn—soybean—alfalfa	7.80	3.48	103	12.5
Continuous alfalfa	4.26	1.90	53	6.5

Source: R. R. Klein and D. A. Miller, "Allelopathy and its role in agriculture, *Commun. in Soil Sci. and Plant Anal.* 11(1)43–56 (1980).

One should prepare a fine, firm seedbed and select a high-yielding, pest-resistant cultivar that is adapted to the area. Seed should be inoculated; then, one should use a roller-type seeder, which compacts the soil and seed so that there is excellent contact between the two. In the northern one-half of the United States, seeding should be done in the early spring, at the rate of at least 20 kg/ha (18 lb/A). Late-summer seedings may be done in the southern one-half of the United States (34). Optimum seeding depth is 0.8 cm (1/4 in).

Weed Control

The later that one seeds in the spring, or as the soil temperature rises, the greater the problem of weed control. Most rapid and vigorous alfalfa seedling

FIGURE 16-4
Allelopathy effects of alfalfa forage exudate on growth of alfalfa.

FIGURE 16-5
Allelopathy effects of alfalfa seed exudate on alfalfa establishment.

emergence occurs when the daily mean air and soil temperatures are near 25°C (77°F). Seedling emergence and growth are minimal when the soil temperature is below 10°C (38°F) or above 35°C (95°F). Most summer annual weeds require temperatures of 25 to 35°C (77 to 95°F) for germination (34). For example, in central Illinois, if one were to seed prior to April 20, one may not need to use a preplant herbicide for weed control. But if seeding is done after May 1, it is recommended that a preplant herbicide be used because the soil and air temperatures are sufficiently high to encourage summer-annual weed germination. Therefore, as a rule of thumb, a certain date related to soil temperature can be set as to when or when not a preplant herbicide should be used. Common preplant herbicides used include Eptam and Balan.

Broadleaf weeds can be controlled by several different postemergence herbicides applied about 1 month after seeding or when weeds are 5 to 7.5 cm (2 to 3 in) tall. One of the most popular postemergence herbicides is 2,3-DB. Generally, weeds will not be a problem in late-summer seedings.

Insect Control

In the midwest, leafhoppers should be sprayed about 30 days after seeding, and then again within 3 to 5 days after each harvest. If no control is practiced, leafhoppers will invade new seedlings and stunt the plants. Leafhoppers are not as serious a problem in the west and southwest as they are in the midwest. It is important to control these pests and so get a new stand off to a fast and healthy start so that if a droughty period is experienced, then the plants will have well-developed roots going into the fall, with sufficient food reserves for winter survival (2).

Harvest Schedule

The first harvest should be taken at the first-flower to one-tenth flower stage (28a). All subsequent harvests should be taken every 28 to 30 days in the west, and every 30 to 35 days in the midwest. The last harvest in the fall should occur about 30 to 35 days prior to fall dormancy in the south or the average killing frost in the midwest (32).

The first-flower stage occurs about 60 to 65 days after seeding in regions with four or more harvests per year. There should be 30 to 35 days of growth following the last harvest in the fall so that sufficient food reserves are stored for the winter.

Cutting height should be as close to the soil surface as possible, or 5 to 7 cm (2 to 3 in). This encourages even regrowth and removes all accumulated top growth. It also, as a result, ensures the removal of all top growth that may have diseases or sites for harboring insects (31).

One should never remove late-fall growth from a stand established in the fall. In the midwest, one should also refrain from taking a late-fall harvest on a stand seeded the preceding spring, although if there is considerable regrowth, one may take a late harvest after a killing frost provided that the soil is well-drained (27). In the southwest, one may get 2 to 4 harvests before winter dormancy occurs. When freezing and thawing may occur, however, it is important to leave late-fall growth on poorly drained soils; this leaves a healthy root system, which helps reduce spring heaving. Assuming good drainage and sufficient root reserves for winter survival and heaving resistance, then, one may harvest after a killing frost (Fig. 16-3). It is important to leave a 10- to 12-cm (4- to 5-in) stubble. This helps insulate the crown during the winter, helps collect snow cover, and helps penetrate ice sheets that may form (27). After a killing frost there will be essentially no regrowth to reduce the food reserves, thereby getting one additional harvest and not lowering the next year's yield (Table 16-3) (31).

Following the year of establishment, the first harvest should be taken at the late-bud or first-flower stage. The remaining portion of the season should be divided into a harvest schedule of every 30 to 35 days in the midwest, and 28 to 30 days in the west (31).

Fertility

The fertility level of an alfalfa field should be maintained by annually applying the amount of P and K removed by harvest. With every ton of forage dry matter removed from the field, one removes 6 to 7 kg (12 to 15 lb) of P_2O_5 and 27 kg (60 lb) of K_2O for a 10-ton yield; this generally amounts to 135 to 168 kg/ha (120 to 150 lb/A) of P_2O_5 and 673 kg/ha (600 lb/A) of K_2O. It is suggested that this

TABLE 16-3
INFLUENCE OF FALL CUTTING
OF ALFALFA ON
FIRST-CUTTING YIELD THE
FOLLOWING SPRING AT
MADISON, WI

Fall harvest	% relative yield
None	100
Sept. 1	90
Sept. 15	69
Oct. 1	83
Oct. 15	92
Nov. 1	97

Source: D. Smith, "Cutting schedules and maintaining pure stands," in C. H. Hanson (ed.), *Alfalfa Science and Technology, Agronomy* 15:481–496 (1972).

application be split, one-half after the last harvest in early fall and one-half after the first harvest in the spring. In this manner, the nutrients are more evenly supplied for growth throughout the year and are sufficient for excellent winter survival. One should apply approximately 2.5 to 3.5 kg/ha (2 to 3 lb/A) of actual boron per year (26).

Insect Control after Year of Establishment

If needed, one should spray for alfalfa weevil control prior to the first harvest or possible second growth. Potato leafhopper control should be practiced 3 to 5 days after each harvest. Leafhoppers and other insects are present throughout most of the growing season, until a killing frost. Controlling the leafhopper also helps control other insects, such as aphids, and reduces the spread of some diseases (2).

Food Reserves

Food reserves were discussed in Chapter 4, but it is important to mention the subject again. Continuous grazing of alfalfa or harvesting it at an immature stage, such as the prebud or vegetative stage, reduces vigor and total dry matter production, and causes early death of the stand. It is also important not to harvest alfalfa during the critical fall period (31). As shown in Table 16-3, if alfalfa is harvested in the fall with less than 30 to 35 days of regrowth before a killing frost, or fall dormancy, the dry matter yield will be significantly reduced the next spring. This period of growth is necessary to permit sufficient top growth and translocation of food storage into the roots before the final freeze (24).

COMPOSITION AND QUALITY

The nutritive value of any forage crop is related to both its physical and chemical composition. Since the leaf fraction contains higher levels of essential nutrients than does the stem, the relative proportion of leaves to stems is important, but more significantly, one must be able to harvest a forage with a high percentage of leaves maintained (4).

Leafiness

There are cultivar differences related to leafiness. In a study conducted in West Virginia in 1969, five cultivars harvested at the late-bud stage ranged from 40.1 to 46.4% leaves (3). Recently released cultivars tend to be higher in leaf content. At the University of Illinois,[1] it was found that various cultivars ranged from 44.5 to 57.8% leaves. Generally, harvests later in the season tend to have a higher percentage of leaves than that in earlier harvests (3). This is associated with higher temperatures and fewer leaf diseases. As the temperature increases, the percentage of leaves in alfalfa increases (22).

Insects often affect the leaf content, and thus the protein content, of alfalfa. Leafhoppers, alfalfa weevils, and pea aphids are probably the most damaging insects in terms of leaf production or leaf damage. It has been shown that when leafhoppers are prevalent and fertility varies, forage production can be significantly reduced (Table 16-4). Leafhoppers reduce protein production by 29% under high fertility and by 43% under low fertility. Leafhoppers, weevils, and

[1] Unpublished data, University of Illinois, 1979.

TABLE 16-4
YIELD AND PROTEIN PERCENTAGES OF "VERNAL" ALFALFA AS INFLUENCED BY LEAFHOPPER DAMAGE AND SOIL FERTILITY

Treatment	D.M. production kg/ha (lb/A)	%	Protein kg/ha (lb/A)	Reduction due to leafhopper, %
High fertility, no leafhopper	2780 (2480)	17.5	486 (434)	—
High fertility plus leafhopper*	2186 (1950)	15.8	345 (308)	29
Low fertility, no leafhopper	2477 (2209)	19.3	478 (426)	—
Low fertility plus leafhopper	1771 (1580)	15.3	271 (242)	43.3

*Average of 80 leafhoppers per sweep (52% nymphs) with a 37.5-cm net.
Source: D. Smith and J.T. Medler, "Influence of leafhoppers on the yield and chemical composition of alfalfa hay," *Agron. J.* 51:118–119 (1959).

pea aphids can be controlled readily by timely spraying. When yellowing of the alfalfa appears, it is too late to spray, for the leafhoppers have already done their damage. Soil fertility also greatly affects dry matter production and total protein produced, as shown in Table 16-4.

Lodging reduces dry matter production and causes leaf droppage. Lodged forage can not always be clipped closely enough to remove all of the top growth, so that yields are reduced and regrowth is slower. Lodging also reduces nitrogen fixation (26, 40).

Cutting at early growth stages provides forage of high quality and high nutrient content. The percentage of constituents important for animal nutrition, such as digestible nutrients, protein, sugars, carotene, minerals, and amino acids, decreases as the plant advances in maturity. Cutting at late maturity causes lodging, loss of leaves, and a higher percentage of stems, with lower protein and digestible nutrient content and a higher fiber content (Table 16-5). Maturity also affects total net energy (Table 16-6).

Leaf diseases play a very important role in leaf production and, therefore, forage quality. Any disease that causes leaf droppage or interferes with photosynthesis affects protein production and percent protein. It is very important to produce disease-free alfalfa.

DISEASES

It is often thought that bacterial wilt, *Corynebacterium insidiosum* (McCull) H.L. Jens., is the most important alfalfa disease. This disease is found in each of

TABLE 16-5
CHANGES IN PERCENT TDN, FIBER, AND PROTEIN WITH
DIFFERENT STAGES OF MATURITY OF AN ALFALFA-GRASS
COMBINATION

Stage of maturity	TDN, %	Fiber, %	Protein, %
First cut			
Prebud	68.4	25.8	20.0
Late bud to bloom	60.3	32.2	16.6
½ bloom	57.4	34.5	15.1
Full bloom–mature	54.0	37.0	13.2
Second cut			
35 days' regrowth	59.0	32.3	16.7
Third cut			
35 days' regrowth	67.0	27.1	19.2

Summary after prebud stage of alfalfa	
Crude protein	Decreased 0.22% per day
TDN	Decreased 0.48% per day
Crude fiber	Increased 0.37% per day

TABLE 16-6
CHANGES IN PERCENT DIGESTIBLE PROTEIN AND NET ENERGY WITH ADVANCES
IN MATURITY OF ALFALFA AND ALFALFA–SMOOTH BROMEGRASS AT ST. PAUL, MN

	Harvest date	Percent digestible protein	Net energy, therms/ton
Alfalfa	May 18	20.0	1240
	June 3	16.9	980
Alfalfa–	June 15	13.2	900
smooth bromegrass	May 17	16.9	1240
	May 24	14.1	1120
	May 31	11.3	1040
	June 7	9.5	960
	June 14	8.6	900
	June 21	5.8	760

Sources: R. F. Barnes and C. H. Gordon, "Feeding value and on-farm feeding," in C. H. Hanson (ed.), *Alfalfa Science and Technology, Agronomy* 15:601–630 (1972); and A. G. Matches, W. F. Wedin, G. C. Marten, D. Smith, and B. R. Baumgardt, "Forage quality of Vernal and DuPuits alfalfa harvested by calendar date and plant maturity schedules in Missouri, Iowa, Wisconsin, and Minnesota," *Wis. Agr. Sta. Res. Rep. 73,* 1973.

the United States. The first symptom is a yellowish-brown discoloration in the woody cylinder of the taproot. Affected plants are stunted, and die within three years. The disease organism enters the root through wounds within a few weeks after germination. This disease is controlled by growing wilt-resistant cultivars (16, 16).

Phytophthora root rot is caused by the fungus *Phytophthora megasperma* Drechs. f. *medicaginis*. It develops rapidly in fields where there is an excess of soil moisture caused by inadequate drainage or excess irrigation. This disease attacks the taproot as a rot, appearing as a yellowish-brown to black girdling lesion (Fig. 16-6). In contrast to bacterial wilt, phytophthora root rot may appear within the first year or whenever excess moisture is present. It occurs over the entire United States. Resistant cultivars and good drainage are good control methods (17).

Fusarium wilt, *Fusarium oxysporum* Schlect. f. *medicaginis* (Weim), attacks alfalfa in the warmer regions of the United States (15a). This disease infects the vascular tissue, first causing stems on one side, and later the entire plant, to die. Brown to dark-red streaks appear in the taproot. Resistant cultivars and crop rotation are effective controls (17).

Fusarium root rot, *Fusarium* spp., occurs most often in the eastern half of the United States, It causes brown, rotted areas on the taproot and crown. This disease progresses rather slowly. Again, resistant cultivars and crop rotations seem to provide adequate control (17).

In late summer, following prolonged periods of hot, wet weather, anthracnose, *Colletotrichum trifolii* Bain, appears as scattered, dead, bleached stems with the shoot tips curved downward, forming a "shepherd's crook." Diamond-

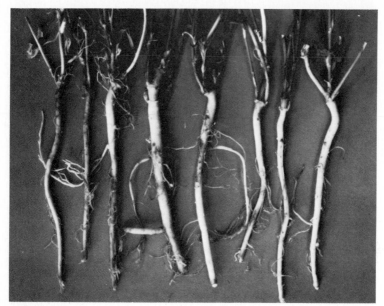

FIGURE 16-6
Phytophthora root rot of alfalfa, which occurs in poorly drained soils.

shaped, tan lesions with dark borders form near the base of the stem. The fungus eventually girdles the stem, causing sudden death or the "shepherd's crook" appearance. Resistant cultivars and crop rotation help control this disease (10, 17).

Other diseases of alfalfa include sclerotinia crown and stem rot, *Sclerotinia trifoliorum* Eriks.; rhizoctonia stem blight, *Rhizoctonia solani* Kuehn.; common leaf spot, *Pseudopeziza medicaginis* (Lib.) Sacc.; lepto leaf spot, *Leptosphearulina briosiana* (Poll.) Graham and Luttrell Sym.; rust, *Uromyces striatus* Schroet.; downy mildew, *Peronospora trifoliorum* D. By.; spring black stem and leafspot, *Phoma medicaginis* Malbr. and Roum. Syn.; summer black stem and leafspot, *Cercospora medicaginis* Ell. and Ev.; bacterial leafspot, *Xanthomonas alfalfae* (Riker, Jones, and Davis) Dows. All of these cause leaf drop, ultimately to death of the stand. There is also an alfalfa mosaic virus (AMV), which creates yellow mottling of the leaflets in no set pattern (17).

Nematodes, which are microscopic roundworms, cause many diseases. Generally, they attack the stems or roots of alfalfa, but they may also attack the crowns, leaves, or seeds. Symptoms are dwarfing of the plant, weakened growth, and swollen areas where the nematodes attack or live in the plant. The stem nematode, *Ditylenchus dipsaci* (Kühn) Filipjer, and the root knot nematode, *Meloidogyne* spp., are found on alfalfa most often in the far west and other dry regions of the United States. Resistant cultivars provide the means of control (17).

INSECTS

There are many insects that attack alfalfa. The alfalfa plant has no pubescence, and it is thought that this is the reason insects attack it more readily than other species. However, hairy peruvian alfalfa and its derivatives do possess pubescence, which aids in insect resistance.

The potato leafhopper, *Empoasca fabae* (Harris), probably causes more yield reduction in the United States than any other insect does (Fig. 16-7). This insect is very small and as an adult is pale green and wedge-shaped. The nymphs are pale yellow and characteristically walk sideways since they cannot fly until they become adults. The nymphs hatch from eggs the adults lay on the plant. The nymphs feed on the plant by sucking its juices and releasing a toxic substance into it. This causes the appearance of a yellow wedge on the tips of the alfalfa leaflets, and then severe stunting until the entire plant turns pale green or yellow, and eventually a reddish color. If alfalfa is not cut and sprayed to control the next hatch, the plants will eventually die (Fig. 16-8). Adult leafhoppers survive in the gulf states and Mexico during the winter. They are transported northward each spring via air currents. Resistant cultivars and spraying are the best controls (2).

A widely publicized insect of alfalfa is the alfalfa weevil, *Hypera postica* (Gyllenhal). This is a damaging insect that defoliates plants by shredding and eating leaves, giving the field a grayish cast (Fig. 16-9). Most of the damage from this insect occurs in the first crop in the upper midwest, to the first two growth periods in the south. Damage occurs in the spring, when the larvae emerge (Fig. 16-10). The larvae have black heads and a white stripe down the back with a dark-green halo. Following the larva stage, the weevils pupate in netlike cocoons either on old plants or debris left in the field (Figs. 16-11, 16-12, 16-13). Control

FIGURE 16-7
Potato leafhopper on an alfalfa leaflet and the chlorotic effect of its feeding.

FIGURE 16-8
Proper leafhopper control results in rapid regrowth and higher yields.

methods include tolerant cultivars, insecticide application, and clean, timely harvest methods (2, 33, 35, 38).

Other prevalent insects that attack alfalfa are the pea aphid, *Acyrthosiphon pisum* (Harris), and the blue alfalfa aphid, *A. kondoi* (Shinji). These insects suck plant juices, which causes the plant to wilt and yellow. They attack the terminal buds and stems during the cool, wet season. It appears that the blue aphid problem will continue to increase in the near future. Resistant cultivars are available for control of each of these insects (2).

FIGURE 16-9
The results of alfalfa-weevil feeding on alfalfa, with individual tall fescue clover unaffected.

FIGURE 16-10
The larvae of an alfalfa weevil,
illustrating its skeletonizing feeding
pattern.

FIGURE 16-11
Pupae stage of alfalfa weevil.

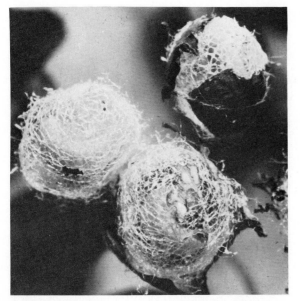

FIGURE 16-12
Adult alfalfa weevil, following the pupae stage. Female on left and male on right.

The spotted alfalfa aphid, *Therioaphis maculata* (Buckton), is most severe in the arid areas of the southwestern and western United States. These insects normally build up during the hot, dry season; however, they can cause extensive damage on first harvest. Plant resistance is the best control (39).

FIGURE 16-13
Cross section of an alfalfa stem, showing the adult alfalfa-weevil puncture of the stem and its eggs on the inside.

Other insects that attack alfalfa include the meadow spittlebug, *Philaenus spumarius* (L.); the variegated cutworm, *Peridroma saucia* (Hubner); the grasshopper, *Melanopus differentialis* (Thomas); and plant bugs, *Lygus lineolaris* (Palisot de Beauvois). Most of these insects can be controlled by insecticides (2, 33).

SUMMARY

Alfalfa is the species to which all others must be compared.

Advantages

1 Assuming good drainage and adequate fertility, alfalfa:
   a has the highest yield of all legumes
   b is the highest protein producer of all legumes
   c is a very good green-manure legume
2 Is resistant to more plant diseases than red clover
3 Is a long-lived perennial
4 Is deep-rooted and drought-resistant
5 Is a good seed producer in the far west
6 Is relatively easy to establish
7 Recovers rapidly after harvest

Disadvantages

1 Does poorly on poorly drained soils
2 Requires high soil pH and fertility
3 Because of lack of pubescence, has more insect problems than other species
4 Heaves more readily than other legumes, except for sweet clover

QUESTIONS

1 Describe the historical importance of alfalfa.
2 Describe the areas of adaptation of alfalfa.
3 How does alfalfa maintain its regrowth?
4 What is the difference between fall dormancy and winterhardiness?
5 Name the different types of alfalfas and why they are important.
6 Describe the value of alfalfa.
7 Describe the establishment of alfalfa.
8 What is the significance of having 30 to 35 days of growth before a killing frost in some parts of the United States?
9 What is the average amount of nutrients removed from a field when harvesting alfalfa?
10 How does the quality of alfalfa change with maturity?
11 Describe the differences between phytophthora root rot and bacterial wilt?
12 Why is the leafhopper considered one of the more serious insect problems of alfalfa?

REFERENCES

1 Ahlgen, G. H. *Forage Crops.* McGraw-Hill, New York, 3d ed., pp. 53–64, 1956.
2 App, B. A. and G. R. Manglitz. Insects and related pests. In C. H. Hanson (ed.), *Alfalfa Science and Technology, Agronomy,* 15:527–554, 1972.
3 Barnes, R. F. and C. H. Gordon. Feeding value and on-farm feeding. In C. H. Hanson (ed.), *Alfalfa Science and Technology, Agronomy,* 15:601–630, 1972.
4 Baumgardt, B. R. and D. Smith. Changes in the estimated nutritive value of the herbage of alfalfa, medium red clover, Ladino clover, and bromegrass due to stage of maturity and year. *Wis. Agr. Exp. Sta. Res. Rept. 10,* 1962.
5 Bolton, J. L. *Alfalfa: Botany, Cultivation, and Utilization.* World Crop Books, Leonard Hill, London, 1962.
6 Bolton, J. L., B. P. Goplen, and H. Baenziger. World distribution and historical developments. In C. H. Hanson (ed.), *Alfalfa Science and Technology, Agronomy,* 15:1–34, 1972.
7 Brown, B. A. Fertilizer experiments with alfalfa, 1915–1960. *Conn. Agr. Exp. Sta. Bull. 363,* 1961.
8 Busbice, T. H., R. R. Hill, Jr., and H. L. Carnahan. Genetics and breeding procedures. In C. H. Hanson (ed.), *Alfalfa Science and Technology, Agronomy,* 15:283–318, 1972.
8 a Collins, Changes in composition of alfalfa, red clover, and birdsfoot trefoil during autumn. *Agron. J.,* 75:287–291, 1983.
9 Davis, R. L. and C. Panton. Combining ability in alfalfa. *Crop Sci.,* 2:35–37, 1962.
10 Devine, T. E., C. H. Hanson, S. A. Ostazeski, and T. A. Campbell. Selection for resistance to anthracnose *(Colletotrichum trifolii)* in four alfalfa populations. *Crop Sci.,* 11:854–855, 1971.
10 a Dutt, T. E., R. G. Harvey, and R. S. Fawcett. Influence of herbicides on yield and botanical composition of alfalfa hay. *Agron. J.,* 75:229–233, 1983.
11 Frakes, R. V., R. L. Davis, and F. L. Patterson. The breeding behavior of yield and related variables in alfalfa. III. General and specific combining ability. *Crop Sci.,* 1:210–212, 1961.
12 Hanson, C. H. and D. K. Barnes. Alfalfa. In *Forages,* Iowa State Univ. Press, Ames, IA, 3d ed., pp. 136–147, 1973.
13 Harte, W. *Essays on Husbandry.* W. Frederick, London, 1770.
14 Heinrichs, D. H. Creeping alfalfas. *Advan. Agron.,* 15:317–337, 1963.
15 Hendry, G. W. Alfalfa in history. *J. Amer. Soc. Agron.,* 15:171–173, 1923.
15 a Hijano, E. H., D. K. Barnes, and F. I. Frosheiser. Inheritance of resistance to fusarium wilt in alfalfa. *Crop Sci.,* 23:31–34, 1983.
15 b Jodari-Karimi, F. V. Watson, H. Hodges, and F. Whisler. Root distribution and water use efficiency of alfalfa as influenced by depth of irrigation. *Agron. J.,* 75:207–211, 1983.
16 Jones, F. R. A new bacterial disease of alfalfa. *Phytopathol.,* 15:243–244, 1925.
17 Kehr, W. R., F. I. Frosheiser, R. D. Wilcoxson, and D. K. Barnes. Breeding for disease resistance. In C. H. Hanson (ed.), *Alfalfa Science and Technology, Agronomy,* 15:335–354, 1972.
17 a Kehr, W. R., G. R. Managlitz, and R. L. Ogden. Insect control on seedling alfalfa by cultivars and soil and foliar insecticides. *Agron. J.,* 74:407–411, 1982.
18 Klinkowski, M. Lucerne: its ecological position and distribution in the world. *Herbage Plants,* 12:1–62, 1933.

19 Kohler, G. O., E. M. Bickoff, and W. M. Beeson. Processed products for feed and food industries. In C. H. Hanson (ed.), *Alfalfa Science and Technology, Agronomy,* 15:659–676, 1972.

19 a Krall, J. M. and R. H. Delaney. Assessment of acetylene reduction by sainfoin and alfalfa over three growing seasons. *Crop Sci.,* 22:762–766, 1982.

20 Krasnuk, M., F. H. Witham, and G. A. Jung. Hydrolytic enzyme differences in cold tolerant and cold sensitive alfalfa. *Agron. J.,* 70:597–604, 1978,

21 Krasnuk, M., F. H. Witham, and G. A. Jung. Dehydrogenase levels in cold tolerant and cold sensitive alfalfa. *Agron. J.,* 70:605–613, 1978.

21 a Leyshon, A. J. Deleterious effects on yields of drilling fertilizer into established alfalfa stands. *Agron. J.,* 74:741–743, 1982.

22 Lowe, C. C., V. L. Marble, and M. D. Rumbaugh. Adaptation, varieties, and usage. In C. H. Hanson (ed.), *Alfalfa Science and Technology, Agronomy,* 15:391–413, 1972.

23 Mainer, A. and K. T. Leath. Foliar diseases alter carbohydrate and protein levels in leaves of alfalfa and orchardgrass. *Phytopathol.,* 68:1252–1255, 1978.

24 Matches, A. G., W. F. Wedin, G. C. Marten, D. Smith, and B. R. Baumgardt. Forage quality of Vernal and DuPuits alfalfa harvested by calendar date and plant maturity schedules in Missouri, Iowa, Wisconsin, and Minnesota. *Wis. Agr. Sta. Res. Rept. 73,* 1973.

25 Niedermeier, R. P., N. A. Jorgensen, C. E. Zehner, D. Smith, and G. P. Barrington. Evaluation of alfalfa-brome forage harvested two and three times annually, stored as low-moisture silage, and fed to lactating dairy cows. *Wis. Agr. Exp. Sta. Res. Rept. 2378,* 1972.

26 Rhykerd, C. L. and C.J. Overdahl. Nutrition and fertilizer use. In C. H. Hanson (ed.), *Alfalfa Science and Technology, Agronomy,* 15:437–468, 1972.

27 Russell, W. E., F. J. Olsen, and J. H. Jones. Frost heaving in alfalfa establishment on soils with different drainage characteristics. *Agron. J.,* 70:869–872, 1978.

28 Schertz, D. L., and D. A. Miller. Nitrate-N accumulation in the soil profile under alfalfa. *Agron. J.,* 64:660–664, 1972.

28 a Sheaffer, C. C. Seeding year harvest management of alfalfa. *Agron. J.,* 75:115–119, 1983.

28 b Sheaffer, C. C. and D. R. Swanson. Seeding rates and grass suppression for sod-seeded red clover and alfalfa. *Agron. J.,* 74:355–358, 1982.

28 c Sheaffer, C. C., D. L. Rabas, F. I. Frosheiser, and D. L. Nelson. Nematicides and fungicides improve legume establishment. *Agron. J.,* 74:536–538, 1982.

29 Sinskaya, E. N. Flora of Cultivated plants of the U.S.S.R. XIII Perennial leguminous plants. Part I. Medic, sweetclover, fenugreek. 1950, translated by Israel Program for Scientific Translation, Jerusalem, 1961.

30 Smith, D. Association of fall growth habit and winter survival in alfalfa. *Can. J. Plant Sci.,* 41:244–251, 1961.

31 Smith, D. Cutting schedules and maintaining pure stands. In C. H. Hanson (ed.), *Alfalfa Science and Technology, Agronomy,* 15:481–496, 1972.

32 Smith, D. The establishment and management of alfalfa. *Wis. Agr. Exp. Sta. Bull. 452,* 1960.

33 Sorensen, E. L., M. C. Wilson, and G. R. Menglitz. Breeding for insect resistance. In C. H. Hanson (ed.), *Alfalfa Science and Technology, Agronomy,* 15:371–390, 1972.

33 a Sumberg, J. E., R. P. Murphy, and C. C. Lowe. Selection for fiber and protein concentration in a diverse alfalfa population. *Crop Sci.,* 23:11–16, 1983.

34 Tesar, M. B. and J. A. Jackobs. Establishing the stand. In C. H. Hanson (ed.), *Alfalfa Science and Technology, Agronomy,* 15:415–435, 1972.

35 Thompson, T. E., R. E. Shade, and J. D. Axtell. Alfalfa weevil resistance mechanism characterized by larval convulsions. *Crop Sci.,* 18:208–209, 1978.

36 Tysdal, H. M., T. A. Kiesselbach, and H. L. Westover. Alfalfa breeding. *Neb. Agr. Exp. Sta. Res. Bull. 124,* 1942.

37 Tysdal, H. M. and H. L. Westover. Alfalfa improvement. *USDA Yearbook Agr.,* pp. 1122–1152, 1937.

38 USDA. The alfalfa weevil: how to control it. *Leaflet 368,* 1971.

39 USDA. The spotted alfalfa aphid: how to control it. *Leaflet 422,* 1957.

40 VanKeuren, R. W. and G. C. Marten. Pasture production and utilization. In C. H. Hanson (ed.), *Alfalfa Science and Technology, Agronomy,* 15:641–658, 1972.

41 Wilcoxson, R. D., D. K. Barnes, F.I. Frosheiser, and D. M. Smith. Evaluating and selecting alfalfa for reaction to crown rot. *Crop Sci.,* 17:93–96, 1977.

CHAPTER 17

RED CLOVER

Red clover, *Trifolium pratense* L., is a very important short-lived perennial legume. It is grown with grasses or in pure stands, primarily in the northern part of the United States and southern Canada. Recently, though, alfalfa has been replacing red clover as a forage legume both as hay and as pasture because of alfalfa's higher yields and longevity. Red clover is widely grown for pasture, hay, and silage in short rotations with small grains and corn. Seed is usually abundant and relatively inexpensive.

HISTORICAL BACKGROUND

It is thought that red clover originated in Asia Minor and southeastern Europe, the same general area as did alfalfa. The ancient Greeks and Romans did not know of red clover as a crop. It was first cultivated in Media and south of the Caspian Sea, approximately where alfalfa was domesticated (35). Red clover was first mentioned as a feed for cows in the thirteenth century. Records indicate that it was cultivated in Spain in 1500, later in Italy, in Holland and Lombardy by 1550, and in France in 1583. By 1645, it was brought to England from Germany. It is believed to have been brought to the United States by English colonists (12, 24). Jared Eliot wrote about the culture of red clover in Massachusetts in 1747 (12).

WHERE RED CLOVER GROWS

Today red clover grows extensively in the United States, eastern Canada, most European countries, New Zealand, and Australia. The main areas of production in the United States are from the midwest to the south central region and to the northeast, and extending into Ontario and Quebec, Canada. This area is best described as humid with summer temperatures moderately cool to warm and with sufficient rainfall during the growing season (35).

Further south, into Arkansas, Tennessee, and North Carolina, red clover grows like a winter annual. In this region, when red clover is sown in the fall, it germinates and grows throughout the winter in a rosette-type pattern (35). In the far west, from western Montana southward through Utah, Nevada, and most of California, red clover grows in lower elevations under irrigation as a biennial (35). In the higher elevations of the far west and southeastern United States, it grows as a short-lived perennial.

Red clover grows best on well-drained soils, but will tolerate soils that are not very well drained. It grows better on poorer-drained soils than does alfalfa, and it is not as sensitive to acidity. Liming is desirable for obtaining the highest yield on acidic soils. Red clover is better adapted to heavier-textured soils than it is to light sandy or gravelly soils (35), and it requires high fertility.

HOW RED CLOVER GROWS

Red clover has a taproot system. The taproot is heavily branched, and a large portion of the plant's root system is concentrated in the upper 30 cm (12 in) of the soil (Fig. 17-1). As a result, red clover does not tolerate drought conditions as well as alfalfa does.

Botanically, red clover is a short-lived perennial, but agronomically it grows as a biennial in the northern portion of the United States and as a winter annual in the south. Its biennial nature is the result of early death, caused by various disease organisms that attack the plants (35). However, new cultivars with improved disease resistance provide the opportunity for more than 2 years of production.

Red clover has an upright, determinate growth habit. The first year's growth is a rosette-type, with numerous leafy stems arising from a crown. The crown of red clover is set more above-ground than is alfalfa's. There are frequently fewer upright stems from each crown after the first year (10, 33).

The flowers of red clover are borne on a head at the tip of the branch. Flowering progresses down the stem, with flowers rising from axillary buds on lateral branches. Flowering of red clover requires a 14-hour photoperiod, but mammoth red clover requires a longer photoperiod, of approximately 18 hours. As many as 125 flowers may be present per head (45) (Fig. 17-2). The flowers are usually reddish-purple, with a few of lilac, rose, or pink. The flowers are self-sterile and must be cross-pollinated (35). The flowers are typical legume flowers, but with a rather long corolla tube (34). Flowers produced early in the year, or following a cool spring, have a longer corolla than do those produced during the warmer months. Generally, the longer corollas produced early in the year are too long for honeybees to collect nectar, resulting in low seed yields. Therefore, the second growth usually produces more seed than does the first growth (34).

FIGURE 17-1
Red clover taproot and crown
(Courtesy University of Illinois)

Following pollination, red clover flowers turn brown. One seed develops per flower. The seeds are short, 2 to 3 mm long, mitten-shaped, long, and varying in color, from purple to yellow (34).

Red clover leaves are palmately trifoliolate, usually with distinctive, light-colored, "crescent" markings, approximately in the center of each leaflet (Fig. 17-3). The leaves and stems of strains found in the United States are pubescent, whereas the European strains are smooth. This trait, pubescence, helps reduce the frequency of insect visits, but may increase disease problems because it helps retain moisture on the stems and leaves. The pubescence tends to create an annoying dusty hay (35).

Established red clover is not as winter-hardy as are alfalfa and sweet clover. Bula and Smith (6) found that compared with those two species, red clover had a greater loss of available carbohydrates during winter dormancy because of its higher level of metabolic activity (Table 17-1). Smith (29, 33) also found that those plants that flowered during the seedling year were more susceptible to winter injury than were those that did not flower. Conversely, red clover seedlings appear to have more tolerance for cold weather than do seedlings of

FIGURE 17-2
Close-up of red clover flowering head.

FIGURE 17-3
Close-up of red clover leaf, showing the crescent marking and pubescence on the leaf.

TABLE 17-1
PERCENT LOSS OF TOTAL AVAILABLE CARBOHYDRATES OVER THE
WINTER IN THE ROOTS AND CROWNS OF RED CLOVER, ALFALFA, AND
SWEET CLOVER AT MADISON, WI

	Red clover	Alfalfa	Biennial sweet clover
2 year average loss	56.6%	48.5%	27.5%

Source: D. Smith, *Forage Management in the north*, Kendall-Hundt, Dubuque, IA, 3d. ed.,
pp. 101–111, 1975.

other legumes. Red clover seedlings are more competitive than are those of
many other legumes and grasses, especially in the early spring, when temperatures are cool and there is sufficient moisture. Red clover is also more
shade-tolerant than are alfalfa and ladino clover (4, 31).

Later in the year, when temperatures increase and moisture decreases, red
clover is adversely affected. It has been shown that when temperatures reach
35/27°C (95/80°F) day/night, its carbohydrate root reserves at the early flowering
stage are very low, but when grown under cooler temperatures, 24/15°C
(75/60°F) the root reserves are much higher (32). Low moisture supply is often
associated with high temperatures and since red clover has a relatively shallow
root system, after it flowers, it cannot absorb sufficient water from the dry soil to
maintain a healthy, vigorous stand. Soon diseases attack the weakened plants,
and they often die.

TYPES OF RED CLOVER

There are two major types of red clover: (1) the early, two-cut type, called
medium red clover, and (2) the late, one-cut type, called mammoth. Most of the
red clover grown in the United States is of the medium type.

Because of its longer photoperiod requirement, mammoth red clover flowers
approximately 2 weeks later than the medium type does. In fact, mammoth red
clover does not flower the seedling year, though it does form a rosette-type
growth. Generally, only one harvest is taken from the mammoth type because it
produces very little regrowth (12). Compared with the medium type, mammoth
red clover is more winter-hardy and more adapted to Canadian conditions.

Red clover cultivars from Canada and European countries do not perform
well in the United States. In fact, most red clover cultivars and strains do not
perform well in areas far from where they were developed.

USES OF RED CLOVER

Red clover is used mostly for hay or pasture. It is often grown with companion
forage grasses and makes a very high quality hay when field-cured without rain

damage and stored inside a barn. Red clover is easily cut and crimped into a windrow for field curing. This reduces leaf losses and shortens the time needed for drying. Red clover and red clover–grass mixtures make excellent pasture crops for livestock. The carrying capacity, however, is less than that of alfalfa or alfalfa-grass mixtures. Red clover grown with forage grasses also makes good silage when slightly wilted.

Red clover is a principal legume used in pasture renovation because of its seedling vigor and ability to withstand shading. It has been used as a forage crop in short rotations more than has alfalfa. Red clover also has great value as a green manure crop for helping to improve the physical conditions of the soil and ensuring high yields of a following cultivated crop.

DESIRABLE CHARACTERISTICS

Producers are always looking for cultivars that have more disease resistance and higher yields. As stated earlier, red clover has excellent seedling vigor, and it works well in pasture renovation. But the regrowth of red clover is not as rapid as that of alfalfa, and it produces only two or perhaps three harvests per year. If a grower wants a cultivar that produces more dry matter per year, he or she should select alfalfa over red clover. Red clover has fairly good resistance to many insects, but it is very susceptible to many diseases because of its inability to form a callus, or to heal, when injured (35).

The fertility requirements for red clover are not as high as those for alfalfa or sweet clover but they are higher than those of lespedeza. In addition, compared with alfalfa, red clover has more resistance to heaving since it has a more branched root system (35).

Locally produced red clover seed is generally sufficient, which reduces seed costs for common red clover. The seed supply for improved cultivars is rather limited.

MANAGEMENT OF RED CLOVER

Establishment

Red clover grows best on well-drained, loamy soils, but it is also adapted to soils that are not very well drained. It is most productive on soils with a pH of 6.0 to 7.5; on acid soils, though, it is more productive than is alfalfa (12). Like alfalfa, red clover removes considerable amounts of P and K from the soil. Low yields of red clover decrease the total amount of P and K removed.

Seeding

A fine, firm seedbed is important when seeding red clover. When seeding any forage crop, in fact, if the soil is loose, it should be packed before and after seeding with a cultipacker or corrugated field roller.

A high percentage of red clover is seeded with a small-grain companion crop such as oats, wheat, or barley. When seeded in this way, it is recommended that the small-grain seeding rate be reduced by 25% compared with the rate used when the small grain is grown alone. This ensures less competition for the red clover.

When red clover is grown in the northern half of the United States, early spring seeding is preferred. When seeding red clover in an established winter small-grain companion crop, it is usually broadcasted over the small grain early in the year, in February or March. Red clover is easily established when there is good seedling cold tolerance and sufficient moisture, in conjunction with alternate freezing and thawing.

Red clover is seeded from October to November 15 in the deep south, but from September 1 to October in the upper south. In the far west, it may be sown as late in the winter as February because of the area's mild winters with sufficient moisture (35, 38).

Seeding rates for red clover seeded alone range from 9 to 11 kg/ha (8 to 10 lb/A) in the midwest; the rate is slightly higher in the eastern United States (12). When seeded with a companion forage grass, the seeding rate is approximately 4.5 to 7 kg/ha (4 to 6 lb/A).

Weed Control

Approximately the same weed control practices used with alfalfa are applicable for red clover.

Insect Control

Insects can be a pest on red clover, although in general there is much less of a problem with strains grown in the United States since pubescence is present. Several insects that do cause some damage on red clover will be discussed later in the chapter. Insecticides, timely harvests, and seeding with forage grasses are the best control practices.

Harvest Schedule

When red clover is sown with a companion small grain, there is generally one harvest the seeding year. This harvest is taken approximately 35 to 40 days before the average killing frost or fall dormancy. In the upper midwest, assuming that winter wheat is harvested around the first week of July, and following straw and stubble removal, there will be from 40 to 55 days of growth before the red clover is harvested. There will be very little bloom the seedling year, but the late fall growth period allows enough time for sufficient root reserves to be built up for winter survival (31).

When a companion crop of wheat or oats is harvested, if the red clover was not heavily seeded, it will be approximately 20 to 36 cm (8 to 14 in) tall, or about stubble height. One can then harvest the remaining stubble and red clover as a hay crop. This type of hay will be approximately 12 to 14% protein, sufficiently high for a beef cow herd or for nonlactating animals.

Red clover regrowth should be cut close to the soil surface, approximately 5 to 10 cm (2 to 4 in). This ensures the removal of all top growth and reduces the incidence of disease buildup (12).

During the year following establishment, red clover should be harvested when it is in about 50% bloom. If harvested too early, weakened stands result because insufficient food reserves are stored for rapid regrowth (12). Conversely, if harvested too late, regrowth occurs during the drier, hotter months, resulting in slower or little regrowth.

It has been shown that red clover dry matter yields increase until the plants are just beyond full bloom (30, 32). When grown in the upper midwest, there are three separate growths of red clover from the crown (31). The second and third harvests are taken when about 50% of the plants are blooming. These harvests may be determined by setting the last harvest approximately 35 to 40 days before the average killing frost or fall dormancy, provided that a third year of growth is desired. If red clover is to be grown for only two years, then the third or last harvest can be taken when the greatest amount of dry matter has been produced. Alfalfa generally produces one more harvest per year than does red clover, and alfalfa lives longer.

Fertility

Red clover removes about the same amount of plant nutrients per unit of dry matter as does alfalfa, although with lower yields, the amount of nutrients removed is reduced. A fertility program similar to that outlined for alfalfa should be followed for red clover, except that boron may not be needed. The only other difference is that one should top dress the fertilizer needed after the last harvest in the fall.

Insect Control after Year of Establishment

As mentioned earlier, there are fewer insect problems with red clover than there are for alfalfa because of the pubescence that is present on most strains grown in the United States (12, 35). Most insects can be controlled by the application of an insecticide after each harvest; low, clean harvesting practices; and crop rotation.

Food Reserves

The food reserves in medium red clover are typical of those in any perennial legume (Fig. 17-4). The highest storage of food reserves is at or just past full bloom. Mammoth red clover follows a very similar cyclic pattern (32). Com-

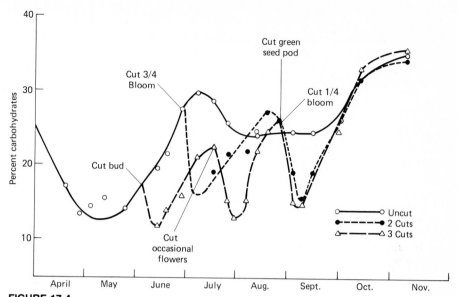

FIGURE 17-4
Seasonal variation of percent carbohydrates stored in red clover under three different cutting schedules at Madison, Wisconsin (31).

pared with alfalfa, red clover has a lower level of stored food reserves when managed in a similar fashion (29).

COMPOSITION AND QUALITY

Quality of red clover depends upon maturity, leafiness, lack of diseases, percent fiber, percent protein, and digestibility. Red clover quality decreases with maturity, as shown in Table 17-2, but not as rapidly as that of alfalfa beyond the bloom stage.

SEED PRODUCTION

The second growth of red clover is often harvested for seed. Seed production is secondary to forage production in the eastern half of the United States, whereas in the northwestern part of the country, seed production is the primary purpose (1).

Since red clover is self-sterile, pollinating insects are essential for seed production. Honeybees, *Apis mellifera* L., and bumblebees, *Bombus* spp., are the primary pollinators. Honeybees collect nectar primarily, whereas bumblebees collect pollen. The bumblebee population is often low, so honeybees must be brought in to facilitate pollination. It is to one's advantage to eliminate other

TABLE 17-2
CHANGES IN DRY MATTER PRODUCTION, TDN, FIBER, AND PROTEIN WITH
DIFFERENT STAGES OF MATURITY OF RED CLOVER

Stage of maturity	Dry matter, kg/ha	TDN, %	Fiber, %	Protein, %
First growth				
Prebud	1789	68.5	15.5	26.5
Late bud	3603	60.9	22.0	20.5
Full bloom	4493	62.9	26.2	15.9
Early seed pod	5188	59.7	28.4	15.3
Seed set	5159	56.2	29.6	14.6
Second growth				
Prebud	1112	68.8	16.2	24.5
1/2 bloom	1491	68.5	20.6	21.4
Full bloom	1525	65.2	24.2	17.8
Early seed pod	1948	60.9	28.2	16.6
Seed set	1769	55.1	26.9	16.8

Source: D. Smith, "Chemical composition of herbage with advance in maturity of alfalfa, medium red clover, ladino clover and birdsfoot trefoil," *Wis. Agr. Exp. Sta. Res. Rep. 16,* pp. 3–10, 1964.

flowering plants in the vincinity so that the honeybees will work only the red clover plants (12).

Seed may be combined standing or from a windrow. When red clover seed heads mature, they turn brown and the stems turn brownish-yellow in color. This generally occurs approximately 25 to 30 days after full bloom. Following harvest, the seed is cleaned to remove trash and immature seeds (46).

Hybrid Seed Production

In recent years, the self-incompatability mechanism in red clover has broken, so that self-seed can be produced by means of a process called pseudo self-compatability (22). Inbred lines can be produced, thus allowing for single crosses and double-cross-hybrid seed production (2, 8, 39). Hybrid red clover is now available on a limited basis.

DISEASES

Red clover is plagued with diseases, which makes the species grow as a biennial. Northern anthracnose, *Kabatiella caulivora* (Kirch.) Korak., is one of the most serious diseases of red clover (35) (Fig. 17-5). It causes small, dark lesions on the stems and sometimes on the leaves. The lesions increase in size until they encircle the stem, causing the plant to wilt and die, creating a droop. This disease is most serious in wet climates with temperatures of 20 to 25°C (68 to 77°C). A number of cultivars are now resistant to it.

Southern anthracnose, *Colletotrichum trifolii* Bain, is more important in red

FIGURE 17-5
Northern anthracnose of red clover, with healthy, normal plant on left.

clover's southern regions of adaptation (35). This disease requires a higher temperature than does northern anthracnose, but it results in similar symptoms. Southern anthracnose may develop as far north as southern Wisconsin, and northern anthracnose as far south as southern Illinois. There are resistant cultivars available.

Root and crown rots are caused by several different fungi, *Fusarium*, *Sclerotinia*, and *Rhizoctonia*, (19, 25). These fungi are present in the area where red clover is grown and, depending upon plant or winter injury, soil type, and climatic conditions, these diseases cause severe thinning of the stand (35).

Viruses also attack red clover. Those viruses found most often are bean yellow mosaic virus, pea streak virus, and vein mosaic virus. Viruses are more common following periods of stress (25).

Other diseases of red clover include powdery mildew, *Erysiphe polygoni* DC; spring black stem, *Phoma trifolii* Johnson and Valleau; summer black stem, *Cercospora zebrina* Pass.; target spot, *Stemphylium sarcinaeforme* (Cov.) Wiltshire; and pepper spot, *Pseudoplea trifolii* (Rostr.) Petr. Crop rotation, resistant cultivars, good fertility, and harvest at proper stage of maturity are various appropriate control measures (35).

INSECTS

Insects cause stand loss and reduced yields in red clover. The clover rootborer, *Hylastinus obscurus* (Marsham) damages plants during the seedling year by tunneling through the root (27). Crop rotation and some insecticides are control methods that may be applied.

The potato leafhopper, *Empoasca fabae* (Harris), damages strains of red clover that lack pubescence (16). The meadow spittlebug, *Philaenus spumarius* (L.); the pea aphid, *Acrythosiphon pisum* (Harris); and the yellow clover aphid, *Therioaphis trifolii* (Monell) are some additional insects that may attack red clover (35).

In seed-producing fields, one may look for the clover seed chalcid, *Bruchophagus platyptera* (Walker), which attacks developing legume seeds. The larvae feed on the young seed. The lesser cloverleaf weevil, *Hypera nigrirostris* (F.), also attacks the developing seed (35).

Insecticides, crop rotation, clean harvests, and resistant cultivars are the best control practices for insects that attack red clover.

SUMMARY

The advantages and disadvantages of red clover are as follows:

Advantages

1 It is adapted to a wide range of climatic conditions.
2 It is easy to establish.
3 It may be better than alfalfa when needed for short duration in pasture renovation programs.
4 It is a high-quality forage.
5 Its quality remains higher beyond maturity than does alfalfa's.
6 It has a more fibrous root system than does alfalfa.
7 It is resistant to more insects than alfalfa is.
8 Its regrowth is fairly rapid.
9 Compared with alfalfa, it will grow on more-acid, less-fertile, and wetter soils.

Disadvantages

1 It is a short-lived perennial.
2 Many diseases attack red clover.
3 It is not as high-yielding as alfalfa is.
4 It produces a dusty hay because of its pubescence.

QUESTIONS

1 Describe the areas where red clover is adapted.
2 Compare the growth of red clover with that of alfalfa.

3 Why does the second growth of red clover produce more seed than does the first? Why does mammoth red clover produce the most seed on the first growth the second year?

4 Why is red clover more competitive in a small-grain companion crop than is alfalfa?

5 Discuss the insect resistance of red clover and the control of those insects that do attack it.

6 Compare the harvest schedule of red clover with that of alfalfa.

7 Compare the quality of red clover hay with that of alfalfa in relation to maturity and time of harvest.

8 Why are food reserves important in red clover?

9 Why do diseases of red clover play such an important role in its culture?

10 What are the advantages and disadvantages of red clover?

REFERENCES

1 Ahlgren, G. H. *Forage Crops.* McGraw-Hill, New York, 2d ed., pp. 77–93, 1956.

2 Anderson, N. K., N. L. Taylor, and R. Kirthithavip. Development and performance of double-cross hybrid red clover. *Crop Sci.,* 12:240–242, 1972.

3 Baumgardt, B. R. and D. Smith. Changes in the estimated nutritive value of the herbage of alfalfa, medium red clover, ladino clover, and bromegrass due to stage of maturity and year. *Wis. Agr. Exp. Sta. Res. Rept. 10,* 1962.

4 Bula, R. J. Vegetative and floral development in red clover as affected by duration and intensity of illumination. *Agron. J.,* 52:74–77, 1960.

5 Bula, R. J., R. G. May, C. S. Garrison, C. M. Rincker, and J. G. Dean. Floral response, winter survival, and leaf mark frequency of advanced generation seed increases of 'Dollard' red clover, *Trifolium pratense* L. *Crop Sci.,* 9:181–184, 1969.

6 Bula, R. J. and D. Smith. Cold resistance and chemical composition in overwintering alfalfa, red clover, and sweet clover. *Agron. J.,* 46:397–401, 1954.

7 Clark, R. U. and J. H. Reynolds. Changes in stand density and carbohydrate root reserves of two varieties of red clover with several cutting managements. *Tenn. Farm and Home Sci. Prog. Rep. 82,* 1972.

7 a Collins, M. Changes in composition of alfalfa, red clover, and birdsfoot trefoil during autumn. *Agron. J.,* 75:287–291, 1983.

8 Cornelius, P. L., N. L. Taylor, and M. K. Anderson. Combining ability in I. single crosses of red clover. *Crop Sci.,* 17:709–713, 1977.

9 Cressman, R. M. Internal breakdown and persistence of red clover. *Crop Sci.,* 7:357–371, 1967.

10 Cumimings, B. G. The control of growth and development in red clover (*Trifolium pratense* L.) I. Recording and analysis of developmental patterns in morphogenesis. *Can. J. Plant Sci.,* 39:9–24, 1959.

11 Dischum, S. and L. Henson. Symptom reaction of individual red clover plants to yellow bean mosaic virus. *Phytopathol.,* 46:150–152, 1956.

12 Fergus, E. N. and E. A. Hollowell. Red clover. *Advan. Agron.,* 12:365–436, 1960.

13 Gorman, L. W. Effect of photoperiod on varieties of red clover, *Trifolium pratense* L. *N. Z. J. Sci. Tech.,* 37:40–54, 1956.

14 Goth, R. W. and R. D. Wilcoxson. Effect of bean yellow mosaic virus on survival and flower formation in red clover. *Crop Sci.,* 2:426–429, 1962.

15 Herron, J. C. Biology of sweet clover weevil and notes on the biology of the clover root curculio. *Ohio J. Sci.,* 53:105–112, 1953.

16 Hollowell, E. A., J. Wontieth, Jr., and W. P. Flint. Leafhopper injury to clover. *Phytopathol.,* 17:399–404, 1927.

17 Kendall, W. A. The persistence of red clover and carbohydrate concentration in the roots at different temperatures. *Agron. J.,* 50:657–659, 1958.

18 Kendall, W. A. and E. A. Hollowell. Effect of stage of development on carbohydrate content, growth, and survival of red clover. *Agron J.,* 51:685–686, 1959.

19 Kilpatrick, R. A., E. W. Hanson, and J. G. Dickson. Root and crown rots of red clover in Wisconsin and the relative prevalence of associated fungi. *Phytopathol.,* 44:252–259, 1954.

20 Klebesadel, L. J. and D. Smith. The influence of oats stubble management on the establishment of alfalfa and red clover. *Agron J.,* 50:68–83, 1958.

21 Kreitlow, K. W., J. H. Graham, and R. J. Garber. Diseases of forage grasses and legumes in the northeastern states. *Pa. Agr. Exp. Sta. Bull. 573,* 1953.

22 Leffel, R. C. Pseudo self-compatibility and segregation of gametophytic self-incompatibility alleles in red clover, *Trifolium pratense* L. *Crop Sci.,* 3:377–380, 1963.

23 Ludwig, R. A., H. G. Barrales, and H. Steppler. Studies on the effects of light on the growth and development of red clover. *Can. J. Agr. Sci.,* 33:274–287, 1953.

24 Merkenschlager, F. Migration and distribution of red clover in Europe. *Herb. Rev.,* 2:88–92, 1934.

25 Newton, R. C. and J. H. Graham. Incidence of root-feeding weevils, root rot, internal breakdown, and virus and their effect on longevity in red clover. *J. Econ. Entomol.,* 53:865–867, 1960.

26 Ohio Agr. Exp. Sta. Handbook of experiments in agronomy. *Spec. Cir. 53,* pp. 48–49, 1938.

27 Pruess, K. P. and C. R. Weaver. Estimation of red clover yield losses caused by clover root borer. *J. Econ. Entomol.,* 51:491–492, 1958.

28 Rincker, C. M., J. G. Dean, C. S. Garrison, and R. G. May. Influence of environment and clipping on the seed-yield potential of three red clover cultivars. *Crop Sci.,* 17:58–60, 1977.

28 a Sheaffer, C. C. and D. R. Swanson. Seeding rates and grass suppression for sod-seeded red clover and alfalfa. *Agron. J.,* 74:355–358, 1982.

29 Smith, D. Carbohydrate root reserves in alfalfa, red clover, and birdsfoot trefoil under several management schedules. *Crop Sci.,* 2:75–78, 1962.

30 Smith, D. Chemical composition of herbage with advance in maturity of alfalfa, medium red clover, ladino clover and birdsfoot trefoil. *Wis. Agr. Exp. Sta. Res. Rept. 16,* pp. 3–10, 1964.

31 Smith, D. Forage management in the north. Kendall-Hund, Dubuque, IA, 3d ed., pp. 101–111, 1975.

32 Smith, D. Influence of temperature on yield and chemical composition of five legume species. *Agron. J.,* 62:520–521, 1970.

33 Smith, D. Reliability of flowering as an indication of winter survival in red clover. *Can. J. Plant Sci.,* 43:386–389, 1963.

34 Starling, T. M., C. P. Wilsie, and N. W. Gilbert. Corolla tube length studies in red clover. *Agron. J.,* 42:1–8, 1950.

35 Taylor, N. L. Red clover and alsike clover. In *Forages,* Iowa State Univ. Press, Ames, IA, 3d ed., pp. 148–158, 1973.

36 Taylor, N. L., M. K. Anderson, and D. M. TeKrony. Producing red clover seed in Kentucky. *Ky. Agr. Ext. Serv. AGR-2,* 1972.

37 Taylor, N. L., E. Dade, and C. A. Garrison. Factors involved in seed production of

red clover clones and their polycross progenies at two diverse locations. *Crop Sci.,* 6:535–538, 1966.

38 Taylor, N. L., J. K. Evans, and G. Lacefield. Growing red clover in Kentucky. *Ky. Agr. Ext. Serv. AGR-33,* 1975.

39 Taylor, N. L., K. Johnston, M. K. Anderson, and J. C. Williams. Inbreeding and heterosis in red clover. *Crop Sci.,* 10:522–525, 1970.

40 Taylor, T. H., W. C. Templeton, Jr., and E. M. Smith. History and concept of grassland renovation in Kentucky. *Proc. 334d. So. Pasture and Forage Crop Improv. Conf.,* Miss. State, Miss. 1976.

41 Thomas, H. L. Breeding potential for forage yield and seed yield in tetraploid versus diploid strains of red clover *(Trifolium pratense). Crop Sci.,* 9:365–366, 1969.

42 Thomas, H. L. Inbreeding and selection of self-fertilized lines of red clover, *Trifolium pratense. Agron. J.,* 47:487–489, 1955.

43 Torrie, J. H. and E. W. Hanson. Effects of cutting first year red clover on stand and yield in the second year. *Agron. J.,* 47:224–228, 1955.

44 Van Riper, G. E. and D. Smith. Changes in the chemical composition of the herbage of alfalfa, medium red clover, ladino clover, and bromegrass with advance in maturity. *Wis. Agr. Exp. Sta. Res. Rept. 4,* pp. 1–5, 1959.

45 Wilsie, C. P. Producing alfalfa and red clover in Iowa. *Agron. J.,* 41:545–545, 1949.

46 Winch, J. E. and W. E. Tossell. Management of medium red clover for seed and hay production. *Can. J. Plant Sci.,* 40:21–28, 1960.

CHAPTER 18

LADINO CLOVER

Ladino clover, or white clover, *Trifolium repens* L., is one of the most nutritious and widely grown forage legumes in the world. When grown under favorable conditions, it is a long-lived perennial. Some strains or cultivars react more like annuals or biennials depending upon drought, diseases, insects, and management practices.

There are three major types of white clover: small, intermediate, and large. The small type is often considered a weed or weedy-type white clover, whereas the intermediate type is called common white clover. The large or giant type, called ladino, is an autotetraploid, that is, 2X chromosome number of common white clover. Ladino clover is often grown with other forage grass species.

HISTORICAL BACKGROUND

White clover may have originated in the Near East (12). Migratory grazing animals and animals used for transportation helped spread it throughout western Asia and Europe. White clover was first cultivated in the Netherlands in the late sixteenth century (3, 14). Later it moved to England (27).

It is believed that the British colonists brought white clover seed to the east coast of America, but French settlers and missionaries brought it into the Ohio River valley. In 1891, the North Carolina experiment station raised the first ladino clover crop introduced from Italy (6). Ladino clover was not really recognized as a forage in the northeast until the 1930s. Several years later it moved into the southeastern and north central United States (1). Since 1940, ladino clover has been a very popular pasture forage.

WHERE LADINO CLOVER GROWS

White clover grows in the temperate regions of the world. In the United States, it is grown primarily from Minnesota to Louisiana and eastward. In the Pacific Coast area, the northwest to the southwest, it is grown under irrigation.

Ladino clover is well-adapted to very fertile soils with sufficient moisture. Since it has a very shallow, fibrous root system, it needs considerable moisture. If grown on a sandy soil, it needs irrigation. A soil pH of 6 to 7 is preferred for ladino clover grown in the United States, whereas in Canada strains have been developed that tolerate acid soils with a pH of 4.5 (19). White clover will not tolerate highly alkaline soils.

HOW LADINO CLOVER GROWS

Ladino clover is seldom grown in pure stands because its very succulent, high moisture content makes it difficult to cure when harvesting it as hay, and also creates a great possibility of bloat when grazing. It is longer-lived as a perennial in the north than in the subtropical south. The intermediate type grows as a winter annual in the south. It has a prostrate growth habit. Unlike red clover, white clover is glabrous. The individual plant may spread in diameter up to more than 1 m. Its height may be from 30 to 45 cm (15 to 18 in) for the ladino type (14) (Fig. 18-1).

White clover seedlings have a rosette-type leaf growth. The plant develops from a small crown, from which creeping fleshy stems or stolons arise and spread outward (Fig. 18-2). The young seedling develops a short primary stem with several internodes. The primary stem stops growing and the stolons develop in about 2 months. Leaflets are heart-shaped to elliptical and borne on long petioles. Adventitious roots may develop at the nodes of the stolon; these generally lie on the soil surface (14).

The root system of ladino clover starts as a deep, primary taproot of approximately 1 m in depth, but it dies during the second year of growth (29). Then the secondary root system, which comprises the roots developed from the nodes of the stolon, acts as the main root system. This root system is very shallow and fibrous.

Ladino clover flowers develop following long days of 14 hours and 15 minutes, whereas vegetative growth has been shown to be promoted following short days and cool temperatures (14). The intermediate type will flower at a much shorter daylength. Flowers develop on peduncles, which may be longer or taller than the leaves produced on petioles. The flower head is round and white, and composed of 20 to 150 florets. The flowers are perfect and borne on a short pedicel.

White clover seeds mature approximately 30 days after pollination. The seeds are small and heart-shaped to oval. Seed color ranges from mainly yellow to reddish and brownish with age. White clover seed has a very hard seedcoat. When the seedheads are consumed by cattle, the hard seed may pass through the digestive system intact, which allows the cattle to help spread the species.

FIGURE 18-1
Ladino clover in flowering stage.
(Courtesy University of Illinois)

TYPES OF WHITE CLOVER

As mentioned earlier in the chapter, there are three types of white clover: large, intermediate, and small. The large type, generally grown in the northern region, is known primarily as 'Ladino' clover. Ladino grows to about two to four times the size of the small or common white clover. Size alone is the main morphological difference among the three types (14).

This size difference means that the ladino type makes greater demands upon soil nutrients. To obtain maximum growth and production from ladino clover, one needs to fertilize the crop adequately and add the proper amounts of lime to the soil. When ladino clover is grown with a companion forage grass, proper management is very important if one wants to maintain the clover (1). Ladino clover cannot be distinguished from the other white clover types by seed appearance. Therefore, it is important to use certified seed.

The intermediate type of white clover is grown primarily in the south, where it is widely used in permanent pastures as a reseeding annual. There are

FIGURE 18-2
Ladino clover stolon. *(Courtesy University of Illinois)*

numerous local ecotypes. The intermediate types blend nicely with dallisgrass, and are generally more heat-tolerant and produce more seed than do ladino clovers.

Small white clover grows freely throughout the United States and much of Canada. It grows voluntarily in pastures and lawns, where it is not considered a particularly desirable plant. When grown in areas of its adaptation, production of small white clover is generally much lower than that of either the ladino or intermediate type (2).

USES OF WHITE CLOVER

White clover is probably one of the most important pasture legumes in the temperate zone. It is grown with forages grasses to help reduce the hazard of bloat in grazing cattle (26). It has been seeded alone for swine or poultry pasture. Very little white clover is ensiled. Some white clover–grass combinations are harvested as hay crops, but they are difficult to field cure (2).

Ladino makes a highly palatable forage because only the leaves and petioles are removed under average grazing conditions. This makes it a highly concentrated, nutritive forage. Since only the leaves and petioles are grazed, ladino can be grazed early and closely without damage to the plant. The only time that stems or stolons might be grazed is during overgrazing (7, 11). Ladino is a relatively high-yielding legume and is excelled only by birdsfoot trefoil and subclover in its tolerance to close grazing (14).

Ladino responds to irrigation. It grows well on soils with a high water table and on poorly drained soils, and can withstand higher-rainfall conditions. White clover–grass mixtures are excellent in soil conservation and watershed management programs. White clovers are also used a great deal in pasture renovation when legumes are completely missing in a permanent pasture. Earlier research estimated that when white clover is added to a grass pasture, approximately 225 kg N/ha (200 lb/A) is returned to the soil in the United States (twice this amount in a 10-month growing season in New Zealand) by transfer of N from the movement of nitrogenous compounds from the nodules, the decay of roots and top growth, and the return of animal wastes. Documentation of these figures has been obtained by evaluating the production of pastures receiving up to the estimated amount of N fertilizer per hectare (14). Recently researchers have questioned whether it is really N released from the nodules by means of using labeled N. They found that legumes are utilizing considerable amounts of soil N.

Ladino clover is an excellent source of protein and minerals because only the leaves are grazed (Table 18-1). The forage is low in fiber and very succulent. Its feeding value changes very little with maturity. Protein percentages range from 20 to 30% on a dry-weight basis. It has a rapid recovery when there is sufficient moisture.

DESIRABLE CHARACTERISTICS

Farmers like ladino clover in grass-legume grazing situations because it is long-lived and highly nutritious. Ladino clover acts as a perennial, not by the continued life of the original plant, but by means of reseeding and stolons that develop new root systems (7). The old, original plant often dies in 1 or 2 years. The perennial nature results from the production of new stolon growth each year.

Regrowth of ladino clover is rapid and very succulent. In fact, it may be too succulent, which hinders field curing if the crop is harvested for hay. Cattle may

TABLE 18-1
COMPOSITION OF LADINO CLOVER AT VARIOUS STAGES OF MATURITY

Stage of growth	Percent						
	TDN	Protein	Fiber	Cellulose	Ca	P	K
1/10 bloom	78.6	27.6	15.8	23.7	1.7	0.30	2.01
Full bloom	76.4	23.3	16.5	25.1	1.5	0.28	1.95
Early seed set	70.3	20.4	18.2	27.4	1.5	0.27	1.85

Source: Unpublished data, University of Illinois, Urbana.

selectively graze ladino from a mixture, which may result in bloat. Many producers use poloxalene to prevent bloat (7). A bloat-resistant strain is obviously desirable.

MANAGEMENT OF WHITE CLOVER

Establishment

White clovers prefer silt loams or clay soils with a pH of 6 to 7 and with a sufficient water-holding capacity.

Seeding

Seedbed preparation for white clovers should be similar to that for alfalfa and red clover; that is, a fine, firm seedbed is necessary. Ladino may be seeded with a companion forage grass, but severe shading will result in a reduction of the clover stand or its complete removal (14). In the midwest, ladino is usually sown in pasture mixtures at rates of 0.28 to 0.56 kg/ha (1/4 to 1/2 lb/A), resulting in a 2:1 grass-to-clover ratio (16). When sown as a monoculture for swine and poultry pastures, the seeding rate is usually 3.4 kg/ha (3 lb/A).

In elevated areas in the south, intermediate white clover is sown in grass mixtures at a rate of 1.1 to 2.2 kg/ha (1 to 2 lb/A), but in the deep south a rate of 3.4 kg/ha (3 lb/A) is recommended with grass. In a pure stand, a rate of 5.6 kg/ha (5 lb/A) is recommended (7).

Seeding dates vary according to location. In the upper midwest, ladino is planted in the early spring; in the upper south, from late August to late October; and in the deep south, from mid-October to mid-November.

Weed Control

The best method of weed control in white clover is to maintain a healthy, well-fertilized stand of grass and clover. Repeated tillage prior to seeding largely eliminates any weed problem. The most common postemergence herbicide used to control broadleaf weeds, 2,4-DB, does not injure grass or ladino clover. One can apply up to 2.2 kg/ha (2 lb/A) of 2,4-DB without injury to the white clover. In stands of intermediate white clover, many producers use 2,4-D amine at a rate of up to 1.1 kg/ha (1 lb/A) since white clover is very tolerant to this herbicide.

Insect Control

Since white clover has glabrous leaves, the potato leafhopper, *Empoasca fabae* (Harris), causes a lot of damage. Simple insecticide programs control this insect. Most insecticides must be applied repeatedly throughout the growing season, since there are continuous generations of the leafhopper. Most insects that attack white clover can be controlled by insecticides, whether they feed on leaves or invade the flowers or seeds.

Disease Control

White clover has very little resistance to root and stolon rots. Several strains have some tolerance to the various diseases of white clover. The best method of disease control involves proper fertility, crop rotation, and proper harvesting schedules.

Management

Management of ladino clover involves maintaining the proper grass-ladino mixture in a pasture. Keeping ladino clover in a grass mixture is difficult for several reasons: (1) the shading effect of tall-growing companion grasses; (2) the fertility level, especially that of N; (3) the need to keep the moisture level high enough so that droughty conditions do not develop and the ladino does not die out because of its shallow root system; and (4) insect and disease problems.

Although grasses need N, white clover is adversely affected by the addition of N (14).The amount of white clover in a mixture is affected by the amount of inorganic N present in the soil. As the N level increases, grasses tend to dominate. But if the N level is lower than that sufficient for the grass, and the levels of other major elements are adequate, white clover becomes dominant. Grasses will not always be maintained in ladino-grass mixtures without additional N fertilizer.

To maintain the correct clover content, one should maintain a high level of P and K in the soil. Annual applications of P and K should be applied according to a soil test and the amount of dry matter removed. The rates of P and K removal are the same as those for alfalfa. In the midwest, P and K should be applied after the last harvest in late summer or in the early fall. In the south, where white clover is often grown with bermudagrass, dallisgrass, or tall fescue, nitrogen is often applied to bermuda or dallisgrass–white clover mixtures in the summer or to mixtures with tall fescue in the late spring. If one wants to encourage the white clover, no N should be applied, but if one wants to favor winter grasses, N should be applied in the fall or early spring. To encourage the growth of warm-season grasses, one should apply N in late spring and summer.

Frequent harvesting during the period of greatest growth encourages white clover domination by minimizing the competition from grasses. The correct balance between the two forage components may be reached by changing the frequency and stage of harvest (14). Forage produced by a white clover–grass mixture should be utilized prior to the greatest degree of competition from the grass because tall-growing grasses will competitively shade out the white clover.

Ladino clover is best maintained in a grass mixture by rotational or controlled continuous grazing. The grazing height should be controlled or maintained at approximately 5 to 15 cm (2 to 6 in). One should graze the forage down to 5 to 8 cm (2 to 3 in), move the cattle to another pasture, and rotate them back to the first pasture when it is approximately 15 to 25 cm (6 to 10 in) tall. Managing the pasture in this manner allows the plants to recover and ensures grazing at the most digestible stage. Excess forage produced may be harvested for hay before the seed heads develop.

Bloat Control

Bloat is always a potential problem in pure ladino stands or mixtures where it constitutes a high percentage, but one can control bloat with poloxalene blocks or a liquid feed supplement containing poloxalene (7, 13, 26). Cattle should be placed on poloxalene several days before turning them into a ladino stand or grass mixture. One to 2 g of poloxalene per 45 kg (100 lb) of body weight per day controls bloat in cattle (7, 13). It is advisable to fill the cattle with dry hay or some roughage before they are turned in to the pasture the first day. Lick blocks or molasses liquid feeders should be placed near the water source or rest area, and the grazing cattle observed closely during the first few days for signs of bloat. If bloat develops, the affected animal should be drenched with a concentrate containing poloxalene for quick relief. If one is feeding a grain supplement containing poloxalene, approximately 0.22 kg (1/2 lb) of the supplement should be given twice a day.

Food Reserves

Food reserves of white clover are stored in the crown and taproot of the original plant and stolon regrowth (22, 29). The food reserves are rapidly depleted during regrowth the following spring (Table 18-2). It is critical that white clover have a rest period of approximately 20 to 35 days before fall dormancy or a killing frost; otherwise, it will go into the winter with a very low food reserve.

TABLE 18-2
PERCENT OF TOTAL SOLUBLE CARBOHYDRATES STORED IN VARIOUS PARTS OF LADINO CLOVER OVER THE WINTER AND THROUGH THE FOLLOWING SPRING, MADISON, WI

	Sampling date			
Plant part	Sept. 10	Nov. 10	June 1	June 10
Basal area of stolon	28.6%	37.7%	17.3%	9.3%
Tip area of stolon	26.6	35.6	23.5	16.0
Complete stolon	26.8	39.0	20.4	12.7
Root	20.2	—	11.0	—

Source: O.C. Ruelke and D. Smith, "Overwintering trends of cold resistance and carbohydrates in medium red, ladino, and common white clover," *Plant Physiol.* 31: 364–368 (1956).

Even with a rest period, roughly half of the white clover stolons are killed during winter in the north central states. There is very little protection of the stolons from winter injury unless the companion grasses help protect them.

SEED PRODUCTION

Very little seed of common white clover is produced in the United States; the main area of this limited production is in Louisiana. The seed is harvested from pastures, and special management of the pasture is required when a seed crop is to be harvested (15). This involves competitively reducing the grasses by early, heavy grazing, allowing the white clover regrowth to produce flowering heads for seed.

Most ladino seed is produced in the irrigated northwest. Under good management, seed yields of approximately 700 kg/ha (560 lb/A) are possible. Management involves excellent soil fertility, sufficient moisture via irrigation, insect control, sufficient honeybee pollinators, and excellent harvesting and threshing of the seed (3). Roughly 3 to 10 hives of honeybees are needed per hectare. Lush growth, along with the proper photoperiods in the northern region of the United States, can then produce many flowering heads for seed production.

DISEASES

Root and stolon rots are the major disease problems of white clover. *Sclerotium rolfsii* Sacc. attacks white clover in the humid south, whereas *Sclerotinia trifoliorum* Eriks. attacks thick stands in the northern regions and in the early spring in the southeast. Root and stolon rots involve species of *Rhizoctonia, Fusarium, Leptodiscus, Curvularia,* and *Colletotrichum* (14). There are also virus diseases of white clover, which are spread by mowing, insects, and grazing. Many of the foliar diseases of white clover are controlled by plant resistance. Nematodes cause considerable damage in the south. The most damage comes from the root knot nematode, *Meloidogyne* Spp. The best controls are plant resistance, crop rotation, control of insects that may aid in the spread of a disease, good fertility, and a proper harvest schedule.

INSECTS

The most important insect problem of white clover is the potato leafhopper. This insect feeds upon white clover by sucking plant juices, causing a stunting and yellowing of the leaves; the leaves then turn reddish or brownish (3, 11, 20). Resistant plants and proper use of insecticides help control the leafhopper.

Other insects that attack white clover are the meadow spittlebug, *Philaenus spumarius* (L.); clover leaf weevil, *Hypera punctata* (F.), and alfalfa weevil, *H. postica* (Gyllenhal). Seed fields may be attacked by the lygus bugs, *Lygus hesperus* Knight and *L. elisus* Van D., as well as by the clover seed weevil,

Miccotrogus picirostris (F.), and the clover seed midge, *Dasineura gentneri* Pritchard (14).

Control of most insects attacking white clover involves some plant resistance, insecticides, clean harvests, including weed control around fields that may harbor insects, and crop rotation.

SUMMARY

The advantages and disadvantages of ladino clover are as follows:

Advantages

1 Ladino clover makes an excellent, nutritious pasture for hogs and poultry.
2 White clovers are high-quality forages—high in protein and minerals, highly digestible, and low in fiber.
3 They respond to irrigation.
4 Due to its shallow root system, white clover tolerates and requires moist soil conditions (also a disadvantage).
5 It will perpetuate permanent pasture conditions.
6 The quality of the forage does not decline rapidly with maturity.

Disadvantages

1 The main disadvantage of ladino clover is that it is more bloat-prone than any other legume.
2 It can be crowded out when grown with tall-growing, sod-forming grasses—it is not shade-tolerant.
3 It is not drought-tolerant.
4 It is susceptible to winter kill.
5 In the southeast, most stand losses occur during the late summer.

QUESTIONS

1 What characteristics of ladino clover make it adapted to the deep south?
2 Compare the perennial aspects of ladino clover with those of alfalfa.
3 Why does the quality of ladino clover remain relatively high through the various stages of maturity?
4 Where do the three different types of white clover grow in the United States?
5 Why is white clover grown in pasture situations?
6 How can one help control bloat when grazing white clover?
7 Why is the northwest United States a favorable place to produce ladino clover seed?

REFERENCES

1 Ahlgren, G. H. *Forage Crops.* McGraw-Hill, New York, 2d ed., pp. 94–105, 1956.
2 Ahlgren, G. H and R. F. Fuelleman. Ladino clover. *Advan. Agron.*, 2:207–232, 1950.

3 Bacon, O. G., V. E. Burton, and J. E. Swift. Pest and disease control programs for ladino clover seed production. *Calif. Agr. Exp. Sta. Ext. Serv. Leaflet,* 1970.

4 Baumgardt, B. R. and D. Smith. Changes in the estimated nutritive value of the herbage of alfalfa, medium red clover, ladino clover, and bromegrass due to stage of maturity and year. *Wis. Agr. Exp. Sta. Res. Rept. 10,* 1962.

5 Brougham, R. W., P. R. Ball, and W. M. Williams. The ecology and management of white clover-based pastures. In *Plant Relations in Pastures,* CSIRO, Melbourne, Australia, pp. 309–324, 1978.

6 Carrier, L. and K. S. Bort. The history of Kentucky bluegrass and white clover in the United States. *J. Amer. Soc. Agron.,* 8:256–266, 1916.

7 Chessmore, R. A. Grazing pure legume pastures. *Proc. of a Forage Legume Conf.,* Noble Found., Ardmore, Okla., 1977.

8 Daday, H. Gene frequencies in wild populations of *Trifolium repens* L. III. World distribution. *Heredity,* 12:169–184, 1958.

9 Daday, H. Gene frequencies in wild populations of *Trifolium repens* L. IV. Mechanisms of natural selection. *Heredity,* 20:355–365, 1965.

10 Erith, A. G. *White Clover (Trifolium repens* L.), Duckworth, London, 1924.

11 Gibson, P. B. and E. A. Hollowell. White clover. *USDA Agr. Handbook 314,* 1966.

12 Harlan, J. R. Agricultural origins: centers and noncenters. *Science,* 174:468–474, 1971.

13 Johnston, J. E., L. E. Foote, J. Rainey, R. E. Girouard, L. Guthrie, P. B. Brown, and W. H. Willis. Clover bloat control in grazing cattle. *La. Agr. Exp. Sta,* 1966.

14 Leffel, R. C. and P. B. Gibson. White clover. In *Forages,* Iowa State Univ. Press, Ames, IA, 3d ed., pp. 167–176, 1973.

15 Marble, V. G., L. G. Jones, J. R. Goss, R. B. Jeter, V. E. Burton, and D. H. Hall. Ladino clover seed production in California. *Calif. Agr. Exp. Sta. Ext. Serv. Circ. 554,* 1970.

16 Martin, T. W. The role of white clover in grassland. *Herb. Abstr.,* 30:159–164, 1960.

17 McCarthy, G. and F. E. Emery. Some leguminous plants. *N.C. Agr. Exp. Sta. Bull. 98,* 1894.

18 McCloud, D. E., W. A. Cole, and B. A. Barton. Photoperiodic minimum for blossoming of white clover as related to variety and seed source. *Agron. Abstr.,* p. 56, 1958.

19 McConkey, O. The origin and ecological adaptation of the agricultural grasses, clovers, and alfalfa of eastern Canada. *Herb. Rev.,* 3:185–192, 1935.

20 Monroe, W. E. White clover—the legume supreme. *La. Agr. Ext. Serv. Pub. 1777,* 1974.

21 Ronningen, T. S. Susceptibility to winter injury and some other characteristics in ladino and common white clovers. *Agron. J.,* 45:114–117, 1953.

22 Ruelke, O. C. and D. Smith. Overwintering trends of cold resistance and carbohydrates in medium red, ladino, and common white clover. *Plant Physiol.,* 31:364–368, 1956.

23 Sprague, V. G. and R. J. Garber. Effect of time and height of cutting and nitrogen fertilization on the persistence of the legume and production of orchardgrass-ladino and bromegrass-ladino association. *Agron. J.,* 42:586–593, 1950.

24 Stewart, J. and F. E. Baer. Ladino clover, its mineral requirements and chemical composition. *N.J. Agr. Exp. Sta. Bull. 759,* 1951.

25 Thomson, D. J. and W. F. Raymond. White clover in animal production—nutritional factors (review). In J. Lowe (ed.), *White Clover Research,* Brit. Grassl. Soc. Hurley, *Occasional Symp.,* 6:277–284, 1970.

26 Todd, J. R. White clover in animal production—animal health factors (review). In J. Lowe (ed.), *White Clover Research,* Brit. Grassl. Soc. Hurley, *Occasional Symp.,* 6:297–307, 1970.

27 Ware, W. M. White clover. *Min. Agr. Fish. Misc. Publ. 46,* London, 1925.

28 Wedin, W. F., A. W. Burger, and H. L. Ahlgren. Effect of soil type, fertilization, and stage of growth on yield, chemical composition, and biological value of ladino clover (*Trifolium repens* L.) and alfalfa (*Medicago sativa*). *Agron. J.,* 48:147–152, 1956.

29 Westbrooks, F. E. and M. B. Tesar. Tap root survival of ladino clover. *Agron. J.,* 47:403–410, 1955.

CHAPTER 19

BIRDSFOOT TREFOIL

Birdsfoot trefoil (*Lotus corniculatus* L.) is a long-lived perennial forage legume, grown primarily for pasture and hay. It is fine-stemmed, leafy, slightly decumbent, and does not cause bloat when grazed.

HISTORICAL BACKGROUND

Birdsfoot trefoil has not attracted much attention. It probably originated in the Mediterranean region and is native to Europe and parts of Asia. The plant was described in Europe in about 1597. Birdsfoot trefoil was first cultivated in Europe at about the turn of this century (17). It was probably brought to the United States as an impurity in other forage seed imported from Europe. Between 1885 and 1900 seed samples from Europe were tested at various places in the U.S. Birdsfoot trefoil received its first major recognition in 1934 via Cornell University and the state of New York (29).

WHERE BIRDSFOOT TREFOIL GROWS

Birdsfoot trefoil is generally adapted to the temperate climate of the northeastern states and to the corn belt area of the north central states, as well as to the northern part of the Pacific coast. It is also grown in southeastern and central Canada. Worldwide, birdsfoot trefoil is grown in Europe, Great Britain, Brazil, Chile, Uruguay, Australia, and New Zealand. Birdsfoot trefoil grows best on fertile soils, and produces well on sandy loams or poorly drained clay or silty clay loams. However, it prefers a well-drained soil with a pH of 6.2 to 6.5. It grows on a wide range of soil types and conditions, including moderately alkaline soils, shallow soils, and moderately acid or infertile soils (28). Birdsfoot trefoil is considered to be winter-hardy, and thus provides another good perennial legume for the northern regions of the United States (29).

HOW BIRDSFOOT TREFOIL GROWS

Birdsfoot trefoil has a well-developed taproot with many branches in the upper 30 to 60 cm (15 to 24 in) of soil, which allows the plant to grow on shallow soils as well as deep, well-drained soils (Fig. 19-1). The taproot is not as deep as that of sweet clover or alfalfa. Branch roots of birdsfoot trefoil have the ability to develop new shoots. Birdsfoot trefoil also has the ability to develop new plants and root systems from callus tissue of the internode and from axillary buds of the node. This enables it to survive when its roots have been cut or when heaving occurs during freezing and thawing in the spring (29).

The aboveground portion of birdsfoot trefoil has many branches developing from one crown (Fig. 19-2). Plant height may reach 60 to 90 cm (24 to 35 in) under ideal conditions. The stems are slender and not as rigid as those of alfalfa, creating a more decumbent growth habit.

FIGURE 19-1
Birdsfoot trefoil plant showing the taproot with its many branches.
(Courtesy University of Illinois)

FIGURE 19-2
Birdsfoot trefoil growth pattern. *(Courtesy University of Illinois)*

The leaves of birdsfoot trefoil are compound, with five leaflets. Three of the leaflets are at the terminal point of the petiole, and two leafletlike sepals are located at the petiole base. A diurnal movement of the leaflets occurs during darkness, so that they close around the stem (17).

Flowers are clustered in a typical umbel having four to eight florets on a long peduncle (Fig. 19-3). Flower colors vary from yellow to yellow-orange. Following cross-pollination and fertilization by insects, slender, brown to brownish-purple seed pods develop. Ten to 15 seeds develop in each pod. Five to six pods 2.5 to 4 cm (1 to 1½ in) long develop at right angles to the end of the peduncle, giving the appearance of a bird's foot (Fig. 19-3). The pods ripen 25 to 30 days after fertilization. The seeds are oval to spherical in shape and very small, 825,000 to 925,000/kg (17). Seedcoat colors range from light to dark brown, and the seeds are usually speckled with dark spots. Self-incompatibility exists in birdsfoot trefoil, thus reducing self-seed. A very high percentage of the seed is cross-pollinated.

Birdsfoot trefoil is considered a long-day plant, requiring approximately 16 hours for complete flowering. With shorter day lengths, flowering is reduced, and a more prostrate growth habit is exhibited (12). Birdsfoot trefoil is widely

FIGURE 19-3
Flower and seed pod
arrangement of birdsfoot trefoil.
(Courtesy University of Illinois)

known as a nonbloating legume, and because of this one trait alone, birdsfoot trefoil makes an excellent pasture forage crop.

Birdsfoot trefoil plants usually contain a certain amount of a cyanogenetic glucoside that, when hydrolyzed, produces hydrocyanic acid (HCN). Many of the new cultivars or selections do not contain the glucoside. There has been no reported HCN poisoning of cattle when grazing birdsfoot trefoil *(L. corniculatus)*, though some species of *Lotus* may be poisonous (29).

TYPES OF TREFOIL

There are three different species of trefoil growing in the United States and Canada: broadleaf birdsfoot trefoil (*L. corniculatus* L.); narrowleaf trefoil (*L. tenuis* Wald et Kit.), and big trefoil (*L. uliginosus* Schkuhr.), sometimes referred to as *L. major* (29).

Within the broadleaf type of birdsfoot trefoil there are two subtypes: the low-growing, or Empire, type; and the erect-growing, or European, type.

Compared with the erect strain, the low-growing strain blooms later and is weaker-stemmed, causing it to be more prostrate and to have a slower spring growth and slower recovery following harvest. The erect type is earlier-flowering and semi-erect; it starts growth earlier in the spring and has a more rapid regrowth following harvest (4, 37).

Narrowleaf trefoil has narrow, linear-lanceolate leaflets on a more slender, weaker stem with longer internodes. This species has a more shallow root system. Flowers are smaller and fewer in number and, at maturity, usually change in color from yellow to orange-red (30).

Big trefoil looks very similar to broadleaf trefoil, though big trefoil has rhizomes and more flowers per umbel than the other two types of trefoil. It has a shallow root system, is not drought-resistant, and grows best in moist areas. Seed size is much smaller than in the other two species (30).

USES OF BIRDSFOOT TREFOIL

Birdsfoot trefoil is a very desirable long-lived perennial legume, and it is used a great deal in permanent pastures. Because of its nonbloating nature, there is no need to grow it with a grass when grazing.

The low-grazing, Empire-type cultivars persist better in pastures than do the erect, European-type cultivars. One can use the erect-type trefoil for pasture but it must be rotationally grazed, just as one would graze alfalfa or red clover. Both erect and low-growing trefoils persist under closer and more frequent grazing than do either alfalfa or red clover.

Birdsfoot trefoil is often used in renovating old permanent grass pastures. Added to a grass sward, it increases animal production more than does the addition of N alone (34).

Trefoil can be harvested when it begins to bloom, and it makes a good-quality hay. Since the stems are fine-textured with many leaflets, the quality of birdsfoot trefoil is usually very high. Most of the cultivars used for hay are of the European type. Two to three harvests can be taken per year (1, 11, 17, 30, 34, 38). Yields range from 4 to 10 t/ha (1.78 to 4.46 T/A). Since trefoil does not have a rapid regrowth, approximately 6 to 8 weeks should elapse between harvests. No late fall harvest should be taken so that food reserves can build up for winter survival (36).

In order to maintain a good, healthy stand, birdsfoot trefoil should not be harvested close to the soil surface since it needs to maintain some axillary buds for regrowth (29). The cutting height should be approximately 5 to 10 cm (2 to 4 in). Regrowth depends upon the carbohydrate reserves generated by the leaves or top growth remaining after harvest (9, 27).

Trefoil maintains itself as a perennial because of its reseeding ability, in addition to the inherent perennial nature of the plant. Therefore, some remaining growth needs to be maintained for reseeding the species. The original plant often dies as a result of insects, diseases, or environmental stresses (30).

Some silage is made of pure stands of trefoil, but more often trefoil is harvested as a silage when grown with a grass. The forage should be cut, wilted to at least 60 to 65% moisture, finely chopped, and ensiled to exclude any air. It has been shown in New York that trefoil silage and other legume silages exhibited similar animal results and consumption patterns (29, 35).

DESIRABLE CHARACTERISTICS

Producers like birdsfoot trefoil because it is an excellent, nonbloating pasture legume. Trefoil is widely adapted to many different soil types and soil conditions. It is long-lived, primarily because of its reseeding or seed-shattering characteristic and its ability to reroot and establish a new plant. It is also very winter-hardy, and it does not require as high a fertility level as does alfalfa.

Growers would like to have a trefoil with more seedling vigor and faster regrowth following harvest. Trefoil is also slow in getting established. With a faster establishment and rapid regrowth, one could increase the dry matter production of trefoil by 50% and obtain one or two harvests during the year of establishment. In addition, seed production is rather low due to the shattering characteristic. And there are several diseases of trefoil that tend to limit dry matter production and stand persistence.

MANAGEMENT OF BIRDSFOOT TREFOIL

Establishment

Because of trefoil's slow seedling establishment, seedbed preparation is even more important than for alfalfa or red clover. A firm seedbed and proper placement of the seed is most important because seedling establishment is closely related to depth of seeding. Earlier results have shown that when trefoil was seeded at depths of 0.6, 1.3, 1.9, and 2.5 cm (0.2, 0.5, 0.7, and 1 in) in a silt loam, the percent emergence was 31, 18, 8, and 3, respectively (30).

Preparation of a firm seedbed, followed by broadcast seeding and cultipacking or rolling, greatly improves seedling establishment. It is important to have a good soil-to-seed contact. This provides for better moisture uptake and a higher percent emergence, resulting in a faster establishment with a good root system.

As stated earlier, trefoil can tolerate a lower pH and less-fertile and poorer-drained soils than can alfalfa. It is important to have a soil pH of 6.2 to 6.5. This ensures a higher availability of nutrients and proper media for the *Rhizobium* bacteria to function. It is advisable to have the same level of available P and K in the soil as for alfalfa. High levels of P ensure rapid root development (30).

Seeding Rate

High-quality seed is important to ensure a rapid establishment of trefoil. Many seed lots may have small, immature, shriveled seeds because the seeds are harvested early, prior to pod shattering. Most seeding rates are from 5.5 to 12 kg/ha (5 to 11 lb/A). Early researchers found that increasing the seeding rate does not increase trefoil forage production once it is established (17).

If trefoil is to be seeded with a companion forage grass, it is best established with a slow-growing grass, such as Kentucky bluegrass or red fescue (30). When seeded with a rapid-establishing grass, such as ryegrass, or a companion crop, such as oats, the competition may be too great to ensure a good stand of birdsfoot trefoil.

Inoculation of Seed

Since birdsfoot trefoil may be new to an area, one should always inoculate the seed with a special strain of *Rhizobium* bacteria specific for trefoil. The bacteria that inoculate other legumes, such as red clover or alfalfa, do not inoculate the trefoil species. Sufficient amounts of inoculant should be applied to the seed immediately prior to seeding (30).

Weed Control

Trefoil seedlings grow slowly and furnish little competition to rapidly growing weeds. Proper weed control at time of establishment is most important. Lightly seeded nurse crops that provide little competition may be used, such as oats seeded at about one-half of the normal seeding rate.

Herbicides that are used for alfalfa establishment are recommended for trefoil as well: EPTC or Benefin incorporated into the soil surface at a depth of 5 to 7.5 cm (2 to 3 in) prior to seeding. EPTC may be better than Benefin for controlling or suppressing yellow nutsedge growth. Both of these herbicides are effective in annual-grass control (13, 16, 29). One should remember that a companion forage grass or grass nurse crop cannot be established along with trefoil when these preplant herbicides are used.

Postemergence broadleaf-weed control can be applied to stands of trefoil using 2,4-DB or dinoseb. Trefoil may show some curling or cupping of the leaves when 2,4-DB is used, but the plant will soon recover and continue to grow. Dinoseb is not recommended in all cases because of the injury observed when applications are made during periods of high temperature (16).

Harvest Schedule

Trefoil should be cut for hay when it first begins to bloom. It can be harvested every 6 to 8 weeks until the last harvest, which should be 40 days prior to fall dormancy or a killing frost. This allows for two harvests in Canada and the northern regions of the United States, and three harvests in the lower north central region of the United States and the corn belt area.

If trefoil is used for pasture, it should be grazed rotationally, with rest periods of about 4 weeks. The low-growing type may be grazed lightly and continuously

TABLE 19-1
FORAGE YIELDS OF BIRDSFOOT TREFOIL WITH VARIOUS HARVEST
TREATMENTS AT URBANA, IL

Harvest treatment for 2 years	Hay production year			Plants per 0.28 sq.m‡
	1	2	3†	
	t/ha (t/A)	t/ha (t/A)	t/ha (t/A)	
2.5 cm (1 in) every 3 wks to late Aug.	4.62 (2.06)	6.01 (2.68)	8.36 (3.72)	6.75
10 cm (4 in) every 3 wks to late Aug.	2.33 (1.05)	5.29 (2.35)	8.43 (3.75)	7.50
2.5 cm (1 in) every 3 wks to early Oct.	5.22 (2.32)	5.58 (2.48)	5.67 (2.52)	6.25
10 cm (4 in) every 3 wks to early Oct.	3.56 (1.58)	5.65 (2.51)	7.56 (3.37)	8.00
2.5 cm (1 in) when 10 cm (4 in) high to early Oct.	4.33 (1.93)	5.27 (2.35)	6.57 (2.93)	7.50
2.5 cm (1 in) when 5–7.5 cm (2–3 in) high to late Aug.	3.16 (1.40)	4.37 (1.94)	7.24 (3.32)	8.75
2.5 cm (1 in) when 5–7.5 cm (2–3 in) high to early Oct.	3.92 (1.75)	4.30 (1.91)	6.91 (3.08)	8.75
Prebloom*	4.98 (2.22)	6.39 (2.84)	6.91 (3.08)	7.25
1/10 bloom*	5.47 (2.43)	6.52 (2.90)	6.46 (2.88)	5.50
Full bloom*	5.99 (2.67)	8.39 (3.78)	7.80 (3.47)	9.25
Mature*	4.19 (1.86)	6.39 (2.84)	5.81 (2.59)	3.75
LSD at 5%	0.21	0.33	0.55	—

*Harvested for a second time on Aug. 30 regardless of stage of growth.
†Harvested alike 3d year, measure of residual effect of first 2 years.
‡Measured Oct. of 3d year.

if the entire top is not removed at any one time. Some food reserves are stored in the stem portion and leaves and are needed to manufacture soluble carbohydrates for regrowth, which also allows for some flowering and reseeding.

Results of various cutting treatments of trefoil are shown in Table 19-1. Duell and Gausman (7) also found that the highest dry matter production of trefoil occurs when harvests are taken three times at 1/10 bloom rather than waiting for the mature stage. The best cutting height is around 7.5 to 10 cm (3 to 4 in). This results in maintenance of stand and, therefore, higher yields (7, 9, 27). It is more important to leave taller stubble in trefoil than in alfalfa.

Food Reserves

The soluble carbohydrate food reserves of trefoil follow a cyclic pattern similar to that in alfalfa. Studies at Wisconsin indicate that during the growing season, percent carbohydrate is at a lower level in trefoil than in alfalfa and red clover (31). During the winter, trefoil food reserves are stored in the roots, but during the growing season regrowth depends upon the carbohydrate synthesized by the top growth rather than the amount stored in the roots (9, 14).

Fall management is as critical for trefoil as it is for alfalfa. Trefoil should have at least 40 days of fall regrowth after the last harvest before fall dormancy or a killing frost so that sufficient soluble carbohydrates are stored for winter survival.

COMPOSITION AND QUALITY

When harvested at the flowering stage, birdsfoot trefoil is equal in quality to alfalfa and red clover. Trefoil declines in protein and increases in fiber with maturity, as does alfalfa (32). Protein content decreases from approximately 28% for young, immature growth, to 9% in plants with mature seed pods (32, 35). There is no difference in feeding value between the low-growing and erect-type trefoils (35).

When trefoil is harvested and fed as silage or hay, consumption and milk production are similar (7). Trials at Minnesota showed that when trefoil was substituted for alfalfa-grass as one-third of the pasture, daily gains in lambs increased by 22 to 24% over a 3-year period (18). Where trefoil rather than alfalfa was added to a pasture, the forage contained more legumes, greater digestibility, higher crude protein, and lower percent cell walls than did the alfalfa-grass mixture (18). In earlier trials at Purdue, it was found that trefoil had a higher carotene content than did alfalfa and ladino clover (23).

SEED PRODUCTION

Compared with other legumes, seed yields of birdsfoot trefoil are usually rather low. The shattering of seed pods when mature is the major reason. Pollinating insects are also very important, especially the species of *Hymenoptera*. Seed producers must bring in the pollinating bees to help ensure good seed set. Seed yields range from a low of 50 to 150 kg/ha (45 to 134 lb/A) to a high average of 200 to 550 kg/ha (180 to 490 lb/A) (29). Potential seed yields range from 500 to over 1000 kg/ha (445 to 890 lb/A) (21).

The pollinating insect forces itself into the base of the trefoil flower, which causes the extrusion of the pistil through the tip of the keel. Then, the insect collects and transfers pollen from its body to another flower with an extruded pistil. Each floret has many ovules but only 5 to 8 of them develop within 24 to 48 hours after pollination. Within 3 weeks, the seed pods are physiologically mature and turn light brown to black, depending on climatic conditions (8).

Seed Harvest

Most trefoil is mowed, windrowed, and harvested with a combine from the windrow. An indeterminate flowering habit makes it difficult to decide when to harvest for seed. Most growers cut and windrow trefoil for seed harvest when the maximum number of pods are light green to light brown in color (2, 8). If one waits too long, the pods shatter, creating a significant loss of seed.

DISEASES

The most destructive disease of birdsfoot trefoil species is caused by the soil-inhabiting fungus *Rhizoctonia solani* Kuehn., which causes not only foliar blight but also crown and root rots. This disease is most destructive in dense stands during hot, humid weather. *R. solani* is best controlled by harvesting when the foliage first shows symptoms of foliar blight (29).

There is also a root rot complex of trefoil that is caused by *Fusarium* spp., *Verticillium* spp., *Leptodiscus terrestris* Gerdemann, *Sclerotinia* spp., *Rhizoctonia* spp., *Mycoleptodiscus* spp., and *Macrophomina* spp. (26, 29). The upper center portion of the taproot and crown is affected by extensive decay, causing death of the plant (10). There are now cultivars available that are relatively resistant to this root rot complex.

INSECTS

Trefoil has fewer insect problems than it has diseases. The potato leafhopper, *Empoasca fabae* (Harris), and the alfalfa plant bug, *Adelphocoris lineolatus* (Goeze), are two sucking insects that do attack trefoil; injury occurs in trefoil just as it does in alfalfa. The meadow spittlebug, *Philaenus spumarius* (L.) is considered one of the most important pests in the midwestern and eastern areas of trefoil production. They also suck the sap from the plant, and create a white, foamy mass on the leaves and stems. All of these insects cause stunting and lack of flower production (abortion) (25).

In seed fields of birdsfoot trefoil, the trefoil seed chalcid, *Bruchophagus kolobovae* Fed., causes a great deal of seed loss. This insect lays its eggs in the seed pods, and when they hatch, the larvae feed upon the developing ovule, leaving a hollow seedcoat. Insecticides do not always control the chalcid (25). But in most cases, a good insecticide program, along with proper harvest schedules and clean harvests, adequately controls most of the insects that attack trefoil.

SUMMARY

The advantages and disadvantages of birdsfoot trefoil are as follows:

Advantages

1 Trefoil does not cause bloat when grazed.
2 It is a long-lived perennial.
3 It reseeds itself as a result of pod shattering.

4 The quality of forage is similar to that of alfalfa.

5 It persists longer under heavier grazing than alfalfa and red clover do.

6 It can grow on a wider range of soil types and conditions than can alfalfa; that is, birdsfoot trefoil can tolerate a lower pH, a lower general fertility, and more moisture.

Disadvantages

1 Birdsfoot trefoil is more difficult to get established than is alfalfa.

2 It has less seedling vigor than alfalfa and red clover have.

3 Its regrowth is slower than that of alfalfa and red clover.

4 Seed production of trefoil is rather low because of pod shattering.

5 Less nitrogen is fixed by trefoil than by alfalfa.

6 Trefoil is not tall-growing.

QUESTIONS

1 How do the area of adaptation and soil preference of birdsfoot trefoil differ from those of alfalfa?

2 Describe the morphological characteristics of birdsfoot trefoil.

3 Why is there interest in using trefoil as a pasture?

4 Describe the different types of trefoil.

5 How does the establishment rate of trefoil compare with that of alfalfa and red clover?

6 How does trefoil store its food reserves?

7 Does trefoil respond differently to different cutting heights and intervals between harvests?

8 What is unique about trefoil seed production?

9 Are there any diseases and insect problems in trefoil production? Name the most serious of each.

10 What are the advantages and disadvantages of birdsfoot trefoil?

REFERENCES

1 Ahlgren, B. H. *Forage Crops.* McGraw-Hill, New York, 2d ed., pp. 106–113, 1956.

2 Anderson, S. R. Development of pod and seeds of birdsfoot trefoil, *Lotus corniculatus* L., as related to maturity and to seed yields. *Agron. J.,* 47:483–487, 1955.

3 Anderson, S. R. and D. S. Metcalfe. Seed yields of birdsfoot trefoil (*Lotus corniculatus* L.) as affected by preharvest clipping and by growing in association with three adapted grasses. *Agron. J.,* 49:52–55, 1957.

3 a Collins, M. Changes in composition of alfalfa, red clover, and birdsfoot trefoil during autumn. *Agron. J.,* 75:287–291, 1983.

3 b Collins, M. Yield and quality of birdsfoot trefoil stockpiled for summer utilization. *Agron. J.,* 74:1036–1041, 1982.

4 Davis, R. R. and D. A. Bell. A comparison of birdsfoot trefoil and ladino clover-bluegrass for pasture. I. Response of lambs. *Agron. J.,* 49:436–440, 1957.

5 Decker, A. M., H. J. Retzer, M. L. Sarna, and H. D. Kerr. Permanent pastures improved with sod-seeding and fertilization. *Agron. J.*, 61:243–247, 1969.

6 Drake, C. R. Diseases of birdsfoot trefoil in six southern states in 1956. *Plant Dis. Reptr.*, 42:145–146, 1958.

7 Duell, R. W. and H. W. Gausman. The effect of differential cutting on the yield, persistence, protein, and mineral content of birdsfoot trefoil. *Agron. J.*, 49:318–319, 1957.

8 Giles, W. L. *The morphological aspects of self-sterility in Lotus corniculatus L.* Ph.D. thesis, Univ. of Mo., Columbia, 1949.

9 Greub, L. J. and W. F. Wedin. Leaf area, dry matter production, and carbohydrate reserve levels of birdsfoot trefoil as influenced by cutting height. *Crop Sci.*, 11:734–738, 1971.

10 Henson, P. R. Breeding for resistance to crown and root rots in birdsfoot trefoil, *Lotus corniculatus* L. *Crop Sci.*, 2:429–432, 1962.

11 Henson, P. R. and H. A. Schath. The trefoils—adaptation and culture. *USDA Agr. Handbook 223*, 1962.

12 Joffee, A. The effect of photoperiod and temperature on the growth and flowering of birdsfoot trefoil. *S. African J. Agr. Sci.*, 1:435–450, 1958.

13 Kerr, H. D. and D. L. Klingman. Weed control in establishing birdsfoot trefoil. *Weeds*, 8:157–167, 1960.

14 Langille, J. E. L. B. MacLeod, and F. S. Warren. Influences of harvesting management on yield, carbohydrate reserves, etiolated regrowth and potassium uptake of birdsfoot trefoil. *Can. J. Plant Sci.*, 48:575–580, 1968.

15 Laskey, B. C. and R. C. Wakefield. Competitive effects of several grass species and weeds on establishment of birdsfoot trefoil. *Agron. J.*, 70:146–148, 1978.

16 Linscott, D. L. and R. D. Hagin. Interaction of EPTC and DNBP on seedlings of alfalfa and birdsfoot trefoil. *Weeds*, 16:182–184, 1968.

16 a Luu, K. T., A. G. Matches, and E. J. Peters. Allelopathic effects of tall fescue on birdsfoot trefoil as influenced by N fertilization and seasonal changes. *Agron. J.*, 74:805–808, 1982.

17 MacDonald, H. A. Birdsfoot trefoil (*Lotus corniculatus* L.): its characteristics and potentialities as a forage legume. *Cornell Univ. Agr. Exp. Sta. Mem. 261*, 1946.

18 Marten, G. C. and R. M. Jordan. Substitution value of birdsfoot trefoil for alfalfa-grass in pasture systems. *Progress Report: Clovers and Special Purpose Legumes Research*, 11:40–41, 1978.

19 Mays, D. S. and J. B. Washko. The possibility of stock piling legume-grass pasture. *Agron. J.*, 52:190–193, 1960.

20 McKee, G. W. Some effects of liming, fertilization, and soil moisture on seedling growth and nodulation in birdsfoot trefoil. *Agron. J.*, 53:237–240, 1961.

21 Miller, D. A., L. J. Elling, J. D. Baldridge, P. C. Sandal, S. G. Carmer, and C. P. Wilsie. Predicting seed yield of birdsfoot trefoil clones. *North Central Reg. Res. Bull. 227*, 1975.

22 Morse, R. A. The pollination of birdsfoot trefoil (*Lotus corniculatus* L.). *Diss. Abstr.*, 15:946, 1955.

23 Mott, G. O., R. E. Smith, W. M. McVey, W. M. Beeson. Grazing trials with beef cattle. *Purdue Agr. Exp. Sta. Bull. 581*, 1952.

24 Nelson, C. J. and D. Smith. Growth of birdsfoot trefoil and alfalfa. II. Morphological development and dry matter distribution. III. Changes in carbohydrate reserves and growth analysis under field conditions. *Crop Sci.*, 8:21–28, 1968.

25 Neunzig, H. H. and G. G. Gyrisco. Some insects injurious to birdsfoot trefoil in New York. *J. Econ. Entomol.*, 48:447–450, 1955.

26 Ostazeski, S. A. An undescribed fungus associated with a root and crown rot of birdsfoot trefoil (*Lotus corniculatus* L.). *Mycologia*, 59:970–975, 1967.

27 Pierre, J. J. and J. A. Jackobs. The effect of cutting treatments on birdsfoot trefoil. *Agron. J.*, 45:463–468, 1953.

28 Rachie, K. O. and A. R. Schmid. Winter-hardiness of birdsfoot strains and varieties. *Agron. J.*, 47:155–157, 1955.

29 Seaney, R. R. Birdsfoot trefoil. In *Forages,* Iowa State Univ. Press, Ames, IA, 3d ed., pp. 177–188, 1973.

30 Seaney, R. R. and P. R. Henson. Birdsfoot trefoil. *Advan. Agron.*, 22:119–157, 1970.

31 Smith, D. Carbohydrate root reserves in alfalfa, red clover, and birdsfoot trefoil under several management schedules. *Crop Sci.*, 2:75–78, 1962.

32 Smith, D. Chemical composition of herbage with advance in maturity of alfalfa, medium red clover, ladino clover, and birdsfoot trefoil. *Wis. Agr. Exp. Sta. Res. Rep. 16,* 1964.

33 Smith, D. Influence of temperature on the yield and chemical composition of five forage legume species. *Agron. J.*, 62:520–523, 1970.

34 Templeton, W. C., Jr., C. F. Buck, and D. W. Watenbarger. Persistence of birdsfoot trefoil under pasture conditions. *Agron. J.*, 59:385–386, 1967.

35 Trimberger, G. W., W. K. Kennedy, J. T. Reid, J. K. Loosli, K. L. Turk, and V. N. Krukovsky. Feeding value and digestibility of birdsfoot trefoil hay harvested at different stages. *N. Y. Agr. Exp. Sta. Bull. 974,* 1962.

36 VanKeuren, R. W. and R. R. Davis. Persistence of birdsfoot trefoil (*Lotus corniculatus* L.) as influenced by plant growth habit and grazing management. *Agron. J.*, 60:92–95, 1968.

37 VanKeuren, R. W., R. R. Davis, D. A. Bell, and E. W. Klosterman. Effect of grazing management on the animal production from birdsfoot trefoil pastures. *Agron. J.*, 61:422–425, 1969.

38 Wakefield, R. C. and N. Skaland. Effects of seeding rate and chemical weed control on establishment and subsequent growth of alfalfa and birdsfoot trefoil. *Agron. J.*, 57:547–550, 1965.

39 Wittwer, L. S., W. K. Kennedy, G. W. Trimberger, and K. L. Turk. Effects of storage methods upon nutrient losses and feeding value of ensiled legume and grass forage. *N. Y. Agr. Exp. Sta. Bull. 931,* 1958.

CHAPTER 20

CROWNVETCH

Crownvetch, *Coronilla varia* L., is a member of the pea family of leguminous plants. It gets its name from its "vetchlike" leaves.

HISTORICAL BACKGROUND

Crownvetch is native to middle and southern Europe, Asia Minor, and northern Africa; it has not been reported in northern Europe. Its habitat has been more or less adjacent to the Mediterranean area. It has been introduced into Belgium, the Netherlands, Spain, Portugal, Switzerland, and Czechoslovakia, and has been distributed quite widely in France and Hungary, especially where droughty limestone soils are prevalent. It is common in northern and central Italy, Rumania, and Greece, and is abundant in the central and southern Crimean regions of the U.S.S.R. It is also common in Syria, Iran, Turkistan, and adjacent areas (10).

Crownvetch was first made available in the United States around 1890 (10). Early introductions were used as ornamentals. It is found primarily north of the 35th parallel and east of the 97th meridian in the United States.

WHERE CROWNVETCH GROWS

Crownvetch grows best on permeable, well-drained, calcareous soils that have only moderate amounts of clay, good supplies of phosphorus and potassium, and a pH of 4.5 or higher. The most vigorous stands have been observed on well-drained sites with a pH of 6 or higher. On wet soils, stands fail or produce very little growth. A shortage of moisture sometimes causes thin stands on sandy soils, but once crownvetch is established, it is quite drought-tolerant (10). Crownvetch grows on soils that are too droughty for birdsfoot trefoil, and it grows well in partly shaded areas. The crowns may be used for propagation, but seed is preferable.

HOW CROWNVETCH GROWS

Crownvetch is a long-lived, deep-rooted, winter-hardy, drought-tolerant, perennial legume. It is a herbaceous plant with weak, hollow stems that can tolerate light shading. The creeping stems may grow to 2 m or more under favorable conditions (13).

The leaves of crownvetch are pinnately compound, with 6 to 13 pairs of oblong leaflets per leaf, arranged on each of the petioles (Fig. 20-1). The roots are profuse and penetrate deeply. New shoots or plants arise from the creeping rootstalks or underground rhizomes. The plant does not climb or vine, but spreads laterally (1).

Flower colors vary from purplish-pink to whitish-pink. A cylindrical seed pod

FIGURE 20-1
Crownvetch, illustrating its growth habit, which provides excellent soil conservation and wildlife cover.

FIGURE 20-2
Close-up of crownvetch, illustrating its flower and pod arrangement.

develops, which is 2.5 to 5 cm (1 to 2 in) long and divided into five to seven segments (Fig. 20-2). The seed pods shatter very easily. When dry, the pods break into the individual segments, each about 0.6 cm (0.2 in) long. Each segment contains a single seed (1, 11) (Fig. 20-3).

USES OF CROWNVETCH

Because of its creeping stems, strong rhizomes, and decumbent growth habit, crownvetch appears to be the answer to some erosion problems. It is an ideal plant for controlling erosion on steep banks, worn-out land that needs improving, gullied areas, rock outcrop areas, and especially roadside banks. Crownvetch has gained wide acceptance in bank stabilization for soil and water conservation programs (11) (Fig. 20-2).

Crownvetch is often used for pasture. It has been used in pasture improvement programs in relation to sod-seeding (10). There is no potential problem of bloat when grazing crownvetch, nor are there any toxic substances present in the foliage that would make it unpalatable for ruminants. However, when crownvetch is fed to nonruminants, there appear to be some toxicity problems. Some *Coronilla* species contain cardiac glycosides in the seed and plant tissue (7).

Recent studies have shown that the digestibility of crownvetch is as high or higher than that of alfalfa (3, 14). This is especially true at the older stages of maturity. The intake of crownvetch, though, may be slightly lower than that of alfalfa (6, 16).

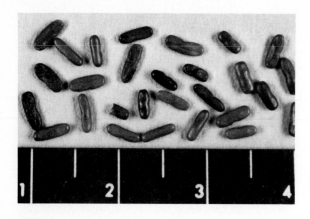

FIGURE 20-3
Crownvetch seeds. *(Courtesy University of Illinois)*

ADDITIONAL CHARACTERISTICS

Crownvetch has a slower germination and seeding vigor than do red clover and alfalfa. Because of this slower seedling growth, control of weeds and competition from other crops is very important (13).

Regrowth of crownvetch following harvest is slower than that of alfalfa, so that the harvest frequency for crownvetch is every 8 weeks (11).

A special *Rhizobium* inoculum strain is needed for crownvetch. Since many areas have not been planted with crownvetch, it is advisable to double the usual rate of inoculum when seeding.

Crownvetch is a long-lived perennial, spread by rhizomes and seed. The seed pods shatter easily when dry, thus allowing seed dispersal. But seed shattering creates a seed production problem related to harvest time. One of the limiting factors in crownvetch seed production is the plant's excessive vegetative growth between the peak of bloom and the time of normal seed harvest. The effect is to retard maturation of the partly mature pods to the extent that they do not fully develop. If the amount of regrowth is excessive, practically no seed is available for harvest. Low seed yield is a major problem in crownvetch culture (13).

MANAGEMENT OF CROWNVETCH

Establishment

Spring seedings of crownvetch in a clean, firm seedbed have proved successful. In critical areas, dormant-season seedings made from December to February, followed by mulching, have proved effective at the Soil Conservation Service (SCS) Plant Materials Center in Elsberry, Missouri (13). Mulch has been helpful in establishing crownvetch on road banks and critical areas. Around 4 t/ha (2 T/A) of straw tied down with net materials is sufficient to stabilize these areas.

Depth of seeding is critical when crownvetch is grown in a heavy clay soil. It has been shown that germination rates may be decreased one-half for each

3.2-mm (1/8-in) increase in depth of seeding (1). Seedings should not be greater than 6 mm (1/4 in) deep in clay soils. Viable seeds usually average 40 to 50% germination, with approximately 22 to 40% hard seed. High-quality crownvetch seed has a minimum purity of 98% and a germination rate of 70%, and should include no more than 35% hard seed (13).

Crownvetch can be established with the aid of preplant herbicides such as EPTC, Benefin, and Treflan.

Fertility Requirements

Crownvetch can withstand a lower soil pH than can alfalfa, but it prefers a pH of at least 5.5. Recent studies have shown that dry matter yields increased as soil pH increased from 4.2 to 5.4, and continued to increase up to a pH of 6.3 (13, 17). Earlier studies indicated that crownvetch would continue to increase in dry matter production up to a pH of 6.5, but this is not true. It now appears that the higher pH reduces the amount of available P and causes an excess of exchangeable soil Mn. It was found that P application resulted in a lower incidence of leaf black spot disease, and also enhanced root penetration (17). The potassium and phosphorus needs for crownvetch are similar to those for red clover and alfalfa.

Companion Crop

Since crownvetch is characteristically slow to establish and poor in seedling vigor, 12 kg/ha (10 lb/A) of ryegrass, tall fescue, or field brome may be used as a companion crop to help control erosion until the crownvetch is established. Crownvetch seeding rates range from 12 to 17 kg/ha (10 to 15 lb/A). Overseeding crownvetch in oats in April or early May has produced good stands. The oats should be seeded at 36 kg/ha (1 bu/A) and clipped in the early-head growth stage. Clipping retards growth of summer-annual grasses and allows the crownvetch time to become established.

Management of Stand

The mowing of banks or critical areas is not necessary or desirable after the first year of establishment. Crownvetch should be first harvested when it is in 1/10 to 1/2 bloom. After crownvetch is well-established, the harvest schedule should be in intervals of approximately 8 weeks after the first harvest because of its slow regrowth.

Weed Control

Weed control in crownvetch is associated more with seed production than with forage production. Field trials conducted recently in New Jersey have shown that the best herbicide treatment in crownvetch seed fields is terbacil (0.25 lb/A a.i.), simazine (1.0 lb/A a.i.) or diuron (1.0 lb/A a.i.) (12). None of these herbicides controlled all weeds, but they gave relatively good weed control with no crop injury (12). An early-spring application is better than an application after mid-April.

Insect and Disease Control

Crownvetch seems to be free of most insect and disease problems. The main insect problem is the lygus bug, which infects the blossoms and causes the flowers to abort or produce shriveled seeds. Stem rot is the main disease problem; it attacks the base of the stems, causing a reduction in dry matter and seed production.

QUALITY OF CROWNVETCH

Feeding trials, including grazing, have shown crownvetch to be a very acceptable forage from the standpoint of animal performance (3, 4, 5, 6). However, crownvetch has not been reliable as a high-producing forage when continuously grazed (3). There has been some evidence that free-grazing animals may periodically reject crownvetch (4).

Laboratory evaluation has shown that the total phenol concentration is a reasonable index of crownvetch's nutritive value, being negatively correlated (r [=] -0.76) with IVDMD (2).

Crownvetch hay is similar in protein and crude fiber content to other legume hays. The tannin content is 3.2 to 3.8%, about half that of sericea lespedeza. In dry matter digestibility, crownvetch is similar to alfalfa when harvested on the same date (2, 14).

Mineral composition and IVDMD indicate that crownvetch has an acceptable nutritive level and digestibility for cattle (2, 14). IVDMD values for 17 different clones over two different dates ranged from 45.0 to 67.1%, with average values for two different stages of maturity of 53.1 and 62.5%.

SUMMARY

Following are the advantages and disadvantages of crownvetch:

1 Crownvetch produces no bloat problem.
2 It provides excellent erosion and gully control.
3 The soil pH requirement is lower for crownvetch than for most legumes.
4 It is a long-lived perennial.
5 It does not climb or twine in growth.
6 It is tolerant to most diseases and insects.
7 It is drought-resistant.

Disadvantages

1 Crownvetch is slow in establishment.
2 Its regrowth is slow.
3 Its seed shatters.
4 Its seed is expensive.
5 Dry matter production is lower in crownvetch than in alfalfa.
6 Animals may be slow to utilize it.

QUESTIONS

1 Describe the area in which crownvetch is best adapted.
2 Describe the structure and appearance of crownvetch.
3 What are the uses of crownvetch?
4 How is crownvetch best established?
5 Compare crownvetch with alfalfa.

REFERENCES

1 Alex, J. F. Botanical aspects of crownvetch. In E. M. Watkin (ed.), *The Potential for Crownvetch in Ontario,* Univ. of Guelph, Ont., Canada, pp. 35–42, 1971.
2 Burns, J. C. and W. A. Cope. The nutrutive value of crownvetch forage as influenced by structural constituents and phenolic and tannin compounds. *Agron. J.,* 66:195–200, 1974.
3 Burns, J. C., W. A. Cope, and E. R. Barrick. Cow and calf performance per hectare productivity and persistence of crownvetch under grazing. *Agron. J.,* 69:77–81, 1977.
4 Burns, J. C., W. A. Cope, L. Goode, R. W. Harvey, and H. D. Gross. Evaluation of crownvetch (*Coronilla varia* L.) by performance of beef cattle. *Agron. J.,* 61:480–481, 1969.
5 Burns, J. C., R. D. Mochrie, and W. A. Cope. Intake of dairy heifers fed mixed crownvetch-tall fescue hays harvested in May and July. *Agron. J.,* 70:886–888, 1978.
6 Burns, J. C., R. D. Mochrie, and W. A. Cope. Responses of dairy heifers to crownvetch, sericea lespedeza, and alfalfa forages. *Agron. J.,* 64:193–195, 1972.
7 Cassady, J. M. Toxic constituents of *Coronilla* species. In *Proc. 2d Crownvetch Symp., Pa. State Univ. Agron. Mimeo. 6,* pp. 99–100, 1968.
8 Decker, A. M., H. J. Retzer, M. L. Larna, and H. D. Kerr. Permanent pastures improved with sod-seeding and fertilization. *Agron. J.,* 61:243–247, 1969.
9 Hart, R. H. Frequency and severity of defoliation of crownvetch stems by grazing sheep. *Agron. J.,* 62:626–627, 1970.
10 Henson, P. R. Crownvetch—soil conserving legume and a potential pasture and hay plant. *USDA Publ. ARS 34-63,* 1963.
11 Leffel, R. C. Other legumes. In *Forages,* Iowa State Univ. Press, Ames, IA, 3d ed., pp. 208–209, 1973.
12 Meade, J. A. and C. R. Belcher. Weed control in crownvetch and flatpea. *Progress Report: Clovers and Special Purpose Legumes Research,* 11:45–49, 1978.
13 McKee, G. W. Crownvetch in the United States—present status, future developments. In E. M. Watkin (ed.), *The Potential for Crownvetch in Ontario,* Univ. of Guelph, Ont., Canada, pp. 49–62, 1971.

14 Miller, J. D. Digestibility and mineral content of crownvetch clones. *Progress Report: Clovers and Special Purpose Legumes Research*, 11:80–89, 1978.

15 Reid, R. L., G. A. Jung, and R. O. Thomas. Nutritive value of crownvetch hay using sheep and cattle. In *Proc. 2d Crownvetch Symp., Pa. State Univ. Agron. Mimeo. 6*, pp. 86–98, 1968.

16 Reynolds, P. J., C. Jackson, Jr., and P. R. Henson. Comparisons of the effects of crownvetch (*Coronilla varia* L.) and alfalfa hays on the liveweight gain of sheep. *Agron. J.*, 61:187–190, 1969.

17 Vickers, J. C. and J. M. Zak. Effects of pH, P, and Al on the growth and chemical composition of crownvetch. *Agron. J.*, 70:748–751, 1978.

CHAPTER 21

LESPEDEZAS

Lespedezas are warm-season legumes which grow during the summer months. They play an important role in overall forage production in the southern and southeastern areas of the United States. Lespedezas are generally grown on soils low in fertility, as lespedeza yields are usually lower than those of other legumes.

There are many summer annual species of lespedeza, but the two most important species are Korean lespedeza, *Lespedeza stipulacea* Maxim., and common lespedeza, *L. striata* (Thumb) H. and A. In addition, there is one major perennial species of lespedeza, sericea, *L. cuneata* (Dumont) G. Don (8).

HISTORICAL BACKGROUND

Lespedezas are native to Asia and the eastern United States. Both common and Korean lespedezas were introduced to the United States from the Far East (12).

Common lespedeza was first grown in the United States in Georgia in 1846. An improved strain, known as 'Kobe,' was introduced from Japan in 1919, and grown in South Carolina. In that same year, Korean lespedeza was brought into the United States from Korea. By 1921, seed was produced in this country on a small scale and, within two years, a large quantity of seed was distributed (16).

Sericea lespedeza was first evaluated in North Carolina in 1896. Improved strains were introduced until 1924, and seed subsequently was distributed throughout the southern United States.

WHERE LESPEDEZA GROWS

Lespedezas was grown from eastern Kansas, Oklahoma, and Texas eastward to the Atlantic Coast. Korean lespedeza is best adapted to the upper south, and the better soils of the lower south. Common lespedeza is grown in the upper and lower south on soils of low fertility that are too acid to grow other legumes. Common lespedeza requires a longer growing season than the Korean type. Sericea lespedeza grows over this entire southern region (8).

Korean lespedeza requires the highest soil pH—a range of 6.0 to 6.5—of the species described. It is less tolerant of acid soils and more tolerant of alkaline soils than is common lespedeza. Common lespedeza tolerates soils with a pH of 5.5 to 6.5. Sericea lespedeza tolerates the lowest general soil fertility and pH of the three species, a pH range of 5.0 to 6.5. Sericea is generally considered an excellent soil builder and was the first legume to be grown on soils of very low fertility where no other legumes would grow (15).

TYPES OF LESPEDEZA

Annual Species

Although these lespedeza species are somewhat similar, they can readily be distinguished by their vegetative characteristics (8, 14). Koren lespedeza is coarser and matures earlier than common lespedeza. Korean lespedeza has broad, heart-shaped leaflets, while common lespedeza has leaves that are longer and less rounded on the ends. Korean has larger stipules, or bracts, at the base of the leaves. The common lespedeza cotyledons, upon emergence, are bluish-green in color and have an indentation on one edge near the outer end, while Korean cotyledons are dark green, almost elliptical, and have pronounced veins. The seeds of Korean lespedeza grow in clusters at the terminal end of the branch, while those of common lespedeza are set in leaf axils along the entire length of the stem. The small hairs on the stems of Korean are slanted upward, in contrast to those on common lespedeza, which are slanted downward (14) (Fig. 21-1).

FIGURE 21-1
Korean lespedeza flower and leaf. *(Courtesy University of Illinois)*

Flowers for both annual species range in color from purple to light pink. After flowering in initiated, leaves of Korean lespedeza begin turning forward, around the seed pod. This folding around the seed pod helps reduce seed shattering. Leaves of common lespedeza do not fold around the seed pod, thus increasing susceptibility to seed shattering (14).

Common lespedeza requires a longer growing season than the Korean strains. Therefore, common lespedeza is more adapted to the climate of the lower south, and Korean to the upper two-thirds of the lespedeza region. Day length and temperature affect the vegetative growth, flowering, and seed set in both of the annual species. Each species remains vegetative while the day length is longer than the critical photoperiod. Flowering occurs when the photoperiod begins to shorten. In fact, if the species is grown too far north and the shorter photoperiod occurs during the growing season, a killing frost might occur before the seed matures. Higher temperatures in the south hasten flowering and shorten the time needed for seed maturation (12, 18). The leaves will remain on the plant after frost.

Both annual lespedezas are thin-stemmed and leafy. If the plant density is low, the annual lespedezas tend to grow more prostrate (Fig. 21-2). The annuals have a rather shallow root system.

Perennial Species

Sericea lespedeza is a long-lived, perennial species. It has an erect growth habit and reaches heights of 60 to 100 cm (24 to 40 in) (Fig. 21-3). It is a leafy plant

FIGURE 21-2
Mature common lespedeza in a rowed nursery.

FIGURE 21-3
Sericea lespedeza nursery illustrating both upright and spreading growth habit.

with coarse stems. The leaves are trifoliate, with long, narrow leaflets that are squared off at the end. Flowers, ranging in color from purple to cream (8, 14), are produced on very short pedicels in the leaf axils along the stem (Fig. 21-4).

Sericea lespedeza will die back to the ground after a killing frost. Regrowth the following spring occurs from crownal buds. During the growing season, regrowth occurs from axillary buds, since there are no crownal buds regenerated during the growing season. Sericea has a well-branched, deep taproot.

FIGURE 21-4
Close-up view of sericea lespedeza showing the flower sturcture and leaf arrangement. *(Courtesy University of Illinois)*

Sericea has a high tannin content, which makes it lower in palatability and digestibility than either of the annual species. Sericea usually yields less than an equal planting of the annual species. Sericea tolerates lower soil fertility than either of the annual species, but does respond to lime and fertilizer applications on poor acid soils (1, 3, 4, 7).

USES OF LESPEDEZA

Annual lespedezas are used primarily for pasture and hay. Leaf shattering, which occurs soon after cutting, reduces the acreage that is baled. If one cuts, crimps, and windrows the annual lespedeza in one operation, baled hay can be harvested with little leaf shatter. It can also be cut and baled the same day, before it has dried completely. This helps reduce leaf shatter.

Considerable hectarage of the lespedezas, especially sericea, is used for soil conservation. Many seedings of perennial lespedeza are established along the sides of interstate highways to reduce soil erosion (8, 17) (Fig. 21-5). Lespedezas are also used along roadsides for wildlife cover, and the seed is used for feed by wildlife such as deer and quail.

Lespedezas are not grown as widely as they once were. With the existence of other improved legumes and better fertility practices, lespedezas have been replaced in many mixtures. Arrowleaf clover has replaced some of the lespedeza acreage in the south. Hop clover has become naturalized in some areas, and clover is now grown with bermuda pastures in the south.

FIGURE 21-5
Use of sericea lespedeza on roadsides.

CHARACTERISTICS OF LESPEDEZA

Lespedezas will tolerate soils lower in fertility than most other legumes. Lespedezas can be grown with both cool- or warm-season grasses and with small grains. Both of the annual lespedezas are warm-season species and are good producers of forage late in the summer. This extends the carrying capacity of the pasture through the sumer, when the productivity of most grasses is tapering off (3).

"Common" lespedeza in general usage refers to the small, prostrate, unproductive type of *L. striata* also called "jap clover." It has become naturalized and occurs widely as a "wild" plant. This strain is generally not planted or cultivated; the 'Kobe' cultivar is the only *L. striata* that is planted and cultivated. Annual lespedeza is one of the best plants for wild bird food (seeds) and is widely planted for this purpose, especially for use as quail food (1, 3, 8, 14).

Probably the best characteristic of lespedezas is that there is no bloat problem associated with grazing. The annual types are easy to establish, and will reseed themselves each year; thus stands persist as if the plants were perennial. Lespedezas resist heat and drought conditions, therefore providing good summer grazing (3, 14). Seed is rather inexpensive.

Four drawbacks to the utilization of sericea lespedeza are: (1) possession of a tannin which reduces its palatability; (2) low yields; (3) inability to compete well with weeds; and (4) a lack of tolerance for wet soil conditions (3, 14). Recently 'Au Lotan' was released. It is a cultivar of sericea that has about one-half the tannin of original sericeas. It is more palatable, more digestible, and produces higher annual gains than other sericea. The yield of 'Au Lotan', however, is only about 85% of the other sericeas'. Seed is now being commercially produced.

MANAGEMENT OF ANNUAL LESPEDEZA

Establishment

Annual lespedezas grow on almost any soil type except deep sands. They are very drought-resistant and will grow on eroded soils which have a pH of 5.5 to 6.5 and a low P level. They are not adapted to high pH soil. Korean lespedeza prefers a soil with a higher pH than common lespedeza. Both annual lespedezas respond to fertilizer and somewhat to increased pH (15). Averages of many trials over the years have shown yields to increase from 1745 to 4048 kg/ha (1557 to 3611 lb/A) with increased fertilizer and lime. Research has shown that raising the pH to 6.5 with annual applications of 11 to 20 kg of P/ha (9.8 to 18 lb/A) will greatly increase production. Sometimes high rates of fertilizer will lower the quality of lespedeza because of the increase in stem proportion to leaf proportion (15).

Seeding

Annual lespedezas are easy to establish. They are sometimes seeded on the snow in winter, or early in the spring; the seed germinates as the snow melts. They often reseed themselves each year, so the stands appear to be perennial in nature. Common lespedeza is often sown earlier than the Korean type; Korean is planted in March or April, while common lespedeza is sown about one month earlier. Annual lespedezas are often broadcast into a grass pasture. If this is the case the lespedeza should not be grazed until it is at least 16 cm (6 in) tall (14).

Seeding rates for annual lespedezas range from 22 to 33 kg/ha (20 to 30 lb/A) of unhulled seed. Lespedeza seed should be inoculated before it is seeded in an area where lespedeza has not grown before. To ensure a full stand in the second year, 11 to 17 kg/ha (10 to 15 lb/A) should be seeded during the second year to supplement the natural reseeding (14). Annual lespedeza seed often contains up to 40 to 60% hard seed.

Harvest

Since annual lespedezas are warm-season producers, their period of growth occurs at around the end of the most productive time for most cool-season grasses. This extends the carrying capacity of the pasture through mid-summer when productivity of most grasses is tapering off.

The highest quality forage from annual lespedezas occurs at early bloom. If harvested at this stage it will approach alfalfa in quality. There will be sufficient regrowth after an early-bloom harvest to allow enough flowering so that reseeding may occur, or enough regrowth for grazing. If harvested at a more mature stage, there will be very little regrowth, or even death of the stand. Common lespedeza can be grazed closer than Korean and still survive (14).

If annual lespedezas are harvested for hay, they should be cut, crimped and windrowed to prevent leaf shatter. Annual lespedeza should be harvested before the plants are completely dry. Leaves do remain on the plants after frost, making it a good crop to graze late into the fall.

Weed Control

Heavy seeding rates and excellent seedbed preparation help reduce weed invasion. An excellent herbicide for controlling broadleaf weeds is 2,4-D, amine form, at the rate of 1.12 kg/ha (1 lb/A). Applying this herbicide in May or June effectively controls weed without injuring the lespedezas (11).

MANAGEMENT OF PERENNIAL LESPEDEZA

Establishment

Perennial or sericea lespedeza will grow on lower-fertility soils than the annual types. It also responds to fertilizer and lime application but will not grow on high pH soils. A well-prepared seedbed is important, and the seed should be covered with a shallow layer of soil to ensure a good stand (3) (Figs. 21-6 and 21-7).

FIGURE 21-6
Lespedeza sown in wheat with no nitrogen added.

Seeding

Sericea lespedeza is usually seeded at 22 to 33 kg/ha (20 to 30 lb/A) or more of well-scarified seed. Generally no forage is harvested during the year of establishment. It has been shown that with the herbicide Vernolate incorporated as a

FIGURE 21-7
Small grains interseeded with lespedeza. Only oats has been chopped.

preplant treatment at the rate of 3.4 kg/ha (3 lb/A), the seeding rate can be reduced to 11 kg/ha (10 lb/A). With this method of establishment one can harvest in the year of seeding (11).

Managemant

Sericea lespedeza has a coarser stem and is more upright in growth habit than are annuals. Increasing the stand density will increase quality by reducing stem size without a corresponding reduction in leaves. Sericea must be harvested early to ensure quality forage (Fig. 21-8). Grazing should occur when the plant height reaches 15 to 20 cm (6 to 8 in). Stands will thin and be lost if grazing is not terminated in late summer to allow recovery before frost. In the early spring cattle gains may reach around 0.9 kg (2 lb) per day, while in July and August there may be no gains from grazing sericea (2).

The grazing period of sericea lespedeza can be extended by two months if the sericea is overseeded with rye in the fall. The added cost may not be justified for a cow-calf program, but the rye provides a high-quality forage for the cows before sericea begins growth in early spring.

If sericea is harvested for hay, it should be cut when it is 30 to 35 cm (12 to 14 in) high, while the stems are still very pliable and the tannin content is low. There is more tannin in the leaves than in the stems. Sericea harvested at this stage will contain from 10 to 15% crude protein and 2 to 7% digestible protein. The net energy derived from sericea lespedeza is about 75% that of alfalfa hay (10). Sericea hay is not recommended for milk cows.

One important consideration in managing the growth of sericea lespedeza is that it should not be harvested below the lowest internode, as regrowth is from axillary, not crownal buds. Crownal buds are generated during the winter for the regrowth the next spring. Sericea should not be harvested more than twice during the year before a seed crop, nor should there be more than three hay harvests per year. Adherence to this practice will ensure regrowth the following year (14).

FIGURE 21-8
Sericea lespedeza must be grazed at a rather early stage of growth, prior to 30 cm, to ensure high quality.

Root Reserves

Sericea lespedeza should not be grazed late in the fall (that is, 30 to 40 days before dormancy or a killing frost), because it needs to develop a high root reserve for winter survival.

Weed Control

The control practices are about the same for sericea as for the annual lespedeza. Most growers prefer to have some weeds rather than take a chance with 2,4-D. If 2,4-D is applied at the rate of 0.23 to 0.45 kg/ha (0.25 to 0.5 lb/A) at the bud stage, many of the broadleaf weeds can be controlled, with little injury to the sericea. The main use of 2,4-D on sericea would be as a spot treatment for dodder control (11).

DISEASES

There are more diseases of Korean lespedeza than the other lespedeza species. Bacterial wilt, *Zanthomonas lespedezae* (Ayers, et al.) Starr, is probably the most serious disease (13). Tar spot, *Phyllachora lespedezae* (Schw.) Sacc. attacks the leaves, causing leaf shatter and a reduction in yield (6). There are several other diseases periodically attacking lespedeza.

INSECTS

Lespedezas are relatively free of insects (14). This may be due to the tannin content of the plant, although the correlation is unclear, as annuals are relatively low in tannin.

SUMMARY

The advantages and disadvantages of growing lespedezas are as follows:

Advantages:

1 They tolerate acidic, infertile soils.
2 The annual species are easy to establish.
3 Annuals reseed themselves.
4 They resist drought and heat.
5 They provide good summer grazing.
6 Seed is relatively inexpensive.
7 They are resistant to most diseases and insects.
8 They do not cause bloat.
9 Sericea begins growth rather early in the spring, as compared to the annuals.

Disadvantages:

1 Their yields are relatively low.
2 Sericea contains tannin, which reduces its palatability and digestibility.
3 Quality of sericea is lower than that of alfalfa or red clover.
4 Regrowth is slow and low in yield.
5 Late growth in the spring requires supplementary pastures when growing the annuals.
6 They do not tolerate wet soil conditions.
7 They do not compete well with weeds.
8 There is the risk of rapid leaf shatter when harvesting for hay.
9 Annuals fail to reseed in hot, dry weather.

QUESTIONS

1 Describe and compare the three major species of lespedeza.
2 Compare the adaptation of lespedezas with that of alfalfa.
3 Which species of lespedeza is the most palatable? Which is the least palatable? What accounts for the differences in palatability?
4 Compare the seeding practice of the three species.
5 What caution must one exhibit when harvesting sericea if the stand is to remain productive?

REFERENCES

1 Anderson, K. L. Lespedeza. *Kan. Agr. Exp. Sta. Circ. 251,* 1956.
2 Anthony, W. B., R. R. Harris, C. S. Hoveland, E. L. Mayton, and J. K. Boseck. Serala sericea as a grazing crop for beef cattle. *Ala. Agr. Exp. Sta. Highlights of Agr. Res. 14,* No. 4, 1967.
3 Baldridge, J. D. The lespedezas. II. Culture and utilization. *Advan. Agron.,* 9:122–142, 1957.
4 Donnelly, E. D. and G. E. Hawkins. The effects of stem type on some feeding qualities of sericea lespedeza, *L. cuneata,* as indicated in a digestion trial with rabbits. *Agron. J.,* 51:293–294, 1959.
5 Elder, W. C. Utilization of sericea lespedeza in Oklahoma. *Okla. Agr. Exp. Sta. Processed Series* P-557, 1967.
6 Hanson, C. H., W. A. Cope, and J. L. Allison. Tar spot of Korean lespedeza caused by *Phyllachora* spp.: Losses in yield and differential susceptibility of strains. *Agron. J.,* 48:369–370, 1956.
7 Hawkins, G. E. Composition and digestibility of lespedeza sericea hay and alfalfa hay plus gallotannin. *J. Dairy Sci.,* 38:237–243, 1955.
8 Henson, P. R. The lespedezas I. *Advan. Agron.,* 9:113–122, 1957.
9 Henson, P. R. and C. H. Hanson. Annual lespedezas culture and use. *USDA Farmers Bull. 2113,* 1958.
10 Holdaway, C. W., W. B. Ellelt, J. F. Eheart, and A. O. Pratt. Korean lespedeza and sericea lespedeza hays for producing milk. *Va. Agr. Exp. Sta. Bull. 305,* 1936.
11 Hoveland, C. S., G. A. Buchanan, and E. D. Donnelly. Establishment of sericea lespedeza. *Weed Sci.,* 19:21–24, 1971.

12 Offutt, M. S. Some effects of photoperiod on the performance of Korean lespedeza. *Crop Sci.,* 8:308–313, 1968.

13 Offutt, M. S. and J. D. Baldridge. Inoculation studies related to breeding for resistance to bacterial wilt in lespedeza. *Mo. Agr. Exp. Sta. Res. Bull. 603,* 1956,

14 Offutt, M. S. and J. D. Baldridge. The lespedezas. In *Forages,* Iowa State Univ. Press, Ames, IA, 3d ed., pp. 189–198, 1973.

15 Offutt, M. S., L. H. Hileman, and O. T. Stallcup. Effect of lime, phosphorus, and potassium on yield and composition of Korean lespedeza. *Ark. Agr. Exp. Sta. Bull. 717,* 1966.

16 Pieters, A. J. The annual lespedezas as forage and soil-conserving crops. *USDA Circ. 536,* 1939,

17 Pieters, A. J., P. R. Henson, W. E. Adams, and A. P. Barnett. Sericea and other perennial lespedezas for forage and soil conservation. *USDA Circ. 863,* 1950.

18 Smith, S. E. The effect of photoperiod on the growth of lespedeza. *J. Amer. Soc. Agron.,* 33:231–236, 1941.

CHAPTER 22

OTHER LEGUMES

There are many other native or introduced legumes in the United States, some of which have limited distribution but are nevertheless very important in local situations or for restricted purposes. Various legume species have been used for soil improvement, and other species for soil conservation, hay, pastures, or emergency crops.

SWEET CLOVER

Sweet clover, *Melilotus* spp., is a prime example of a legume species that was important in its day as a forage and as a soil-improvement crop. It is one of the oldest legume species and was brought to the United States from Europe in the early 18th century. It was found in Virginia in 1739 (37). Sweet clover probably originated in the same general regions of the world as did alfalfa and red clover, and is a native of the temperate areas of Europe and Asia. Having a wide range of adaptation, it has been widely distributed over the world, and is found throughout the United States and much of Canada. It does not grow well on acid soils, since it requires a pH of 6.0 or higher for good growth. Sweet clover is one of the best soil-building legumes because of its large, deep taproot system. It is drought-resistant, making it well-suited to the Great Plains of the United States. Sweet clovers prefer well-drained soils that are relatively high in organic matter (37).

Most of the sweet clovers grown in the United States are of two species: the yellow-flowered, *M. officinalis* (L.) Lam., and the white-flowered, *M. alba* Desr. The yellow-flowered sweet clovers are biennial in growth habit, while the white-flowered sweet clovers include both annual and biennial types. Most hectarage in the United States is of the biennial white-flowered sweet clover (11, 36, 37).

Sweet clovers differ from alfalfa in several morphological characteristics. Sweet clover has a more definite taproot, which is less branched than that of alfalfa. Leaves are trifoliate as in alfalfa, but the margins of the sweet clover leaflets are completely serrated, while in alfalfa only the upper one-third of the leaflet margin is lightly serrated. Alfalfa stipules are larger and serrated, while sweet clover stipules are smooth, small, and narrow. The sweet clover flowers are either yellow or white, much smaller than red clover and alfalfa florets, and are loosely arranged on a slender raceme (Fig. 22-1). Yellow sweet clover flowers 10 to 14 days earlier than white sweet clover.

Sweet clover seed pods shatter easily, potentially causing a weed problem over the years. Generally only one seed is produced per pod, compared with many seeds per pod in alfalfa (11). One can distinguish germinating alfalfa seeds from those of sweet clover by viewing them under ultraviolet light, as sweet clover seeds do not fluoresce (4). Yellow sweet clover also has finer stems, is leafier, has less top growth, and is somewhat shorter than white. Yellow sweet clover is less sensitive than alfalfa to competition from companion crops, weeds, and drought (11).

The biennial strain of white sweet clover is preferred in the corn belt because it matures later and produces more growth, and has soil-improvement properties on a par with the yellow-flowered types (36). Both yellow- and white-flowered types have about the same number of stems per plant and similar root systems (Fig. 22-2). Both are winter-hardy, although sweet clovers as a whole may be subject to some winter injury (11, 36).

Annual white sweet clover, such as the cultivar 'Hubam', does not over-winter. It is an excellent green-manure crop. If sweet clover is allowed to shatter and is plowed down, there will be volunteer sweet clover for some 20 to 25 years due to the high percentage of hard seed, and this can be a problem. It must be plowed down prior to flowering (11). Photoperiod requirements limit the

FIGURE 22-1
Close-up of sweet clover raceme showing leaf morphology.
(Courtesy University of Illinois)

FIGURE 22-2
Flowering plant of white sweet clover.
(Courtesy University of Illinois)

flowering of some of the recent cultivars in the upper midwest, so flowering and seed shatter are not a problem in this area. When sweet clover is properly inoculated, it may equal or exceed alfalfa in its ability to fix and utilize atmospheric nitrogen (16, 36).

Cultural practices are essentially the same as for alfalfa. The soil pH must be above 6.0 for good growth. The seed must be well-scarified before seeding, or the hard-seed carryover will be encountered, lowering plant density because of low germination. Seeding rates of 11 to 17 kg/ha (10 to 15 lb/A) are recommended for pure stands (11).

Biennial sweet clover produces a single main stem for the first year of growth. Regrowth must come from axillary buds on the main stem after the plant is cut off. There is no regrowth from crownal buds, as is also the case with alfalfa. If the sweet clover is cut too close to soil surface during the first year, a severe reduction of the root system will result. This reduces the ability of the plant to overwinter and decreases its vigor during the second year, because the root system serves as a food storage area for the overwintering plant (11, 36, 38).

In addition, biennial sweet clover can become a weed problem in succeeding years if the grower plows it down in the fall of the seedling year instead of waiting for growth to begin the next spring and then plowing down. Biennial sweet clover develops several fleshy crownal buds early in the fall of the seedling year. These buds are not initiated during the fall unless all of the top growth is removed below the first leaf node. If sweet clover is cut below the first node, the crownal buds will initiate growth because no axillary bud remains. Once the crownal buds are initiated, no new crownal buds will develop, thereby stopping regrowth from the crown. If the fall growth were plowed down with the fleshy buds intact, then the following spring, growth would occur either from the crownal buds that were plowed down or the crownal rhizomes that had overwintered. But biennial sweet clover will not flower the seedling year. Therefore, biennial sweet clover can be killed easily by plowing it down in the spring of the second year, after 10 to 15 cm (4 to 6 in) growth has occurred (11, 45).

Sweet clover possesses coumarin, a chemical that reduces palatability of the forage to the point that cattle may refuse to consume the crop, especially if there is an alternative forage (11). There is a characteristic odor associated with coumarin. If sweet clover hay is harvested when it has a very high moisture content and spoilage or heating occurs, the coumarin changes into a compound called dicoumarol. Dicoumarol reduces the ability of blood to clot (5). Cattle have been known to bleed to death after consuming spoiled sweet clover hay, from an open wound or internal hemorrhaging, a condition often called the "bleeding disease." It is therefore very important to watch for spoiled sweet clover hay in livestock feed. Dicoumarol is used commercially in a poison to kill rats and mice (36).

If sweet clover is to be used as a pasture crop, the top growth should reach around 20 to 25 cm (8 to 10 in) in height before grazing it down to 10 to 15 cm (4 to 6 in), to prevent rank growth (11, 25). A bloat problem, similar to that with alfalfa or red clover, exists in sweet clover, but to a lesser degree.

There are a few diseases associated with sweet clover: black stem, stem canker, and root rots. The only major insect problem is the sweet clover weevil, which has restricted widespread use of sweet clover as a green-manure crop. Some of the advantages of sweet clover are:

1 It is an excellent soil-improvement crop.
2 It has a very deep root system.
3 It is more resistant to drought and heat than alfalfa.
4 It is an excellent source of nectar and pollen for honeybees.
5 It grows in a wide range of deep soils and climatic conditions.

A few of the disadvantages are:

1 It may become a weed problem due to the hard seed content.
2 It contains coumarin, which may cause a "bleeding disease" in cattle.
3 It makes poor-quality hay because it is quite stemmy.
4 It is low in palatability as a pasture crop.

ALSIKE CLOVER

Alsike clover, *Trifolium hybridum* L., is a short-lived perennial legume. It was brought into the United States in 1839, and is adapted to the upper Great Lakes region, and Oregon, western Idaho, and northern California. These areas have cool climates and wet soils. Alsike tolerates more acid soil and alkaline conditions than red clover (39). It is managed in almost the same way as red clover.

The seeds of alsike clover are greenish-yellow and much smaller than red clover or alfalfa seeds. Flowers range in color from pink to pinkish-white. The flowering head, or inflorescence, is smaller than that of red clover (39) (Fig. 22-3). Alsike is finer stemmed than red clover and much more decumbent in growth. It is usually grown with grasses, which tends to help keep the plants more erect.

Alsike may be injured by prolonged hot weather. It can tolerate flooded conditions and withstand wet soil for longer periods than ladino clover, and has been used successfully in irrigated mountain meadows. Alsike responds to irrigation and does well in naturally wet grass pastures.

Because of its small seeds, the recommended seeding rate is only 4.5 to 7 kg/ha (4 to 6 lb/A). Alsike clover usually produces only one harvest per year. The recommended stage of maturity for harvesting is full bloom. However, it may be difficult to determine when full bloom occurs because of alsike clover's indeterminate growth type and flowering habit. One cannot harvest both a hay crop and a seed crop in the same year (39).

Alsike clover is susceptible to about the same diseases and insects that attack red clover, but is resistant to northern and southern anthracnose (39).

CRIMSON CLOVER

Crimson clover, *T. incarnatum* L., is a winter annual legume native to southeastern Europe. It was introduced to the United States from Italy in 1819.

FIGURE 22-3
Inflorescence of alsike clover. *(Courtesy University of Illinois)*

Crimson clover is widely adapted to the southeastern United States and to the southern portions of Ohio, Indiana, Illinois, and Missouri, as well as along the Pacific coastal area. In northern Maine it is considered a summer annual and is seeded in late May or early June (25).

Crimson clover grows best on soils with a pH of 5.7 or above (25) and responds to P, K, and B fertilization for seed production. Crimson clover is grown primarily for pasture but can be used as hay or silage. It produces more growth in midwinter than most legumes (25). It can be grown in combination with small grains and ryegrass or overseeded in bermudagrass to increase forage quality and production. Crimson clover has increased the forage yield by 60% of 'Coastal' bermudagrass fertilized with 224 kg/ha N (200 lb/A) (26). Crimson clover provides early spring forage while the 'Coastal' bermudagrass is dormant, as well as supplying residual N for the bermudagrass.

Crimson clover seeds are yellow, larger, and rounder than red clover seeds. Florets are a bright crimson, formed on a pointed, conical flower head with 75 to 125 florets per head (Fig. 22-4). Flowering is initiated when the photoperiod exceeds 12 hours (24). Flowers are self-fertile, but insect visitation is needed for effective pollination. After the seed matures, about 25 to 30 days after pollination, the plant dies. Seed shattering perpetuates crimson clover in a pasture. Leaves are trifoliate and the plant is densely covered with hairs (25).

Seeding rates vary from 22 to 34 kg/ha (20 to 30 lb/A) when seeded in a mixture, and should be at least 34 kg/ha (30 lb/A) when seeded alone (21). There are new cultivars that are earlier-maturing, and have larger seeds with more seedling vigor, which will reseed themselves more readily than some of the older cultivars. There are now cultivars that have no hard seed (22). In the older

FIGURE 22-4
Crimson clover in bloom.

cultivars the hard-seed content may vary from 30 to 75%, which ensures reseeding of the crimson clover.

Crimson clover may be seeded from July to November, depending on the growing region and its ultimate use. If used primarily as an early winter or early spring grazing crop, the clover should be seeded early enough to allow sufficient growth before the advent of winter (25). An excellent time to seed crimson clover is following or just preceding a rain, because the moisture will ensure good germination (25).

Crimson clover may be grazed when it is 10 to 15 cm (4 to 6 in) tall but should not be grazed to below 5 cm (2 in). During the winter, it can be stocked in moderately heavy amounts, but stocks should be much heavier in the spring, when there is rapid growth. If crimson clover is harvested for hay, it should be harvested in the early-bloom stage to ensure high-quality forage (25).

Crimson clover does not have many serious disease problems, probably because it is a winter annual. It is susceptible to some crown and stem rots under wet conditions and high temperatures, and occasionally to anthracnose. There are very few insect problems on crimson clover, possibly due to the heavy pubescence on the plants. There are some insect problems when seed is being produced.

ARROWLEAF CLOVER

Arrowleaf clover, *T. vesiculosum* Savi, is one of the newest reseeding winter annual legumes cultured in the United States. It was first brought into the United States from Italy in 1930. More recently, in 1956, several more introductions were made into the United States, and were grown in Georgia (25). Arrowleaf is grown on well-drained, sandy soils and even clay soils, from South Carolina to Texas, and from Tennessee and Arkansas south to the Gulf of Mexico. Arrowleaf clover does not tolerate high lime soils, and is less tolerant of acid soils than crimson clover. The soil pH should be between 6.0 and 7.0. An iron chlorosis may appear in soils above a 7.5 pH. Arrowleaf clover requires more P and K than crimson clover. If P and K additions are needed, they should be applied in August, as indicated by soil tests (22, 25).

Arrowleaf clover is replacing some of the crimson clover hectarage in the lower south and the hop clover hectarage in the upper south because it has a longer growing season and there is very little bloat hazard (3). Regrowth of arrowleaf clover is rapid after a rain. The clover is also fairly drought-tolerant. The forage quality is high, with a digestible dry-matter content of from 80% in late winter to 70% in the early-bloom stage of growth. Plant height may reach 150 cm (60 in) if the plants are not grazed or harvested.

Arrowleaf clover blooms over a long period. Stems are hollow, thick, and smooth. A long, white, pointed stipule is present at the base of the petiole. The leaflets have pronounced veins. Leaf markings may or may not be present. The flowering head is conically shaped, and there are 150 or more florets per head. The flower color ranges from white, at first emergence, to pink or purple (25).

There are several cultivars now available that differ in winterhardiness, maturity, and spring growth. Seeding practice is very important because costs are relatively high. Arrowleaf should be seeded on a well-prepared seedbed along with small grains or ryegrass. It should be seeded very shallow, no more than 0.6 cm (¼ in) deep, with a grain drill or, when seeding into an existing sod, with a sod seeder. Seeding rate recommended is 6 to 9 kg/ha (5 to 8 lb/A) of well-scarified seed. Hard-seed content may range from 75 to 80% (25). Arrowleaf is sown from August until November, depending on the region where it is grown. Since arrowleaf will germinate at low temperatures, if it is planted into sod it should be sown after the grass goes dormant, thus reducing competition for moisture (21, 22). A special *Rhizobium* inoculum is required for arrowleaf (25, 32a). If one is overseeding a sod such as tall fescue or bermudagrass with arrowleaf clover, the grass should be closely grazed in August or harvested for hay. The sod should be disked lightly and seeded in late August or September. Then the seed is harrowed or lightly covered (22).

Arrowleaf clover produces most of its growth during the spring or early summer. One should start grazing the plants when they are around 15 cm (6 in) tall and grazing should remain moderately heavy throughout the growing season (22). If a seed crop or reseeding is desired, grazing should be discontinued in early April. High digestibility is retained for longer than with many legumes. Dry matter yields of over 6.7 t/ha (3 T/A) have been reported (18).

Arrowleaf clover is nonbloating clover. Grazing trails of arrowleaf clover in combination with wheat, rye, or ryegrass have produced excellent results with feeder and stocker cattle. In an Alabama trial using arrowleaf clover in combination with rye and ryegrass, the average daily gain of beef cattle was 0.985 kg (2.17 lb) and the average over a 2-year test was 461 kg/ha (412 lb/A) during a 177-day grazing period (2). Other tests with 'Coastal' bermudagrass showed gains of 345 kg/ha (308 lb/A) as compared with a gain of 295 kg/ha (263 lb/A) using dallisgrass–ladino clover (18). These trials showed a grazing season of 222 cow-calf days for the arrowleaf 'Coastal' bermudagrass combination, compared with only 145 days for the dallisgrass–ladino clover mixture. Other trials have shown similar results (22).

VETCHES

There are over 150 species of the genus *Vicia*. These legume species are referred to as vetch. Vetches are widely grown in temperate areas throughout the world. Within the genus *Vicia,* the most widely cultivated species are hairy vetch, *V. villosa* Roth; common vetch, *V. sativa* L.; and purple vetch, *V. benghalensis* L. All three species are considered winter annuals and used primarily as ground cover (12, 13, 28).

Hairy Vetch

Hairy vetch is a winter annual and is the most winter-hardy of the three species. Hairy vetch is adapted to a wide range of soils. One would think it gets its name

FIGURE 22-5
Hairy vetch inflorescence. *(Courtesy University of Illinois)*

from the pubescence of some of the plants, but some are nearly completely hairless (Fig. 22-5). There are many subspecies of hairy vetch, including smooth vetch and winter vetch (Fig. 22-6). In general, hairy vetch is less winter-hardy than alfalfa and red clover but sufficiently hardy to be grown throughout most of the United States. It has a viney growth habit. It is used as a green-manure crop and is mixed with small grains such as rye for grazing (12, 13). If it is grown with

FIGURE 22-6
Seed pod of hairy vetch. *(Courtesy University of Illinois)*

rye as a pasture mixture, one should seed around 100 to 112 kg/ha (90 to 100 lb/A) of rye and 17 to 22 kg/ha (15 to 20 lb/A) of hairy vetch. For a green-manure crop, seed 28 to 34 kg/ha (25 to 30 lb/A) of vetch. It has hard seed and will reseed itself. The flowers are purple.

Common Vetch

This species is less winter-hardy than hairy vetch and is therefore restricted to the southern region of the United States. It is better adapted to well-drained and higher-fertile soils than hairy vetch (13). Common vetch has more winter growth and flowers earlier than hairy vetch. Old cultivars of common vetch will not reseed themselves, but some new cultivars have hard seed, enabling reseeding. Common vetch is resistant to the root knot nematode, making it a good crop to rotate with crops susceptible to the nematode.

Purple Vetch

Purple vetch is the least winter-hardy of the three species; therefore, as a winter annual, it is grown in only the most southerly regions. Flowers are reddish-purple or white in color and have pubescent pods (28).

Cicer Milkvetch

Cicer milkvetch, *Astragulus cicer,* is not related to hairy or common vetch, or to crownvetch. Cicer is a long-lived perennial adapted to the dryland conditions of the Great Plains and the northwest (41).

Cicer exhibits a rhizomatous, decumbent-type growth. Cicer will often spread some 3 m (10 ft) under good growing conditions. The stems are hollow, coarse, and very succulent. The leaves are pinnately compound with one terminal leaflet. Flower color ranges from light yellow to white. Flowers are arranged on a raceme from the leaf axils (6).

Cicer has a wide range of adaptation, from dryland conditions to cool, moist areas, and also to high elevations. Cicer milkvetch is more tolerant of acid or alkaline soils than alfalfa (6). It is also resistant to drought and frost damage. Cicer grows more slowly and its seeds emerge later than those of alfalfa. It does not cause bloat. There are no major insect or disease problems.

KUDZU

Kudzu, *Pueraria lobata* (Willd.) Ohwi, is adapted to the humid south and other areas where it will not be killed over the winter. It is a long-lived perennial used primarily for controlling soil erosion. It can also be used for hay or pasture.

Kudzu grows rapidly as a coarse, hairy vine with long runners (33). Once it becomes established, it will persist for years, Kudzu vines may reach over 20 m (65 ft) in 1 year. Kudzu can become a pest if it is not controlled or managed properly, as it may climb utility poles, trees, or shrubs to the point where it will weaken or kill the trees or shrubs.

Kudzu will not persist if continuous or heavy grazing is practiced. It is best used to accumulate forage during a drought or other emergency situation. It is very palatable and cattle performance is very good when grazing is properly managed (29).

Kudzu roots easily at the nodes. Propagation is through rooted cuttings, a division of the crown, or transplanted seedlings (29). It very seldom sets seed in the United States, but does bloom. Flowers are dark purple in color.

SAINFOIN

Sainfoin, *Onobrychis viciifolia* Scop., is a perennial legume adapted to the dry regions of the northern United States and to soils with a high pH. On irrigated lands, it is a short-lived perennial. Sainfoin is a deep-root species which will persist on soils low in P (28). The plant has an upright growth habit. Stems arise from a crown. Leaves are pinnately compound. Flowers arise from a raceme and are pink. The seeds, larger than those of red clover, are kidney-shaped, and range in color from brown to yellowish-green. Sainfoin seeds shatter readily when mature.

The protein content of sainfoin is lower than for alfalfa, but sainfoin proves to be higher in total digestible nutrients than alfalfa when harvests at the same stages of maturity are compared (7, 9). It is a nonbloating legume. One of the main disadvantages of sainfoin is that it has a slow regrowth and is not as productive as alfalfa when both are adapted.

OTHER CLOVERS

Strawberry clover, *Trifolium fragiferum* L., has an inflorescence that resembles a strawberry, with color from pink to white. It is grown mainly in temperate climates on wet, alkaline soils. Strawberry clover is especially well-adapted to areas which may flood for several weeks at a time. Strawberry clover will probably survive prolonged flooding better than other legumes. It is grown primarily in the northwestern United States in areas that are too alkaline or wet to support other legumes (15).

Hop clover has two species, large hop clover, *T. campestre* Schub., and small hop clover, *T. dubium* Sibth. They have small round flowering heads with yellow flowers. The ripe seedheads are brown. They are adapted to poor soils and, in the southern United States, to higher elevations. They only produce forage during a short period of growth in the spring (28).

Ball clover, *T. nigrescens* Vi., is a winter annual legume which looks like ladino clover but has no stolons. It also grows taller than ladino. It is grown on the fertile loamy soils of the lower southeastern United States, and begins to grow about 1 week after crimson clover. One can use heavy grazing for ball clover. It reseeds itself (17).

Persian clover, *T. resupinatum* L., is another winter annual legume which grows in the southeastern United States on heavy, moist soils. It has an upright growth habit. The forage is very palatable and nutritious, but cattle will bloat rather easily when grazing persian clover unless a bloat-preventive chemical is utilized. Persian clover can also be used for soil improvement or as hay. Seed shattering ensures reseeding (43). The only desirable feature about this clover is that it will grow on wetter soils than other winter annual clovers.

Berseem clover, *T. alexandrinum* L., a winter annual legume native to the Mediterranean area, has been adapted to the Gulf Coast area and southwestern United States. It grows vigorously in the fall and early winter, and will generally produce more forage during the fall than most legumes. It has an erect growth habit. Berseem clover has hollow stems, narrow leaves, with yellowish flower inflorescence. It winter kills at temperatures only slightly below freezing (20).

Rose clover, *T. hirtum* All., is grown on rangeland in California (46). It has pubescent stems, leaves, and flower heads and rose-colored flowers. Rose clover prefers well-drained soils. Under grazing conditions, it will reseed itself. Rose clover is a very palatable species even after it has matured (28).

Subterranean clover, *T. subterraneum* L., is another legume that reseeds itself readily through seed shattering, its high percentage of hard seed, and its high-temperature seed dormancy. Burrs are formed on the seed pod, which in turn buries itself in the soil. This method of regeneration gives subterranean clover its name. It is grown as a rangeland species with grasses in western Oregon and California and is a very important legume species in Australia. It has very small flowers that are white to pink in color. It is adapted to acid soils and its main use is for pasture (30).

LUPINES

Lupines, in general, contain a high percentage of alkaloids, which makes them bitter in taste. It is thought that the alkaloids have made the plants resistant to insects and unappetizing to cattle, as there has been a lack of grazing. Alkaloids are contained in both the plant tissue and seeds. When cattle are quite hungry, they have freely grazed lupines and been killed by the high alkaloid content (10).

Lupines have been cultivated for over 3000 years. In 1928, an alkaloid-free lupine, sweet lupine, was discovered, which proved to be nontoxic to cattle (10, 31). The presence or absence of an insect, the thrip, is a good method for determining if a legume is bitter or sweet, as alkaloids repel thrips.

The growth habit of lupines is upright; lupines reach heights of around 1 m (3 ft) on coarse stems. They are adapted to the southern United States on well-drained soils. In the south, lupine is considered a winter annual, but when grown in the upper midwest on sandy soils, it is considered a summer annual (28).

There are three major large-seeded lupines grown in the United States. These are:

1 The blue lupine, *Lupinus angustifolius* L.: The blue lupine is the most common species and grows on the coastal plains of the southeast on soils that generally are neutral. Blue lupine is more winter-hardy than yellow lupine. The main use of blue lupine cultivars is soil improvement, although the sweet cultivars have been used for grazing (10).

2 Yellow lupine, *L. luteus* Kell.: Yellow lupines can grow on more acid soils, or on sandy soils with low fertility, more readily than can blue lupines. Yellow lupines are limited to Florida because of their lack of winterhardiness. Yellow lupine seed shatters more easily than blue lupine seed. Yellow lupine seed is flat, light in color, and usually has black speckles, while blue lupine seed is not as flat and is gray to brownish-gray in color (28).

3 White lupine, *L. albus* L.: White lupines are the most winter-hardy of the lupines. They grow in the lower midsouth region of the United States around the Mississippi Delta (Fig. 22-7). They require probably the most fertile soils of the three and exhibit the least amount of seed shattering. White lupine seeds are large, flat, and white to cream in color. White lupine is used primarily for soil improvement.

FIELD PEA

Field pea, *Pisum sativum* subsp. arvense (L.) Poir., is considered a winter annual in the south and a summer annual in the north. It is used as a forage when grown with a small grain, and for soil improvement when grown in pure stands.

FIGURE 22-7
Growth characteristics of white lupine. *(Courtesy North Carolina State University)*

When grown with a small grain it can be harvested for hay or silage, or sometimes used as an emergency pasture. The garden pea has a more delicate, sweet flavor than the field pea. There are several winter-hardy cultivars that are adapted to the south (28).

ROUGHPEA

Roughpea, *Lathyrus hirsutus* L., is adapted to the areas of the south which have mild winters and heavy clay or loamy soils. Roughpea is used for soil improvement, as the seed is poisonous to livestock. If grazing is allowed, cattle should be removed when pods begin to develop. The seed is hard and persists in the soil for many years; therefore reseeding occurs. Roughpea can be sown with summer grasses, such as johnsongrass or bermudagrass, in the south, or with tall fescue further north (28).

COWPEA

Cowpea, *Vigna sinensis* (L.) Savi ex Hassk., has a viney top growth. It is grown as a summer annual in the southeastern United States on a wide range of soil types. Cowpeas have large leaves and curved beanlike pods (Fig. 22-8). The seeds are kidney-shaped. Cowpeas have been used for soil improvement, hay, pasture, silage, and human food. They should be harvested for hay as the pods begin to turn yellow and before defoliation (28).

SOYBEAN

Soybean, *Glycine max* (L.) Merr., was first introduced into the United States as a forage crop in 1804. Soybeans did not gain any major interest until the early 1900s. After the 1920s, soybeans were harvested as a hay crop (1). Following this period, interest developed in using soybeans as a grain crop, so that presently

FIGURE 22-8
Cowpea seed pod. *(Courtesy University of Illinois)*

99% of the crop is harvested for the seed and processed for oil and as a protein supplement.

Soybeans have a wide range of maturity and photoperiod requirements. They are grown east of a line that runs lengthwise through the middle United States, as far south as the Rio Grande Valley of Texas. Soybeans are erect in growth habit. Plants are pubescent, which makes them rather resistant to many insects (28). Very few hay cultivars have been developed in recent years; therefore, the taller, more lodging-resistant grain cultivars have been grown as an emergency hay or silage crop. Soybeans have been grown as a double crop, following the harvest of wheat. Sometimes this hectarage will be caught by an early frost or must be harvested for hay or silage. If soybeans are to be harvested for hay, they should be cut by midbloom to early pod fill (30a). Mature soybean seeds contain large quantities of oil, which makes it an undesirable hay crop. Once soybeans begin to mature, leaves turn yellow and dehisce. Soybeans have been inter-cropped with grain sorghum and corn to increase the protein content when these plants are harvested as a silage crop.

PEANUT

The vines of peanut, *Arachis hypogaea* L., are sometimes used for hay after nuts are harvested. Peanuts are combined from windrows, so very little peanut-vine hay is produced today. Recently, however, the practice of salvaging the vines for hay has increased, purely for economic reasons. Excellent hay may be made if it is not weathered and if some leaves are conserved. One must be careful of remaining pesticide residues. Recently peanut cultivars have been developed solely for the purpose of making hay. These cultivars do not produce large quantities of nuts. Peanuts are grown primarily in the southeastern United States.

REFERENCES

1 Ahlgren, G. H. *Forage Crops*. McGraw-Hill, New York, 2d ed., pp. 113–123, 1956.
2 Anthony, W. B., C. S. Hoveland, E. L. Mayton, and H. E. Burgess. Rye-ryegrass–Yuchi arrowleaf clover for production of slaughter cattle. *Auburn Univ. Agr. Exp. Stat. Circ. 182*, 1971.
3 Baxter, A. TP & L report on forage fed beef program. Texo. Power and Light Co., Dallas, TX, 1976.
4 Brink, V. C. Note on a fluorescence test to distinguish seeds of alfalfa and sweet clover in mixtures. *Can. J. Plant Sci.* 38:120–121, 1958.
5 Campbell, H. A. and K. P. Link. Studies on the hemorrhagic sweetclover disease. IV. The isolation and crystalization of the hemorrhagic agent. *J. Biol. Chem.* 138:21–33, 1941.
6 Carleton, A. E., R. D. Austin, J. R. Stroh, L. E. Wiesner, and J. G. Scheetz. Cicer milkvetch (*Astragalus cicer* L.) seed germination, scarification and field emergence studies. *Mont. Agr. Exp. Sta. Bull. 655*, 1971.

7 Carleton, A. E., C. S. Cooper, R. H. Delaney, A. L. Dubbs, and R. F. Eslick. Growth and forage quality comparisons of sainfoin (*Onobrychis viciifolia* Scop.) and alfalfa (*Medicago sativa* L.). *Agron. J.,* 60:630–632, 1968.

8 Daniel, T. W., F. B. Wolberg, U. L. Miller, J. H. Alswager, M. E. Ensminger, and A. A. Spielman. Chemical composition and digestibility of flat pea forage in three stages of maturity. *J. Anim. Sci.* 5:80–86, 1946.

9 Eslick, R. F. Sainfoin—Its possible role as a forage legume in the West. In C. S. Cooper and A. E. Carleton (eds.), Sainfoin Symp., pp. 1–2. *Mont. Agr. Exp. Stat. Bull. 627,* 1968.

10 Gladstones, J. S. Lupines as crop plants. *Field Crop Abstr.* 23:123–148, 1970.

11 Gorz, H. J., and W. F. Smith. Sweet clover. In *Forages,* Iowa State Univ. Ames, Press, IA, 3d ed., pp. 159–166, 1973.

12 Gunn, C. R. Seeds of native and naturalized vetches of North America. *USDA Agr. Handbook 392,* 1971.

13 Henson, P. R., and H. A. Schoth. Vetch culture and uses. *USDA Farmers' Bull. 1740,* 1968.

14 Hermann, F. J. A botanical synopsis of the cultivated clovers *(Trifolium). USDA Monogr. 22,* 1963.

15 Hollowell, E. A. Strawberry clover: A legume for the West. *USDA Leaflet 464,* 1960.

16 Hollowell, E. A. Sweet clover. *USDA Leaflet 23,* 1959.

17 Hoveland, C. S. Ball clover. *Auburn Univ. Agr. Exp. Stat. Leaflet 64,* 1960.

18 Hoveland, C. S. W. B. Anthony, E. L. Mayton, and H. E. Burgess. Pastures for beef cattle in the Piedmont. *Auburn Univ. Agr. Exp. Stat. Circ. 196,* 1972.

19 Johnson, H. W., J. L. Cartter, and E. E. Hartwig. Growing soybeans. *USDA Farmers' Bull. 2129,* 1967.

20 Kaddah, M. T. Tolerance of berseem clover to salt. *Agron. J.,* 54:421–425, 1962.

21 Knight, W. E. Effect of seeding rate, fall disking, and nitrogen level on stand establishment of crimson clover in a grass sod. *Agron. J.* 59:33–36, 1967.

22 Knight, W. E. Productivity of crimson and arrowleaf clovers grown in a Coastal bermudagrass sod. *Agron. J.* 62:773, 1970.

23 Knight, W. E., E. D. Donnelly, J. M. Elrod, and E. A. Hollowell. Persistence of the reseeding characteristics in crimson clover, *Trifolium incarnatum. Crop Sci.* 4:190–193, 1964.

24 Knight, W. E., and E. A. Hollowell. The influence of temperature and photoperiod on growth and flowering of crimson clover. *Agron. J.* 50:295–298, 1958.

25 Knight, W. E., and C. S. Hoveland. Crimson Clover and Arrowleaf clover. In *Forages,* Iowa State Univ. Press, Ames, IA, 3d ed., 199–207, 1973.

26 Knight, W. E., and V. H. Watson. Subclover, *Trifolium subteranean* L., a high quality, persistent, reseeding, winter annual clover for the Southeast. *Proc. 70th Assoc. Southern Agr. Workers,* 1973.

27 Kretschmer, A. E. Berseem clover—A new winter annual for Florids. *Fla. Agr. Exp. Stat. Circ. S-163,* 1964.

28 Leffel, R. C. Other legumes. In *Forages,* Iowa State Univ. Press Ames, IA, 3d ed., pp. 159–166, 1973.

29 McKee, R., and J. L. Stephens. Kudzu as a farm crop. *USDA Farmers' Bull. 1923,* 1948.

30 Morley, F. H. W. Subterranean clover. *Advan. Agron.* 13:57–123, 1961.

30 a Munoz, A. E., E. C. Holt, and R. W. Weaver. Yield and quality of soybean hay as influenced by stage of growth and plant density. *Agron. J.* 75:147–149, 1983.

31 Nowacki, E. Inheritance and biosynthesis of alkaloids in lupine. *Genet. Pol.* 4:161–202, 1963.

32 Piper, C. V. Forage plants and their culture. MacMillan, New York, 2d ed., 1924.

32 a Rich, P. A., E. C. Holt, and R. W. Weaver. Establishment and nodulation of arrowleaf clover. *Agron. J.* 75:83–86, 1983.

33 Romme, H. J. The development and structure of the vegetative and reproductive organs of kudzu, *Pueraria thubergiana* (Sieb. and Zucc.) Benth. *Iowa State Coll. J. Sci.* 27:407–419, 1953.

34 Rumberg, C. B. Grazing of grassland cicer milkvetch—grass pastures with yearling calves. *Agron. J.* 70:850–852, 1978.

35 Smith, Dale. Influence of temperature on the yield and chemical composition of five forage legume species. *Agron. J.* 62:520–523, 1970.

36 Smith, Dale. Sweetclover. In *Forage Management in the North,* Kendall-Hunt, Dubuque, IA 3d ed., pp. 125–132, 1975.

37 Smith, W. K., and H. J. Gorz. Sweet clover improvement. *Advan. Agron.* 17:163–231, 1965.

38 Stickler, F. C., and K. J. Johnson. The comparative value of annual and biennial sweet clover varieties for green manure. *Agron. J.* 51:184, 1959.

39 Taylor, N. L. Red clover and Alsike clover. In *Forages,* Iowa State Univ. Press Ames, IA, 3d ed., pp. 157–158, 1973.

39 a Touchton, J. T., W. A. Gardner, W. F. Hargrove, and R. R. Duncan. Reseeding crimson clover as a N source for no-tillage grain sorghum production. *Agron. J.* 74:283–287, 1982.

40 Townsend, C. E. Phenotypic diversity for agronomic characters in *Astragalus cicer* L. *Crop Sci.* 10:691–692, 1970.

41 Townsend, C. E., D. K. Christensen, and A. D. Dotzenko. Yield and quality of cicer milkvetch forage as influenced by cutting frequency. *Agron. J.* 70:109–113, 1978.

42 USDA. Growing crimson clover. Leaflet 482, pp. 1–10, 1971.

43 USDA. Persian clover—A legume for the south. Leaflet 484, 1960.

44 Wiggans, S. C. The relationship of photoperiod to growth and flowering of *Melilotus* species. *Iowa Acad. Sci.* 60:278–284, 1953.

45 Wilkens, F. S., and E. V. Collins. Effect of time, depth, and method of plowing upon yield and eradication of biennial sweetclovers. *Iowa Agr. Exp. Stat. Res. Bull. 162,* 1933.

46 Williams, W. A., R. M. Love, and L. J. Berry. Production of range clovers. *Calif. Agr. Exp. Stat. Circ. 458,* 1957.

FOUR

FORAGE GRASSES

CHAPTER 23

TIMOTHY

Timothy, *Phleum pratense* L., was brought to the United States by the early settlers in New England (21). It is a short-lived, perennial, cool-season, bunch-type grass which has a shallow root system.

Timothy is grown primarily for hay and pasture. It is often grown with legumes and sometimes with other grasses in pasture mixtures. Timothy has traditionally been used as the standard for comparing the quality of other grasses. Timothy has been declining in popularity due to low yields, lack of vigor, and stand loss when it is cut frequently.

HISTORICAL BACKGROUND

The early history of timothy is not well known. In the early 1700s, timothy was grown in New England and initially called "Herd's grass," following its discovery by John Herd. Herd found it growing wild in the Piscataqua River area, near Portsmouth, New Hampshire. Timothy Hanson promoted the use of this grsss, and it now carries his name. He introduced timothy to Maryland from New England in around 1720 (21). From that time, timothy has occurred naturally in pastures and has fulfilled a very important role in forage production in the northeastern United States.

WHERE TIMOTHY GROWS

Timothy is adapted to cool, humid climates. Since it has a shallow root system that will not tolerate droughts or hot, dry areas, timothy is grown in the upper midwest, from Minnesota south through Missouri and east to the Atlantic coast.

It is also grown in the Pacific northwest and in the northern intermountain area. Timothy growth is lusher in the northern part of this region. The various timothy cultivars show marked differences in their responses to changes in day length because of north-south photoperiod differences. For example, when early cultivars are grown in the south, they will flower up to 24 days earlier than the same cultivars grown in the north. Late-flowering cultivars will flower 6 to 16 days earlier in the north than when grown in the south (9). The optimum day/night temperature for timothy is close to 21/15°C (70/60°F) (2, 6).

HOW TIMOTHY GROWS

Timothy may be considered a short-lived perennial. It is a bunchgrass with erect stems that grow to a height of 51 to 102 cm (20 to 40 in) with a dense cylindrical, spikelike inflorescence (22) (Fig. 23-1). Timothy is unusual among the grasses in that one of the lower internodes remains relatively short and enlarges to form a haplocorm. A haplocorm looks like a bulb and serves as a storage organ until the plant blooms. Buds on the haplocorm, or just below it, develop into new shoots (Fig. 23-2). New haplocorms develop from these new culms, or shoots, and the old haplocorm dies. Therefore, the individual timothy shoots are biennial in nature; each new shoot develops vegetatively from the haplocorm, which contains itself as a perennial, growing at the base of the older shoots (15). After timothy is harvested new sets of buds form and timothy overwinters as tertiary shoots (25). It does not spread, nor is it a sod-forming species (15) (Fig. 23-3). Timothy does not have a vernalization requirement for the initiation of flowering (26).

FIGURE 23-1
Timothy, an important forage grass that remains palatable until mature.

FIGURE 23-2
Timothy haplocorm and tiller. Left, one month following a harvest in May; right, uncut.

FIGURE 23-3
Close-up of timothy root system, haplocorms, and inflorescences. *(Courtesy University of Illinois)*

Normally timothy is cross-pollinated, as the flowers are primarily self-sterile. The spikelike panicle of timothy is longer than most other species of *Phleum,* but the glume awns are short (4, 22).

Seedling vigor of timothy, relative to its seed size, occurs more rapidly than in most other grasses, resulting in easy establishment from which a hay or pasture crop can be attained the year of seeding (22).

USES OF TIMOTHY

Most of the timothy grown is used for hay production. Considerable hectarage is used for pasture, especially after hay harvest, and for horses or when the grazing pressure is not intense. Timothy is often grown with alsike clover under moist soil conditions, with red clover for short durations, and with alfalfa or birdsfoot trefoil for longer durations. During the early spring growth period, when all of the dry matter may not be profitably consumed, the first growth is often harvested as a hay crop and the regrowth is pastured.

Use of timothy as a hay or pasture plant has declined in recent years due to the advent of more productive grasses available for forage use, and timothy's lack of regrowth when it is harvested frequently. The regrowth of timothy is slower than that of orchardgrass, tall fescue, reed canarygrass, and even smooth bromegrass. More improvements have been made in these grasses than in timothy. Primary factors for the decline in timothy utilization are: (1) limited regrowth when moisture is limited; (2) a shallow root system which brings on a lack of tolerance for droughty conditions; and (3) lack of regrowth, which limits its availability prior to its optimum stage of maturity. Timothy is therefore being replaced by more productive grasses either in monoculture systems or in forage mixtures.

DESIRABLE CHARACTERISTICS

Since timothy is easily established, producers would like to see more persistent and higher producing cultivars. Timothy hay is bright, clean, and highly palatable. Timothy hay is highly desirable for horses because it is so clean.

Timothy produces more dry matter the further north it is grown. Dry matter yield increases as the plant matures until early or full flowering (Table 23-1). The digestibility of timothy decreases with maturity, as it does in all grass species, but probably not at the same rate as with any other cool-season grass (13). It is important to note the decline in crude and digestible protein that accompanies advancing maturity. In a study comparing the digestibility of timothy at two different locations, Maine and West Virginia, the digestibility was always higher in Maine than West Virginia (3) (Table 23-2). The quality of timothy, as measured by digestibility, is higher when grown in cooler, moist climates. When comparing timothy with smooth bromegrass, both grown under temperatures of 19/16°C (65/50°F), timothy proved higher in IVDDM and soluble carbohydrates than the bromegrass. When both were grown at a higher day/night temperature,

TABLE 23-1

DRY MATTER PRODUCTION AND CHEMICAL COMPOSITION OF TIMOTHY AT VARIOUS STAGES OF MATURITY

Characteristic	Preheading	Heading	Early flower	Full flower	Beginning of seed maturity
Dry matter (t/ha)	3.21	5.29	6.59	6.91	6.39
Crude protein (%)	14.5	12.2	9.6	7.2	5.1
Crude fiber (%)	24.7	27.6	29.2	29.4	28.5
Ash (%)	7.2	6.8	5.7	5.4	4.7
Lignin (%)	4.5	5.5	6.5	7.1	7.4
Digestible dry matter coefficient	75	70	64	59	57
Digestible crude protein coefficient	76	68	60	52	44
Digestible crude fiber coefficient	79	73	65	58	54
Digestible energy coefficient	70	66	58	58	58

Source: Kivimäe, A., "Estimation of digestibility and feeding value of timothy," *Proc. 10th Intl. Grassl. Congr.,* Finland, pp. 389–392, 1966.

30/24°C (90/75°F), timothy was higher in crude protein, crude fiber, ash, and ether extract (27).

Growers like timothy because it is very easy to maintain alfalfa or another legume grown as a companion crop as timothy exhibits nonaggressive and bunch-type growth (Fig. 23-4). Present-day growers would like faster regrowth of timothy so that a higher percentage of grass could be maintained in an alfalfa mixture harvested frequently. The rate of regrowth is similar to that of red

TABLE 23-2

DIGESTIBILITY OF TIMOTHY AT TWO DIFFERENT LOCATIONS, MAINE AND WEST VIRGINIA, AT DIFFERENT STAGES OF MATURITY

Harvest	Location	Prejoint	Early head	Early bloom	Past bloom
			% digestibility		
First	Maine	84.9	74.5	68.9	56.3
	W. Va.	76.4	62.1	58.9	55.3
Second	Maine	73.7	77.6	76.7	78.8
	W. Va.	65.5	73.6	70.1	73.4
Third	Maine	77.4	—	—	—
	W. Va.	66.4	—	—	—

Source: C. S. Brown, G. A. Jung, K. A. Varney, R. C. Wakefield, and J. B. Washko, "Management and productivity of perennial grasses in the northeast," IV, "Timothy," *West Vir. Agr. Exp. Sta. Bull.* 570T, 1968.

FIGURE 23-4
Alfalfa-timothy mixture makes an excellent forage. *(Courtesy University of Wisconsin)*

clover; both can be harvested two to three times a year. Because red clover begins to die out the third year, red clover and timothy should be considered short-time stands.

MANAGEMENT OF TIMOTHY

Establishment

Timothy is easily established. It has a rather rapid seeding growth. Timothy prefers loamy or heavy soils with good fertility. It is not well adapted to alkaline soil conditions. Timothy prefers cool, humid climates with ample moisture during the growing season (22). The seedbed should be prepared as for alfalfa or red clover. If seeded with a legume, no more than 11 to 22 kg/ha (10 to 20 lb/A) of N should be applied at seeding.

Seeding

Timothy and its companion legume should be seeded at the same time. Early spring seeding is generally preferred. This will enable a harvest during the year of establishment. Timothy can be seeded alone either in early spring or late summer. It should be seeded around 1.25 cm (0.5 in) deep in moist soil. Sometimes timothy is broadcasted on the surface of frozen ground, in which case the freezing and thawing will provide the proper depth and moisture for quick germination (22).

Seeding rates for timothy, when seeded with a companion legume, range from 2 to 4 kg/ha (2 to 3 lb/A).If timothy is seeded alone, a typical seeding rate would be from 4 to 8 kg/ha (3 to 7 lb/A) (22). Timothy seed is extremely small—there are approximately 2,422,900 to 2,860,400 seeds per kg (1,100,000 to 1,300,000 seeds per lb); therefore, the seeding rates are rather low compared to some other forage crops.

Harvest Schedule

Differences in cultivars of timothy dictate when each particular cultivar should be harvested. In general, timothy should be harvested for hay at the early-bloom stage (Tables 23-1 and 23-2) to maintain high quality. In a study in the northeastern United States, it was found that the dry matter yield increases until past bloom (3). Quality of timothy hay declines rapidly beyond the full-bloom stage (Table 23-1). Early-cut timothy will produce 3.2 times more digestible protein and 1.25 as much metabolizable energy as late-bloom hay (5).

When timothy is grown in the northern regions of the United States, without any nitrogen, it may produce more dry matter than smooth bromegrass or orchardgrass, whether harvested for hay or in a pasture-type management (Table 23-3). Orchardgrass is more productive than timothy when fertilized with nitrogen and harvested as a pasture. Timothy will exceed orchardgrass production when grown further north, and in an association with alfalfa and ladino, or red clover and birdsfoot trefoil, harvests will exceed orchardgrass production (24).

Fertility

When timothy is grown with a legume and the percentage of legume in the stand goes lower than 33%, then N should be added for profitable yields. If timothy is grown alone, or if the legume has disappeared, profitable rates of N applications range from 67 to 135 kg/ha (60 to 120 lb/A) (17).

Timothy removes about the same amount of plant nutrients per unit of dry matter as alfalfa, but with lower yields there are fewer total nutrients removed. The P and K fertility program is similar to that outlined for alfalfa.

Food Reserves

Food reserves in timothy are stored in the haplocorm and stem bases. Timothy, or any perennial forage crop, must store food reserves for winter survival. The maximum level of soluble carbohydrate reserve is reached when the plants are in the flowering to dough stage of growth. Following this stage the plants decline somewhat in reserves and then level off (23).

Winter survival is affected by the level of soil fertility and the amount of carbohydrates stored in the roots and haplocorm. Excessive rates of N fertilizer cause a decrease in the dry weight of the primary haplocorm and increase the dry weight of the secondary haplocorm, which in turn lowers the weight of the tillers, or shoots (3, 18). Split applications of N are therefore advisable.

TABLE 23-3
THREE-YEAR AVERAGE DRY MATTER YIELDS OF VARIOUS GRASSES HARVESTED AS HAY
(Grown in pure stands with and without N, and with 2 different legume combinations, in Wisconsin)

Species	Hay—2 harvests (D.M. kg/ha)				Pasture—3 harvests (D.M. kg/ha)			
	no N	90.8 kg N	Alfalfa-ladino	Birdsfoot trefoil-red clover	no N	90.8 kg N	Alfalfa-ladino	Birdsfoot trefoil-red clover
Timothy	4,560	10,821	8,466	7,182	2,920	7,175	5,759	4,358
Smooth bromegrass	3,794	11,053	8,485	5,660	2,296	6,295	6,133	3,809
Orchardgrass	3,795	10,982	8,050	5,445	2,910	8,063	5,641	3,783

Source: D. R. Schmidt, and G. H. Tempas, "Comparison of the performance of bromegrass, orchardgrass, and timothy in Northern Wisconsin," Agron J., 52:689–691, 1960.

The last harvest should be made 30 to 35 days prior to a killing frost. If a late fall harvest is taken, after a killing frost, at least 8 to 14 cm (3 to 5 in) of stubble growth should be left (14). This will provide some insulation, aid in collecting snow cover, and, most important, store some of the food reserves.

Owing to the regrowth from the haplocorm, and overall slowness in regrowth, timothy, if harvested at the wrong stage of regrowth, will not persist long in a stand when grown with alfalfa. Most of the present-day cultivars of alfalfa have a faster regrowth than timothy and are harvested more frequently than timothy can withstand. The total yield is greatly reduced if harvests are too frequent or made during the jointing stage (20). If harvest occurs at early-heading instead of bloom or seed-set stage, yield, as well as vigor, is reduced (10).

DISEASES

There are several diseases of timothy, but none is extremely important. Most diseases can easily be controlled (28). Probably the most significant disease of timothy is stem rust, *Puccinia graminis* var. *phlei-pratensis* (Eriks. and E. Henn.) Stakman and Piem. Stem rust has been controlled by developing rust-resistant cultivars (28).

SUMMARY

The advantages and disadvantages of timothy are as follows:

Advantages:

1 Timothy is easily established.
2 It has very good seedling vigor.
3 It is a palatable grass until full-bloom stage.
4 It is not competitive with legumes in a mixture.
5 It produces a clean hay.
6 It produces well further north, but is not as high-yielding as other cool-season perennial grasses.

Disadvantages:

1 Timothy is lower-yielding than orchardgrass, smoothbrome, tall fescue, or reed canarygrass.
2 It is susceptible to drought and heat.
3 It will not tolerate heavy grazing or frequent harvesting.
4 It is a short-lived perennial.
5 Regrowth is greatly affected by the stage of growth at the first harvest.
6 It is not aggressive in forming a good sod.

QUESTIONS

1 Describe the growth characteristics of timothy and explain how these might limit its production.
2 What is a haplocorm?
3 Why is timothy often used as the basis for quality comparisons?
4 Why is the hectarage of timothy decreasing?
5 In what way does excessive N affect the persistence of timothy?
6 Describe the role of food reserves and their storage in timothy.

REFERENCES

1 Ahlgren, G. H. *Forage Crops*. McGraw-Hill, New York, 2d ed., pp. 171–181, 1956.
2 Baker, B. S., and G. A. Jung. Effect of environmental conditions on the growth of four perennial grasses. I. Response to controlled temperature. II. Response to fertility, water, and temperature. *Agron. J.*, 60:155–162, 1968.
3 Brown, C. S., G. A. Jung, K. A. Varney, R. C. Wakefield, and J. B. Washko. Management and productivity of perennial grasses in the northeast. IV. Timothy. *West Vir. Agr. Exp. Sta. Bull. 570T,* 1968.
4 Clarke, S. E. Self-fertilization in timothy. *Sci. Agr.* 7:409–439, 1927.
5 Colovos, N. F., H. A. Kenner, J. R. Prescott, and A. E. Terri. The nutritive value of timothy hay at different stages of maturity as compared with second cutting clover hay. *J. Dairy Sci.* 32:659–664, 1949.
6 Cooper, J. P., and N. M. Tainton. Light and temperature requirements for the growth of tropical and temperate grasses. *Herb. Abstr.* 38:167–176, 1968.
7 Evans, M. W. Improvement of timothy. *USDA Yearbook of Agriculture,* pp. 1103–1121, 1968.
8 Evans, N. W. The life history of timothy. *USDA Bull. 1450,* 1927.
9 Evans, M. W. Relation of latitude to certain phases of the growth of timothy. *Amer. J. Bot.* 26:212–218, 1939.
10 Evans, M. W., H. A. Allard, and O. McConkey. Time of heading and flowering of early, medium, and late timothy plants at different latitudes. *Sci. Agr.* 15:573–579, 1935.
11 Evans, M. W., and L. E. Thatcher. Comparative study of an early, a medium, and a late strain of timothy harvested at various stages of development. *J. Agr. Res.* 56:347–364, 1938.
12 Hanson, A. A., and V. G. Sprague. Heading of perennial grasses under greenhouse conditions. *Agron. J.* 45:248–251, 1953.
13 Kivimäe, A. Estimation of digestibility and feeding value of timothy. *Proc. 10th Intl. Grassl. Congr.,* Finland, pp. 389–392, 1966.
14 Knoblauch, J. C., G. H. Ahlgren, and H. W. Gausnan. Persistence of timothy as determined by physiological response to different management systems. *Agron. J.* 47:434–439, 1955.
15 Langer, R. H. M. Growth and nutrition of timothy *(Phleum pratense)*. I. The life history of individual tillers. *Ann. Appl. Biol.* 44:166–187, 1956.
16 Lloyd, L. E., H. F. M. Jeffers, E. Donefer, and E. W. Crampton. Effect of four maturity stages of timothy hay on its chemical composition, nutrient digestibility and nutritive value index. *J. Anim. Sci.* 30:468–473, 1961.

17 Mack, A. R., and B. J. Finn. Differential response of timothy clonal lines and cultivars to soil temperature, moisture and fertility. *Can. J. Plant Sci.* 50:295–305, 1970.

18 Mislevy, P., J. B. Washko, and J. D. Harrington. Plant maturity and cutting frequency effect on total nonstructural carbohydrate percentages in the stubble and crown of timothy and orchardgrass. *Agron. J.* 70:907–912, 1978.

19 Nath, J., and E. L. Nielsen. Interrelations of reproductive stages of timothy. *Crop Sci.* 2:49–51, 1962.

20 Peters, E. J. The influence of several managerial treatments upon the grass morphology of timothy. *Agron. J.* 50:653–656, 1958.

21 Piper, C. V., and K. S. Bort. The early agricultural history of timothy. *J. Amer. Soc. Agron.* 7:1–4, 1915.

22 Powell, J. B., and A. A. Hanson. Timothy. In *Forages,* Iowa State Univ. Press, Ames, IA, 3d ed., pp. 277–284, 1973.

23 Reynolds, J. H., and D. Smith. Trend of carbohydrate reserves in alfalfa, smooth bromegrass and timothy grown under various cutting schedules. *Crop Sci.* 2:333–336, 1962.

24 Schmidt, D. R., and G. H. Tenpas. Comparison of the performance of bromegrass, orchardgrass, and timothy in Northern Wisconsin. *Agron. J.* 52:689–691, 1960.

25 Sheard, R. W. Relationship of carbohydrate and nitrogen compounds in the haplocorm to the growth of timothy. *Crop Sci.* 8:658–663, 1968.

26 Smith, D. Effect of day-night temperature regimes on growth and morphological development of timothy plants derived from winter and summer tillers. *J. Brit. Grassl. Soc.* 27:107–110, 1972.

27 Smith, D. Influence of cool and warm temperatures and temperature reversal at inflorescence emergence on yield and chemical composition of timothy and bromegrass at anthesis. *Proc. 11th Intl. Grassl. Congr.,* Australia, pp. 510–514, 1970.

28 Sprague, R., and G. W. Fischer. Check list of the diseases of grasses and cereals in the western United States and Alaska. *Wash. Agr. Exp. Sta. Circ. 194,* 1952.

CHAPTER 24

SMOOTH BROMEGRASS

Smooth bromegrass, *Bromus inermis* Leyss., is a long-lived, sod-forming, cool-season grass. It is often grown with alfalfa as a hay or pasture crop in the northern half of the United States. It is noted for its palatability to all livestock. Some people refer to smooth bromegrass as the "king of the forages" in the same manner they refer to alfalfa as the "queen of the forages."

HISTORICAL BACKGROUND

Smooth bromegrass is native to western Europe and northern Asia. It was first grown as a forage crop in Hungary. In 1884, smooth bromegrass was brought into the United States via California from Hungary and Russia (15). Seed was later distributed throughout the western portion of the United States and spread to the midwest by the 1890s. Smooth bromegrass never gained great prominence as a forage crop until the drought period in the mid-1930s. Since that period, smooth bromegrass has had an important role as a forage crop in American agriculture (15). In moister areas, it is seldom grown alone, and usually is grown with alfalfa (13). In the drier regions of Iowa, Nebraska, and Kansas, however, it is often grown alone.

WHERE SMOOTH BROMEGRASS GROWS

Smooth bromegrass is adapted to a wide range of soil types. It prefers a deep, well-drained, fertile soil. Smooth bromegrass will tolerate alkaline conditions, drought situations, and extreme temperature variations. Its deep root system makes it drought-tolerant. Smooth bromegrass spreads by seeds and rhizomes. Within a few years, bromegrass forms a complete sod (15). It may have a tendency to become sod-bound under certain conditions: (1) high light intensity; (2) frequent rainfall; (3) low fertility; and (4) frequent, low-application rates of nitrogen (3).

Smooth bromegrass is grown predominantly in the upper midwestern United States and in adjacent areas northward into Canada. It has been grown as far south as the panhandle of Texas (15). There are many selections or strains, especially of the southern types of smooth bromegrass (8).

TYPES OF SMOOTH BROMEGRASS

There are two major types of smooth bromegrass: (1) northern and (2) southern (15). The northern type is grown predominatly in Canada and, to a limited degree, in the Dakotas, where it is also called common bromegrass. Northern-type bromegrass does not spread as rapidly as the southern type, and therefore produces an open sod. Legumes are grown more easily with the northern type because this type is less competitive and less productive. It is also slower to establish itself than southern types (1, 15, 24).

The southern-type smooth bromegrass is more aggressive in its growth habit. Southern bromegrass is superior to northern bromegrass in the following ways: (1) it is more aggressive in growth habit; (2) it produces greater yields in the corn belt; (3) it has more seedling vigor; (4) it is easier to establish; (5) it exhibits faster spring growth; (6) it forms a solid sod rather quickly; and (7) it possesses more drought tolerance because of its deeper root system (25). Because it produces a more complete sod, the southern type is better for soil-conservation purposes and use in waterways.

HOW SMOOTH BROMEGRASS GROWS

Smooth bromegrass is a cool-season, leafy, sod-forming perennial (Fig. 24-1). The young smooth bromegrass seedlings tiller shortly after emergence and continue to increase their number of tillers until late in the summer (14). Rhizome formation begins 3 weeks to 6 months after emergence. Buds are formed on the rhizomes, which become aerial shoots and new rhizomes. Smooth bromegrass has a deep root system, although a high percentage of the root mass is concentrated in the upper few centimeters of the soil (14).

Smooth bromegrass not only spreads by rhizomes but also is spread readily by seeds. Flowering is initiated by cool, short days (14); flowers are borne on an open panicle in late spring to early summer. Many florets make up the numerous spikelets of the inflorescence (Fig. 24-2). Smooth bromegrass is generally cross-fertilized because it is weakly self-fertile. The seeds are paper-thin with a very short awn, and are light in weight (15).

Smooth bromegrass has an open panicle inflorescence. Smooth bromegrass has auricles and a solid leaf sheath (Fig. 24-3). On the underside of the leaf blade, about midway from the tip to the collar, there is a W-shaped constriction, which is useful in identifying smooth bromegrass.

FIGURE 24-1
Smooth bromegrass is a leafy,
highly palatable, sod-forming,
cool-season grass.

Smooth bromegrass matures like any grass—emergence, tillering, elongation, inflorescence emergence, anthesis, and seed set. New tillers usually do not develop until near the anthesis stage. The food reserves are sufficiently high at anthesis that basal tillers are initiated. These tillers will elongate but generally do not produce flowering heads. These shoots are usually leafy and produce

FIGURE 24-2
Close-up of a smooth bromegrass
inflorescence. *(Courtesy University of
Illinois)*

FIGURE 24-3
Close-up of a smooth bromegrass showing the collar area and auricles. *(Courtesy University of Illinois)*

shortened internodes (15).

The quality of smooth bromegrass is generally considered very good, when bromegrass is compared with other cool-season grasses. The quality of smooth bromegrass is maintained at a higher level throughout maturity than in orchard-grass, for example. Digestible protein declines with maturity as with any forage, but the total digestible dry matter increases until seed set (24). Smooth bromegrass responds to N application, which in turn increases crude protein production and total dry matter production. In terms of dry matter production, most other cool-season grasses are more responsive to N application than smooth bromegrass (21).

In a study of bromegrass grown in the northeast, it was shown that the greatest dry matter production of smooth bromegrass will occur when the first growth is harvested at past-bloom stage (Table 24-1). It was also shown that when the first growth was harvested at the same stage of maturity for three years, the advantage of harvesting past bloom for increased dry matter yield was very apparent by the third year. The quality of forage, however, is lower at the past-bloom stage than in an early-bloom harvest; legumes will generally disappear from the stand by the end of the third year (25).

The time to cut or graze is guided by the stage of growth of new tillers from the crown area. When tillers are initiated, harvest should occur. Early spring grazing should leave enough vegetation intact to protect the growing point (25). Once smooth bromegrass heads out and flowering has occurred, tillers are initiated and the food reserves are at a very high level, allowing closer cutting or grazing. The second regrowth continues and the growth cycle is repeated. This may be repeated for a third or fourth time depending on available moisture and adequate fertility (15). Since bromegrass is most often grown with alfalfa, this mixture is usually harvested when the alfalfa is ready to be harvested. When

TABLE 24-1
DRY MATTER PRODUCTION OF SMOOTH BROMEGRASS FOR FIRST-CUTTING
HARVESTS AT VARIOUS STAGES OF MATURITY

Maturity of 1st harvest each yr	Forage Production			3-year average
	1st yr	2d yr	3d yr	
	metric ton/ha (T/A)			
Prejoint	7.62b (3.40)	6.95bc (3.09)	2.05bc (0.91)	5.54c (2.47)
Early head	9.06ab (4.04)	6.73c (3.00)	0.85c (0.37)	5.55c (2.48)
Early bloom	9.69a (4.32)	8.79ab (3.92)	3.27b (1.45)	7.25b (3.23)
Past-bloom	9.84a (4.38)	10.22a (4.55)	5.36a (2.39)	8.47a (3.77)

Source: M. J. Wright, G. A. Jung, C. S. Brown, A. M. Decker, K. E. Varney, and R. C. Wakefield, "Management and productivity of perennial grasses in the northeast," II, "Smooth bromegrass," *West Vir. Agr. Exp. Sta. Bull. 554T*, 1967.

alfalfa is at the first-flower stage of growth, smooth bromegrass may be at the early-head stage. Harvesting of smooth bromegrass–legume mixture is usually conducted so as to maintain the legume in the sward.

USES OF SMOOTH BROMEGRASS

Smooth bromegrass is used for hay production and grazing. It is often seeded alone for grazing, but much more hectarage is seeded in combination with a legume, especially alfalfa. Smooth bromegrass makes higher quality hay than orchardgrass, but orchardgrass may be preferred as pasture. When properly grazed or managed, alfalfa can remain in the mixture for many years before the smooth bromegrass predominates.

An alfalfa–smooth bromegrass mixture will produce high yields of high-quality forage when grazed. The alfalfa component will ensure palatability and production during the summer slump period or dry summer months, when pure grass stands are very low in production. Present-day alfalfa cultivars are more aggressive in their growth and can be harvested more frequently, thus ensuring their prolonged presence in a stand (16).

Smooth bromegrass–alfalfa mixture makes a high-quality hay product. The presence of smooth bromegrass in a legume hay will enable outside storage of large round-bales because the grass helps reduce rain penetration.

Considerable amounts of smooth bromegrass and smooth bromegrass–alfalfa mixtures are harvested as silage or haylage. Smooth bromegrass is one of the best sod-forming, soil-conserving crops available, providing there is good drainage and highly fertile soil.

DESIRABLE CHARACTERISTICS

Smooth bromegrass has a deep root system, tolerates heat and drought, will grow on highly alkaline soils, develops a dense sod, and makes a high-quality, highly palatable forage. Slightly more seedling vigor is desired. It is very long-lived and winter-hardy, and is one of the more productive forage grasses in the northern United States.

The dense sod may be an advantage, but it may also be an undesirable characteristic in the event that a sod-bound condition develops. Disking to cut or open up the sod or using some type of light tillage and additional N can reduce this problem.

MANAGEMENT OF SMOOTH BROMEGRASS

Establishment

Smooth bromegrass is rather difficult to establish. Its fluffy seed adds to the seeding problem. A moist, firm seedbed is important. Smooth bromegrass requires a high soil pH of 6.5 to 7.0 for good establishment. It is adapted to a wide range of soil types but prefers a well-drained soil (15). Nitrogen is important to its establishment. If a legume is not sown with smooth brome-grass, a starter fertilizer containing N should be applied at the time of seed-ing (21).

Seeding

Smooth bromegrass can be broadcasted or drilled. Because the seed is fluffy, a drill is preferred, to control the seeding rate and place the seed at a desired depth. A drill with a special grass-seed box is best because the feeding devices inside the box aid in getting the fluffy seed into the drill spouts (13). Fertilizer spreaders are also used to broadcast smooth bromegrass seed; then a device or field roller is needed to cover the seed. Smooth bromegrass should not be seeded any deeper than 1.1 to 1.3 cm (0.5 in) (11, 13).

Smooth bromegrass should be seeded in the early spring with a companion legume crop. Often it is sown with a small grain nurse crop. Soil is usually moister in the spring. Late summer seedings may become established but suffer more winter injury if a lack of soil moisture prevents the root system from developing well. Further south in its region of adaptation, smooth bromegrass can be seeded in the late summer with considerable reliability because there are more fall rains and a longer growing season. There is usually no weed problem when it is sown in the late summer. In many of the drier areas, smooth bromegrass is seeded without a nurse crop. A nurse crop may reduce the chances for establishing the smooth bromegrass because the two crops may compete for soil moisture (10, 17).

Seeding rates for smooth bromegrass vary according to the mixture and location. In the drier regions of the midwest, Kansas and Nebraska, when smooth bromegrass is seeded alone, it is seeded at the rate of 11.2 to 16.8 kg/ha (10 to 15 lb/A), in mixtures at 9 to 11.2 kg/ha (8 to 10 lb/A), and alfalfa seeded at 3.4 to 5.6 kg/ha (3 to 5 lb/A) (15). Further north, the seeding rate for smooth bromegrass decreases and the legume seeding rate increases. For example, in

Illinois, the recommended seeding rate for smooth bromegrass is 7 kg/ha (5 lb/A), and for alfalfa 9 to 11 kg/ha (8 to 10 lb/A). Comparable seeding rates are suggested for Indiana and Michigan.

Harvest Schedule

The harvest schedule, or management, of smooth bromegrass is very important to maintain a productive, long-lived stand, but it may involve a compromise of quality. It is easy to maintain vigorous, productive smooth bromegrass stands if the bromegrass is harvested infrequently and at the seed-set stage (15). If harvested too early, dry matter yield is depressed.

The critical stage for harvesting smooth bromegrass is when the young shoots begin to elongate, because then the growing point is developing upward in the culm. When one removes or harvests these young shoot apices close to the surface, regrowth must come from crownal buds, which do not develop until the flowering stage. Therefore, if cut too early, the recovery is very slow, because new crownal buds or tillers have not developed enough to begin growth (15, 23). Excessive early grazing will reduce the overall vigor and regrowth as described above.

After the first growth has been removed, regrowth originates from the basal tillers. These shoots eventually elongate in the same manner in which the first growth developed. Therefore, this period of growth is a critical time, and smooth bromegrass should not be harvested during this phase. It is much better to harvest or graze smooth bromegrass after the new basal tillers have been initiated, because this allows the growth cycle to repeat itself. Flowering heads are not generally produced during the second or later growth cycles (15).

Management of a smooth bromegrass–legume mixture is dictated by the legume. Usually the companion legume is alfalfa. It is extremely important to maintain alfalfa in the stand since these two forage species are both upright in growth habit, highly productive, and very digestible. Frequent removal will eventually eliminate the alfalfa, and the stand will become one of only smooth bromegrass (1).

The best method of managing smooth bromegrass–alfalfa is the use of rotational grazing or harvesting for hay (15). In this manner, food reserves are allowed to build up, periodically, before removal and both crownal buds of alfalfa and basal tillers of smooth bromegrass are allowed to develop for the next growth cycle (20). When alfalfa is grown with smooth bromegrass, the mixture should be harvested when the alfalfa is within the bud to first-flower stages of maturity. Alfalfa is the indicator plant as to proper harvest time.

Since bromegrass has a deeper root system than timothy or bluegrass, it will provide more growth during the drier summer months. With proper management of the spring growth of an alfalfa–smooth bromegrass mixture, greater summer growth will occur. This is accomplished by allowing the spring growth to reach 15 to 20 cm (6 to 8 in) and then permitting moderate grazing until around mid-May, removing animals, and not resuming grazing of the stand until early or mid-July. Good production will occur from mid-July through August. Again, alfalfa will be the determining plant as to when the last harvest should occur, so that soluble carbohydrates will build up for winter. The fall management practiced for alfalfa will hold true here. If a late fall harvest is needed, or there is grazing after a killing frost, forage should not be removed to any shorter than 13 to 21 cm (5 to 8 in). This will allow some insulation and a high enough stubble to help collect snow during the winter. Smooth bromegrass retains its palatability for some time after a killing frost (24).

If alfalfa–smooth bromegrass is harvested when the bromegrass is in the critical stage, the bromegrass will not persist much longer than 1 year. To maintain smooth bromegrass with alfalfa, it is suggested that an early maturing cultivar of alfalfa be used with the bromegrass, so that the mixture is harvested when the smooth bromegrass is in a very young leafy rosette stage. Conversely, smooth bromegrass grown with a very late maturing cultivar of alfalfa should be harvested when the first growth of bromegrass is more mature, or in a fully headed stage.

Fertility

Smooth bromegrass responds to N fertilization, but not as much as orchardgrass, tall fescue, or reed canarygrass. Nitrogen should be applied annually, in the early spring, at the rate of 67 to 140 kg/ha (60 to 125 lb/A). Nitrogen increases dry matter production and protein percentage.

A sod-bound condition may develop when nitrogen is limiting. Since bromegrass is a cool-season grass, more N is needed earlier in the growing season, and especially during the cool-season, for its growth. Nitrogen is not released from decaying vegetation during the cool portion of the season, limiting N release and creating a dense, shallow root system or a sod-bound condition. Adding N early in the growing season will alleviate this condition. N-demand by smooth bromegrass is greatest early in the season, during the period of maximum growth.

Additional N is needed for each growth cycle in order to provide top production. In a trial in Wisconsin, it was found that cool-season grasses fertilized with a split application of 224 kg/ha (200 lb/A) produced significantly more dry matter than when supplied with the same N-level in early spring (21). In another trial, when N was applied in three split applications, up to an annual amount of 940 kg/ha (840 lb/A), it was found that smooth bromegrass dry matter production increased up to a maximum of 313 kg/ha (280 lb/A) (Table 24-2) (19). When compared with yields after no N application was made, dry matter production increased by 276% and protein by 412%. Dry matter production began to decrease when the N rate exceeded 450 kg/ha (400 lb/A). The highest percent stand was found when 78 to 156 kg/ha (70 to 140 lb/A) of N was applied. Several researchers have found that it takes around 270 kg/ha (240 lb/A) of N on pure grass stands to equal the yield of alfalfa-grass mixtures that receive no N (3, 10, 16).

TABLE 24-2
DRY MATTER YIELDS AND PROTEIN PRODUCTION
FROM SMOOTH BROMEGRASS
(With various rates of nitrogen and three harvests per year)

Total annual N applied*	Dry matter yield	Protein	Stand†
kg/ha (lb/A)	t/ha (T/A)	t/ha (T/A)	%
0 (0)	2.40 (1.07)	0.36 (0.16)	89
78 (25)	3.45 (1.53)	0.52 (0.23)	92
156 (140)	6.61 (2.95)	1.08 (0.48)	93
234 (210)	7.80 (3.47)	1.37 (0.61)	85
313 (280)	9.01 (4.01)	1.84 (0.82)	75
470 (420)	8.88 (3.96)	2.04 (0.90)	69
628 (560)	8.72 (3.88)	2.11 (0.94)	63
785 (700)	8.23 (3.67)	2.11 (0.94)	55
942 (840)	8.09 (4.60)	2.02 (0.90)	—

*One-third applied in early spring, the remaining thirds after first and second harvest.
†% stand after 2 years of treatment.
Source: D. Smith, "Influence of nitrogen fertilization on the performance of an alfalfa-bromegrass mixture and bromegrass grown alone," *Wis. Agr. Exp. Sta. Res. Rep. R2384,* 1972.

P and K should be applied at the rate of nutrient removal from the soil by the forage; soil tests are relied on for accurate rates of application.

Food Reserves

Food reserves are stored in the rhizomes in smooth bromegrass. The stored carbohydrates are needed for winter survival (20). Before a killing frost there must be enough top growth left to provide a net accumulation of soluble carbohydrates and a long enough period to allow a buildup of carbohydrates. This period generally falls about 30 to 35 days prior to a frost. Since storage is in the rhizomes, smooth bromegrass can be moderately grazed in the fall and grazed closely after frost, because there is essentially no aboveground storage (20).

A proper balance of plant nutrients is needed for the buildup of food reserves in smooth bromegrass, as is also true for any grass species. After N, P and K are probably the two most important nutrients ensuring proper soluble carbohydrate buildup and good, healthy cells for winter survival.

COMPOSITION AND QUALITY

The first growth of smooth bromegrass cultivars grown in the northern region is generally more digestible than of the same cultivars grown in the southern region

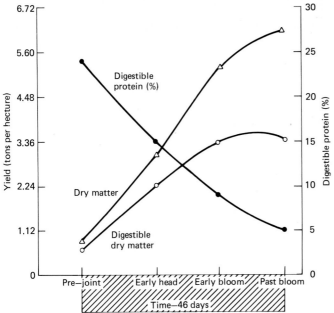

FIGURE 24-4
Yield, digestible dry matter, and digestible protein of smooth,
bromegrass harvested at various stages of maturity.

(24). The nutritive value of first-growth smooth bromegrass is very high; it
steadily declines with maturity (Fig. 24-4) (25). Digestible protein declines with
maturity while total digestible dry matter increases up to about midbloom. Since
smooth bromegrass does not usually flower again once it has matured, growth in
the aftermath of maturity is primarily vegetative; estimates of digestibility equal
those of the first-growth boot stage (25) (Fig. 24-5).

SEED PRODUCTION

Most of the smooth bromegrass seed produced in the United States is of the
southern type, while seed produced in Canada is of the northern type.
Southern-type smooth bromegrass seed is produced in many of the northern
states and in the northwest, especially Oregon, Washington, and Idaho. Seed
production depends on the amount of tillering, as well as the panicle formation
in relation to photoperiod, temperature, and the amount of N available (15).
Nitrogen is important for panicle formation because N must be readily available
during the cool spring days when flowering is initiated. Excessive N may cause
severe lodging. Older stands need more N. Generally around 56 to 112 kg N/ha
(50 to 100 lb/A) is recommended (9).

FIGURE 24-5
Smooth bromegrass cultivars showing different maturity and growth habits. *(Courtesy University of Wisconsin)*

DISEASES

There are a number of foliar diseases of smooth bromegrass, which are particularly prevalent in the humid regions. Plant breeders and pathologists are working to develop cultivars which would be resistant to the various diseases. Bacterial blight, caused by *Pseudomonas coronafaciens* var. *atropurpurea* (Reddy and Godkin) Stapp, is prevalent on smooth bromegrass. It appears as a light-colored halo around a dark lesion. *Pythium* spp. is common on the heavy soils in the humid midwest. These diseases include seedling blight and damping-off of young seedlings. Brown leaf spot is caused by *Helminthosporium bromi* Dreschsl. and appears as small, dark brown, oblong spots with a yellow margin, that develop later in the season. There are many different leaf spots to which plant resistance is readily found (15).

Plant resistance and good cultural practices are the two best control methods for plant diseases. Close, clean, and early harvests, and excellent fertility program, along with good drainage, are the main cultural practices used to reduce the incidence of most diseases of smooth bromegrass.

INSECTS

Grasshoppers are probably the most serious insect problem in smooth brome-grass, especially in the year of establishment. A simple insecticide program will help control this insect. In seed fields, the bromegrass seed midge, *Stenodiplosis bromicola* Marikovsky and Agafonova, will sometimes have a serious negative effect on seed production. Again, a simple insecticide spray program is the best control method.

SUMMARY

There are numerous advantages and disadvantages of smooth bromegrass.

Advantages:

1 It is deep-rooted.
2 It has good drought and heat tolerance.
3 It is long-lived.
4 It forms a dense sod.
5 It is winter-hardy.
6 It is a palatable species throughout its maturity.
7 It makes a high-quality hay and good pasture.
8 It can be grown with alfalfa and is fairly easily managed.
9 It serves as a good waterway because of its dense sod.
10 It responds to N, less than will reed canarygrass or tall fescue, but more than timothy and bluegrass.
11 It remains green at mature stage longer than most grasses.
12 It is a good seed producer.
13 Seed is generally inexpensive.

Disadvantages:

1 It is slow in establishment.
2 It may be difficult to seed, due to the light, fluffy seed.
3 It may become sod-bound if there is insufficient nitrogen.
4 It will not withstand heavy grazing.
5 Without N or a companion legume, it is low in production.
6 Aftermath results are usually lower than with orchardgrass, but depend upon harvest schedule.
7 Fall regrowth is less than in orchardgrass but more than in timothy.
8 It may be too competitive and crowd out some legumes.
9 The southern strain of smooth bromegrass is much more productive than the northern strain. (Common seed produced in Canada is probably a northern strain.)
10 It requires a high pH.
11 It does not tolerate poorly drained soils.

QUESTIONS

1 Describe the areas where smooth bromegrass is grown.
2 What are the various types of smooth bromegrass? What are the characteristics of each type?

3 Describe the growth habit of smooth bromegrass. Are there any characteristics of the smooth bromegrass grown in the United States that make it unique?

4 What precautions must one observe in the harvest of smooth bromegrass, its regrowth, and its growth with alfalfa?

5 What is meant by an "indicator plant" in a smooth bromegrass–alfalfa mixture?

6 How can one alleviate a sod-bound condition of smooth bromegrass?

7 When should N be applied to smooth bromegrass? Why?

8 What are the advantages and disadvantages of growing smooth bromegrass?

REFERENCES

1 Ahlgren, G. H. *Forage Crops.* McGraw-Hill, New York, 2d ed., pp. 181–191, 1956.

2 Baumgardt, B. R., and D. Smith. Changes in the estimated nutritive value of the herbage of alfalfa, medium red clover, ladino clover, and bromegrass due to stage of maturity and year. *Wis. Agr. Exp. Sta. Res. Rep. 10,* 1962.

3 Carter, L. P., and J. M. Scholl. Effectiveness of inorganic nitrogen as a replacement for legumes grown in association with forage grasses. I. Dry matter production and botanical composition. *Agron. J.* 54:161–163, 1962.

4 Eastin, J., M. R. Teel, and R. Langston. Growth and development of six varieties of smooth bromegrass (*Bromus inermis* Leyss.) with observations on seasonal variations of fructosan and growth regulators. *Crop. Sci.* 4:555–559, 1964.

5 Evans, M., and C. P. Wilsie. Flowering of bromegrass, *Bromus inermis,* with greenhouse as influenced by length of day, temperature, and time of fertility. *J. Amer. Soc. Agron.* 38:923–932, 1946.

6 Fortmann, H. R. Responses of varieties of bromegrass (*Bromus inermis* Leyss.) to nitrogen fertilization and cutting treatments. *Cornell Univ. Agr. Exp. Sta. Memo. 322,* 1953.

7 Gist, G. R., and R. M. Smith. Root development of several common forage grasses to a depth of eighteen inches. *J. Amer. Soc. Agron.* 40:1036–1042, 1948.

8 Gould, F. *Grass Systematics.* McGraw-Hill, New York, 1968.

9 Knowles, R. P., D. A. Cooke, and C. R. Elliott. Producing certified seed of bromegrass. *Can. Dept. Agr. Publ. 866,* 1969.

10 Krueger, C. R., and J. M. Scholl. Performance of bromegrass, orchardgrass, and reed canarygrass grown at five nitrogen levels and with alfalfa. *Wis. Agr. Exp. Sta. Res. Rep. 69,* 1970.

11 Lueck, A. G., V. G. Sprague, and R. J. Garber. The effects of a companion crop and depth of planting on the establishment of smooth bromegrass, *Bromus inermis* Leyss. *Agron. J.* 41:137–140, 1949.

12 Moore, R. A. Symposium on pasture methods for maximum production in beef cattle pasture systems for a cow-calf operation. *J. Anim. Sci.* 30:133–137, 1970.

13 Moore, R. A., and G. B. Harwick. Establishing pasture and forage crops. *S. Dak. Farm and Home Res.* 20(2):6–9, 1969.

14 Newell, L. C. Controlled life cycles of bromegrass, *Bromus inermis* Leyss., used in improvement. *Agron. J.* 43:417–424, 1951.

15 Newell, L. C. Smooth bromegrass. In *Forages,* Iowa State Univ. Press, Ames, IA, 3d ed., pp. 254–262, 1973.

16 Newman, R. C., and D. Smith. Influence of two seeding patterns, nitrogen fertilization and three alfalfa varieties on dry matter and protein yields and persistence of alfalfa-grass mixtures. *Wis. Agr. Exp. Sta. Res. Rep. 2377, 1972.*

17 Rehm, G. W., W. J. Moline, E. J. Schwartz, and R. S. Moomaw. Effect of fertilization and management on the production of bromegrass in northern Nebraska. *Nebr. Agr. Exp. Sta. Res. Bull. 247, 1971*

18 Reynolds, J. H., and D. Smith. Trend of carbohydrate reserves in alfalfa, smooth bromegrass, and timothy grown under different cutting schedules. *Crop Sci.* 2:333–336, 1962.

19 Smith, D. Influence of nitrogen fertilization on the performance of an alfalfa-bromegrass mixture and bromegrass grown alone. *Wis. Agr. Exp. Sta. Res. Rep. R2384, 1972.*

20 Smith, D., and R. D. Grotelueschen. Carbohydrates in grasses. I. Sugar and fructosan composition of the stem base of several northern-adapted grasses at seed maturity. *Crop. Sci.* 6:263–266, 1966.

21 Smith, D., and A. V. A. Jacques. Influence of alfalfa stand patterns and nitrogen fertilization on the yield and persistence of grasses grown with alfalfa. *Wis. Agr. Exp. Sta. Res. Rep. R2480, 1973.*

22 Sotola, J. The chemical composition and apparent digestibility of nutrients in smooth bromegrass harvested in three stages of maturity. *J. Agr. Res.* 63:427–432, 1941.

23 Sprague, V. G., and R. J. Garber. Effect of time and height of cutting and nitrogen fertilization on the persistence of the legume and production of orchardgrass-ladino and bromegrass-ladino associations. *Agron. J.* 42:586–593, 1950.

24 Watkins, J. M. The growth habits and chemical composition of bromegrass, *Bromus inermis* Leyss., as affected by different environmental conditions. *J. Amer. Soc. Agron.* 32:527–538, 1970.

25 Wright, M. J., G. A. Jung, C. S. Brown, A. M. Decker, K. E. Varney, and R. C. Wakefield. Management and productivity of perennial grasses in the northeast. II. Smooth bromegrass. *West Vir. Agr. Exp. Sta. Bull. 554T, 1967.*

CHAPTER 25

ORCHARDGRASS

Another popular long-lived, cool-season forage grass is orchardgrass, *Dactylis glomerata* L. It is an upright-growing, perennial, bunch-type grass, forming an open or bunchy sod. Orchardgrass has grown in the United States for over 200 years. Orchardgrass gets its name from its ability to grow readily under orchard conditions or under 30% shade. Another common name, used in Europe for orchardgrass, is "cocksfoot." This name comes from the shape of the inflorescence.

Orchardgrass is grown primarily for pasture but also for hay and silage. It is found along roadways and on less productive lands.

HISTORICAL BACKGROUND

It is known that orchardgrass was introduced into the United States before 1760 because records state that orchardgrass seed was produced in Virginia in 1760 and sent to England (12). Orchardgrass is native to western and central Europe, and was introduced into the United States along the Atlantic coast by colonists. Records show that in 1830, Philip Henshaw collected orchardgrass seed in an orchard in Virginia and seeded it on a farm in Kentucky. The Kentucky field was harvested for seed and planted around Goshen, Kentucky. Commercial seed was sold from this field (12).

Orchardgrass was grown and tested at many different locations over the years, but it did not gain a great deal of acceptance until about 1940.

WHERE ORCHARDGRASS GROWS

Orchardgrass is adapted to the well-drained, medium-textured soils of the humid regions of the United States. It has a deep root system, developing from a bunch-type growth that never forms a dense sod. It is a persistent, fairly drought-resistant, shade-tolerant, cold-resistant, cool-season grass. Orchardgrass grows best on fertile soils that are calcareous, neutral, or more than medium in acidity. It is grown under irrigation throughout the West and in high-rainfall mountain areas. Orchardgrass is grown primarily in the eastern United States, from southeastern Canada, west throughout Iowa, south to the northern portion of the Gulf states, and east to the Atlantic Ocean (12).

The northern limit of adaptation for orchardgrass is determined by its moderate winterhardiness. It is not as winter-hardy as smooth bromegrass and some of the other cool-season grasses (5, 7, 12). Where snow cover is very predictable over the years, orchardgrass will survive in more northerly locations.

Orchardgrass is a very popular grass, grown on many hectares throughout the world. Orchardgrass is grown in all of Europe, except the northern regions; in northern portions of Asia; in New Zealand and Australia; and in the temperate regions of South America (12).

TYPES OF ORCHARDGRASS

There are four generally recognized types of orchardgrass among the many introductions and cultivars grown in the United States. The four types are: (1) tall, stemmy, early; (2) tall, leafy, late; (3) medium-tall, leafy, medium-late; and (4) dwarf, medium-late. All but the tall, leafy, late type produce high dry matter yields and good seed yields. A great deal of the commercial seed is from the tall, stemmy, early type (12).

HOW ORCHARDGRASS GROWS

The growth habit of orchardgrass is erect (Fig. 25-1); it reaches a height of 1 to 1.4 m (3 to 4.5 ft), with an inflorescence 8 to 16 cm (3 to 6 in) long. At the base of the culm, which bears the inflorescence, only a few leaves are present. The inflorescence is made up of numerous crowded spikelets, each of which bears 2 to 5 florets (1, 12) (Fig.25-2). The lower branches of the inflorescence are longer and exhibit more branching than the upper branches. This gives the "cocksfoot" appearance in the flowering and seed-set stages.

Orchardgrass has a faster rate of establishment than timothy or smooth bromegrass. This characteristic allows one or more harvests during the year of establishment. Since it is shade-tolerant, orchardgrass is established more easily than timothy or smooth bromegrass when grown in competition with other species (6).

Orchardgrass is one of the earliest and fastest-growing grass species in the early spring, possibly surpassed only by tall fescue. It develops rapidly and flowers from mid-May, in Illinois, to early June, in Wisconsin, depending on day length, temperature, and cultivar growth (27). Orchardgrass flowers about 1 week before tall fescue or reed canarygrass, 2 weeks before smooth bromegrass, and about 4 weeks before timothy. The optimum day-night temperature is

FIGURE 25-1
Flowering plant of orchardgrass. *(Courtesy University of Illinois)*

FIGURE 25-2
Close-up of an orchardgrass panicle. *(Courtesy University of Illinois)*

22/12°C (73/54°F) (5, 7). The optimum daytime temperature is the same for orchardgrass, timothy, and bluegrass, while the optimum daytime temperature for smooth bromegrass is 24°C (76°F) (27). As daytime temperatures increase to 35°C (95°F), dry matter yields decrease for all cool-season grass species, although smooth bromegrass yields do not decrease as rapidly as these other species'. Orchardgrass is more heat-tolerant than timothy or bluegrass (5). Orchardgrass has good fall growth, but is less productive than tall fescue (27).

Orchardgrass is known for its tolerance to shade. It has been shown that it can grow with 33% reduced light for 3 years without this affecting the yield or stand (6). Detached leaves may be saturated at a light intensity of 32,000 lx, but intact leaves, with a leaf-area index (LAI) between 3 and 8, require 48,000 lx (19). Researchers have also found that a LAI of 5 is optimum and can be attained 2 weeks after harvest, providing adequate moisture and nutrients are available (19). The leaves at this stage can intercept 95% of the sunlight. They also found that orchardgrass can maintain a high level of photosynthesis for 2 to 3 weeks before photosynthesis decreases (32). Therefore, the optimum frequency for harvesting is every 4 to 5 weeks. Light intensity and the longer photoperiod increase tillering and buildup of food reserves (4).

Soon after emergence, tillering occurs, followed by elongation of the stems. The inflorescence differentiates at the base of the plant and grows upward by means of elongation of the internodes. Orchardgrass produces tillers continuously throughout the year (27). Once orchardgrass produces inflorescence, all of the regrowth is vegetative and leafy (27). Orchardgrass reproduces sexually by means of seeds and asexually by means of tillers. Orchardgrass is considered weakly self-fertile.

The leaves of orchardgrass are folded in the emerging tiller, or bud, and reach a total length of 1 m (3 ft). There is no auricle present at the collar. The leaf sheath is flattened (12). With the continuous regrowth of primarily aboveground tillers, recovery is faster than for a sod-forming grass such as smooth bromegrass.

Food reserves are stored in the basal portion of the leaf blade, in the tillers, and in the roots. The largest percentage of stored carbohydrates is located near the soil surface in the plant (7). Therefore, one should not harvest too closely in late fall, or food reserves will be removed.

Orchardgrass is often grown with alfalfa. Since some cultivars of orchardgrass mature rather early, it is important to pick a cultivar that matures when alfalfa is at first flower. With species that mature at about the same rate, it is much easier to maintain both species in the sward, and the quality of both species will be at approximately the same point, even as quality changes with maturity. Quality of orchardgrass decreases rapidly with maturity in the early spring, as measured by the percent of protein and dry matter digestibility, while the cell wall components increase (27). As quality decreases, so does the consumption level, which affects the total amount of energy available for animal production. Researchers have found that the energy level of orchardgrass forage may be rather low when compared with other grasses (33). It has been shown that the energy level of orchardgrass is low during the summer and when it is highly fertilized with N (33).

When orchardgrass is grown with alfalfa, it is often overly mature for a harvest just when alfalfa is in the first flower. When alfalfa is at the first-flower stage, orchardgrass is often fully headed or at the anthesis stage. When orchardgrass is in the vegetative stage, its nutritive value is close to alfalfa's at the vegetative stage, but when it is in full bloom its nutritive value has fallen to approximately one-half of its former level (Fig. 25-3). To avoid a significant drop in quality, it is suggested that an alfalfa-orchardgrass mixture be harvested when the orchardgrass is within the boot to full-head stages.

Orchardgrass regrowth is approximately equal in nutritive value to alfalfa at the early-bud stage (Fig. 25-3). All of its regrowth is leaves. Leaves of orchardgrass senesce approximately every 35 days, so regrowth can be harvested safely within this time span, with no fear of harvest damage that would affect recovery (14).

Since orchardgrass is a bunch-forming grass, with proper management one can establish and maintain various legume species in an orchardgrass stand. Ladino clover is often grown with orchardgrass in the southern United States, but in the north alfalfa is more commonly grown with orchardgrass. Since

FIGURE 25-3
Nutritive value index of alfalfa (A) at vegetative, flower-bud, early-bloom, one-fourth-bloom, full-bloom and seed-set stage, plus orchardgrass (O) at vegetative, late-boot, headed, early-bloom, late-bloom, seed-set and vegetative-regrowth stage (14).

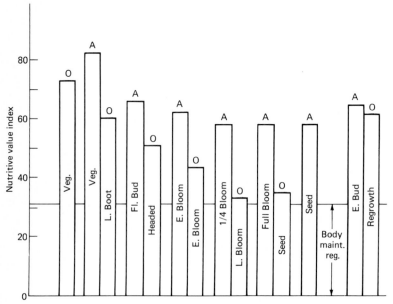

orchardgrass matures rather early and since its nutritive value drops rapidly with maturity, an early harvest of orchardgrass will ensure a high-quality forage, as well as eliminate the shading effect on ladino clover. It has been found that one should harvest or graze orchardgrass-ladino when the orchardgrass is 20 to 25 cm tall (8 to 10 in) and when the regrowth is about 20 cm (8 in) (11, 24, 28). A close clipping of 5 cm (2 in) is more satisfactory than a 7.5 cm (3 in), owing to fast regrowth and the location of the growing points in both ladino and orchardgrass. Therefore, this combination makes a good pasture. Controlled grazing or rotational grazing will keep orchardgrass in a leafy stage of growth and, at the same time, maintain the ladino clover.

When high rates of N are applied to an orchardgrass-legume mixture, the legume will usually diminish to the point that the stand will eventually shift to all grass. In general, N applied to pure stands of orchardgrass results in greater response than when applied to timothy, or even smooth bromegrass, under the same management system (25, 28). Dry matter yields of orchardgrass may exceed 13.5 t/ha (6 T/A) when the crop is properly fertilized and sufficient moisture is provided (23). It is one of the most productive cool-season grasses.

USES OF ORCHARDGRASS

Orchardgrass, because of its growth characteristic, lends itself to grazing. Growth begins early in the spring, and if properly managed through controlled grazing, will be maintained in a young, leafy stage and will not be overly competitive with the companion legume, especially if it is ladino clover. Rotational grazing is highly recommended for orchardgrass, even with its rapid regrowth (Fig. 25-4).

DESIRABLE CHARACTERISTICS

The desirable characteristics of orchardgrass are: (1) it is one of the earliest-maturing grass species in the midwest; (2) 1t has a denser root system than smooth bromegrass, timothy, or bluegrass; (3) it is easy to establish; (4) it has fast recovery; and (5) it is shade-tolerant. Orchardgrass will grow on a wider range of soil types with slightly lower fertility than will smooth bromegrass. Growers would like to see the quality of orchardgrass maintained at a higher level throughout the growing season than is presently attained, and desire more winterhardiness. In general, orchardgrass is classified as moderately winter-hardy. If snow cover is sufficient and complete for the winter, orchardgrass will consistently overwinter in Wisconsin and Minnesota.

Since orchardgrass is a bunch-type grass, it produces an open sod; thus, a legume can be maintained fairly easily in combination with the orchardgrass. Low-growing legumes may be smothered if the orchardgrass is allowed to grow too tall in the early spring and if the regrowth is not properly managed.

FIGURE 25-4
'Boone' orchardgrass, a very important cool-season, bunch-forming grass that is often grown in combination with alfalfa or other legumes.

MANAGEMENT OF ORCHARDGRASS

Establishment

Orchardgrass is easily established. A firm, moist seedbed is important. Seed should be seeded no deeper than 1.2 cm (0.5 in). Quick establishment is aided by using a field roller or cultipacker to press the seed into the soil (12, 27).

Seeding

In a procedure similar to that used in the establishment of smooth bromegrass, often orchardgrass is seeded with a drill along with a companion legume, plus a nurse crop of oats in the early spring. The oats should be harvested as silage or hay in the early summer, leaving time for possibly one forage harvest before September. The seeding rate for orchardgrass sown alone is about 11 kg/ha (10 lb/A). When sown with a legume, the seeding rate for orchardgrass is often reduced to 4 to 7 kg/ha (4 to 6 lb/A). Late summer seedings can become established further south in the area of orchardgrass adaptation (12, 27), but very late seedings, 45 days prior to a killing frost, may not survive the winter.

Newly harvested orchardgrass seed will not germinate as well as seed that has gone through an after-ripening period or dormancy (12). There appears to be a growth inhibitor in the lemma and palea, as well as the caryopsis, of dormant seed. The inhibitor is reduced following storage of the seed at temperatures near freezing (12). If temperatures exceed 30°C (76°F), germination is reduced (12). Moderate temperatures ensure good emergence.

Harvest Schedule

The harvest schedule for orchardgrass is very important, because it affects the quality of forage produced and ease of maintenance of a legume within the stand. Since orchardgrass is a bunch grass, the competition is not as great as with a sod-forming grass. If the first growth of orchardgrass is harvested late each consecutive year, the stand thins out and becomes bunchy.

Alfalfa-orchardgrass should be harvested when the orchardgrass is in the boot to the fully headed stage. This will not only maintain both species but will ensure harvest of a high-quality forage. When orchardgrass is grown with ladino, it should be harvested no later than when the orchardgrass is in the boot stage. This will reduce the above-ground competition for the ladino and ensure a high-quality forage.

If orchardgrass is grown in a monoculture or with alfalfa, it should be harvested at least every 30 to 35 days. With ladino, it should be harvested or grazed at least every 28 to 30 days because of the growth habit of ladino.

The last fall harvest should be made 30 to 35 days before a killing frost or fall dormancy. This will allow enough time for the food reserve to be stored in the crown and basal tillers.

Soil Fertility

Orchardgrass responds to N fertilization, but probably not as readily as tall fescue or reed canarygrass. Sufficient soil fertility is needed for uniform dry matter production throughout the growing season. If soil fertility is low, the greatest portion of the total seasonal yield occurs in the spring. Early studies found that N rates of 140 kg/ha (125 lb/A) would give good dry matter yields (12). More recently, N-fertility studies at Purdue University indicated that dry matter yields increased up to approximately 560 kg of N/ha (500 lb/A) and decreased at higher rates (Table 25-1) (23). Other elements must be in proper balance or in sufficient amounts to maintain these high yields. Nitrogen should not be applied to legume-orchardgrass mixture, as the grass will increase competitively, to the point of eliminating the legume.

Phosphorus is needed for good root development early in the plant's life cycle. In general, N fertilizer will increase the plant content of K, Ca, and Mg, but it may decrease the P content. Occasionally N fertilizer may decrease the uptake of K (21).

Potassium should be applied annually to orchardgrass at the rate it is removed. High rates of N fertilizer will increase N uptake and, in general, increase K uptake, which in turn may reduce Mg uptake (4, 17, 26). Orchardgrass has been known to be a low accumulator of Mg under cool temperatures

TABLE 25-1
FORAGE YIELDS OF POTOMAC ORCHARDGRASS, KENTUCKY 31 TALL FESCUE, LINCOLN SMOOTH BROMEGRASS, AND REED CANARYGRASS AT LAFAYETTE, INDIANA
(Averaged over 2 years with increasing levels of nitrogen fertilization)*

N applied†		Orchard grass,‡ t/ha (T/A)	Tall fescue,‡ t/ha (T/A)	N applied		Smooth bromegrass,‡ t/ha (T/a)	Reed canarygrass,‡ t/ha (T/A)
kg/ha	lb/A			kg/ha	lb/A		
0	0	4.5 (2.0)	5.6 (2.5)	0	0	6.5 (2.9)	5.4 (2.4)
70	62.5	7.6 (3.8)	8.3 (3.7)	56	50	7.6 (3.4)	7.4 (3.3)
140	125	11.0 (4.9)	12.1 (5.4)	112	100	10.8 (4.8)	9.6 (4.3)
280	250	13.7 (6.1)	14.8 (6.6)	224	200	11.9 (5.3)	13.2 (5.9)
560	500	15.0 (6.7)	16.6 (7.4)	448	400	13.5 (6.0)	16.4 (7.3)
1121	1000	14.3 (6.4)	15.7 (7.0)	896	800	13.0 (5.8)	15.2 (6.8)
2242	2000	13.2 (5.9)	13.9 (6.2)	1794	1600	11.9 (5.3)	15.0 (6.7)

*Plots received 112 kg P/ha and 336 kg K/ha the first year and twice this amount the second year.
†Nitrogen applied in split amounts in the spring and after each harvest.
‡Orchardgrass and tall fescue were harvested five times per year, while smooth bromegrass and reed canarygrass were harvested four times per year.
Source: C. L. Rhykerd, C. H. Noller, K. L. Washburn, S. J. Donohue, K. L. Collins, L. H. Smith, and M. W. Phillips, "Fertilizing grasses with nitrogen," *Purdue Univ. Agronomy Guide AY-176,* 1969.

(27). The subsequent deficiency of Mg may lead to grass tetany, a nutritional disease.

Food Reserves

Food reserves are stored in the basal portion of the stems or tillers of orchardgrass. Orchardgrass should not be grazed or harvested too closely as winter approaches because of the location of the storage structures (20, 27). Similar to smooth bromegrass, orchardgrass should have 30 to 35 days of fall growth prior to a killing frost, so that enough food reserves are built up for winter survival. Orchardgrass is not quite as winter-hardy as smooth bromegrass, reed canarygrass, or timothy (27).

COMPOSITION AND QUALITY

The quality of orchardgrass depends on the amount and balance of fertilizer applied. Many growers consider the nutritive value of orchardgrass to be poor, based on two factors other than the soil-fertility level. Orchardgrass quality declines a little more rapidly with the maturity of the first growth than smooth bromegrass or timothy. As shown in Fig. 25-2, when orchardgrass is cut after the early bloom, it provides little more than a maintenance level of energy for livestock. The other aspect that growers often overlook is that orchardgrass is a very early-maturing species. This makes it more difficult to harvest orchardgrass at the proper stage of early growth because inclement weather and other interfering factors must be taken into account. Despite its early maturity, growers often harvest orchardgrass at a later growth stage than smooth bromegrass or timothy.

Orchardgrass regrowth is equal in nutritive value to its first growth or to the nutritive value, almost equal to that of alfalfa, in the late-boot stage (14). Therefore, once orchardgrass is harvested at the proper stage during the early spring, the regrowth is very high in quality.

SEED PRODUCTION

Most of the United States orchardgrass seed is produced in the northwest, primarily in Oregon (12). Most of the cultural practices for seed production in orchardgrass are the same as for smooth bromegrass.

DISEASES

Since orchardgrass is grown in many of the humid regions of the United States, there are many foliar diseases that attack it. Foliar diseases cause leaf death and reduce the quality and quantity of both hay and pasture. Weakened plants are also less likely to withstand droughts and severe winters. There are leaf spots and blotches caused by species of the fungus *Helminthosporium,* which occurs in lush, dense stands during wet weather. Small to large spots, or lesions, usually oval to oblong or elongated in shape, form on the leaves and leaf sheaths. The lesions may be yellowish-tan to brown, and are often surrounded by a yellowish border or halo (8). Brown stripe, or leaf streak, caused by the fungus *Scolecotrichum graminis* Fckl., occurs throughout the growing season. Lesions on the leaf blades are, at first, small, round to elongated, and appear water-soaked. Later the lesions turn gray to purplish-brown and extend many centimeters, leaving a grayish-white to light brown center (8).

Selenophoma leaf spot, also called speckle blotch or eye spot, is caused by the fungi *Selenophoma bromigeva* (Sacc.) Sprague and A. G. Johnson and *S. donacis* (Pass.) Sprague and A. G. Johnson. The common names describe the appearance of the spot with brown, red, or purple borders. It is more common in cool, moist weather in the spring or fall (8). Another leaf spot, or blotch, is caused by several species of *Stagonospora*. Other diseases are scald, *Rhynchosprium orthosporum* Cald., plus the various rusts (8).

Plant resistance and good cultural practices are the best methods of reducing the buildup of these diseases. Stands in excellent health with high fertility, and control of weeds help control diseases.

INSECTS

There are numerous insects that attack both the foliage and root system of orchardgrass. The grasshopper is probably one of the most destructive insects to orchardgrass. In addition, sawflies, *Dolerus* spp., consume many leaves. Those insects attacking the root systems are the grubs of Japanese beetles, *Popillia japonica* Newman and the June beetles, *Catinis nitida* (L.) (12). Control programs involve insecticide application, timely and clean harvests, and crop rotation.

SUMMARY

Following are the advantages and disadvantages of orchardgrass:

Advantages:

1 It is a long-lived perennial bunch grass.
2 A legume can very easily be maintained as a companion crop since orchardgrass has a bunch-type growth.
3 It is easily established.
4 It has rapid regrowth following harvest.
5 It is shade-tolerant.
6 It produces well under high summer temperatures.
7 It is very early in maturity.
8 Its life cycle matches well with alfalfa's.

9 All of its regrowth is vegetative.

10 It is high-yielding with adequate N applications.

11 It makes an excellent pasture.

12 It is more heat-tolerant than timothy or Kentucky bluegrass, but less so than tall fescue or smooth bromegrass.

13 It exhibits a relatively uniform growth rate throughout the season.

Disadvantages:

1 It is moderately winter-hardy.

2 Quality declines rapidly with maturity.

3 If not properly managed and under high N conditions, it may be too competitive for the legume.

4 It may produce an open, bunchy growth pattern.

5 It is not as drought-resistant as smooth bromegrass.

6 It requires high fertility, especially N, in pure stands.

QUESTIONS

1 From where does orchardgrass get its name?

2 Compare the growth habit, environmental adaptation, nutritive quality, and management of orchardgrass with these features in smooth bromegrass.

3 What are problems of management in growing alfalfa or ladino with orchardgrass?

4 Compare the storage of food reserves of orchardgrass with smooth bromegrass.

5 Why does orchardgrass have a fast regrowth?

6 Discuss the fertility requirements of orchardgrass.

7 Why is orchardgrass a popular forage grass? What are some of its limitations?

REFERENCES

1 Ahlgren, G. H. *Forage Crops,* McGraw-Hill, New York, 2d ed., pp. 205–212, 1956.

2 Archer, K. A., and A. M. Decker. Autumn-accumulated tall fescue and orchardgrass. I. Growth and quality as influenced by nitrogen and soil temperature. *Agron. J.* 69:601–605, 1977.

3 Archer, K. A., and A. M. Decker. Autumn-accumulated tall fescue and orchardgrass. II. Effects of leaf death on fiber components and quality parameters. *Agron. J.* 69:605–612, 1977.

4 Auda, H. R., R. E. Blaser, and R. H. Brown. Tillering and carbohydrate contents of orchardgrass as influenced by environmental factors. *Crop Sci.* 6:139–143, 1966.

5 Baker, B. S., and G. A. Jung. Effect of environmental conditions on the growth of four perennial grasses. I. Response to controlled temperature. II. Response to fertility, water, and temperature. *Agron. J.* 60:155–162, 1968.

6 Blake, C. T., D. S. Chamblee, and W. W. Woodhouse, Jr. Influence of some environmental and management factors on the persistence of ladino clover in association with orchardgrass, *Agron. J.* 58:487–489, 1966.

7 Davidson, J. L., and F. L. Milthorpe. Carbohydrate reserves in the regrowth of cocksfoot. *J. Brit. Grassl. Soc.* 20:15–18, 1965.

8 Dickson, J. G. *Diseases of Field Crops.* McGraw-Hill, New York, 2d ed., 1956.

9 Fleming, G. A., and W. E. Murphy. The uptake of some major and trace elements by grasses as affected by season and stage of maturity. *J. Brit. Grassl. Soc.* 23:174–184, 1968.

10 Gardner, F. P., and W. E. Lomis. Floral induction and development of orchardgrass. *Plant Physiol.* 28:201–212, 1953.

11 Harrison, C. M., and C. W. Hodgson. Response of certain perennial grasses to cutting treatments. *J. Amer. Soc. Agron.* 31:418–430, 1939.

12 Jung, G. A., and B. S. Baker. Orchardgrass. In *Forages,* Iowa State Univ. Press, Ames, IA, 3d ed., pp. 285–296, 1973.

13 Jung, G. A., and R. E. Kocher. Influence of nitrogen and clipping treatments on winter survival of perennial cool-season grasses. *Agron. J.* 66:62–65, 1974.

14 Jung, G. A., R. L. Reid, and J. A. Balasko. Studies on yield, management, persistence, and nutritive value of alfalfa in West Virginia. *W. Va. Agr. Exp. Sta. Bull. 581T,* 1969.

15 Knievel, D. P., and D. Smith. Influence of cool and warm temperatures and temperature reversal at inflorescence emergence on growth of timothy, orchardgrass, and tall fescue. *Agron. J.* 65:378–383, 1973.

16 Krueger, C. R., and J. M. Scholl. Performance of bromegrass, orchardgrass, and reed canarygrass grown at five nitrogen levels and with alfalfa. *Wis. Agr. Exp. Sta. Res. Rep. 69,* 1970.

17 MacLeod, L. B. Effect of nitrogen and potassium on the yield and chemical composition of alfalfa, bromegrass, orchardgrass, and timothy grown as pure stands. *Agron. J.* 57:261–266, 1965.

18 Nittler, L. W., T. J. Kenny, and E. Osborne. Response of seedlings of orchardgrass to photoperiod, light intensity, and temperature. *Crop Sci.* 3:125–128, 1963.

19 Pearse, R. B., R. H. Brown, and R. E. Blaser. Relationships between leaf-area index, light interception and net photosynthesis in orchardgrass. *Crop Sci.* 5:553–556, 1965.

20 Phillips, T. G., J. T. Sullivan, M. E. Loughlin, and V. G. Sprague. Chemical composition of some forage grasses. I. Changes with plant maturity. *Agron. J.* 46:361–369, 1954.

21 Reid, R. L., A. J. Post, and G. A. Jung. Mineral composition of forages. *W. Va. Agr. Exp. Sta. Bull. 589T,* 1970.

22 Reynolds, J. H., C. R. Lewis, and K. F. Leaker. Chemical composition and yield of orchardgrass forage grown under high rates of nitrogen fertilization and several cutting managements. *Tenn. Agr. Exp. Sta. Bull. 479,* 1971.

23 Rhykerd, C. L., C. H. Noller, K. L. Washburn, Jr., S. J. Donohue, K. L. Collins, L. H. Smith, and M. W. Phillips. Fertilizing grasses with nitrogen. *Purdue Univ. Agronomy Guide AY-176,* 1969.

24 Robinson, R. R., and V. G. Sprague. Responses of orchardgrass-ladino clover to irrigation and nitrogen fertilization. *Agron. J.* 44:244–247, 1952.

25 Scholl, J. M., T. H. McIntosh, and L. R. Frederick. Response of orchardgrass, *Dactylis glomerata* L., to nitrogen fertilization and time of cutting. *Agron. J.* 52:587–589, 1960.

25 a Shenk, J. S. and M. O. Westerhaus. Selection for yield and quality in orchardgrass. *Crop Sci.* 22:422–425, 1982.

26 Singh, R. N., D. C. Martens, S. S. Obenshain, and G. D. Jones. Yield and nutrient

uptake by orchardgrass as affected by 14 annual applications of N, P, and K. *Agron. J.* 59:51–53, 1967.

27 Smith, D. Orchardgrass. In *Forage Management in the North,* Kendall-Hunt, Dubuque, IA, 3d ed., pp. 181–186, 1975.

28 Sprague, V. G. and R. J. Barber. Effect of time and height of cutting and nitrogen fertilization on the persistence of the legume and production of orchardgrass-ladino and bromegrass-ladino association. *Agron. J.* 42:586–593, 1950.

29 Sprague, V. G., and J. T. Sullivan. Reserve carbohydrates in orchardgrass clipped periodically. *Plant Physiol.* 25:92–102, 1950.

30 Taylor, T. H., J. P. Cooper, and K. J. Treharne. Growth response of orchardgrass (*Dactylis glomerata* L.) to different light and temperature environments. I. Leaf development and senescence. *Crop Sci.* 8:437–440, 1968.

31 Templeton, W. C., Jr., J. L. Menees, and T. H. Taylor. Growth of young orchardgrass (*Dactylis glomerata* L.) plants in different environments. *Agron. J.* 61:780–783, 1969.

32 Treharne, K. J., J. P. Cooper, and T. H. Taylor. Growth response of orchardgrass (*Dactylis glomerata* L.) to different light and temperature environments. II. Leaf age photosynthetic activity. *Crop Sci.* 8:441–445, 1968.

33 Washko, J. B., G. A. Jung, A. M. Decker, R. C. Wakefield, D. D. Wolf, and M. J. Wright. Management and productivity of perennial grasses in the northeast. III. Orchardgrass. *West Vir. Agr. Exp. Sta. Bull. 557T,* 1967.

CHAPTER 26

REED CANARYGRASS

One of the most productive perennial, cool-season, sod-forming grasses is reed canarygrass, *Phalaris arundinacea* L. This grass is often grown on poorly drained fertile soils or on sites that are too wet for most other crops. Reed canarygrass is very responsive to nitrogen fertilization and is drought-tolerant. It is usually grown in a monoculture system as a pasture or hay crop.

HISTORICAL BACKGROUND

Reed canarygrass was first grown as a forage crop in Sweden in 1749 (3, 38). It is native to the temperate regions of Europe, Asia, and the United States. The aggressive type grown in Europe is the strain presently cultivated in the United States, instead of the type native to the United States. It was reportedly grown in England as a forage crop by 1824, and in Germany by 1850. In approximately 1835, reed canarygrass was grown in the United States, and it was planted in Oregon by 1885 (3, 29).

WHERE REED CANARYGRASS GROWS

Reed canarygrass is adapted to a wide range of poorly drained soils ranging in pH from 4.9 to 8.2 (13, 16). Reed canarygrass has a very deep root system, which enables it to tolerate droughty conditions. It has more drought tolerance than many of the cool-season grasses (29).

Reed canarygrass is grown in the humid and subhumid regions of the upper half of the United States and into southern Canada. The plant may die over the winter, especially on drier soils in Canada with insufficient snow cover (16).

FIGURE 26-1
Flowering plant of reed canarygrass. *(Courtesy University of Illinois)*

HOW REED CANARYGRASS GROWS

Reed canarygrass is a tall-growing, sod-forming, cool-season grass (Fig. 26-1). It spreads by short rhizomes, thus forming a dense sod (29). Seeds also shatter readily after ripening, which helps in its establishment and maintenance (15).

Buds are formed in the axils of the scale leaves of the rhizomes near the main shoot from early in the year through August. The aboveground tillers develop from these axillary buds on the rhizomes during the fall and the following spring. Roots develop at the nodal region on the rhizome and from the lowest nodes of the tillers. Tiller growth and elongation occur in mid-April. Plant height ranges from 60 to 240 cm (24 to 105 in) (29). The inflorescence begins to develop from the growing points of each shoot or tiller in about mid-April. Flowers develop in June to early July on a spikelike panicle 5 to 20 cm (2 to 8 in) in length (29) (Fig. 26-2). The mature seed is waxy gray to dark gray in color, and around 3 mm in length. The flowers mature from the top of the panicle downward, and the seed shatters readily. Soon after seed shatter, the panicle dies back, leaving only the remaining green plant. It is therefore very difficult to harvest reed canarygrass seed at the proper stage before seed shatter. Once reed canarygrass flowers, most all of the regrowth is vegetative.

Individual plants have many coarse leaves and are fairly resistant to lodging. Leaves of reed canarygrass are killed by low temperatures earlier in their growth cycles than those of Kentucky bluegrass or timothy (15). Smooth bromegrass

FIGURE 26-2
Close-up of an inflorescence of reed canarygrass.
(Courtesy University of Illinois)

will outyield reed canarygrass at lower soil temperatures, from 10 to 21°C (50 to 70°F) (33). Maximum photosynthesis occurs at air temperatures of about 20°C (68°F), and lessens when the air temperatures reach 38°C (100°F) (29).

Reed canarygrass responds to N application in production of dry matter, protein, and IVDDM (Table 26-1). In an Ohio study, the crude protein of reed canarygrass harvested at the boot stage increased from 13 to 19% as the N level increased from 84 to 672 kg/ha (75 to 600lb/A) (31). Similar results have been found in Minnesota and New Jersey (11, 29).

Earlier work has shown a relationship between digestible dry matter produc-

TABLE 26-1
YIELDS OF REED CANARYGRASS IN OHIO UNDER A HAY MANAGEMENT TREATMENT (3 HARVESTS PER YEAR) WITH VARIOUS LEVELS OF N APPLIED OVER A 3-YEAR AVERAGE

N applied,* kg/ha (lb/A)	Dry matter,† t/ha (T/A)	Protein, t/ha (T/A)	IVDDM,‡ t/ha (T/A)
84 (75)	6.44 (2.89)	0.81 (0.36)	4.22 (1.88)
168 (150)	9.26 (4.12)	1.23 (0.54)	6.01 (2.68)
336 (300)	11.95 (5.32)	1.95 (0.86)	7.44 (3.31)
672 (600)	13.12 (5.85)	2.51 (1.11)	8.72 (3.88)

*Split applications, one half in early spring and one half after the first harvest.
†First harvest taken at boot stage.
‡*In vitro* digestible dry matter.
Source: M. H. Niehous, "Effect of N fertilizer on yield, crude protein content, and in vitro dry matter disappearance in *Phalan's arundinacea* L.," *Agron. J.,* 63:793-794, 1971.

tion, which increases with maturity, and percent digestible protein (Fig. 26-1). In an Illinois study it was found that the crude protein concentration of reed canarygrass is about equal to that of smooth bromegrass and orchardgrass, but higher than that of tall fescue, all compared when harvested at similar stages of growth (Fig. 26-3) (14). Reed canarygrass has been found to have as much or more crude protein than orchardgrass, smooth bromegrass, meadow foxtail, or meadow fescue. Reed canarygrass definitely contains more crude protein than tall fescue, timothy, or tall oatgrass. Crested wheatgrass contains more crude protein than reed canarygrass. When reed canarygrass is highly fertilized with N, it may contain the same amount of crude protein as alfalfa when harvested at the same point in their growth cycles (5, 25, 27, 32).

The highest digestible dry matter production of reed canarygrass is obtained when the grass is harvested at the early-head stage (Fig. 26-4). Since the

FIGURE 26-3
Changes in the crude protein content of various forage crops with advancing stages of maturity (14).

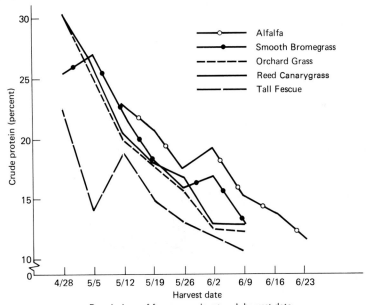

Descriptions of forage samples at each harvest date

Harvest date	Alfalfa	Smooth bromegrass	Orchard grass	Tall fescue	Reed Canarygrass
4/28	...	Vegetative	Vegetative	Vegetative	Vegetative
5/5	...	Early boot	Boot stage	Early boot	Vegetative
5/12	Bud stage	Late boot	Heading	Boot stage	Vegetative
5/19	Bud stage	Early heading	Fully headed	Late boot	Flag leaf
5/26	5% bloom	Fully headed	Flowering	Fully headed	Early heading
6/2	20% bloom	Early flowering	Early seed	Early flowering	Flowering
6/9	50% bloom	Late flowering	Milk stage	Early seed	Late flowering
6/16	Full bloom	...	...	...	...
6/23	Early seed	...	...	...	...

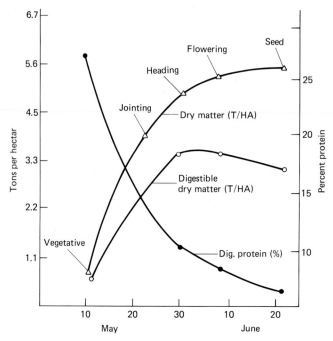

FIGURE 26-4
Dry matter yields and digestible dry matter and protein as related to spring growth of reed canarygrass (13).

regrowth consists primarily of leaves, it is suggested that the first growth be cut early. This will allow at least two or three more harvests and ensure the production of high-quality hay for the year.

Reed canarygrass has been known to be low in palatability and to contribute to poor animal performance. Some studies have indicated that animals prefer smooth bromegrass or mixtures of various grasses and legumes to reed canarygrass, but data indicate no significant difference in animal performance (6, 18). It has been shown, in a Minnesota study, that a contributing factor to the lower palatability of reed canarygrass may be the presence of several toxic alkaloids in the forage (41). Forage intake was low and the grazing cattle displayed a rough haircoat, along with profuse watering of the eyes (37). In Alabama it was found that addition of N fertilizer increased the palatability of reed canarygrass (29). Reed canarygrass is similar to most grasses in that its palatability decreases with maturity.

Several studies have been conducted on the actual intake of reed canarygrass by livestock, with other forages. Using sheep in Michigan, it was found that animal intakes of different forage species (based on hay dry matter in grams per kilogram of body weight) were, in descending popularity: (1) alfalfa, (2) smooth bromegrass, (3) common reed canarygrass, and (4) a Siberian source of reed canarygrass (6).

On in vitro digestibility tests, alfalfa and birdsfoot trefoil scored higher in IVDDM than reed canarygrass (5). Reed canarygrass is more digestible than orchardgrass, but equal in digestibility to smooth bromegrass (39).

In an Iowa trial using beef cattle, it was found that reed canarygrass provided more grazing days than all forage grasses, with the exception of tall fescue. Reed canarygrass produces well in the early part of the growing season, but the late fall or summer deferred growth is not as digestible or productive (in terms of beef produced per hectare) as tall fescue (Table 26-2) (9, 39). In fact, reed canarygrass equals smooth bromegrass and orchardgrass in deferred grazing characteristics and beef gain per hectare.

USES OF REED CANARYGRASS

Reed canarygrass is used primarily for grazing, but it is also used as a hay and silage crop. As a pasture crop, it is usually grown as a monoculture, without a legume. Due to the extreme competitiveness of the sod-forming grasses, most legumes are rather difficult to maintain in a stand with reed canarygrass. With proper management techniques, alfalfa and ladino clover, plus alsike clover under extremely wet conditions, may be grown with reed canarygrass. Dry matter yields increase until seed production is initiated (Fig. 26-4) (31).

TABLE 26-2
CATTLE GAINS, DAILY AND PER SEASON, FOR 4 COOL-SEASON GRASSES IN IOWA

	Reed canarygrass	Tall fescue	Smooth bromegrass	Orchardgrass
	kg/ha (lb/A)			
Liveweight gain				
May, June, July*	368 (328)	336 (300)	464 (414)	415 (370)
Oct. 15 to Dec. 15†	197 (176)	335 (299)	206 (184)	815 (165)
Season total	565 (504)	671 (599)	670 (598)	600 (535)
	kg (lb)			
Daily gain				
May, June, July	.58 (1.27)	.46 (101)	.73 (1.61)	.56 (1.23)
Oct. 15 to Dec. 15	.56 (1.23)	.72 (1.59)	.84 (1.85)	.64 (1.41)

*Alternate grazing, 2 paddocks per pasture.
†Deferred grazing occurred to complete utilization of paddock 1 before moving to paddock 2. Split applications of 269 kg/ha (240 lb/A) N were given annually.
Source: W. B. Bryan, W. F. Wedin, and R. L. Vetter, "Evaluation of reed canarygrass and tall fescue as spring-summer and fall-saved pasture," *Agron. J.,* 62:75-80, 1970; and W. F. Wedin, I. T. Carlson, and R. L. Vetter, "Studies on nutritive value of fall-saved forage, using rumen fermentation and chemical analyses," *Proc, 10th Intl. Grassl. Congr.,* pp. 424-428, 1966.

Where drainage is sufficient, the hay can dry, and reed canarygrass can be harvested as a hay crop.First harvest, taken when the heads first appear, gives the best quality hay. When harvested at this time, a very leafy, productive second crop results, and an excellent hay can be made from this (13).

Reed canarygrass is frequently harvested as silage or haylage. It should be harvested when the heads first begin to appear or emerge, and when field drying conditions allow placement of the haylage at the 40 to 50% moisture level. The result should be an excellent quality haylage. Wilted silage is made at about the 65% moisture level. Silage adds the possibility of saving or reducing the number of days for which the forage must lie in the field, providing it is crimped.

Reed canarygrass is often used for soil conservation, especially for gully control, stabilizing waterways, and the control of erosion of the edges of farm ponds. Probably no other cool-season grass equals reed canarygrass in terms of its usefulness in soil conservation and gully repair (23).

DESIRABLE CHARACTERISTICS

Probably the greatest attribute of reed canarygrass is that it is very tolerant of poorly drained soil conditions. Reed canarygrass has a very deep root system, which aids in its ability to withstand droughty conditions (29). It is a very long-lived perennial; in fact, in some irrigated areas in the west, it may be considered a noxious weed around irrigation ditches or in slowly moving waterways.

Reed canarygrass is very easily established by vegetative propagation using the following procedure: (1) disk an established sward several times to chop up the grass into small sections; (2) place plant sections in a manure spreader or some other type of spreader; (3) scatter the plant sections over the area to be planted; and (4) disk the cuttings into the soil and roll for added soil compacting and contact. Rooting occurs at the nodes of the stem (Fig. 26-5). In this manner, reed canarygrass is very easily established in a waterway. In fact, reed canarygrass will survive more silting-over than any known crop or grass. It may be covered by 7.5 to 10 cm (3 to 4 in) of silt and permitted to continue to grow (Fig. 26-6) (23).

Reed canarygrass can also be established from mature green hay, by immediately tramping it into a wet soil. This procedure requires a very wet, muddy area for quick establishment. It has been found that within 30 days, plant height may approach 15 cm (6 in) (19).

Reed canarygrass is known for its early growth and great production in the spring (29). Approximately 60% of its total season production will occur in May, June, and July (13, 15, 20, 29). Annual gain per hectare is equal to that for smooth bromegrass and orchardgrass; tall fescue is probably the best fall-saved or summer-deferred grazing grass.

Producers would like to see improvement in two aspects of reed canarygrass: palatability, and digestible dry matter production over the entire season, when

FIGURE 26-5
Reed canarygrass established by rooted
nodes. *(Courtesy University of Wisconsin)*

compared with the other cool-season grasses. Research indicates that the
digestibility among fall-saved reed canarygrass can be greatly improved since it
is a fairly highly heritable trait (8, 10, 41).

MANAGEMENT OF REED CANARYGRASS

Establishment

Reed canarygrass is rather difficult to establish by seed. It is slower in its
establishment rate than most of the other cool-season grasses (38). Establishment may fail when weed competition is very severe in the spring. In a poorly
drained area, a late summer seeding should be considered, as stand establishment is more successful at this time and weeds are only a minor problem.
Germination of the seed of reed canarygrass is slow and irregular (1, 15, 16, 29).
New seed is often immature and sometimes dormancy may occur (38).

A moist, firm seedbed is very important for stand establishment of reed
canarygrass, as it is with any other grass. Reed canarygrass will tolerate a wide
range of pH, as indicated earlier, and is rather easily established in a poorly
drained area. Nitrogen is important for its establishment even if the reed
canarygrass is established with a legume. It is suggested that a starter fertilizer
containing nitrogen be applied at the time of seeding, due to the slower
establishment of the grass.

FIGURE 26-6
Reed canarygrass established by
vegetative propagation. Note the
new regrowth from the rhizome.
*(Courtesy University of
Wisconsin)*

Seeding

Seeding practices for reed canarygrass are almost identical to those for the other grasses. A very firm seedbed is highly desirable. Reed canarygrass, however, has a lower percentage of germination, so one should attain at least 80% germination or adjust the seeding rate accordingly. Suggested seeding rates range from 6 to 10 kg/ha (5 to 9 lb/A) (29). As with all forage crops, it is essential that one use a field roller, a cultipacker, or some type of press wheel to ensure close contact of the soil and seed. The depth of seeding should be no more than 1.1 to 1.3 cm (0.5 in) (29).

If a late summer seeding is planned, the seedbed should be prepared approximately 2 to 4 weeks before seeding. This will allow enough time for the soil to become firm and will provide the opportunity to accumulate good seedbed moisture. The best time for seeding in the upper midwest is approximately September 1 through 15 for Illinois, and as one proceeds north, into Wisconsin or northern Iowa, seeding should be carried out between August 5 and 25. Late summer or early fall seeding may be required on extremely poorly drained areas, especially where one cannot seed in the early spring. Even if reed canarygrass seed does not germinate in the fall, it will overwinter and germinate the next spring (29). As mentioned earlier, both the reed canarygrass seedling and seed have a great tolerance for flooding conditions (16, 20, 30). Reed canarygrass seedlings may not be as drought-tolerant as orchardgrass, and they are somewhat more susceptible to freezing temperatures than most of the other cool-season grasses (29, 43).

When seeding reed canarygrass with companion legumes, the seeding rate of the reed canarygrass should be approximately two-thirds of the rate that would be used to seed it in a monoculture. The appropriate companion legumes can be then added at suggested rates, as follows: Ladino clover and alsike clover should be seeded in combination with reed canarygrass at 1.1 and 2.2 kg/ha (1 and 2 lb/A), respectively. These seeding rates are for a poorly drained area (29). For an upland soil, or where drainage is adequate and a hay harvest is anticipated, the suggested rate is reed canarygrass at two-thirds the regular seeding rate, plus 9 kg/ha (8 lb/A) of alfalfa or 7 kg/ha (6 lb/A) of red clover (29).

A companion crop or nurse crop may be used for spring seedings, but is generally not used for a late summer seeding. Spring oats is the most commonly used companion crop. It is suggested that the oats be removed very early as either oatlage or hay, or alternately grazed during the season, so that competition is reduced. This will also reduce competition from weeds, and will allow sufficient amounts of light and moisture for good establishment of the reed canarygrass (29).

Management of Reed Canarygrass as a Pasture

Reed canarygrass is one of the highest-yielding perennial forage grasses used for grazing in the northern United States. Early spring growth facilitates its use as a pasture for the entire growing season (29). Reed canarygrass yields extremely high amounts of dry matter and has a high carrying capacity in terms of animal units per hectare (9, 34, 39). It is very important, when grazing reed canarygrass, to keep it in a very palatable condition throughout the growing season. For the highest quality pasture, reed canarygrass should be grazed when it is from 15 to 60 cm (6 to 24 in) in height. Animal performance and intake decrease very rapidly as the plant grows beyond this height (28). Cattle will selectively graze the young, tender tillers, leaving the more mature plants.

Stocking or light grazing constitutes underutilization. The stocking rate should be maintained at a level such that the plant growth does not exceed 30 cm (12 in) in height (28). This will keep the plant growth in a very palatable stage throughout the growing season. Once ruminants become accustomed to reed canarygrass, they will graze it readily. Acceptability is governed by the absence of alkaloids. Some reports state that lambs do not perform as well on reed canarygrass as other ruminants (29). Rotational grazing for short periods of time with high stocking rates provides the most efficient utilization of reed canarygrass. Cattle will selectively graze the young, tender tillers if the crop is understocked, and will leave the older growth, which must therefore be removed about halfway through the summer by very close clipping. The clipping should occur when the reed canarygrass is at a height of 5 to 7 cm (2 to 2.5 in). Regrowth will then be uniform, and highly palatable forage will be produced (28). If the reed canarygrass is growing in a very poorly drained area, this extremely close clipping should not be used. Under such conditions, the harvest should not be clipped below a height of 10 to 12 cm (4 to 5 in). This stubble height will allow a good sod to develop and will prevent the cattle's hooves from cutting through the sod (29).

The regrowth of reed canarygrass is very similar to that of orchardgrass, in that it produces a very leafy regrowth throughout the pasture season. Tall fescue is a better fall-saved winter pasture crop than reed canarygrass because the quality of reed canarygrass declines rather rapidly after a frost.

Cattle and sheep performance are about equal on reed canarygrass and smooth bromegrass (24). A comparison of reed canarygrass with birdsfoot trefoil in Iowa showed that there were more gains for beef cattle on reed canarygrass than on birdsfoot trefoil pasture (24). Satisfactory milk yields were obtained in Washington when cattle grazed reed canarygrass (18).

Reed canarygrass is sometimes grown with a companion legume such as alfalfa, ladino clover, or alsike clover. Management is a very important aspect in maintaining the legume with a grass as competitive as reed canarygrass. The first growth should be harvested when the reed canarygrass has less than 40 cm (16 in) of growth. Rotational grazing or controlled grazing is very important. Carrying capacity should be increased to the point where all of the forage is grazed or removed at a uniform rate, thus reducing the chances for selective grazing. The animal will then have a very palatable, uniform forage to graze, which, in turn, will ensure the maintenance of a legume in the mixture (29).

In poorly drained areas, alsike clover and ladino clover are often grown with reed canarygrass, instead of alfalfa. Two additional legumes commonly grown with reed canarygrass are red clover and birdsfoot trefoil. The red clover will last approximately 2 years before dying. Especially with these two low-growing species, birdsfoot trefoil and ladino clover, it is very important to manage or control the aboveground competition of the reed canarygrass, so that the legumes are maintained.

Management of Reed Canarygrass as a Hay Crop

It is a little more difficult to make high-quality hay from reed canarygrass than silage. One reason is that reed canarygrass is often grown in a poorly drained or a wet area, and the drying out of the hay, once it is cut and crimped, is slower than with other cool-season grasses. The highest yields of digestible dry matter are obtained when the first growth is harvested for hay at the early-heading stage (Fig. 26-1). The regrowth of the second and third crops is primarily leaves. Since reed canarygrass is a rather coarse-stemmed, wide-leafed species, it is recommended that crimping be done at the time of cutting. Some growers often make silage from the first growth and use the remaining two or three harvests as a hay crop.

Management of Reed Canarygrass as a Silage Crop

A high-quality, nutritious silage can easily be harvested from reed canarygrass. The suggested times of harvest are the same as for a hay harvest. The feeding value of reed canarygrass silage for both dairy and beef cattle has been shown to be of high quality (13, 29). The advantage of making silage from reed canarygrass is the possibility of reducing the drying time in the field and harvesting the grass at a more appropriate stage of growth, ensuring a high-quality product.

Soil Fertility

Reed canarygrass probably responds as well to nitrogen fertilizer as any of the other cool-season grasses. As shown in Table 25-1, reed canarygrass continues to increase in yield up to approximately 448 kg/ha (400 lb/A) of nitrogen per year (7, 31). Split applications are required with high N fertilization, a division of the total amount of nitrogen applied over the year for the first three growth periods. Approximately 60% of the total year's production occurs at the first harvest; therefore, the corresponding amount of nitrogen should be applied prior to early spring growth.

Nitrogen is not the only element required for top production of reed canarygrass; considerable amounts of phosphorus and potassium are also needed. Approximately the same amount of P and K is removed in a ton of reed canarygrass dry matter as in alfalfa. The potassium should be applied annually to reed canarygrass at the same rate as its removal (see Table 10-16). Phosphorus can be applied annually, or larger amounts can be applied every 2 years.

Reed canarygrass can become sod-bound like other sod-forming grasses, such as smooth bromegrass. This is the result of the low levels of nitrogen. Adequate annual applications of nitrogen will therefore reduce the tendency toward sod-bound conditions.

Food Reserves

Food reserves are stored in the crownal area, roots, and rhizomes in reed canarygrass (12, 20, 26). A higher concentration of the soluble carbohydrates is found in the crowns than in the roots; soluble carbohydrates are lowest in the stubble, or tiller area (20, 21). Typically, there is a cyclic production of food reserves throughout the growing season. Carbohydrate reserves are the lowest in the middle to the end of summer, and increase until approximately the first killing frost (20, 21). Grazing pressure should be reduced for the last 30 to 35 days of the growing season so that adequate supplies of food reserves are built up in preparation for winter. Reed canarygrass is not as winter-hardy as smooth bromegrass or intermediate wheat grass (29).

COMPOSITION AND QUALITY

Reed canarygrass may not be as palatable or digestible as some of the other cool-season grasses. Palatability declines rapidly with maturity, while the dry matter yield may increase (13, 14). Reed canarygrass in the vegetative stage has been found to be extremely high in crude protein, as found in research in

Minnesota (28). The amount of crude protein is often associated with the rate of nitrogen application. Crude protein increased in an Ohio study from 13 to 19% as nitrogen rates increased from 75 to 600 kg/ha (67 to 53 lb/A) (31).

Palatability of reed canarygrass appears to be correlated with the alkaloid content and intake of the grass (35, 41). It is, however, not entirely associated with the alkaloid content, as the amount of nitrogen fertilizer is often linked to palatability (29). There are conflicting reports on the performance of sheep on reed canarygrass but, in general, most ruminants will perform very well on reed canarygrass hay or pasture (29). Generally, ruminants prefer smooth brome-grass to reed canarygrass when given a free choice.

Animal intake is the ultimate factor in determining any forage utilization. Most studies have shown that the intake of reed canarygrass is lower than alfalfa or smooth bromegrass (28). If livestock do not have a free choice of different species, however, reed canarygrass will be grazed uniformly. Livestock tend to graze selectively, against reed canarygrass, when the canarygrass is grown in combination with other forage species or legumes. Increasing the carrying capacity or number of animal units per area will ensure uniform removal of the forage and prevent selective grazing.

SEED PRODUCTION

Seed of reed canarygrass shatters very rapidly upon maturity (38). The seeds ripen from the top of the head downward, and fall soon after ripening. Therefore, it is difficult to harvest reed canarygrass seed. The most successful method of harvesting seed is to combine the standing crop. It should be cut as high as possible above the leafy portion. Harvest should occur when most of the seeds are ripe or beginning to turn brown, and prior to shatter. When harvesting reed canarygrass in this manner, there is a great deal of green plant tissue harvested with the ripe seed. Seed must never be allowed to heat before it is completely dry as heating of the seed seriously lowers germination (29). The preferred methods for drying are to immediately dry the seed commercially, thinly spread the combined material over a floor, and stir it repeatedly until dry. Seed yields may range from 60 to over 500 kg/ha (54 to 446 lb/A), but yields have reached 785 kg/ha (700 lb/A) in the midwest (8, 38). Plant breeders have found heritability to be extremely high for reducing seed shatter and increasing seed yield and the number of heads per plant. Therefore, the seed yield could easily be increased by proper breeding and/or selection procedures (8).

DISEASES

One of the unique characteristics of reed canarygrass is that it is not susceptible to very many diseases or insects. The only major disease that is found on reed canarygrass is a leaf disease, *Helminthosporium giganteum* Heald and Wolf. Developing plant resistance and good cultural practices are the two best control methods for this particular disease (29).

INSECTS

The major insect that attacks reed canarygrass is the fritfly, *Oscinella frit* (L.). The fritfly attacks the very young shoot of the plant and that central leaf often dies, while the other, surrounding leaves remain green. It appears that adequate nitrogen fertilization can greatly reduce fritfly infestation of reed canarygrass (42).

SUMMARY

The advantages and disadvantages of reed canarygrass are as follows:

Advantages

1 It grows extremely well in wet areas and can withstand considerable flooding.
2 It is an excellent species for soil conservation, especially for waterways and gully control.
3 It develops a dense sod.
4 It is a very productive grass, providing sufficient nutrients are added.
5 It has a deep root system, which aids in drought resistance.
6 It provides good pasture during the dry summer months.
7 It requires a high carrying capacity to maintain quality forage.
8 It is a long-lived perennial.
9 It responds better to nitrogen than most cool-season grasses.

Disadvantages

1 Its palatability decreases rapidly with maturity.
2 It is slow in establishment.
3 Alkaloids are present, which may reduce palatability and animal intake.
4 The seed shatters upon ripening.
5 Cattle may selectively graze, avoiding consumption of reed canarygrass, if it is grown in mixtures.
6 It cannot be seeded in slowly moving waterways or where silting may occur, because this reduces water runoff.
7 Management is extremely important in order to maintain quality forage.
8 It dies back rapidly upon a killing frost.

QUESTIONS

1 Describe the characteristics unique to reed canarygrass that make it a desirable cool-season forage grass.
2 In what ways does nitrogen application to reed canarygrass influence some of its characteristics?

3 Discuss the palatability, digestibility, antiquality, and animal intake of reed canary-grass.

4 Why is reed canarygrass considered so valuable for soil conservation?

5 Discuss the management of reed canarygrass as a pasture crop and as a hay crop.

6 Compare reed canarygrass to smooth bromegrass and orchardgrass in terms of forage quality and production.

7 Why would you consider reed canarygrass a desirable forage crop?

REFERENCES

1 Ahlgren, G. H. *Forage Crops*. McGraw-Hill, New York, 2d ed., pp. 195–204, 1956.

2 Allinson, D. W., M. B. Tesar, and J. W. Thomas. Influence of cutting frequency, species, and nitrogen fertilization on forage nutritional value. *Crop Sci*. 9:504–508, 1969.

3 Alway, F. J. Early trials and use of reed canarygrass as a forage plant. *J. Amer. Soc. Agron*. 23:64–66, 1931.

4 Arakeri, H. R., and A. R. Schmid. Cold resistance of various legumes and grasses in early stages of growth. *Agron. J*. 41:182–185, 1949.

5 Barnes, R. F., and G. O. Mott. Evaluation of selected clones of *Phalaris arundinacea* L. I. *In vitro* digestibility and intake. *Agron. J*. 62:719–722, 1970.

6 Blakeslee, L. H., C. M. Harrison, and J. F. Davis. Ewe and lamb gains on brome and reed canarygrass pasture. *Mich. Agr. Exp. Sta. Quart. Bull*. 39:230–235, 1956.

7 Bonin, S. G., and D. C. Tomlin. Effects of nitrogen on herbage yields of reed canarygrass harvested at various development stages. *Can. J. Plant Sci*. 48:511–517, 1968.

8 Bonin, S. G., and B. P. Goplen. Heritability of seed yield components and some visually evaluated characters in reed canarygrass. *Can. J. Plant Sci*. 46:51–58, 1966.

9 Bryan, W. B., W. F. Wedin, and R. L. Vetter. Evaluation of reed canarygrass and tall fescue as spring-summer and fall-saved pasture. *Agron. J*. 62:75–80, 1970.

10 Carlson, I. T., K. H. Asay, W. F. Wedin, and R. L. Vetter. Genetic variability in *in vitro* dry matter digestibility of fall-saved reed canarygrass, *Phalaris arundinacea* L. *Crop Sci*. 9:162–164, 1969.

11 Colovos, N. F., R. M. Koes, J. B. Holter, J. R. Mitchell, and H. A. Davis. Digestibility, nutritive value and intake of reed canarygrass (*Phalaris arundinacea* L.). *Agron. J*. 61:503–505, 1969.

12 Davis, W.E.P. Effect of clipping at various heights on characteristics of regrowth in reed canarygrass. *Can. J. Plant Sci*. 40:452–456, 1960.

13 Decker, A. M., G. A. Jung, J. B. Washko, D. D. Wolf, and M. J. Wright. Management and productivity of perennial grasses in the northeast. I. Reed canary-grass. *West Vir. Agr. Exp. Sta. Bull. 550T,* 1967.

14 Dvorak, R. A. and D. A. Miller. The nutritional value of maturing forage crops. *Ill. Research, Univ. of Ill. Agr. Exp. Sta*. 18:16–17, 1976.

15 Evans, M. W., and J. E. Ely. Growth habits of reed canarygrass. *J. Amer. Soc. Agron*. 33:1017–1027, 1941.

16 Goplan, B. P., S. G. Bonin, W. E. P. Davis, and R. M. MacVicar. Reed canarygrass. *Can. Dept. Agr. Publ. 805,* 1963.

17 Hanson, C. L., J. F. Power, and C. J. Erickson. Forage yield and fertilization recovery by three irrigated perennial grasses as affected by N fertilization. *Agron. J*. 70:373–375, 1978.

18 Hodgson, R. E., M. S. Grunder, and J. C. Knott. Carrying capacity of pure stands of reed canarygrass. *Wash. Agr. Exp. Sta. Bull. 275,* p. 29, 1932.
19 Holmberg, G. Vegetating critical areas. *J. Soil Water Conserv.* 14:165–168, 1959.
20 Horrocks, R. D., and J. B. Washko. Spring growth of reed canarygrass and Climax timothy under different harvesting systems. *Crop Sci.* 9:716–719, 1969.
21 Horrocks, R. D., and J. B. Washko. Studies of tiller formation in reed canarygrass and Climax timothy. *Crop Sci.* 11:41–45, 1971.
22 Hoveland, C. S., E. M. Evans, and D. A. Mays. Cool-season perennial grass species for forage in Alabama. *Ala. Agr. Exp. Sta. Bull. 397,* 1970.
23 Hughes, H. D., and V. B. Hawk. Reed canary stops gullies. *Crops and Soils* 6(8):14–15 and 35, 1954.
24 Ingels, J. R., J. W. Thomas, E. J. Benne, and M. Tesar. Competitive response of wether lambs to several cuttings of alfalfa, birdsfoot trefoil, bromegrass, and reed canarygrass. *J. Anim. Sci.* 24:1159–1164, 1965.
25 Krueger, C. R., and J. M. Scholl. Performance of bromegrass, orchardgrass, and reed canarygrass grown at five nitrogen levels and with alfalfa. *Wis. Agr. Exp. Sta. Res. Rep. 69,* 1970.
26 Lawrence, T., and R. Ashford. Effect of stage and height of cutting on the dry matter yield and persistence of intermediate wheatgrass, bromegrass, and reed canarygrass. *Can. J. Plant Sci.* 49:321–332, 1969.
27 Lawrence, T., F. G. Warder, and R. Ashford. Effect of stage and height of cutting on the crude protein content and crude protein yield of intermediate wheatgrass, bromegrass, and reed canarygrass. *Can. J. Plant Sci.* 51:41–48, 1971.
28 Marten, G. C., and J. D. Donker. Determination of pasture value of *Phalaris arundinacea* L. vs. *Bromus inermis* Leyss. *Agron. J.* 60:703–705, 1968.
29 Marten, G. C. and M. E. Heath. Reed canarygrass. In *Forages,* Iowa State Univ. Press, Ames, IA, 3d ed., pp. 263–276, 1973.
30 McKenzie, R. E. The ability of forage plants to survive early spring flooding. *Sci. Agr.* 31:358–367, 1951.
31 Niehous, M. H. Effect of N fertilizer on yield, crude protein content, and *in vitro* dry matter disappearance in *Phalaris arundinacea* L. *Agron. J.* 63:793–794, 1971.
32 Phillips, T. G., J. T. Sullivan, M. E. Laughlin, and V. G. Sprague. Chemical composition of some forage grasses. I. Changes with plant maturity. *Agron. J.* 46:361–369, 1954.
33 Read, D.W.L., and R. Ashford. Effect of varying levels of soil and fertilizer phosphorus and soil temperature on the growth and nutrient content of bromegrass and reed canarygrass. *Agron. J.* 60:680–682, 1968.
34 Schaller, F. W., W. F. Wedin, and E. T. Carlson. Reed canarygrass. *Iowa State Univ. Coop. Ext. Serv. Pm-538,* 1972.
35 Simons, A. B., and G. C. Marten. Relationship of indole alkaloids to palatability of *Phalaris arundinacea* L. *Agron. J.* 63:915–919, 1971.
36 Smith, D., A.V.A. Jacques, and J. A. Balasko. Persistence of several temperate grasses grown with alfalfa and harvested two, three, or four times annually at two stubble heights. *Crop Sci.* 13:553–556, 1973.
37 VanArsdell, W. J., G. A. Branaman, C. M. Harrison, and J. F. Davis. Pasture results with steers on reed canarygrass. *Mich. Agr. Exp. Sta. Quart. Bull.* 37:125–131, 1954.
38 Vose, P. B. The agronomic potentialities and problems of the canarygrass, *Phalaris arundinacea* L. and *Phalaris tuberosa* L. *Herb. Abstr.* 29:77–83, 1959.

39 Wedin, W. F., I. T. Carlson, and R. L. Vetter. Studies on nutritive value of fall-saved forage, using rumen fermentation and chemical analyses. *Proc. 10th Intl. Grassl. Congr.,* pp. 424–428, 1966.

40 Wedin, W. F., and R. L. Vetter. Grain feeding on birdsfoot trefoil or reed canarygrass pastures. *Agron. Abstr.* pp. 77–78, 1970.

41 Williams, M., R. F. Barnes, and J. M. Cassady. Characterization of alkaloids in palatable and unpalatable clones of *Phalaris arundinacea* L. *Crop Sci.* 11:213–217, 1971.

42 Wolf, D. D. Yield reductions reed canarygrass caused by fruit fly infestation. *Crop Sci.* 7:239–240, 1967.

43 Wood, G. M., and P. A. Kingsbury. Emergence and survival of cool-season grasses under drought stress. *Agron. J.* 63:949–951, 1971.

CHAPTER 27

TALL FESCUE

Tall fescue, *Festuca arundinacea* Schreb., is a very popular long-lived, bunch-forming, cool-season grass. It has gained popularity in the United States in the last 20 years. This popularity is due to its high palatability during the fall months. It is the best fall-saved, cool-season grass. Spring growth also possesses good quality, but the summer forage is probably one of the most unpalatable of all grasses grown in the United States. Tall fescue is an upright-growing perennial with a deep root system with the ability to spread and form a dense sod by means of short rhizomes. This species probably resists wear and tear better than any other cool-season grass.

HISTORICAL BACKGROUND

Tall fescue is native to Europe. It was first introduced into the United States in about 1850 (9, 14). The early history of tall fescue is a little unclear, because many writers confused it with meadow fescue, *F. elatior* L. In 1950, the scientific name for tall fescue, *Festuca arundinacea,* was given to denote a separate species, instead of a subspecies of meadow fescue (9, 11).

Morphologically, tall fescue and meadow fescue are very similar and very difficult to distinguish from each other. Tall fescue is more aggressive in its growth and has a pubescence on its auricles, while meadow fescue does not. Both species have a distinct midrib in the leaf blade with a very waxy appearance (15). Tall fescue has 42 pairs of chromosomes, while meadow fescue has 14 (15).

WHERE TALL FESCUE GROWS

Tall fescue is adapted to a wide range of soils from the midsouth to Canada. It is well-adapted to low, wet, poorly drained areas, and will persist during the cold

427

winter months (9, 11, 26). It tolerates soils with a pH in the range of 4.5 to 9.5. Even though tall fescue exhibits primarily bunch-type growth, it also has the ability to spread and form a dense sod by means of the short rhizomes (9, 11). The root system is very deep. Its rhizomes are less vigorous than those of reed canarygrass or smooth bromegrass. The ability to grow in wet areas and its tolerance to a wide pH range make it an excellent turf grass in areas of heavy traffic (31).

Tall fescue prefers a relatively cool environment, but will persist during the hot summer months and will continue to grow during the cool winter months. The optimum growing temperature for fescue is from 20 to 25°C (68 to 77°F), but it will grow in temperature ranges from 5 to 29°C (40 to 85°F). Seed germinates from 12 to 34°C (55 to 91°F), but will not germinate below 5°C (40°F) (25). Therefore, with these cool temperature requirements, tall fescue will not persist in the deep south, coastal states, or in Florida under high summer temperatures. In these areas it performs more like a winter annual than a perennial (3, 9, 26, 32).

Tall fescue is somewhat resistant to shade but probably not as resistant as orchardgrass (35a). Small seedlings of tall fescue will tolerate 65 to 75% shade. Established fescue will continue to grow even in 50% shade (5, 14).

TYPES OF FESCUES

There are actually about 100 species of *Festuca*. Each of the species varies in plant height, leaf width, growth habit, longevity, and many other agronomic characteristics (9, 11). One of the more famous broadleaf species of fescue is meadow fescue, *F. elatior* L. There are several narrowleaf species of fescue, such as red fescue, *F. rubra* L., Idaho fescue, *F. idahoensis* Elmer., and sheep fescue, *F. ovina* L.

Meadow fescue is adapted to the same area as timothy. It is used as a hay crop and, especially in Europe, as a pasture crop. Meadow fescue has never been extremely popular in the United States (11).

Red fescue is quite similar to sheep fescue in that both have very fine leaves. The leaves are very dark green. Red fescue does spread somewhat by very short rhizomes (11).

Sheep fescue is native to North America. Its leaves are rather sharp, and bluish-gray in color (11). Chewing fescue is closely related to red fescue in that neither spreads by rhizomes. Chewing fescue and red fescue are often used in lawns and turfs, and especially in shaded, dry areas. Idaho fescue is grown primarily in the northwestern United States (11).

HOW TALL FESCUE GROWS

Tall fescue is a very aggressive cool-season perennial grass (Fig. 27-1). It is very tolerant of poorly drained areas and is a very palatable grass in early spring and fall (11, 14). It is a deep-rooted bunch grass, although it does produce some

FIGURE 27-1
Flowering plant of tall fescue. *(Courtesy University of Illinois)*

short rhizomes (Fig. 27-2). Continuous mowing or grazing causes this species to form a dense sod (10).

Tall fescue tolerates a wide range of soil pH from 4.5 to 9.5 (14). It will persist for many years in low, wet areas, and continues to grow during the cool winter months. Another characteristic of tall fescue that aids in its adaptation is that it tolerates droughty conditions and shallow soils (11).

Tall fescue is easily established and has excellent seedling vigor (14). There are many studies that indicate that tall fescue is at first somewhat slow in its establishment, but becomes very vigorous in the later stages of plant development (9, 11, 31, 43, 44). Tall fescue compares in aggression with other seedling forage grasses as follows (from most to least aggressive): (1) annual and perennial ryegrass; (2) tall oat grass; (3) orchardgrass; (4) tall fescue; (5) smooth bromegrass; (6) timothy; and (7) Kentucky bluegrass (9, 11, 31).

Tall fescue produces a very leafy growth, but somewhat less so than orchardgrass at similar stages of maturity (Table 27-1). The leaves possess a very distinct midrib, and are a very dark green, with a shiny or waxy appearance. Plant height may reach 105 to 150 cm (40 to 60 in) (9, 11, 14).

Seeds are produced on a branched, panicle-type head that ranges from 10 to 30 cm (4 to 10 in) in length (Fig. 27-3). Seed heads are very numerous. There are five to seven seeds per spikelet. The seeds are somewhat darker in color than those of ryegrass and have a purple cast on the glumes (9, 11). In the United States, flowering occurs from mid-May to early June (9, 11), making tall fescue

FIGURE 27-2
Close-up of tall fescue tillers and
short rhizomes.

one of the earliest-flowering species of the cool-season grasses in the United
States.

Tall fescue is adapted to a wide range of climatic conditions. Probably its best
growth occurs under relatively cool growing conditions, but it will also persist
under very hot summer temperatures (9, 11, 26). Dormancy is not complete
even when the temperatures reach about 1.1°C (34°F); active growth occurs
when the daily or weekly temperatures reach 4.4°C (39°F) (26).

Tall fescue has been shown to be of very poor quality during the late spring or
summer months (37). One of the outstanding attributes of tall fescue is the
extremely high quality of fall-produced forage, which, incidentally, is probably
the best of any cool-season grass, because of the higher concentration of sugar
and soluble carbohydrates in tall fescue (6). It was found in a Kentucky study

TABLE 27-1
LEAF BLADE TO STEM PLUS SHEATH RATIOS
OF TALL FESCUE AND ORCHARDGRASS AT
VARIOUS STAGE OF MATURITY

Maturity stage	Tall fescue	Orchardgrass
Boot	1.97	2.16
Headed	0.43	1.04
Early bloom	0.30	0.68
Late bloom	0.29	0.56
Seed	0.26	0.47

FIGURE 27-3
Close-up of tall fescue inflorescence. *(Courtesy University of Illinois)*

that the digestibility, total sugar content, and palatability of tall fescue is lowest during summer, intermediate in the early spring, and highest during the fall months (6). Protein averaged 22%, 17%, and 19% for the spring, summer, and fall, respectively; while the sugar content averaged 9%, 8%, and 19%, respectively. The digestibility averaged 69% in the early spring, 66% during the summer, and more than 73% in the fall (6). Some stations have reported that cattle lose weight when grazing tall fescue in midsummer (29).

Tall fescue is as responsive to nitrogen fertilizer as any of the cool-season grasses (19, 25, 36, 40) (Table 27-2). The application of nitrogen may, in some

TABLE 27-2
THE EFFECT OF N APPLICATIONS ON THE YIELDS OF DRY MATTER AND PROTEIN IN TALL FESCUE

N	Dry matter		Protein	Additional protein per kg of N,
kg/ha (lb/A)	kg/ha (lb/A)	%	kg/ha (lb/A)	kg (lb)
0 (0)	1299 (1158)	9.9	129 (115)	— —
90 (80)	3694 (3295)	13.6	502 (447)	4.1 (9.0)
180 (160)	5441 (4853)	16.4	892 (795)	4.3 (9.5)
360 (320)	7227 (6446)	20.3	1,467 (1308)	3.2 (7.0)
720 (640)	7869 (7019)	23.3	1,834 (1635)	1.0 (2.2)

Source: W. F. Wedin, "Fertilization of cool-season grasses," In *Forage Fertilization*, Amer. Soc. Agron., pp. 95–118, 1974.

cases, depress the average daily cattle gain, but the carrying capacity may double, and live weight gain may also be greatly increased per hectare (11). Added nitrogen to tall fescue growing in the northerly regions of its adaptation may increase its susceptibility to winter kill (11). Excessive nitrogen tends to encourage growth later in the season and to reduce winterhardiness. Under normal weather conditions, tall fescue possesses sufficient winterhardiness to grow in an area extending from northern Iowa through Pennsylvania.

Most of the recent interest in tall fescue has focused on its ability to produce high-quality forage late into the fall, resulting in more grazing days per year than are possible with most other tall-growing cool-season grasses. Several methods have been developed for managing tall fescue that will encourage more winter grazing for beef cattle (1, 2, 6, 18). This involves a period of deferring summer grazing until the early fall. It also involves the accumulation of early spring, high-quality fescue in large round-bales to be used, along with the standing regrowth, in the late fall and early winter months (27, 30). This is referred to as a "standing hay" crop.

USES OF TALL FESCUE

Tall fescue is used for numerous purposes. These include pasture, hay, silage, soil conservation, and turf. Tall fescue is most often used as a grazing crop, and has been used as a fall-saved crop in recent years. It is very tolerant of continuous and close grazing (11, 27). Under such management schemes, it may be possible to maintain a legume such as ladino along with the tall fescue over a period of years. Tall fescue is a very aggressive, early spring–growing species (Fig. 27-4). This aids the grazing of cow-calf herds when growth in the early spring is needed (27). Tall fescue is not widely used for dairy animals.

When tall fescue is properly managed and fertilized, and harvested at the proper stage of maturity, it makes an excellent hay. For highest-quality hay, harvest should be taken when the heads first begin to appear (4, 19, 36, 39). Regrowth in the aftermath of such a harvest is generally very productive when there is sufficient moisture and nitrogen (11, 40). Tall fescue should be harvested for silage at approximately the same stage of maturity as for hay (27, 28).

Tall fescue is well known for its soil-conservation potential. It has a deep root system and is very tolerant of wet or poorly drained soils (3a, 11, 14). It makes an excellent waterway and is ideal in the low wetlands (11, 14). In these low, wet areas, one may be able to harvest the waterways for silage early in the year and, later in the year, take a hay crop, especially if conditions have become drier. Tall fescue makes a rather dense sod, due to its rhizomic-type growth, and can carry water moving at a great velocity once a dense sod has developed (11, 14).

Tall fescue is known in the turf industry as a very durable sod (11). It is used quite frequently in play areas, athletic fields, airfields, and parks, and on the sides of roadways where a lot of traffic or wear and tear can be expected. When grown for turf or lawn purposes, the seeding rate should be greatly increased so that a very fine leaf growth develops (17, 31). One of the characteristics contributing to tall fescue's desirability as a turf grass is its high tolerance to

FIGURE 27-4
Tall fescue is a productive,
vigorous, cool-season forage
grass.

many soil and drainage conditions. It is tolerant of both sunny and shade conditions and quite resistant to diseases, insects, and drought.

DESIRABLE CHARACTERISTICS OF TALL FESCUE

Most of the desirable characteristics have already been mentioned in connection with other properties of tall fescue. It is one of the most productive early spring grazing crops in the United States. It also produces abundant, highly palatable forage in the early fall months. It will tolerate a wide range of soil, climatic, and drainage conditions. Fescue is relatively tolerant of shade and will survive long periods of flooding during the winter. Tall fescue may form a dense sod, and it has a root system that lends itself to waterways, lanes, and high-traffic areas. The deep-rooted bunch grass is also very tolerant of droughty conditions and possesses good winterhardiness—not as good as that of smooth bromegrass or reed canarygrass, but equal to orchardgrass (11) (Fig. 27-5). Growers would like a tall fescue that would be more palatable in the summer months and have a lower alkaloid content (21). Tall fescue produces large quantities of seed, which makes the seed rather inexpensive (8).

MANAGEMENT OF TALL FESCUE

Establishment

Tall fescue is easily established. It has less seedling vigor than orchardgrass and greater seedling vigor than smooth bromegrass (5, 11). It is very important to have a firm, moist seedbed. The seeding depth should be approximately 0.6 cm.

FIGURE 27-5
Winter damage of tall fescue plots. *(Courtesy University of Wisconsin)*

It is recommended that a corrugated roller be used in a well-prepared seedbed. As with all cool-season grasses, it is most important to get the seed in close contact with moist soil to ensure quick establishment (5).

Seeding

Tall fescue is usually seeded alone in the north, whereas smooth bromegrass and orchardgrass are often seeded with a legume. In the south tall fescue is often seeded with a legume. The seeding rate varies, depending on whether a legume is seeded with it, or the fescue is seeded alone (11). When tall fescue is seeded by itself, it is usually sown at the rate of 9 to 11 kg/ha (8 to 10 lb/A). When sown with a legume, the seeding rate is reduced to 4 to 7 kg/ha (4 to 6 lb/A) (22). When tall fescue is sown by itself, it is most often planted in the early spring. When legumes are grown in association with tall fescue, a late summer seeding often favors the legume establishment, because there will be less competition from the tall fescue grass (11, 22). Further south, tall fescue, seeded alone, may also be sown in the late summer or early fall (11).

Maintaining a legume with tall fescue requires very careful management (5, 11, 22, 25, 36, 38). Both tall fescue and the legume require high amounts of P and K to develop a vigorous stand. When nitrogen fertilizer is used, the tall fescue is favored and competition may become too great for the legume to develop and maintain a good, healthy stand (11). Tall fescue can easily be established on soils with a low pH or low general fertility, but the legume will then be greatly hampered. For this reason, good general fertility is required when establishing tall fescue with a legume (11). The establishment of tall fescue by itself is much easier, for one can apply ample amounts of nitrogen to ensure vigorous growth and to maintain a green, succulent, highly productive stand late into the season (26, 35, 36). When a legume is crowded out of a tall fescue stand, renovation is necessary to reestablish the legume (10, 21).

Harvest Schedule

The first year is a very important period to ensure that tall fescue is well established if it is sown with a legume (5, 11, 22). It is essential that a good root system be developed before grazing or harvesting for hay (22). The grazing or harvest should be delayed until the fescue is 15 to 20 cm (6 to 8 in) tall, and then grazing should not go below 7 to 10 cm (3 to 4 in) before rotational grazing is instituted (22). If a legume is established with tall fescue, do not let the legume accumulate too much growth or it will suppress the young fescue seedlings. Tall fescue is a little slower in establishment than some of the other grasses, so a new seeding can be damaged by overgrazing or by grazing too soon (22). Once tall fescue is well established, it can be closely grazed. During the winter months, excessive cattle trampling should not occur when the soil is wet (22). It will withstand heavy grazing pressure for short periods of time, but it should not be continuously grazed (11). Tall fescue should not be grazed closer than 5 to 10 cm (2 to 4 in) to ensure top production and longevity of the stand (28).

The boot stage is the proper time to cut tall fescue for a hay crop. Tall fescue makes a high-quality hay at this stage, but the quality declines drastically after the onset of heading (6, 11). The succeeding harvests, depending on the amount of nitrogen applied, should be made every 30 to 35 days, until approximately 30 days before a killing frost. This allows for sufficient fall growth that one can harvest after a killing frost, which is desirable because the palatability of tall fescue improves during the fall months (1, 2). It is preferred, if tall fescue is to be harvested as a hay crop, that the late summer growth, especially during August, be accumulated and grazed off after a killing frost. The palatability of the tall fescue forage that is grown during August, whether grazed or harvested as hay, is very low (37).

Fertility

Tall fescue probably responds to nitrogen fertilization more effectively than any of the cool-season grasses, except reed canarygrass (3a, 40). Even though tall fescue will grow at a lower soil fertility, it is recommended that the pH be at least 6 to 6.5 for good establishment. If legumes are growing with tall fescue, it is essential to keep the pH close to 6.5, or higher, along with a high level of available phosphorus and potassium. Available phosphorus should be at 40 or more and potassium at 300 or more. If a legume is grown with tall fescue and becomes less than 30% of the stand, it is suggested that additional amounts of nitrogen be applied both in the fall and early spring to stimulate the fescue and eliminate the remaining legume (40). For good cow-calf performance, the stand should be at least 30 to 50% legumes. If a fescue-legume pasture is used for feeder cattle, the legume content should be at least 40 to 60%. When the level of legumes in the stand reaches 60%, it is suggested that poloxalene be used to help control bloat.

Food Reserves

The winterhardiness of tall fescue is essentially the same as that of orchardgrass; therefore, it is important to allow sufficient amounts of food reserves to be stored in the basal portion of the stems and tillers to ensure good winterhardiness (25). Various trials have indicated that tall fescue may have slightly more winterhardiness than orchardgrass (35). A good fall management practice would be to refrain from close utilization prior to the advent of winter because of the depletion of stored food reserves. Similar to the other cool-season grasses, at least 30 to 35 days of fall growth should be allowed prior to the killing frost so enough food reserves are built up for the winter. Tall fescue is used a great deal as a deferred summer grazing crop (1, 2, 30). Therefore, plant growth is allowed to occur in the late summer and early fall so that food reserves can accumulate. Following a killing frost, that forage which has collected during the early fall can be removed as a standing hay crop or be grazed. After a killing frost it would be advisable not to graze tall fescue too closely, in order to leave enough stubble to ensure some insulation and help collect snow.

Tall fescue is very valuable for spring, fall, and winter grazing. If properly fertilized, fescue will probably produce a high-quality forage in great quantities during these seasons (11, 30). Most growers are now managing tall fescue as a pasture plus a hay crop, or as a pasture during the spring and early summer, and then again in the early fall (11). Cattle are usually removed during the middle portion of the summer, and the pasture is then fertilized with N and the growth accumulated until after a frost occurs. Cattle are then returned and grazing continues until all of the growth is removed (2, 11).

Tall fescue provides excellent hay yields when properly fertilized. For the best-quality hay, tall fescue should be harvested when the heads begin to appear, and before flowering (11). The aftermath, when properly fertilized with nitrogen, is very productive. Dry matter yields will reach 7- to 9- plus tons per hectare (3 to 4 T/A) (11).

COMPOSITION AND QUALITY

Tall fescue has a reputation as being a low-quality hay. This is only true during midsummer or August (6). Quality is measured in terms of crude protein, crude fiber, carbohydrates, minerals, and digestibility. Tall fescue will rank as a good to high-quality forage when harvested in the early spring and before heading. Management and fertility are extremely important in maintaining high-quality tall fescue forage. It has been shown that digestibility and palatability of tall

TABLE 27-3
SEASONAL CHANGES IN PERCENT PROTEIN, SUGARS, AND DIGESTIBILITY OF TALL FESCUE

Growth period	Crude protein (%)	Total sugars (%)	In vitro digestibility (%)
Spring	22.2	9.0	69.1
Summer	18.0	8.4	66.2
Fall	19.0	18.8	73.5

Source: R. H. Brown, R. E. Blaser, and J. P. Fontenot, "Digestibility of fall-grown Kentucky 31 fescue," *Agron. J.* 55:321-324, 1955.

fescue are lowest during the summer months, intermediate in the early spring, and highest during the fall (Table 27-3) (6, 7). The quality of tall fescue is associated with the percent of carbohydrates present (6).

Many researchers have studied the quality of tall fescue in relation to the average daily gain of cattle grazing the tall fescue; their results range from superior to inferior (11, 32). It is generally agreed that cattle gains on tall fescue during the midsummer months are very poor. It was shown at Indiana that the poor animal performance on tall fescue during the summer months was associated with the low level of energy intake (29). Recent results indicate that performance of cattle grazing tall fescue grown with legumes will equal or surpass their performance on orchardgrass (31) (Table 27-4).

Nitrogen fertilizer lengthens the grazing period, increases stocking rates, and realizes higher gains per hectare, when tall fescue production is compared with orchardgrass or grass-legume mixtures (4, 36). Researchers have found that excessive nitrogen fertilizer depresses the average daily gain, but the carrying capacity may be doubled, thus increasing the total live weight of gain per hectare (4).

Tall fescue is probably the most digestible cool-season grass during the early fall months (Fig. 27-6) Cattle intake is generally increased during the fall. The

TABLE 27-4
PERFORMANCE OF BEEF COWS AND CALVES GRAZING VARIOUS GRASSES AND LEGUME-GRASS MIXTURES

	Orchardgrass	Tall fescue	Tall fescue–legume*
	kg/day (lb/day)	kg/day (lb/day)	(lb/day)
Calf gains	0.80 (1.76)	0.54 (1.18)	0.83 (1.82)
Cow gains	0.26 (0.62)	0.01 (0.02)	0.26 (0.62)
Cow conception (%)	90	70	92

*Legumes used were primarily ladino clover and red clover.
Source: D. C. Petritz, V. L. Lechtenberg, and W. H. Smith, "Performance of economic returns of beef cows and calves grazing grass-legume herbage," *Agron. J.,* 72:581–584, 1980.

FIGURE 27-6
Tall fescue makes an excellent fall-saved forage, or it can be harvested in a large round-bale for winter feeding.

enhanced digestibility of tall fescue during the fall months is often associated with a great increase in sugar content (6, 7).

Tall fescue has been criticized for having a high concentration of alkaloids. Perloline is the major alkaloid related to animal disorders developed when grazing tall fescue (12, 18). Perloline content depends on various genotypes, the time of the year, and the level of nitrogen fertility (21). It is highest during the months of July and August, and whenever tall fescue is heavily fertilized with nitrogen (21). Perloline was thought to be the alkaloid that caused fescue foot poisoning. It has been found that perloline is associated with the inhibition of certain cellulose digestion by rumen bacteria (12, 13). Perloline has been connected to low animal intake and poor digestibility during the months of July and August (21).

Recently evidence has been amassed that the presence of a seed-borne internal fungus, *Acremonium strictum,* seriously reduces animal gains. Comparison of gains on fungus-free versus fungus-infected fescue shows an amazing disparity. Animals on fungus-infected fields had an average gain of 0.4 kg (0.8 lb) per day and a total gain of 213 kg/ha (180 lb/A), while those grazing the noninfected field had an average daily gain of 0.8 kg (1.8 lb) and a total gain of 424 kg/ha (379 lb/A), a 125% increase in average daily gain and a 100% increase per hectare.

Tall fescue has also been criticized severely because of the incidence of fescue foot and grass tentay. This criticism is often directed at poor management. Fescue foot poisoning, or toxicity, can be reduced by growing legumes, ryegrass, or other small grains with the tall fescue (11). The first stages of foot fescue are manifested in sore and stiff feet of cattle. In the next stage, the cattle breathe rapidly, run a fever, lose weight, and develop a rough haircoat. Later a limp develops in the left hind leg. A dry gangrene develops from poor blood circulation, affecting the hooves or tails and causing them to drop off (23).

Farmers have observed that fescue foot will often develop in cattle grazing tall fescue that has accumulated from the previous year (11). Proper tall fescue management therefore involves complete removal of the preceding year's growth before allowing grazing the following year. One can also feed cattle dry hay or grain while they are grazing tall fescue, as this will greatly reduce the incidence of fescue foot (23). The growing of legumes with fescue will also decrease the incidence of fescue foot (23).

The specific cause of fescue foot toxicity, or poisoning, is not known. Some research has indicated that the causal agent may be butenolide, a toxin that is produced by the snowmold fungus, *Fusarium nivale* (Fr.) Ces., which is often found growing on tall fescue hay (41, 42).

Cattle grazing fescue pastures infected with the fungus *Acremonium strictum* show various typical fescue toxicity symptoms—a rough haircoat which is not shed in hot weather, a highly nervous condition, easy excitability, a constant low-grade fever [an increase of 1°C (2°F), rectal temperature], and excessive salivation and higher water demand. As indicated earlier, cattle grazing such infected fields perform poorly. At present, fescue foot poisoning cannot be associated with this fungus.

The quality of tall fescue is associated with its use in a cow-calf program as a deferred grazing system. The practice of deferring fescue for winter grazing is well suited to a cow-calf operation (38). One may select a certain area or amount of fescue (pasture) for such a program. In the midwest, usually the first two harvests are grazed or taken off as hay. Then from approximately July 1, all of the growth is allowed to accumulate until the fall, or until after a killing frost. One can then graze the cattle in a limited grazing or rotational system, allowing the stocking rate to utilize all of the growth through January or February (7, 37). Following this type of management, it is ideal to renovate the pasture in the spring by putting legumes into the fescue sod. The reason for rotational grazing or strip grazing of the deferred tall fescue is that the cattle will eat all of the green winter growth along with the dry leaves that have been killed by the frost.

Tall fescue leaves begin to turn yellow or brown five to six weeks after a killing frost (7). It is difficult to maintain a legume in a deferred tall fescue grazing program because, during the hot, dry summer months, the competition for the fall growth is too extreme. A good growth of deferred tall fescue should carry four to six cows and their calves per hectare (10 to 15 per acre), or eight to ten dry cows per hectare (20 to 25 per acre) during the winter months. Since the sugar content of the forage gradually decreases, it is advisable to increase grazing time as the winter progresses (7, 27).

SEED PRODUCTION

Tall fescue is a cross-pollinating crop that produces very good seed yields (8). Tall fescue should not be fertilized with more than 68 to 80 kg of nitrogen per hectare (60 to 75 lb/A) (11). Excessive amounts of nitrogen will create a considerable amount of lodging and low seed yields (8, 11). In addition, as N is increased, perloline increases, resulting in poor animal performance. Cattle can graze tall fescue until late March or early April, and then the fescue can be permitted to produce a seed crop (11). Most fescue is combined directly. Harvest should occur when 60 to 70% of the seedheads are mature and brown, and when some of the seed will come off when the head is pulled lightly between thumb and forefinger (8). Up to 30% of the seed crop may still be lost due to shattering (8, 11). After the seed has been removed, it would be advisable to cut the aftermath from the seed crop and bale it or remove it as a hay crop so that the regrowth will be quality forage.

SUMMARY

Advantages and disadvantages of tall fescue are as follows:

Advantages:

1 It exhibits very productive early season growth and is the most palatable of the cool-season grasses in the fall.

2 It responds very well to nitrogen fertilization.

3 Its regrowth is rapid.

4 It is probably the most wear-and-tear resistant cool-season grass.

5 It resists trampling by livestock.

6 It is adapted to a wide range of soil types and climatic conditions.

7 It will tolerate poor drainage.

8 It is resistant to flooding.

9 It will become established in one year.

10 It is shade-tolerant to an extent comparable to orchardgrass.

11 It is probably the best summer-deferred cool-season grass.

12 It is somewhat tolerant of continuous grazing.

13 It is an aggressive grass at the later stages of plant development.

Disadvantages

1 It lacks palatability in the midsummer season.

2 It is not recommended as a pasture crop for dairy animals or finishing animals.

3 It has a high concentration of alkaloids during the midsummer, which affects its palatability.

4 Animal performance may not be as great as with orchardgrass or smooth bromegrass.

5 It is a very competitive grass when grown in combination with legumes.

6 An animal disorder called fescue foot is associated with extensive grazing of tall fescue.

7 It lacks complete winterhardiness.

8 Its seed shatters very easily.

9 When a legume is grown with tall fescue, there appears to be considerable selective grazing of the legume, consequently rapidly reducing the legume content in the stand.

QUESTIONS

1 Why is tall fescue widely adapted?

2 How can one distinguish tall fescue from meadow fescue?

3 What are some of the characteristics of tall fescue that make it a very desirable plant in a forage system?

4 What are some of the characteristics of tall fescue that make it an undesirable plant in a forage system?

5 Discuss seasonal effects on the quality of tall fescue.

6 How can one manage tall fescue and utilize it to its greatest potential, while still maintaining high quality?

7 Describe the fertility requirement of tall fescue in relation to quality.

8 Why would you consider tall fescue a potentially desirable grass in the overall forage program in the United States?

9 Describe the symptoms of fescue foot toxicity. How can the incidence of fescue foot be reduced?

REFERENCES

1 Archer, K. A., and A. M. Decker. Autumn-accumulated tall fescue and orchardgrass. I. Growth and quality as influenced by nitrogen and soil temperatures. *Agron. J.* 69:601–605, 1977.

2 Archer, K. A. and A. M. Decker. Autumn-accumulated tall fescue and orchardgrass. II. Effects of leaf death on fiber components and quality parameters. *Agron. J.* 69:605–609, 1977.

3 Asay, K. H., A. G. Matches, and C. J. Nelson. Effect of leaf width on responses of tall fescue genotypes to defoliation treatment and temperature regimes. *Crop Sci.* 17:816–818, 1977.

3 a Belesky, S. P., S. R. Wilkinson, and J. E. Pallas Jr. Response of four tall fescue cultivars grown at two nitrogen levels to low soil water availability. *Crop Sci.* 22:93–97, 1982.

4 Blaser, R. E., H. T. Bryant, R. C. Hammes Jr., R. L. Boman, J. P. Fontenot, and C. E. Polan. Managing forages for animal production. *Vir. Poly. Instit. Res. Div. Bull.* 45, 1969.

5 Blaser, R. E., W. H. Skrdla, and T. H. Taylor. Ecological and physiological factors in compounding forage seed mixtures. *Advan. Agron.* 4:179–219, 1952.

6 Brown, R. H., R. E. Blaser, and J. P. Tontenot. Digestibility of fall-grown Kentucky 31 fescue. *Agron. J.* 55:321–324, 1955.

7 Bryan, W. B., W. F. Wedin, and R. L. Vetter. Evaluation of reed canarygrass and tall fescue as spring-summer and fall-saved pasture. *Agron. J.* 62:75–80, 1970.

8 Buckner, R. C., and P. B. Burrus Jr. The effect of certain management and fertilization practices on seed production of tall fescue. *Proc. Assoc. Southern Agr. Workers*, pp. 67–68, 1959.

9 Buckner, R. C., and L. P. Bush. Tall fescue. *Agronomy No. 20 ASA, CSSA, and SSA Public,* 1979.

10 Buckner, R. C., L. P. Bush, and P. B. Burrus. Variability and heritability of perloline in *Festuca* sp., *Lolium* sp., and *Lolium-Festuca* hybrids. *Crop Sci.* 13:666–669, 1973.

11 Buckner, R. C., and J. R. Cowan. The Fescues. In *Forages,* Iowa State Univ. Press. Ames, IA, 3d ed., pp. 297–306, 1973.

12 Bush, L., and R. C. Buckner. Tall fescue toxicity. In *Antiquality Components of Forages,* Crop Sci. Soc. Amer., pp. 99–112, 1973.

13 Bush, L., J. A. Boling, G. Allen, and R. C. Buckner. Inhibitory effects of perloline to rumen fermentation *in vitro. Crop Sci.* 12:277–279, 1972.

14 Cowan, J. R. Tall fescue. *Advan. Agron.* 8:283–320, 1956.

15 Crowder, L. V. A simple method for distinguishing tall and meadow fescue. *Agron. J.* 45:453–454, 1953.

16 Daughtry, C.S.T., D. A. Holt., and V. L. Lechtenberg. Concentration, composition, and *in vitro* disappearance of hemicellulose in tall fescue and orchardgrass. *Agron. J.* 70:550–554, 1978.

17 Dobson, J. W., E. R. Beaty, and C. D. Fisher. Tall fescue yield, tillering, and invaders as related to management. *Agron. J.* 70:662–666, 1978.

18 Fribourg, H. A., and R. W. Loveland. Production, digestibility, and perloline content of fescue stockpiled and harvested at different seasons. *Agron. J.* 70:745–747, 1978.

19 Fribourg, H. A., and R. W. Loveland. Seasonal production, perloline content, and quality of fescue after N fertilization. *Agron. J.* 70:741–745, 1978.

20 Fuller, W. W., W. C. Elder, B. B. Tucker, and W. McMurphy. Tall fescue in Oklahoma—a review. *Okla. Agr. Exp. Sta. Progress Rep., P-650,* 1971.

21 Gentry, C. F., R. A. Chapman, L. Henson, and R. C. Buckner. Factors affecting the alkaloid content of tall fescue (*Festuca arundinacea* Schreb.). *Agron. J.* 71:313–316, 1969.

22 Hart, R. H., G. E. Carlson, and H. J. Retzer. Establishment of tall fescue and white clover: effects of seeding methods and weather. *Agron. J.* 60:385–388, 1968.

23 Jacobson, D. R., and R. H. Hatton. The fescue toxicity syndrome. *Proc. Fescue Toxicity Conf.,* Lexington, KY, pp. 4–5, 1973.

24 Jung, G. A., J. A. Balasko, F. L. Alt, and L. P. Stevens. Persistence and yield of ten grasses in response to clipping frequency and applied nitrogen in the Allegheny Highlands. *Agron. J.* 66:517–521, 1974.

25 Jung, G. A. and R. E. Kocker. Influence of applied nitrogen and clipping treatments on winter survival of perennial cool-season grasses. *Agron. J.* 66:62–65, 1974.

26 Leasure, J. K. Growth pattern of mixtures of orchardgrass and tall fescue with ladino clover in relation to temperature. *Proc. Assoc. Southern Agr. Workers* 49:177–178, 1952.

27 Lopez, R. R., A. G. Matches, and J. D. Baldridge. Vegetative development and organic reserves of tall fescue under conditions of accumulated growth. *Crop Sci.* 7:409–712, 1967.

27 a Luu, K. T., A. G. Matches, and E. J. Peters. Allelopathic effects of tall fescue on birdsfoot trefoil as influenced by N fertilization and seasonal changes. *Agron. J.* 74:805–808, 1982.

28 Matches, A. G. Influence of cutting height in darkness on measurement of energy reserves of tall fescue. *Agron. J.* 61:896–898, 1969.

28 a Moser, L. E., J. J. Volenec, and C. J. Nelson. Respiration, carbohydrate content, and leaf growth of tall fescue. *Crop Sci.* 22:781–786, 1982.

29 Mott, G. O., C. J. Kaiser, R. C. Peterson, R. Peterson Jr., and C. L. Rkykerd. Supplemental feeding of steers on *Festuca arundinacea* Schreb. pastures fertilized at three levels of nitrogen. *Agron. J.* 63:751–754, 1971.

30 Ocumpaugh, W. R., and A. G. Matches. Autumn-winter yield and quality of tall fescue. *Agron. J.* 69:639–643, 1977.

31 Parks, O. C. Jr., and P. R. Henderlong. Germination and seedling growth rate of ten common turfgrasses. *Proc. West Vir. Acad. Sci.* 39:132–140, 1967.

32 Petritz, D. C., V. L. Lechtenberg, and W. H. Smith. Performance and economic returns of beef cows and calves grazing grass-legumes herbage. *Agron. J.* 72:581–584, 1980.

32 a Reynolds, J. H., and W. H. Wall III. Concentrations of Mg, Ca, P, K, and crude protein in fertilized tall fescue. *Agron. J.* 74:950–954, 1982.

33 Ryle, G.J.A. A comparison of leaf and tiller growth in seven perennial grasses as influenced by nitrogen and temperature. *J. Brit. Grassl. Soc.* 19:281–290, 1964.

34 Smith, A. E. Influence of temperature on tall fescue forage quality and culm base carbohydrates. *Agron. J.* 69:745–747, 1977.

35 Smith, D., A.V.A. Jacques, and J. A. Balosko. Persistence of several temperature grasses grown with alfalfa and harvested two, three, or four times annually at two stubble heights. *Crop Sci.* 13:553–556, 1973.

35 a Stritzke, J. F. and W. E. McMurphy. Shade and N effects on tall fescue production and quality. *Agron. J.* 74:5–8, 1982.

36 Templeton, W. C. Jr. and T. H. Taylor. Some effects of nitrogen, phosphorus, and potassium fertilization on botanical composition of a tall fescue–white clover sward. *Agron. J.* 58:569–572, 1966.

37 Van Keuren, R. W. Summer pasture for beef cows. *Ohio Rep.* 53(3):43–45, 1970.

38 Van Keuren, R. W. Symposium on pasture methods for maximum production in beef cattle: pasture methods for maximizing beef cattle production in Ohio. *J. Anim. Sci.* 30:138–142, 1970.

39 Vartha, E. E., A. G. Matches, and G. B. Thompson. Yield and quality trends of tall fescue grazed with different subdivisions of pasture. *Agron. J.* 69:1027–1029, 1977.

40 Wedin, W. F. Fertilization of cool-season grasses. In *Forage Fertilization,* Amer. Soc. Agron. pp. 95–118, 1974.

41 Yates, S. G. Toxin-producing fungi from fescue pasture. In *Microbial Toxins,* Academic Press, New York, 8:191–206, 1971.

42 Yates, S. G., H. L. Tookey, J. J. Ellis, and H. J . Burkhardt. Mycotoxins produced by *Fusarium nivale* isolated from tall fescue (*Festuca arundinacea* Schreb.). *Phytochemistry* 7:139–146, 1968.

43 Zarrough, K. M., C. J. Nelson, and J. H. Couetts. Relationship between tillering and forage yield of tall fescue. I. Yield. *Crop Sci.* 23:333–337, 1983.

44 Zarrough, K. M., C. J. Nelson, and J. H. Couetts. Relationship between tillering and forage yield of tall fescue. II. Pattern of tillering. *Crop Sci.* 23:338–342, 1983.

CHAPTER 28

BLUEGRASSES

Kentucky bluegrass, *Poa Pratensis* L., is a long-lived, cool-season perennial grass, grown in the northeastern quarter of the United States. It is used primarily for lawns, parks, cemeteries, turf, and golf courses. Many of the old permanent pastures in the northeastern United States contain a high percentage of Kentucky bluegrass. It is probably one of the oldest grass species known, thought to have originated many millions of years ago (10). In the northwestern United States, it is considered a weedy species when it invades cultivated pastures or croplands (15). Within the genus *Poa* there are more than 200 species, approximately 69 of which are grown in the United States (15). Many of the *Poa* species are native to North America. Kentucky bluegrass may volunteer and eventually take over old permanent pastures in the northeastern United States.

HISTORICAL BACKGROUND

It is generally thought that Kentucky bluegrass originated in Eurasia, although some botanists consider it native to both North America and Eurasia (10, 14). It is difficult to determine how Kentucky bluegrass actually migrated from the northeastern United States, where it was introduced by early colonists, to its present area of occupation. Animals probably carried and distributed the seed throughout the upper midwest. When the early settlers and traders migrated into the Ohio Valley area and the Great Lakes region in the 1800s, Kentucky bluegrass was already growing abundantly.

Kentucky bluegrass was introduced along the eastern coast by the early colonists soon after 1600. Early settlers, traders, and missionaries also introduced Kentucky bluegrass into the Great Lakes region before the 1700s (10, 14).

There is considerable evidence that, by 1775, Kentucky bluegrass was well established, especially in the state of Kentucky (10). It was found in the open areas and near watering grounds where buffalo and other wild animals normally fed or obtained their water. Bluegrass was used as an indicator plant by some of the earlier settlers to locate very productive soils (10).

The modifier "Kentucky" was appended to bluegrass between 1833 and 1859 when the bluegrass was found growing abundantly in Kentucky (10). Prior to that, in 1750, the *Poa* species was simply called "bluegrass." Kentucky bluegrass has a green color to its leaves, but during anthesis there is a very dark bluish cast in the field. One of the earlier *Poa* species found was Canada bluegrass, *P. compressa* L., which has a bluish-green cast to its leaves; therefore, the species got the name "bluegrass" (15).

Kentucky bluegrass seed was readily available soon after 1860 and was sent from Kentucky to many other states in the nation. From then on it was called Kentucky bluegrass, *Poa pratensis*.

WHERE KENTUCKY BLUEGRASS GROWS

Kentucky bluegrass is adapted to well-drained soils with a pH between 6.5 to 7.0, and to climates which are cool and relatively humid (23). It has a rather shallow root system developing from rhizomes and forms a very dense sod that may become sod-bound, similar to sods of other cool-season grasses. Bluegrass is not very shade-tolerant and does best in bright, sunny areas. It will not tolerate poor soil conditions or droughty conditions (23). It grows early in the spring and late into the fall, but often becomes dormant during the hot, dry summer months. It turns brown when moisture is deficient and the temperature high. It does respond to irrigation or frequent watering, especially in turf areas where a dense sod is required, such as golf courses or playground areas. Kentucky bluegrass is grown from southern Canada, west to the eastern portion of the Dakotas, south throughout Kentucky, and northeast to Washington, D.C. Kentucky bluegrass is very popular and found in old permanent pastures where the pH is sufficiently high for continual growth. Those soils, which were derived from limestone, are the best sites for Kentucky bluegrass (1).

TYPES OF BLUEGRASS

The main bluegrass forage type grown in the United States is Kentucky bluegrass, *P. pratensis*. It has greater forage growth and is much more productive than the other species. Growth is erect with unbranched culms usually 0.3 to 0.75 m (12 to 30 in) in height. Kentucky bluegrass matures earlier than Canada bluegrass (8).

Canada bluegrass, *P. compressa* L., is much lower-growing and has flat stems and short leaves. It produces rhizomes, but is slightly slower in establishment. Its adaptation is in areas with poor drainage and lower general fertility. It is often considered a weed in the upper midwest (10, 15).

Another bluegrass, called roughstalk bluegrass, *P. trivialis* L., often grows in the same area and under similar conditions as Kentucky bluegrass. It is more shade-tolerant and will withstand wetter soils than Kentucky bluegrass (10, 15).

Texas bluegrass, *P. arachnifera* Torr., is adapted from Kansas to Arkansas, south to Texas, and east to Florida. It is sometimes considered a winter annual in these areas (10, 15).

Another bluegrass native to the northern United States is mutton bluegrass, *P. fendleriana* Vasey. This species grows in northern Michigan westward to the Rocky Mountain areas. It is much more erect in growth habit and is also a bunch-forming grass (10, 15).

Another type of bluegrass native is big bluegrass, *P. ampla* Merr. It too is a perennial, bunch-forming grass grown primarily in the eastern United States. Nevada bluegrass, *P. nevadensis* Pasey, ex Scribn., is closely related to big bluegrass and is commonly grown in the wetter intermountain regions in the west (10, 15).

Bulbous bluegrass, *P. bulbosa* L., is grown predominately in the southern half of the United States, south of Maryland and North Carolina and west to the Pacific coast. It is a short-lived grass that grows during the moist or high rainfall season. As indicated by its scientific name, it does produce bulblets that develop quickly and germinate very rapidly (15).

An annual bluegrass, *P. annua* L., is grown throughout the United States and is winter-hardy in the southern region. It is considered a weed in lawns (15).

HOW BLUEGRASS GROWS

Kentucky bluegrass is a long-lived, cool-season, sod-forming perennial grass. Its growth habit is erect, reaching heights of 0.3 to 0.75 m (1 to 2.5 ft) with an open panicle inflorescence (Fig. 28-1). The leaves of Kentucky bluegrass are smooth, range from green to dark green, are about 3 mm (0.12 in) wide, and reach lengths of 10 to 30 cm (4 to 10 in) (8). A distinguishing characteristic of Kentucky bluegrass is a keel- or boat-shaped tip at the end of the leaf.

Young seedlings produce two different types of buds—one that tillers, or has aboveground growth, and another that produces rhizomes, or below-ground stems (8, 9). The aboveground tiller buds generally develop culms and inflorescences, although a small number of them will only produce leaves. Rhizomes are produced from the lower nodes of the new tiller growth. The older rhizomes also produce new rhizomes at the nodal areas (9).

Early work in Ohio showed that Kentucky bluegrass has a very specific cyclic growth throughout the year (9). It has a vegetative growth that is controlled by photoperiod and temperature and, to a certain extent, soil moisture, nutrients, and the management of the crop (8, 9).

During the fall months, when days are short and temperatures cool, the inflorescence is initiated (8). The inflorescence begins to develop in the fall, and emerges when the days become longer, usually in early March. A combination of short days and low temperatures allows tillers to be formed during the fall

FIGURE 28-1
Close-up of a bluegrass open panicle
inflorescence. *(Courtesy University of
Illinois)*

months, while with a combination of proper temperature and moisture, the aboveground tillers begin growth in early spring. The tiller buds that produce only leaves are initiated in spring and early summer. Most of the rhizomes are initiated during the summer and fall months (8).

New shoots or tiller growth occurs at the rhizome apices and from the leaf axis of old shoots or tillers. New tillers develop from midautumn until early spring, as the day lengths then are relatively short. The branches that develop from midspring to midautumn, when the day lengths are long, become rhizomes. Rhizomes develop from the axis of the leaves of the aboveground shoots; from this position they grow downward and penetrate the soil. New rhizomes develop from the axis of the scale leaves at the nodal areas on the old rhizomes (6, 10). When the day lengths shorten during the fall season, most apices turn upward and develop into aboveground tillers (6, 8).

Kentucky bluegrass produces a very shallow root system and is very sensitive to low moisture levels and high temperatures (6, 8). The root system develops from the nodes of the rhizomes and from the basal leaf-bearing nodes of the tillers (6, 10).

Temperature has a great effect on the development of Kentucky bluegrass. It has been shown that very little top growth is produced in the early spring when the soil temperature is below 10°C (50°F) at the 1.25 cm (½ in) depth (23). Tiller growth will reach its maximum when the average soil temperature is 15.5 to 19°C (60 to 64°F). When the summer soil temperatures reach 26.6°C (80°F), very little production occurs (6, 23). The optimum air temperature for Kentucky bluegrass is between 15.5 and 32.2°C (60 to 90°F) (18, 23). Very little growth occurs at 4.4°C (40°F). Similarly, when air temperatures reach 32°C (90°F) or higher, very little growth occurs. The best temperature for root and rhizome growth is 15.5°C (60°F) (6, 10, 23). Therefore, very little growth occurs in the heat of the summer. Irrigation and nitrogen fertilization will overcome this lack of growth.

Kentucky bluegrass is slower in establishment than most other cool-season grasses. Within one year, it does, however, establish a solid sod. The food reserves are stored in the rhizome area in Kentucky bluegrass. The soluble carbohydrates are synthesized very rapidly during the cool period of early spring. Later in the spring and in early summer, there is a net loss of soluble carbohydrates from the root system. This loss continues through the latter part of the summer. With the resumption of fall growth, shorter days, and lower temperatures, carbohydrates are rapidly accumulated. The most rapid translocation of soluble carbohydrates into the rhizomes occurs in mid-October, when top growth has ceased (6, 23). If a drought occurs during the summer months, the bluegrass has the ability to go dormant, thereby not utilizing any further soluble carbohydrates or food reserves (23).

Kentucky bluegrass is usually grown with a low-growing legume species, such as white clover or ladino clover, when used as a pasture. Since many of the leaves are produced very close to the soil surface, bluegrass is less sensitive to heavy grazing or frequent harvest than an upright-growing, cool-season grass like orchardgrass or smooth bromegrass (10, 23). Therefore, a low-growing legume species must be grown with it—one that can be frequently harvested, such as ladino clover. To maintain a white clover in a Kentucky bluegrass sod, frequent harvests need to be made. If Kentucky bluegrass growth is allowed to accumulate and high levels of nitrogen fertilization are applied, the resulting dry matter production will crowd out the white clover, in an effect similar to undergrazing (11). The excessive top growth tends to deplete the stand, and weed content increases as the sod becomes thinner (13). Consequently, when a tall, dense stand is removed, the most active buds are weakened because the shading effect of the tall growth has been removed. The young tillers are then subject to greater injury from exposure to strong sunlight (16, 18). In such a case, the new growth must come from the more dormant buds on the older portion of the rhizomes. Recovery from this type of harvest is slow, and weeds will soon become established in such a sod.

Many researchers have studied the nutritive value and productivity of Kentucky bluegrass (10, 23). It is generally concluded that Kentucky bluegrass will maintain its nutritive value longer than will most other cool-season grasses, such as orchardgrass, smooth bromegrass, reed canarygrass, and tall fescue (3, 20). Comparing the overall animal consumption of bluegrass with animal intake of other cool-season grasses, most research, using cows and calves, yearling steers, or dairy animals as the test animals, indicates that orchardgrass and smooth bromegrass will outyield Kentucky bluegrass (3, 5, 22, 26). Comparing Kentucky bluegrass to tall fescue, however, the overall Kentucky bluegrass

TABLE 28-1
NUTRITIVE VALUE DERIVED FROM GRAZING KENTUCKY BLUEGRASS AS COMPARED WITH OTHER COOL-SEASON GRASSES

Research institution	Test animal	Kentucky bluegrass	Orchardgrass	Smooth bromegrass	Tall fescue
Pennsylvania	cow/calf	100%	105%	—	—
Virginia	yearling steers	100	105	—	99%
Missouri	steers	100	98	—	—
Kentucky	dairy cows	100	90	105%	80

Source: R. E. Blaser, H. T. Bryant, R. C. Hammes Jr., R. L. Bowan, J. P. Fontenot, and C. E. Polan, "Managing forages for animal production," *Va. Polytechnic Inst. Res. Div. Bull. 45,* 1969; E. M. Brown, "Improving Missouri pastures," *Mo. Agr. Exp. Sta. Bull. 768,* 1961; D. M. Seath, C. A. Lessiter, J. W. Rust, M. Cole, and C. M. Bastin, "Comparative value of Kentucky bluegrass, Kentucky 31 fescue, orchardgrass, and bromegrass as pastures for milk cows," I., "How kind of grass affected persistency of milk production, TDN yield, and body weight," *J. Dairy Sci.,* 39:574–580, 1956; and J. B. Washko, and L. I. Wilson, "Progress report: productivity of pastures utilized by crossbred Angus-Holstein cows and calves, 1968–1970," *Pa. Anim. Sci. Res. Sum.,* 2:20–26, 1971.

production will often be greater than that of tall fescue (Table 28-1). These data refer to grazing trials and not dry hay forages (3, 22).

USES OF BLUEGRASS

Kentucky bluegrass is used mostly for pasture or turf. It is one of the main species in the old permanent pastures in the upper midwest and northeastern United States (23). There are several reasons Kentucky bluegrass is being replaced by more productive cool-season grasses in many permanent pastures: (1) Kentucky bluegrass usually goes dormant during the heat of the summer; (2) it may become sod-bound; and (3) very high fertility is required to maintain high productivity.

Kentucky bluegrass is one of the more popular lawn grasses grown in the United States. Seed is plentiful and easy to obtain. At present, Kentucky bluegrass usually is not seeded in pasture renovation mixtures. It does naturally volunteer or come back in most of the old pastures via old rhizomes or natural invasion.

DESIRABLE CHARACTERISTICS OF BLUEGRASS

Kentucky bluegrass is probably the oldest native cool-season grass growing in the United States. It naturally occurs in many old permanent pastures that maintain a high level of fertility. It has the ability to form a very dense shallow root system, probably the most shallow root system in any cool-season grass grown in the United States (8, 15). Kentucky bluegrass is a rather aggressive grass and will readily form a sod. Therefore, good management is required to maintain a certain percentage of legumes in the sward. It is not a very shade-tolerant species. It can be easily crowded out when planted with other tall-growing species, under orchard conditions, or under shade trees in lawn areas (23). The quality and the nutritive value of Kentucky bluegrass remain high through seed set; this nutritive value is among the highest of cool-season grasses (10). Kentucky bluegrass responds to irrigation. When temperatures are cool, it exhibits rapid regrowth, but during the high temperatures of the late summer, it has the ability to go dormant and may not grow at all. It is a very winter-hardy species and the supply of seed is plentiful and cheap (23).

MANAGEMENT OF KENTUCKY BLUEGRASS

Establishment

Kentucky bluegrass is slow in establishment. It takes approximately 1 year for it to develop a complete sod (23). It is very important to have a firm, moist seedbed when seeding bluegrass. Seed should not be seeded any deeper than 0.3 to 0.6 cm (⅛ to ¼ in). It is extremely important to use some type of field roller or cultipacker when broadcasting the seed to obtain good, close contact of seed to soil. This will encourage quick establishment.

Seeding

Bluegrass is generally established as a monoculture because of its lack of shade tolerance. It is generally seeded during a cool, moist season and on highly fertile, well-drained soils with a pH of 6.5 to 7.0 (23). In the midwest, many growers will seed Kentucky bluegrass in the late summer to early fall because the rains and cool temperatures aid in a good establishment. Kentucky bluegrass can be seeded in the early spring but it is then especially important to have a good soil-to-seed contact. If a drought occurs or there is little rainfall, a spring seeding can be lost toward the latter part of the summer because of its shallow root system. The seeding rate for Kentucky bluegrass varies from 6 to 12 kg/ha (5 to 10 lb/A) (10).

Harvest or Grazing Schedule

Since a high percentage of Kentucky bluegrass is grown under pasture conditions, pasture management is important to raising a very productive stand along with a companion legume. Fertility is probably the greatest factor in maintaining the legume with Kentucky bluegrass. Both require a high soil pH and high P and K fertility (23). When bluegrass is grown with a legume, little or no nitrogen should be applied. If excessive nitrogen is applied, the legumes generally disappear from the stand. Many old permanent pastures are being renovated with legumes in order to increase the carrying capacity and production per acre in relation to either beef or milk. When an old Kentucky bluegrass permanent pasture is renovated simply by adding a legume to the stand, the overall production usually increases by more than 100% (10).

The grazing schedule of Kentucky bluegrass is important in maintaining a productive stand that will exhibit rapid growth. If the grass is allowed to grow tall or rank, the shading effect will result and regrowth of the dormant buds will be very slow (10, 23). Consequently, it is suggested that the height of the stubble be kept to approximately 5 to 15 cm (2 to 6 in). (10). One can maintain this height by proper grazing pressure, clipping, or both grazing and clipping. If the stand is maintained at a lower height than the suggested minimum, the root reserves will become depleted, the sod will become more sparse, and weeds will invade (10). If allowed to grow taller than the suggested maximum, regrowth is greatly reduced (8).

Fertility

Kentucky bluegrass responds to nitrogen fertilization, as do other cool-season grasses, but probably not as readily as tall fescue or reed canarygrass (23). With the exception of smooth bromegrass, Kentucky bluegrass probably requires both the highest pH and soil fertility of any cool-season grass grown in the United States. It is very important to bring the soil pH up to 6.5 prior to seeding to ensure good establishment and maintenance of a good stand. Once the stand is established, recommended rates for nitrogen fertilizer range from 65 to 90 kg/ha (60 to 80 lb/A) (23, 25). In economic terms, under most grazing conditions Kentucky bluegrass is not as productive (in a monoculture) as other cool-season grasses (19, 20, 23, 25). It does require sufficient amounts of moisture or irrigation to maintain high yields (10, 21).

Kentucky bluegrass requires high levels of P and K. Phosphorus is needed for good root development and early spring growth, while potassium will ensure healthy plant tissue and winterhardiness (23). Occasionally high rates of nitrogen fertilizer may decrease the uptake of K. Phosphorus and potassium fertilizers should be applied annually, as determined by the amount removed from the soil by grazing or harvesting (19, 20, 23, 25).

Food Reserves

The soluble carbohydrates in Kentucky bluegrass are stored in the rhizome system and in the lower basal tillers (6). Food reserves go into storage in the rhizomes most rapidly in October, especially after the aboveground tillers have ceased growth (6, 23). To ensure good winterhardinees or overwintering of Kentucky bluegrass, it is important to graze it rather lightly about 30 to 35 days prior to the killing frost or fall dormancy; light grazing will spur abundant translocation of soluble carbohydrates. Due to its regrowth capabilities, Kentucky bluegrass can be grazed a little closer to the soil surface than many of the other cool-season grasses. The proportion of winterhardiness as derived from soluble stored food reserves equals that of smooth bromegrass (23, 27).

SEED PRODUCTION

A high percentage of Kentucky bluegrass seed is produced and harvested in Kentucky (10). In recent years seed production has been initiated in southeastern Kansas, Missouri, Minnesota, and northward into Canada (10, 16). In the northwest, Kentucky bluegrass seed is produced in Idaho, Washington, and Oregon (10). Much of the Minnesota and western seed production is of turf cultivars.

Kentucky bluegrass seed production has two major problems: (1) there is a cottony filament at the base of each seed which, until its removal, makes it rather difficult to flow in a seeder; (2) growth is left in the field after the seed is harvested. If excessive plant growth is left in the field, seed yields will be greatly reduced the following year (10).

It is important to have sufficient amounts of nitrogen to ensure good seed yields. Nitrogen fertilizer is generally applied in the late fall or very early spring to achieve good plant growth with very productive seedheads. Presently most Kentucky bluegrass seed is harvested by means of combines. The crop may be either swathed and windrowed and then combined, or combined in a standing position. The combines actually remove the seed using rubbing devices, including rotating drums inside the combines that completely remove the seed from the panicle (10). Excessive trash or plant growth must be removed from newly harvested seed so that the seed does not heat up when it is being stored or when waiting to be cleaned. The excessive heat will destroy the viability of the seed (10).

To ensure good seed production year after year, the surplus or growth remaining after a harvest should be removed mechanically or by grazing. In the past, some of the seed fields have been burned in the late summer or early fall (10). This practice is not being followed as often lately because of environmental regulations. One of the main advantages of complete removal by burning is that it helps control the various diseases and destructive insects that may build up in the stand.

DISEASES

There are numerous diseases of Kentucky bluegrass, but none is serious enough to greatly reduce the production of forage or seed (10). In the far northwest, leaf and stem rust and powdery mildew may affect seed production (10). These are often controlled by stubble burning or by use of fungicides.

INSECTS

There are several insects that may cause some damage to Kentucky bluegrass. Probably the most destructive insects to Kentucky bluegrass are white grubs. The adults are the May beetle, *Phyllophaga* species, and the June beetle, *Continis nitida* L. (10, 24). Both of these insects feed upon the root and rhizomes and may kill large areas within the field or in lawns. Control of these insects includes both pasture renovation and inclusion of legumes such as red clover or alfalfa in the planting (10, 23). In lawns, insecticides are used.

Sod web worms, *Crambeus* spp., may also cause damage to Kentucky bluegrass pastures. These insects will consume the top growth during a dry season (23). They are probably more destructive in lawns, but are more easily controlled in pastures.

There are a few insects that will lower seed yields of Kentucky bluegrass. Insects that fall into this category are the meadow plant bug, *Leptopterna dolobarta* L., and the *Amblytylus nasutus* Krisch., which both feed on the developing flowers and prevent a good seed set (10, 24). In the far northwest there also are a few insects that will lower the seed yield. Most of these insects are controlled by insecticides or close removal of the stubble to reduce most of the egg-laying sites and overwintering habitat (10).

SUMMARY

The advantages and disadvantages of Kentucky bluegrass are as follows:

Advantages:

1 It is long-lived, perennial, dense, sod-forming grass.

2 It is a naturally occurring cool-season grass that volunteers in permanent pastures.

3 It responds to nitrogen fertilization but not as readily as some of the other cool-season grasses such as orchardgrass, smooth bromegrass, tall fescue, or reed canarygrass.

4 It produces high-quality forage in the early spring and the late fall.

5 Its nutritive value is rather high until seed set, as compared with other grasses.

6 It will withstand rather close grazing or continuous grazing longer than most of the cool-season grasses.

7 It possesses sufficient winterhardiness for the upper midwest.

Disadvantages:

1 It requires high pH and highly fertile soils for its dry matter production.

2 It requires a high rainfall.

3 It is not shade-tolerant.

4 Its production is exceeded by the yields of many of the cool-season grasses.

5 Under low nitrogen fertility it may become sod-bound.

6 It becomes dormant during high summer temperatures.

7 It is slower in establishment than timothy or orchardgrass.

QUESTIONS

1 How do the bluegrasses get their name?

2 Distinguish between Canada bluegrass and Kentucky bluegrass.

3 Describe the life cycle of Kentucky bluegrass with regard to the photoperiod and temperature requirements.
4 What are the soil requirements for Kentucky bluegrass?
5 Describe the characteristics of Kentucky bluegrass that make it an acceptable forage species.
6 Describe the management of pasture situations.
7 Why is Kentucky bluegrass used more often under pasture situations than as a hay system?

REFERENCES

1 Ahlgren, G. H. *Forage Crops.* McGraw-Hill, New York, 2d ed., pp. 220–230, 1956.
2 Ahlgren, H. L. Effect of fertilization, cutting treatments, and irrigation on yield of forage and chemical composition of the rhizomes of Kentucky bluegrass (*Poa Pratensis* L.), *J. Amer. Soc. Agron.* 30:683–691, 1938.
3 Blaser, R. E., H. T. Bryant, R. C. Hammes Jr., R. L. Bowan, J. P. Fontenot, and C. E. Polan. Managing forages for animal production. *Vir. Polytechnic Inst. Res. Div. Bull. 45,* 1969.
4 Brown, B. A., and R. I. Munsell. An evaluation of Kentucky bluegrsss. *J. Amer. Soc. Agron.* 37:259–267, 1945.
5 Brown, E. M. Improving Missouri pastures. *Mo. Agr. Exp. Sta. Bull. 768,* 1961.
6 Brown, E. M. Seasonal variation in the growth and chemical composition of Kentucky bluegrass. *Mo. Agr. Exp. Sta. Res. Bull. 360,* 1943.
7 Darrow, R. A. Effects of soil temperature, pH, and nitrogen nutrition on the development of *Poa pratensis. Bot. Gaz.* 101:109–127, 1939.
8 Etter, A. G. How Kentucky bluegrass grows. *Ann. Mo. Bot. Garden* 38:293–375, 1951.
9 Evans, M. W. Kentucky bluegrass. *Ohio Agr. Exp. Sta. Res. Bull. 681,* 1949.
10 Fergus, E. N., and R. C. Buckner. The bluegrasses and redtop. In *Forages,* Iowa State Univ. Press, Ames, IA 3d ed., pp. 243–253, 1973.
11 Graber, L. F. Food reserves in relation to other factors limiting the growth of grasses. *Plant Physiol.* 6:43–72, 1931.
12 Hanson, A. A. Grass varieties in the United States. *USDA Agr. Handbook 170,* rev. 1972.
13 Harrison, C. M. Responses of Kentucky bluegrass to variations in temperature, light, cutting, and fertilizing. *Plant Physiol.* 9:83–106, 1934.
14 Hartley, W. Studies on the origin, evolution, and distribution of the Gramineae. IV. The genus *Poa* L. *Aust. J. Bot.* 9:152–161, 1961.
14 a Hickey, V. G., and R. D. Ensign. Kentucky bluegrass seed production characteristics as affected by residue management. *Agron. J.* 75:107–110, 1983.
15 Hitchcock, A. S. Manual of grasses of the United States. *USDA Misc. Publ. 200,* rev. 1951.
16 Lindsey, K. E., and M. L. Peterson. Floral initiation and development in *Poa pratensis* L. *Crop Sci.* 4:540–544, 1964.
17 Lindsey, K. E., and M. L. Peterson. High temperature suppression of flowering in *Poa pratensis* L. *Crop Sci.* 2:71–74, 1962.
18 Peterson, M. L., and W. E. Loomis. Effect of photoperiod and temperature in growth and flowering of Kentucky bluegrass. *Plant Physiol.* 24:31–43, 1949.

19 Rehm, G. W., R. C. Sorensen, and W. J. Moline. Time and rate of fertilizer application for seeded warm-season and bluegrass pastures. I. Yield and botanical composition. *Agron. J.* 68:759–764, 1976.

20 Rehm, G. W., R. C. Sorensen, and W. J. Moline. Time and rate of fertilizer application on seeded warm-season and bluegrass pastures. II. Quality and nutrient content. *Agron. J.* 69:955–961, 1977.

21 Robinson, R. R., and V. C. Sprague. The clover populations and yield of a Kentucky bluegrass sod as affected by nitrogen fertilization, clipping treatments, and irrigation. *J. Amer. Soc. Agron.* 39:107–116, 1947.

22 Seath, D. M., C. A. Lessiter, J. W. Rust, M. Cole, and C. M. Bastin. Comparative value of Kentucky bluegrass, Kentucky 31 fescue, orchardgrass, and bromegrass as pastures for milk cows. I. How kind of grass affected persistency of milk production, TDN yield, and body weight. *J. Dairy Sci.,* 39:574–580, 1956.

23 Smith, D. Kentucky Bluegrass. In *Forage Management in the North,* Kendall-Hunt, Dubuque, IA, 3d ed., pp. 159–165, 1975.

24 Spencer, J. T., H. H. Jewett, and E. N. Fergus, Seed production of Kentucky bluegrass as influenced by insects, fertilizers and sod management. *Ky. Agr. Exp. Sta. Bull. 535,* 1949.

25 Sund, J. M., E. L. Nielson, and H. L. Ahlgren. Response of Kentucky bluegrass strains to fertility and management. *Wis. Agr. Exp. Sta. Res. Rep. 20,* 1965.

26 Washko, J. B., and L. I. Wilson. Progress report: productivity of pastures utilized by crossbred Angus-Holstein cows and calves, 1968–1970. *Pa. Anim. Sci. Res. Sum.* 2:20–26, 1971.

27 Youngner, V. B., and F. J. Nudge. Growth and carbohydrate storage of three *Poa pratensis* strains as influenced by temperature. *Crop Sci.* 8:455–457, 1968.

CHAPTER 29

OTHER COOL-SEASON GRASSES

REDTOP

Redtop, *Agrostis alba* L., is another cool-season grass that is native to Europe. It is grown in the cooler northeastern portion of the United States. Redtop is generally grown in the same area as Kentucky bluegrass (8). It is well adapted to very acid or poorly drained soils, and to clay-type soils that are low in fertility (15). Redtop is a sod-forming grass which rapidly develops a good turf (15). It does produce a good supply of seed that remains viable for approximately 2 years.

Redtop is a perennial grass that is predominantly upright in growth, although it also forms creeping stems (15) (Fig. 29-1). On very fertile soils, it is highly productive and often reaches 1 m (3 ft) in height. The leaves are flat, with sharp, pointed ends. Redtop possesses ligules that are approximately 0.6 cm (0.2 in) long and pointed. Redtop gets it name from the reddish color of the panicles (15).

Redtop is often grown in mixtures with alsike clover or timothy. It is used for pasture and in short rotations when grown on cultivated land. It is less palatable than most of the cool-season grasses, including Kentucky bluegrass. It matures in approximately the same time as timothy, and persists longer in stands (8).

Redtop should be harvested in the early flowering stage (8). When grown in pastures, it will withstand considerable amounts of traffic, and it is a very valuable pasture crop, especially in wetland areas and on poor soils. Redtop is a good grass to consider in erosion control, particularly at or on the edges of running streams. Since it does develop rather rapidly, it is considered useful in the stabilization of gullies or waterways, or in areas where reclamation is needed (15). It becomes stemmy and dies under close clipping or harvest (15). It is not recommended as a lawn grass.

FIGURE 29-1
Redtop flowering plant. *(Courtesy University of Illinois)*

Tall fescue is now replacing redtop in most mixtures for forage purposes (8). Redtop is seldom sown alone, and is usually seeded in mixtures at the rate of 3.3 to 6.6 kg/ha (3 to 6 lb/A). It does respond to lime, N, P, and K fertilization.

Prior to 1945, most redtop seed production was in south central and southeastern Illinois. The declining demand for redtop means that seed production is very low (8). Redtop seed shatters readily (15).

RYEGRASSES

Two very important ryegrass species are annual ryegrass, *Lolium multiflorum* Lam., and perennial ryegrass, *L. perenne* L. Most of the hectarage of ryegrasses is in the south and southeastern United States (25). This region encompasses essentially the eastern one-third of Texas, north to the middle of Missouri, and from the southern tip of Illinois east to Washington, D.C. Another area where ryegrass is extensively produced is the three western states, from central California into the mountain region and northward into Canada (10).

The ryegrasses are adapted to a wide range of soil types and climatic conditions. Even though they are best adapted to well-drained soils, ryegrasses will grow on soils too wet for small grains and will withstand short periods of flooding (15). And, although they will grow on well-fertilized sandy soils, they are best adapted to heavier clay soils or silt soils that have adequate surface drainage (10). Ryegrass is less winter-hardy than many other cool-season grasses, especially timothy or orchardgrass (10). Annual ryegrass may be the most underrated winter pasture grass grown in the south.

Annual ryegrass has exceptionally strong seedling vigor. It provides a very quick and effective ground cover for soil conservation, grass waterways, or flood canals (10). It is used as a winter cover crop where erosion control on cropland is needed, or where the land is subject to overflow (15). When annual ryegrass is seeded in the early fall and fertilized with nitrogen, it very quickly develops a soil-stabilizing ground cover. It is often grown with red clover or alsike clover in the north, and also, in short rotations, as a green manure crop. In the south, annual ryegrass is usually grown with crimson or arrowleaf clover, often in a three-way mixture with these two clovers. In the north, annual ryegrass is usually not sown with a perennial grass or legume if grown for forage because it may be too aggressive and may actually reduce a stand of the slower, more desirable components of the mixture.

Perennial ryegrass has a slightly slower seedling establishment and less seedling vigor than annual ryegrass (10). Perennial ryegrass has the ability to go dormant during the summer months or when the soil moisture is low (15). Sometimes perennial ryegrass is added to long-lived forage mixtures, but it should never represent any more than 25% of the stand. If the perennial ryegrass exceeds this rate, due to its aggressiveness it will reduce the more desirable long-lived legume or grass (5).

Seed of both the annual and perennial ryegrass is very plentiful. Most of the seed production is in the far northwest (29).

Both annual and perennial ryegrasses are considered bunch-forming grasses. Perennial ryegrass is erect in growth, reaching heights of approximately 90 cm (36 in) (Fig. 29-2). The leaf blades of perennial ryegrass are folded in the young shoots and are glabrous. The inflorescence is a spike-type growth. Perennial ryegrass seed generally does not possess any awns (15) (Fig. 29-3).

Occasional plants of annual ryegrass will persist as short-lived perennials in the south. It too is erect in growth habit, and exceeds the height of perennial ryegrass, as it reaches heights of about 130 cm (52 in). The young shoots of the leaf blades of annual ryegrass are rolled and glabrous, while in perennial ryegrass they are folded in. The leaf sheath is also glabrous in annual ryegrass. The inflorescence is a spike with many florets per spikelet. Awns are generally present on annual ryegrass seed (15).

Ryegrasses are used primarily for pasture but may also be used for silage, hay, soil conservation, or quick turf establishment (15). Annual ryegrass is seeded as a winter pasture crop in the southeast (10). Both of the ryegrasses are seeded with other grasses or legumes. Perennial ryegrass is a short-lived perennial. Generally, a stand can be maintained for 3 to 4 years.

Ryegrasses produce high-quality forage that may equal the quality of small-grain forages. The total forage production is usually higher in ryegrass than in small grains (28). Most of this production comes in the early fall or the early spring. In the south, ryegrass is often planted with small grains, thus increasing the length of the spring grazing season. It is very well adapted to seeding with small grains such as winter wheat or rye and, to some extent, oats. Ryegrass and

FIGURE 29-2
Flowering plant of perennial ryegrass.
(Courtesy University of Illinois)

FIGURE 29-3
Close-up of perennial ryegrass inflorescence.
(Courtesy University of Illinois)

oats produce most of their forage at approximately the same point in the spring; therefore they do not complement each other as well as winter wheat or rye and ryegrass. Ladino clover, crimson clover, or arrowleaf clover is sown with ryegrass, which, too, increases the quality of the pasture (20, 28). Seeding ryegrass with other grasses or legumes, instead of in pure stands, results in a longer grazing season and helps reduce the incidence of bloat. When ryegrass is seeded with fescue in the south, the forage quality is increased (20, 26, 28). Also in the south, ryegrass is sometimes interseeded in bermudagrass pasture. This can provide a 200-day grazing period, which greatly improves the return per hectare when fattening steers are pastured, rather than fed a highly concentrated ration (20).

Ryegrass is not only utilized as a pasture crop but is used as turf, especially in the southern states. When it is overseeded in a bermudagrass lawn, it will give a quick green cover and will remain productive and green throughout most of the winter (5, 10, 20, 28), adding to the beauty of the lawn or turf around homesites. Ryegrass has also been made into pellets for feedlot purposes. The pelleting of ryegrass and inclusion of the pellets in rations increases the performance of animals being fed on high-forage rations. In a trial in Louisiana, it was shown that when animals were fed rations containing 40, 60, and 80% ryegrass pellets, instead of ryegrass pasture, over a 140-day period the daily gains by the steers were 994, 1000, and 908 g (2.2, 2.2, and 2.0 lb), respectively. Daily gain on ryegrass pasture alone was 700 g (1.54 lb) (27). The composition of the ryegrass pellets was 13.1% crude protein, 36.7% N-free extract, and 89% dry matter. When the animals were slaughtered, the carcass grades were higher for those fed with pelleted ryegrasses than with ryegrass pasture (27).

DESIRABLE CHARACTERISTICS OF RYEGRASS

Some of the desirable characteristics of ryegrass are: (1) it is an excellent winter and early spring grazing crop; (2) it may increase the quality of the forage when sown with some other companion grass or legume; and (3) it adds more grazing days, especially in the spring when sown with some other crop. Ryegrass is adapted to many different soil conditions but probably responds best when grown on heavy or silty clays that have adequate surface drainage (10). Ryegrass is possibly one of the most underrated winter pasture grazing crops in the United States.

MANAGEMENT OF RYEGRASS

Of all of the forage grasses, ryegrass rates among the highest in terms of rapid germination and corresponding seedling vigor (10). Ryegrass is usually sown with other cool-season grasses or legumes, primarily to extend the grazing period, and also because of its rapid establishment. Ryegrass helps suppress weedy plants and will provide an earlier pasture when seeded with another forage crop than would that forage if seeded in a pure stand. A well-prepared seedbed, whether it is an existing sod or a newly prepared seedbed, is desirable for ryegrass. It should be seeded approximately 1.25 cm (½ in) deep. The seeding rate varies, if sown for a forage in a pure stand, from 20 to 28 kg/ha (18 to 25 lb/A). In the south, the seeding rate for annual ryegrass, when seeded alone, is 22 to 33 kg/ha (20 to 30 lb/A), and 11 to 17 kg/ha (10 to 15 lb/A) when seeded with a small grain or companion legume. When ryegrass is seeded further north with a small grain for pasture, the seeding rate is generally 11 to 12 kg/ha (10 to 11 lb/A); if sown with a legume, 7 to 8 kg/ha (6 to 7 lb/A) of seed is sufficient. Ryegrass is often seeded for purposes of combating soil erosion. The seeding rate for this use would be at least 41 kg/ha (37 lb/A). When ryegrass is sown for turf purposes or lawns, the seeding rate ranges from 314 to 504 kg/ha (280 to 450 lb/A) (10, 29).

After ryegrass has established a good stand, up to 7.5 to 10 cm (3 to 4 in) high, it should be fertilized with 56 to 65 kg/ha (50 to 60 lb/A) of actual nitrogen. After fertilization, it should be grazed to induce stooling or tiller production because this will help form a dense sod. Cattle should then be removed and the ryegrass allowed to grow to 15 to 25 cm (6 to 10 in) before the cattle are returned (29).

If ryegrass is harvested as a hay or silage, it should be harvested soon after anthesis (5). The harvest or grazing height should be no closer than approximately 7.5 cm (3 in) above the soil surface. This allows for more rapid regrowth and good tiller production.

Due to the rapid growth of ryegrass, it can be grazed or harvested within 2 months after seeding. In the spring, when growth is rapid, it is important to maintain heavy stocking rates so that the forage will stay in a succulent condition. Consequently, it is important that rotational grazing be implemented so that the regrowth is of high quality (5, 29).

Ryegrass responds to nitrogen application. In general, as the nitrogen rate increases, dry matter yield and the amount of digestible dry matter are also increased (26). Many trials have been conducted to study the effects of various rates of nitrogen on ryegrass dry matter production. Generally, the yield increases in a linear manner from 0 to 336 kg/ha (300 lb/A) (20, 26). In a summary of many trials, where maximum yields were produced under irrigated conditions, it was found that approximately 2 kg (2 lb) of N/ha/day prior to the flowering emergence and up to 5 kg (4 lb) of N/ha/day for the remainder of the growing season were optimal (5).

It is important to maintain the proper levels of P and K when additional nitrogen is applied. Ryegrass is similar to all other cool-season grasses in that high levels of both P and K are removed when additional nitrogen is applied.

Ryegrass is considered to be a forage of high quality when it is properly fertilized. Digestibility, as measured by in vitro techniques, ranges from 70 to 88% (26). This is one of the more digestible forage crops grown.

Most of the ryegrass seed production is in the far northwest—primarily Oregon (10). Good cultural practices involved with seed production entail: (1) the complete removal of stubble following the seed harvest; (2) the elimination of any weeds; (3) the control of any diseases; and (4) close clipping, which will stimulate good tiller production. Ryegrass seed will shatter readily upon maturity (10), so it is important to harvest or windrow the ryegrass to ensure that seed shattering will be minimal, and to harvest with a combine that has a pickup reel attachment.

There are several diseases that will occur in ryegrass. Most of the diseases occur in the humid areas, and include crown rusts, brown rust, and some *Helminthosporium*. Sometimes a root rot that reduces the sod may develop, and ergot is occasionally found in the seed fields (24).

WHEATGRASSES

Wheatgrasses belong to the genus *Agropyron*. There are some 150 species of wheatgrasses widely distributed over the temperate areas of the world (15). The wheatgrasses growing in the United States include both native and introduced species of bunch and sod-forming types (15).

Wheatgrasses are grown primarily in the semiarid portions of western United States, a region bounded on the east by the 100°W longitude line and extending westward to the intermountain area. All of the wheatgrasses are cool-season perennials (15). They have early spring growth of nutritious, palatable forage that may deteriorate in quality during the summer (22). The wheatgrasses are valuable in controlling wind and water erosion. They are also used in the Great Plains area for terraced outlets (13).

Western Wheatgrass

Western wheatgrass, *A. smithii* Rydb., a native of the western United States, is a very drought-resistant species adapted to alkaline soils (21). Although it germinates slowly, it spreads rapidly after establishment. It is not well adapted to the eastern United States. It is a widely distributed, long-lived, sod-forming grass found primarily in the central and northern Great Plains (22).

Western wheatgrass produces a high percentage of its forage early in the spring (22). It is a highly desirable forage for cattle early in the spring, but as it matures it becomes more fibrous (7). It is sometimes allowed to stand during the latter part of the growing season and will provide good winter grazing (22). Its regrowth is rather slow, so it should not be cut too close to the soil surface. If overgrazed or frequently harvested near the soil surface, it will soon die (21). It can be easily transplanted when used in sodding terraces. It is a rather vigorous, winter-hardy, and drought-resistant species.

Intermediate Wheatgrass

Intermediate wheatgrass, *A. intermedium* (Host.) Beauv., was introduced into the United States from the Soviet Union in 1932. It is adapted from western Oklahoma northward into the Dakotas, where the yearly rainfall is 37 cm (15 in) or more. It is a vigorous, sod-forming species. It is more palatable than western wheatgrass. It is a large-seeded species and is easy to establish (19, 22). Intermediate wheatgrass forms a dense sod on well-drained, fertile soils where sufficient moisture is present. It too makes an excellent pasture crop early in the spring. It is not as winter-hardy as crested wheatgrass or western wheatgrass, and is difficult to maintain in a stand for much longer than 5 to 6 years. Therefore, it is best used in rotational pasture situations (23).

Tall Wheatgrass

Tall wheatgrass, *A. elongatum* (Host.) Beauv., is a tall, coarse, bunch-forming grass, well adapted to wet alkaline soils (15, 21). It is later-maturing than most of the other wheatgrasses. It has been used in reclamation areas where the soils are wet and have a high pH. The quality of forage from tall wheatgrass is not as good as that from other wheatgrasses, but tall wheatgrass does provide a longer grazing period, as it matures later than most other wheatgrasses. Tall wheatgrass is not as palatable as intermediate wheatgrass, and is generally not recommended where fescue, reed canarygrass, or orchardgrass can be grown (23). Its region of adaptation is approximately the same as that of intermediate wheatgrass (15).

Crested Wheatgrass

Crested wheatgrass, *A. desertorum* (Fisch. ex Link) Schult., is an introduced bunch-forming grass. It is very hardy, drought-resistant, perennial forage grass. It was introduced into the United States from Russia around 1906 (7). Its area of adaptation is the northern Great Plains and the Rocky Mountain area.

It is best adapted to productive soils, from sandy loams to heavy clays. It is not as tolerant of alkaline soils, nor will it persist under wet conditions as well as tall wheatgrass (7). Crested wheatgrass has a deep fibrous root system which gives it excellent drought resistance. It has the ability to become semidormant during the hot, dry periods of the summer; its greatest amount of growth is produced in the early spring (21). It is a very productive, palatable grass and makes a highly valued forage pasture crop (23). The numerous cultivars of crested wheatgrass range from very leafy and fine-stemmed, to very coarse with few leaves and stiff stems.

Slender Wheatgrass

Slender wheatgrass, *A. trachycaulum* (Link) Malte, is a bunch grass native to the United States. It is adapted to the northern Great Plains and Canada. Vigorous seedlings facilitate quicker establishment of slender wheatgrass than most other wheatgrasses. It is not as long-lived as many other wheatgrasses; therefore, it must be planted with other species or grown in shorter rotations. Slender wheatgrass is tolerant of alkaline soils, but is less drought-tolerant than western wheatgrass (15).

Other Wheatgrasses

There are numerous other wheatgrass species. The bluebunch wheatgrass, *A. spicatum* (Purch) Scribn. and Smith, is a long-lived, drought-resistant bunch grass. It is a highly palatable, nutritious forage grass, but has less carrying capacity and requires greater management to maintain longevity.

Fairway wheatgrass, *A. cristatum* (L.) Gaertn., is a drought-resistant, bunch-forming grass. It is one of the shorter species of the wheatgrasses and is fine-stemmed, much more so than crested wheatgrass. The seedlings are less vigorous and smaller than those of crested wheatgrass. Its name is derived from the fact that it is adapted to turf plantings and fairway areas (15).

Siberian wheatgrass, *A. sibiricum* (Willd.) Beauv., is a drought-resistant, bunch-forming grass. It is very similar to crested wheatgrass. Its stems are finer and the heads are more compact than those of crested wheatgrass. In some regions, because of wide adaptation, it is replacing crested wheatgrass (13).

Quackgrass, *A. repens* (L.) Beauv., is a widely adapted weedy species of wheatgrass. Its region of adaptation is from the Great Plains, and eastward, under more humid conditions; it grows best in the temperate areas of the United States. It is a very vigorous, aggressive, persistent perennial weed which spreads rapidly by rhizomes. When grown in pastures it provides a high-quality forage. It does develop a very rapid sod and can become a serious problem in cultivated fields or where one is trying to grow other cultivated species (15, 22).

RESCUEGRASS

Rescuegrass, *Bromus catharticus* Vahl., is a short-lived, perennial bunch grass that acts like an annual under most conditions. It is native to South America and is grown extensively in Australia (19). This grass is related to such annual winter grasses as cheat and chess, but is more productive and palatable than either of these two weedy species. Its nutritive value is similar to that of oats when the two are grown on similar soils (19).

Rescuegrass is best adapted to highly fertile soils, but will also grow on many of the poorer soils when sufficient fertility is added. It grows best where winters are mild and humid, and will survive as far north as Nebraska (19). Rescuegrass reseeds itself and is widely naturalized. It is used primarily for pasture as a volunteer in the southern regions of the United States. It is sometimes planted with annual winter legumes such as burclover, hop clover, crimson clover, or arrowleaf clover. Rescuegrass can become a serious weed problem wherever it grows. When the seed is mature it is sufficiently brittle that, if eaten by cattle, the seed may pierce the cow's tongue and the resulting condition may become so serious that the cow refuses to eat. Therefore, it is not seeded very widely anymore.

HARDINGGRASS

Hardinggrass, *Phalaris aquatica* L., is a cool-season, perennial bunch grass with short rhizomes. It is not very tolerant of cold weather and is used only occasionally in the lower coastal plains in the southern United States. It is very drought-tolerant and sometimes grown in the southwest, along with dry-land alfalfa (15).

Hardinggrass is a nutritious, palatable species that responds to fertilization. It makes a high-quality hay and should be cut whenever seedheads begin to appear. It should not continually be closely grazed; it should instead be grazed rotationally to allow sufficient regrowth.

SMALL GRAINS

Small grains are cool-season crops including oats, wheat, rye, barley, and triticale. They are grown in regions where moisture is sufficient. They have a wide range of soil adaptation but grow better on more fertile, well-drained soils. In recent years, *Triticale,* which is a new species, has been developed. *Triticale* is a cross between wheat, *Triticum,* and rye, *Secale,* thus getting its name. Each of the small grain species will be discussed separately (4a, 4b).

Oats

Oats, *Avena sativa* L., are both a winter-annual small grain, in the southern regions of the United States; and a spring-sown small grain, in the upper midwest and northeast. The winter oats have predominantly a prostrate-type growth in the fall and winter and are more cold-hardy than spring oats. Most of their forage production is in the late spring (3, 6, 11).

Spring oats have an erect growth habit, with low winterhardiness, and produce early forage growth when sown in the early spring (12). There is also another intermediate type of oats which exhibits both spring- and winter-type growth characteristics (16).

Oats produce a high-quality forage and, in the south, are usually planted in the fall for winter or early spring grazing. In the midwest oats may be seeded following an early soybean harvest so that some fall grazing may occur. The main advantage of oats, in a dairy program, is early fall grazing. In the south oats do not grow as late into the fall or as early in the spring as rye or wheat (3).

Winter oats should be planted as early as possible in the fall to get good fall growth. In the southern region of the United States, oats will overwinter and cattle can graze until early March and then be removed to a grain crop. In the south, oats are often seeded as a companion crop in the establishment of tall fescue or orchardgrass with ladino clover (16). In South Carolina, where tall fescue was planted in rows along with ladino clover and overseeded with oats, this combination provided three to four months' more grazing in the fall and spring than it would have without the oats. Adding oats reduces the bloat hazard of ladino clover and protects the tall fescue and ladino clover seedlings (6, 16). In the south, oats have also been sod-seeded into bermudagrass with renovation programs and to extend the winter grazing period. Rye or wheat would be more successfully overseeded into bermudagrass than oats because oats grow longer in the spring and would compete with the early growth of bermudagrass (3, 6).

Oats are probably preferred to other small grains as a forage crop because of the oat's finer stem, faster curing properties, and higher palatability (12). It is not quite as drought-resistant or as cold-resistant as the other three species (12). Dry matter yields of the other species, especially wheat and rye, will most likely exceed those of oats (11). Oats produce the highest quality forage of the small grains. Oats make an excellent silage crop and, if harvested at the boot stage, yield a second growth (3). Oats are probably the only small grain put up as a dry hay, because of their fine stems and lower fiber content. Oats as a grain are preferred by breeders of horses for their animals.

Small grains, if they are to be harvested as forage, should be seeded at a slightly higher rate than if to be harvested for grain. This produces a finer stem but also increases the potential for lodging. The heavier seeding rates are necessary for any sufficient early grazing. Lower seeding rates encourage more tillering, but a winter-sown small grain will produce equal amounts of spring growth when seeded at any reasonable rate.

A balanced fertilizer should be applied to small-grain crops. Nitrogen is usually applied at planting and top-dressed in the winter or early spring. Oats respond less to nitrogen fertilization than any other small grain. The upper limit of top-dressed nitrogen would be 84 kg/ha (75 lb/A) in late winter for the winter-sown small grains, and 40 to 50 kg/ha (35 to 45 lb/A) in split applications for both the fall and early spring.

Wheat

Wheat, *Triticum aestivum* L., is a winter-annual, small-grain crop that may be used as a forage crop (1, 3, 12). It is sown in the southern and southwestern regions of the United States alone or in combination with other species for high-quality winter pastures. Wheat is often interseeded into bermudagrass to extend winter or early spring grazing. Wheat produces less fall forage than oats, but is more drought- and cold-tolerant and has more winter and early spring growth. It is probably the second-highest-quality small grain harvested as a silage crop. It should be harvested for silage when in the boot or early-flower stage (2). If wheat is harvested as a silage crop, it will produce 60% more digestible nutrients than when harvested for grain. Thus it is a very popular silage crop for a dairy herd (1, 2).

Winter wheat is seeded in the early fall on a well-prepared seedbed. The seeding rate, if wheat is grown as a forage crop, ranges from 66 to 135 kg/ha (60 to 120 lb/A). Wheat should have 15 to 25 cm (6 to 10 in) of growth before grazing (12). One should top-dress with nitrogen soon after wheat emerges, which will increase the fall and winter forage production provided there is sufficient moisture available. Wheat responds slightly more to nitrogen than do oats. In recent years a number of wheat cultivars have been developed for pasture use that will produce more fall growth and early spring growth, and then make a grain crop. Grazing must not extend beyond the tillering to early-elongation stage of growth.

Rye

Rye, *Secale cereale* L., is a winter-annual small grain that is usually grown for pasture. It is a very valuable small grain that produces extremely well when seeded alone or in combination with other grasses or legumes. It has a greater amount of fall and early spring growth than wheat or oats (1, 2, 3, 12). Rye is the most winter-hardy of all the small grains, thus having higher winter production than the other small grains. It also matures earlier than the other small grains, making management a very important aspect to see that rye is harvested or grazed at the proper stage of maturity (2).

Rye has a much more extensive root system than other small grains (12). It is used to help prevent wind or water erosion and is adapted to the less fertile soils; it is therefore often used as a winter cover crop (13). It is adapted to a wide range of soils, and performs better than wheat or oats in the sandier or lighter soils. It is probably more drought-tolerant than wheat or oats. It does respond to plentiful fertilizer applications (1, 3, 17). Rye is usually taller than wheat and may lodge with heavy fertilizer applications.

The seedbed should be prepared the same way as for other small grains. It should be firm, and the seed should be drilled 2.5 to 5 cm (1 to 2 in) deep in the soil. Seeding rates of 90 to 112 kg/ha (80 to 100 lb/A) are needed to provide a good fall and early spring growth (17). The same nitrogen fertility recommendations mentioned for oats should be applied to rye.

Barley

Barley, *Hordeum vulgare* L., is a winter-annual small grain that is sometimes used for pasture. It is predominantly used for grain production. Its seeding rate and overall production are very similar to those of rye (12).

Triticale

Triticale is a cross between wheat and rye. This cross was first recorded in 1875, but was then primarily a sterile species (2). Later, strains of triticale were developed that proved to be very promising as a fertile small grain. Triticale is considered to be rather winter-hardy. It is similar in productivity to wheat and rye (1).

Triticale is a late-maturing, tall species that has low tillering capacity, some seed shattering, and is somewhat sterile (2). Its seedheads and seeds are larger than those of wheat or rye, but generally the seeds are more shriveled and therefore weigh less per unit volume. Initial cultivars of triticales were higher in protein because of the shriveled seed. Its overall seeding practices and management are very similar to those of rye. Newer cultivars are plumper and their protein is approximately equal that in wheat. Triticale is generally considered inferior to wheat and rye in most other aspects.

In summary, when small grains are used for forages, the season of forage production is the main factor governing choice between species, as quality and yields and are very similar among the small grains.

QUESTIONS

1 Compare redtop to tall fescue.
2 Why are the ryegrasses considered to be important forage crops in the U.S.?
3 Describe the management of ryegrass in an overall forage system.
4 Compare the various wheatgrasses in terms of area of adaptation and growth habit.
5 Compare the various small-grain crops with regard to forage uses.

REFERENCES

1 Bishnoi, U. R. Effect of seeding rates and row spacing on forage and grain production of triticale, wheat, and rye. *Crop. Sci.* 20:107–108, 1980.
2 Bishnoi, U. R., P. Chitapong, J. Hughes, and J. Nishimuta. Quantity and quality of triticale and other small grain silages. *Agron. J.* 69:439–441, 1978.
3 Burns, J. D. Wheat, barley, oats, and rye for forage. *Tenn. Agr. Ext. Serv. Mimeo Pub.,* 1973.
3 a Carman, J. G., and D. D. Briske. Root initiation and root and leaf elongation of dependent little bluestem tillers following defoliation. *Agron. J.* 74:432–435, 1982.
4 Carpenter, J. C. Jr., R. H. Klett, and S. Phillips. Producing slaughter steers with temporary grazing crops and concentrates. *La. Agr. Exp. Sta. Bull. 643,* 1969.
4 a Cherney, J. H., and G. C. Marten. Small grain crop forage potential: I. Biological and chemical determinants of quality, and yield. *Crop Sci.* 22:227–231, 1982.
4 b Cherney, J. H., and G. C. Marten. Small grain crop forage potential: II. Interrelationships among biological, chemical, morphological, and anatomical determinants of quality. *Crop Sci.* 22:240–245, 1982.
5 Cowling, D. W., and D. R. Lockyer. The response of perennial ryegrass to nitrogen in various periods of the growing season. *J. Agr. Sci.* 75:539–546, 1970.
6 Denman, C. E., and J. Arnold. Seasonal forage production for small grains species in Oklahoma. *Okla. Agr. Exp. Sta. Bull. B-680,* 1970.
7 Dillman, A. C. The beginnings of crested wheatgrass in North America. *J. Amer. Soc. Agron.* 38:237–250, 1946.
7 a Fairbourn, M. L. Water use by forage species. *Agron. J.* 74:62–73, 1982.
8 Fergus, E. N., and R. C. Buckner. The bluegrasses and redtop. In *Forages,* Iowa State Univ. Press, Ames, IA, 3d ed., pp. 243–253, 1973.
9 Forsberg, D. E. The response of various forage crops to saline soils. *Can. J. Sci.* 33:532–549, 1953.
10 Frakes, R. V. The ryegrasses. In *Forages,* Iowa State Univ. Press, Ames, IA, 3d ed., pp. 307–313, 1973.
11 Frey, K. J., P. L. Rodgers, W. F. Wedin, L. Walker, W. J. Moline, and J. C. Burns. Yield and composition of oats. *Iowa State J. Sci.* 42:9–18, 1967.
12 Fribourg, H. A. Summer annual grasses and cereals for forage. In *Forages,* Iowa State Univ. Press, Ames, IA, 3d ed., pp. 344–357, 1973.
13 Hafenrichter, A. L., J. L. Schwendiman, H. L. Harris, R. J. McLauchlan, and H. W. Miller. Grasses and legumes for soil conservation in the Pacific Northwest and Great Basin States. *USDA SCS Agr. Handbook 339,* 1968.

14 Harris, R. R., W. B. Anthony, V. L. Brown, J. K. Boseck, H. F. Yates, W. B. Webster, and J. E. Barrett Jr. Cool-season annual grazing crops for stocker calves. *Ala. Agr. Exp. Sta. Bull. 416,* 1971.

15 Hitchcock, A. S. Manual of the grasses of the United States. *USDA Misc. Publ,* 200 Rev. 1951.

16 Holt, E. C., M. J. Norris, and J. A. Lancaster. Production and management of small grains for forage. *Tex. Agr. Exp. Sta. Bull. B-1082,* 1969.

16 a Marten, G. C., J. L. Halgerson, and J. H. Cherney. Quality prediction of small grain forages by near infrared reflectance spectroscopy. *Crop Sci.* 23:94–96, 1983.

17 Morey, D. D., M. E. Walker, W. H. Merchant, and R. S. Lowrey. Small grain forage production and quality as influenced by rates of nitrogen. *Ga. Agr. Exp. Sta. Res. Bull. 70,* 1969.

18 Morrow, L. A., and J. F. Power. Effect of soil temperatures on development of perennial forage grasses. *Agron. J.* 70:7–10, 1979.

19 Newell, L. C. Wheatgrasses in the West. *Crops and Soils* 8:7–9, 1955.

19 a O'Neil, K. J., and R. N. Carrow. Perennial ryegrass growth, water use, and soil aeration status under soil compaction. *Agron. J.* 75:177–180, 1983.

20 Reid, D. The effect of a wide range of nitrogen application rates on the yields from a perennial ryegrass sward with and without white clover. *J. Agr. Sci.* 74:227–240, 1970.

21 Rogler, G. A. Growing crested wheatgrass in the Western States. *USDA Leaflet 469,* 1960.

22 Rogler, G. A. The Wheatgrasses. In *Forages,* Iowa State Univ. Press, Ames, IA, 3d ed., pp. 221–230, 1973.

23 Rogler, G. A., and R. J. Lorenz. Pasture productivity of crested wheatgrass as influenced by nitrogen fertilization and alfalfa. *USDA Tech. Bull. 1402,* 1969.

24 Sampson, K., and J. W. Western. *Diseases of British Grasses and Herbage Legumes,* 2d ed., Cambridge Univ. Press, Cambridge, England, 1954.

25 Terrell, E. E. A taxonomic revision of the genus *Lolium. USDA Tech. Bull. 1392,* 1968.

26 Waite, R. The structural carbohydrates and the *in vitro* digestibility of a ryegrass and cocksfoot at two levels of nitrogenous fertilizer, *J. Agr. Sci.* 74:457–462, 1970.

27 White, T. W. Dehydrated pellets for fattening steers. *La. Agr. Exp. Sta. Bull. 634,* 1969.

28 Wyatt, M. J. Studies on the causes of the differences in pasture quality between perennial ryegrass, short-rotation ryegrass, and white clover. *N. Z. J. Agr.* 14:352–367, 1971.

29 Yungen, J. A. Ryegrass forage production trial, southern Oregon experiment station. *Oreg. Forage and Seed Rep.,* 1969.

CHAPTER 30

WARM-SEASON GRASSES

BERMUDAGRASS

Bermudagrass, *Cynodon dactylon* (L.) Pers., is probably the most popular warm-season sod-forming perennial grass in the United States. It is native to the Mediterranean region and southeastern Asia (8), and probably originated in Africa. Many of the earlier researchers indicated that its origin was in India (8). Today many believe it originated in Africa because most diversity of *Cynodon* is found there. It reproduces by rhizomes, stolons, and seeds. It is very aggressive forage grass and is considered a weed in cultivated fields (9).

Bermudagrass is adapted to a wide range of soils and climatic conditions from the Gulf Coast to areas as far north as Kansas, Missouri, and Kentucky. It prefers a deep, fertile soil but will grow very well on heavy clay soils or on deep sands when properly fertilized (8). Bermudagrass will withstand flooding for several weeks providing there is some wind movement to provide the floodwater with sufficient amounts of oxygen. It does not grow well on water-logged soils. Bermudagrass is fairly salt-tolerant but will grow on both acid and alkaline soils (8).

Bermudagrass makes an ideal pasture crop. It is high in production when heavily fertilized (40) and produces high-quality forage when managed properly. Bermudagrass will withstand close clipping and continuous grazing, but will respond better to rotational grazing. It has been used primarily for grazing, hay, or a combination of the two. Bermudagrass has also been used for silage, pelleting, cubing, and green chopping (8, 35).

Bermudagrass, when properly fertilized and harvested every 3 or 4 weeks, will produce a forage with a protein content of approximately 12%. The digestibility can be 65% or higher, but the way bermudagrass is commonly managed, it is much lower. The grass can be up to 80% leaves (14).

There are several types of bermudagrass. Common bermudagrass is the oldest type of bermudagrass, and has a more prostrate growth and less palatable forage than the other types grown. Digestibility is lower, and common bermudagrass is generally considered to be more of a weedy type than a forage type (8).

'Coastal' bermudagrass is the hybrid that growers in the south refer to as "the bermudagrass." 'Coastal' bermudagrass was developed by Dr. G. W. Burton from a cross between 'Tift,' a common bermudagrass, and a tall-growing bermudagrass from South Africa (9). 'Coastal' bermudagrass, when compared with common bermudagrass, is much taller and has a lighter green, longer leaf. It is more upright in growth, more resistant to diseases, and responds better to nitrogen fertilization (9). 'Coastal' bermudagrass is resistant to root knot nematodes and is well adapted to overseeding with legumes because of its open-sod characteristics. It is moderate in cold tolerance and is adapted for use from southern Oklahoma and Arkansas to the coastal regions of North Carolina and Virginia (15). Research has indicated that 'Coastal' bermudagrass is most productive when harvested every 3 weeks and well fertilized with nitrogen (32).

'Midland' is a hybrid bermudagrass also developed by Dr. Burton, and released jointly by the USDA, ARS, and the Agricultural Experiment Stations of Georgia and Oklahoma in 1948. 'Midland' is a cross between 'Coastal' bermudagrass and a winter-hardy bermudagrass from Indiana (8). It is grown further north into Kansas, Missouri, and southwestern Kentucky, and eastward to Maryland. The main attribute of 'Midland' is its greater coldhardiness than 'Coastal' bermudagrass, probably because it has more rhizomes, which enable it to overwinter better. 'Midland' is fair in disease resistance and drought tolerance. Its spring growth begins approximately 2 weeks before that of 'Coastal,' it is more upright in growth habit, and it is better for overseeding with legumes and small grains for grazing or hay production (1).

'Hardie' is the result of a three-way cross of bermudagrass, developed in Oklahoma in the 1960s and released in 1974. It is more winter-hardy than 'Midland' but also begins growth earlier in the spring than 'Midland.' It spreads very rapidly on fine-textured soil and has a higher digestibility potential than 'Midland' bermudagrass (21).

'Coastcross-1' is a hybrid bermudagrass developed by Dr. Burton in 1967. It is a cross between 'Coastal' and a bermudagrass introduced from Kenya. 'Coastcross-1' is completely sterile and produces no seed; grows taller than 'Coastal'; has broader, softer leaves; and spreads very rapidly by stolons, having very few rhizomes. 'Coastcross-1' is resistant to most foliar diseases and to the sting nematode. It is more digestible than 'Coastal' bermudagrass but produces approximately the same amount of dry matter per acre. 'Coastcross-1' is not as winter-hardy as 'Coastal' bermudagrass, but it is more drought-tolerant and responds better to nitrogen fertilization. Its area of adaptation is further south than that of other bermudagrasses (10).

'Tifton 44,' developed by Dr. Burton, is a more winter-hardy cultivar than the other bermudagrasses. It is a cross between 'Coastal' and a bermudagrass introduced from Germany. 'Tifton 44' produces more forage dry matter per area and is more digestible than either 'Coastal' or 'Midland.' Livestock gains have been approximately 20% higher with 'Tifton 44' than with 'Coastal' bermudagrass (21).

'Callie' is not a 'Coastal' bermudagrass derivative. It was developed at the Mississippi Agricultural Experiment Station, is slightly self-fertile, and can be propagated from seed. Its seed production is low and it is not cold-hardy. This subspecies could create some weed problems in cultivated fields.

Since 'Coastal' bermudagrass and its various derivatives are very similar in establishment, reference will be made to 'Coastal' bermudagrass in the following discussion of overall management. The cost of establishing bermudagrass is usually rather high; therefore every effort must be made to obtain a good stand the first year. It takes good management, not only for the initial establishment, but also of maintenance and utilization, to provide a high-quality forage.

'Coastal' bermudagrass is established by vegetative cuttings or sprigs from either stolons or rhizomes. Since it is relatively sterile, 'Coastal' produces very few viable seeds and these plants do not breed true. One should purchase sprigs from a reputable dealer or a producer who has an excellent stand. Quality sprigs are very important for a successful establishment. The sprigs should not dry out. They are planted in rows or broadcast over the surface and then disked and rolled. In the southern region, where 'Coastal' bermudagrass is adapted, it is sprigged into the soil from February to May, depending on the specific location. If irrigation is available, 'Coastal' bermudagrass can be sprigged in midsummer, providing moisture is sufficient (8).

On a good seedbed with sufficient fertility, a uniform field stand can be obtained in 1 year. Proper lime and fertilizer should be applied before planting, since lime must be worked into the seedbed to adjust the soil pH to an optimum level for good plant growth.

Weed control, along with nitrogen fertilization, is very important for getting a good establishment of the first-year stand (8). If broadleaf weeds are a problem during establishment, 2,4-D can be used as a control. Simazine, applied during or immediately after sprigging, will help control annual grassy weeds and some broadleaf weeds that may germinate in the early spring (20a).

Most hybrid bermudagrass is planted by using mature (8-week-old), fresh clippings in midsummer (July). These are broadcast on prepared land, disked, and rolled. Rate of planting is about 1 propagule per 4 sq. ft. The most critical factors for establishment are viable sprigs and adequate moisture.

When bermudagrass stolons reach a length of approximately 10 to 15 cm (4 to 6 in), one should top-dress the plants with approximately 55 to 85 kg/ha (50 to 75lb/A) of nitrogen. Bermudagrass should be harvested when it is 25 to 38 cm (10 to 15 in) tall and top-dressed again with 55 to 85 kg/ha (50 to 75 lb/A) of nitrogen (8, 9). Once bermudagrass is established, one may plan on a productive stand as long as it is well managed and fertilized.

Bermudagrass may be managed as either a pasture crop, a hay crop, or a combination of the two. Nutrient removal is greater when the forage is removed as a hay crop than as a pasture crop. Much more attention should be given to the fertility needs of a hay crop, especially N, P, K, S, and Mg. Growers often harvest one or two hay crops early and then graze the remainder of the year.

Bermudagrass normally uses N, P, and K in a ration of 4-1-2 or 5-1-2. Most of the soils on which bermudagrass is grown are low in nitrogen and phosphorus, and many of the sandy soils are also low in potassium. Bermudagrass should be grown on a soil with a pH of 6.0 to 6.5 (4a, 15).

Pure stands of bermudagrass will efficiently use large amounts of nitrogen fertilizer. It has been shown that, depending on the amount of moisture, bermudagrass production will increase in dry matter per area and in protein content, up to rates of 900 kg/ha (800 lb/A) of actual nitrogen (Table 30-1). The most profitable rate is approximately 448 kg/ha (400 lb N/A); variations depending on the price of N. Split applications of nitrogen are desirable when large amounts are applied. The frequency of the application of nitrogen also affects the protein content of 'Coastal' bermudagrass forage. Many growers apply approximately 55 kg/ha (50 lb/A) nitrogen every 21 days when bermudagrass is being grazed (11). This will retain a high-quality, nutritious forage. The nitrogen increases forage yields, protein content, carrying capacity, beef gains per acre, and forage palatability (13). Daily animal gains depend on the availability and digestibility of the forage. When soil pH and levels of P and K are adequate, the application of N will reduce the amount of water required to produce a kilogram of dry matter (11, 12).

One of the best methods of increasing forage quality and quantity, as well as extending the length of the grazing season, is to overseed bermudagrass with legumes (20a). Legumes will increase the protein content of the forage (26). With the addition of legumes, digestibility, daily animal gains, and milk production are increased (30, 31, 41). The legumes most commonly added to bermudagrass are the clovers—white clover, red clover, crimson clover, arrowleaf clover, subclover, and vetch (30). The method of establishing a clover in a

TABLE 30-1
THE EFFECT OF VARIOUS N RATES ON HAY YIELD AND PROTEIN CONTENT OF 'COASTAL' BERMUDAGRASS AT TIFTON, GA.

N rate	Hay		Protein		Amount of protein per kg of N
kg/ha (lb/A)	kg/ha (lb/A)	Protein (%)	kg/ha (lb/A)		kg/ha (lb/A)
0 (0)	4,195 (3,742)	7.12	271 (241)		— —
112 (100)	11,857 (10,576)	9.64	995 (888)		72.3 (159)
224 (200)	15,974 (14,249)	11.54	1550 (1383)		55.6 (123)
448 (400)	21,770 (19,419)	13.83	2503 (2233)		47.6 (105)
900 (800)	23,760 (21,194)	16.04	3152 (2812)		16.7 (37)

Source: G. W. Burton, "Coastal bermudagrass," *Ga. Agr. Exp. Sta. Bull. N.S. 2*, 1954.

grass sod was discussed in the chapter on sod seeding and pasture renovation. It is very important to keep the pH near 6.5 and P and K at higher levels when the legumes are put into a bermudagrass sod, in order to encourage the growth of legumes and ensure their survival.

Bermudagrass is a very nutritious grass if maintained in a very young, tender state. Thus, stocking rates and frequency of harvest are very important factors in achieving a quality forage and high level of animal performance. Rotational grazing of bermudagrass during the summer will increase both milk and beef production (8, 9, 30). The length between grazing periods depends on pasture growth rate, stocking rate, and grazing intensity. The proper stocking rate should be at a frequency that would allow the grass to be grazed down to approximately 5 cm (2 in) in height in 7 to 10 days and then to rest for an additional 2 to 3 weeks (8, 13). The pastures should be divided into small, temporary paddocks so that these can be rotationally grazed. Four or more pastures are generally needed. Early in the season, when there is excessive lush spring growth, or during periods in which bermudagrass will exceed the growth necessary to support the number of grazing cattle, it may be necessary to harvest some of the pasture as a hay crop. It is generally advisable to clip the pasture after each grazing period and then apply nitrogen fertilization. In general, it is very important to move the cattle often enough so that young, tender, high-quality bermudagrass is available at all times for their consumption.

Some producers in the south will use bermudagrass as a crop for pelleting and also for green chop. Many beef producers use bermudagrass as silage or as haylage. It has been shown by researchers that approximately 0.9 kg (2 lb) of a well-fertilized bermudagrass hay is equivalent in nutritive value to 0.45 kg (1 lb) of shelled corn (35). Pelleted bermudagrass will result in greater beef gains and milk production than will bermudagrass hay. This is accomplished primarily because of the increase in dry matter intake. The time of cutting bermudagrass for pelleting is very important; for every 2-week delay in cutting bermudagrass and feeding pellets for dairy cattle, the dry matter digestibility will decrease by 5% and milk production will decrease by 0.7 kg (1.7 lb) per cow per day (35, 39). It has also been shown that the average daily gains of beef cattle fed bermudagrass pellets increased by 50% over daily gains of cattle fed bermudagrass hay (35).

Weed control in bermudagrass is very important. Generally, removing forage as hay or grazing the bermudagrass is a very poor method of controlling grassy weeds. Well-fertilized bermudagrass, especially when fertilized with nitrogen, will maintain a rapid, short growth and a well-developed sod, thus controlling the invasion of most grassy weeds.

Most broadleaf weeds can be easily controlled with the use of 2,4-D or the mixture of 2,4-D and Banvel. The major weeds in hybrid bermudagrass are common bermudagrass and bahiagrass. A vigorous stand of bermudagrass is the only control, and this is not very effective for these weed species.

JOHNSONGRASS

Johnsongrass, *Sorghum halepense* (L.) Pers., is a tall, coarse, persistent, warm-season perennial grass. It is native to the Mediterranean countries, Africa, and India. Johnsongrass was introduced to the United States in about 1830 (5). It spreads rapidly through its very vigorous rhizomes or seeds. Most of the rhizome growth develops when the plant nears the head stage, so control measures must be taken prior to this time (24). One plant may produce over 150 stems and over 80,000 seeds. Many of these seeds are dormant at the time of maturity; thus, after shattering, they may continue to germinate in the soil for many years (24). Johnsongrass is best adapted to very heavy, dark clay soils that have good internal and surface drainage. It will grow on most any soil with adequate fertility (48). Johnsongrass is considered a noxious weed when grown in many field crops. It has many good qualities for hay or pasture use (24).

Johnsongrass is very similar to the sorghums, except that it does maintain itself, botanically and agronomically, as a perennial. As a perennial forage grass, johnsongrass is limited to Missouri, Kentucky, and south of those states. The flowering head is an open-panicle type and is very similar to sudangrass inflorescence (5). The rhizome growth is relatively slow in comparison with the growth of the upper portion of the plant, until the plant begins to bloom. At the time of flowering, rhizomes are produced very rapidly. The plant continues to produce rhizomes until a killing frost, at which time the plant overwinters; new plants will begin to grow the following year (5).

Johnsongrass also responds to nitrogen fertilization. Dry matter yields are directly correlated to the application of nitrogen fertilizer, and will increase as nitrogen is applied in rates up to 1075 kg/ha (960 lb/A). At about 538 kg/ha (480 lb/A), it becomes uneconomical to continue nitrogen fertilizer applications (48). Johnsongrass, if cut as hay, should be harvested when it is in the early-boot stage. After this stage, generally three to four harvests per growing season are obtainable (5). Research has found that the feeding value of johnsongrass harvested in the boot stage is approximately one and a half times as valuable, in terms of gain/day per hectare, as that of johnsongrass cut in the milk stage (46). Crude protein is 12% in the boot stage and 6% in the milk stage. Heavy nitrogen-fertilized johnsongrass cut in the boot stage will average about 13.5% crude protein, and digestibility will be about 65.8%, compared with 7.2% crude protein and 42.5% digestibility when unfertilized grass is cut at the same stage (5). If johnsongrass is heavily fertilized with nitrogen, P and K should be added at approximaately 0.5 unit P and 0.8 unit K for every 6 units nitrogen.

Even though johnsongrass is considered a weed in field crops in the south, it will not stand continuous extensive grazing. Johnsongrass has been shown to provide excellent forage for dairy cattle when grazed for 1 week and then allowed to rest for 3 to 4 weeks (5). There is some danger of cattle contracting prussic acid poisoning from grazing johnsongrass after a drought or frost. Caution should be taken following either of these two situations when livestock are grazing johnsongrass.

If johnsongrass is seeded as a forage crop, it is generally seeded in the early spring in a well-prepared seedbed. The average seeding rates will range from 11.2 to 22.4 kg/ha (10 to 20 lb/A) (46). It is rather easy to increase the density of a sparse stand simply by disking it in early spring (46). The rhizomes are cut and spread by the disk operation, which increases the number of new plants or tillers that can become established.

Johnsongrass produces abundant seed. Seed yields range from 188 to over 500 kg/ha (168 to 446 lb/A) (5). Once johnsongrass seed is mature it will shatter and persist in the soil for many years, thus perpetuating itself as a noxious weed in other crops that follow.

DALLISGRASS

Dallisgrass, *Paspalum dilatatum* Poir., is a deep-rooted, warm-season perennial grass. It is native to South America and was introduced into the United States in approximately 1842 (17). It is adapted from North Carolina southward into Florida and westward to eastern Oklahoma and Texas. It is drought-tolerant but prefers a moist, fertile, loamy soil. Dallisgrass will persist on sandy soils if they are well fertilized and properly grazed. Dallisgrass is possibly more palatable than bermudagrass and will produce higher daily gains than bermudagrass, but generally lower gains per hectare (5, 27).

Dallisgrass has a bunch-type growth, which makes it very suitable for growing with other grasses or legumes. It has a deep root system but extremely short rhizomes. It has a rapid recovery rate, probably faster than most grasses adapted to the same area. Its growth habit is a little different from that of other warm-season grasses because it continues to grow late into the fall and is not injured by moderate frosts. It begins growth very early in the spring (5, 28).

Dallisgrass is used primarily as a grazing crop. It is usually not harvested as a hay crop. Due to its early spring and late fall growth, it does provide a longer period for grazing in comparison with most other warm-season grasses (45).

In the region of its adaptation, it is usually seeded in early spring at a seeding rate of 5.6 kg/ha (5 lb/A) pure live seed. The seed is planted 0.3 to 0.6 cm (1/8 to 1/4 in) deep and the seed zone is generally compacted with a cultipacker after seeding (28).

Dallisgrass responds well to nitrogen fertilization but when excessively high rates of nitrogen are applied, it will generally crowd out a legume. Fertilizer recommendations range from 20 to 100 kg/ha of N, plus 67 kg of P, and 34 kg of K (18 to 90 lb/A of N, 60 lb of P, 30 lb of K) to maintain a good grass-legume sod (5, 28). Generally, early spring applications of nitrogen should be omitted if a legume is sown with dallisgrass, which will aid in establishment of the legume. To maintain a productive stand of dallisgrass that recovers rapidly, the grass should not be grazed any shorter than 5 to 7.6 cm (2 to 3 in) (5).

Dallisgrass may become a serious weed problem in lawns because it can survive very close mowing; however, herbicides are available that easily kill dallisgrass in most turf stands.

BAHIAGRASS

Bahiagrass, *Paspalum notatum* Flugge, is a deep-rooted, warm-season perennial grass native to South America (47). It has short, thick, scaly rhizomes, and forms a dense sod. It is very competitive with other plants to the extent that it may crowd out even bermudagrass. Bahiagrass is more shade-tolerant than bermudagrass and is well adapted to soils low in fertility and to poorly drained upland soils (29). It has the ability to form a very firm turf on loose sand.

It was first introduced into New Jersey, but is now grown from North Carolina, south and west to the Gulf Coast area, including Texas. It is not as winter-hardy as bermudagrass.

There are several strains of bahiagrass. The main cultivar grown in the south is 'Pensacola.' It is winter-hardy as far north as the upper portions of Alabama, Mississippi, and Georgia, as well as the southern areas of Arkansas and Oklahoma. 'Pensacola' is more resistant to leaf diseases than some of the other strains. It has good seed germination and responds well to nitrogen fertilization, but not as well as 'Coastal' bermudagrass (31, 36). Another cultivar, or strain, of bahiagrass is 'Argentine bahia.' This strain is more palatable than 'Pensacola' but has less cold tolerance and is grown mainly in the coastal area of the United States. It is very susceptible to ergot seed fungus. The 'Pensacola' strain has narrower leaves and produces abundant seed of good quality but it does shatter easily (36). The bahiagrasses are often grown for erosion control as well as for forage production (47).

Bahiagrass is broadcast-seeded or drilled in the early spring after the last frost. Bahiagrass can be overseeded with sudangrass or millet during the summer or with oats in the fall. If oats are overseeded with bahiagrass, the field must be grazed completely early in the spring so the bahiagrass can get an early establishment. Ryegrass should not be seeded with bahiagrass because of the severe late spring competition for moisture materials and light that would result from the ryegrass. If bahiagrass is drilled, the seeding rates range from 11 to 17 kg/ha (10 to 15 lb/A) and up to 22 kg (20 lb) if broadcasted. It should be seeded approximately 0.6 to 1.2 cm (1/4 to 1/2 in) deep and cultipacked after seeding. Bahiagrass has a waxy seedcoat and germinates very slowly. The seed should be scarified either mechanically or chemically to increase germination. Even if a stand is slow in establishment, it will often increase in density over time because of seed shattering and the lateral spread of short but vigorous rhizomes. The seeds are very hard and remain viable in the soil for many years. Seeds are easily moved into new areas by cattle. It is very difficult to maintain legumes with bahiagrass because of the bahiagrass's dense sod (29, 31). If a legume is interseeded in bahiagrass, one should graze the existing stand very close in late summer, lightly disk the soil, and overseed with legumes such as arrowleaf clover, crimson clover, or vetch. Because of their ability to compete, these legumes will produce greater yields in combination with bahiagrass than will white clovers.

Bahiagrass responds to nitrogen fertilizer. Split applications of up to 224 kg/ha (200 lb/A) of actual nitrogen in early spring, early summer (June), and late August have given good yield response.

Since bahiagrass will withstand extremely close grazing, it should be grazed short, about 5 cm (2 in) in height, to maintain high-quality forage and ensure good daily animal gains. Pastures should be grazed for 7 to 10 days and the cattle then rotated to another pasture, allowing a rest of at least 15 to 20 days before the next grazing period. After each grazing period, one should clip the stubble to encourage even regrowth and to maintain high-quality forage production. Bahiagrass does not make a hay as high in quality as 'Coastal' bermudagrass, because the plant is usually too mature by the time it has attained an optimal cutting height to produce a high-quality hay. Most of its leaves are basal and not cut or harvested for baling. If bahiagrass is cut earlier and closer, the quality of the hay will equal that of bermudagrass, but yields will be lower (29, 31, 47).

RHODESGRASS

Rhodesgrass, *Chloris gayana* Kunth, was first introduced into the United States in 1902. It is native to South Africa. Rhodesgrass is a fine-stemmed, erect, leafy, perennial, warm-season bunch grass. It does not possess extremely good winterhardiness, as it dies when temperatures reach $-9°C$ (15°F). Therefore, it is best adapted to the Gulf Coast areas. One should leave at least 25 to 30 cm (10 to 12 in) of fall growth to enhance its chances for winter survival. Rhodesgrass possesses a very deep, fibrous root system, is rather drought-tolerant, requires a good fertile soil, and prefers a moist soil to a droughty soil. It is rather salt-tolerant, in that it appears to tolerate alkaline soils in the lower valleys of Arizona. Rhodesgrass has also been used in irrigated pasture mixtures in southern California.

Rhodesgrass spreads by both stolons and seeds, but it can usually be controlled in other crops. It is primarily used as a pasture, hay, and erosion-control crop. It is very palatable and readily eaten by livestock. Rhodesgrass can withstand heavy trampling and will recover quickly following close grazing. Cattle should be rotationally grazed, however, because if a rhodesgrass pasture is continually overgrazed, stand losses will occur.

PANGOLAGRASS

Pangolagrass, *Digitaria decumbens* Stent., was introduced into the United States in 1935 from South Africa. Pangolagrass is adapted primarily to the Gulf Coast area of the United States, from Florida to Texas, and in some portions of southern California. It is a creeping, perennial, warm-season grass that resembles crabgrass, is adapted to well-drained fertile soils ranging from sand to clay, and is not tolerant of soils frequently inundated. Pangolagrass prefers full sunlight but is partially shade-tolerant. It prefers a soil with a pH in the range of 4.2 to 8.5. It is a rather poor seed producer and is usually established by vegetative cuttings or stolons (25, 38).

Pangolagrass is not winter-hardy, thus limiting its northern region of adaptation. It is grown primarily for pasture, hay, or silage. The forage is generally higher in protein and dry matter yields than that of bermudagrass or bahiagrass, and it produces higher animal gains. Pangolagrass is one of the more palatable nutritious grasses. It does not withstand excessive trampling but recovers rather rapidly from overgrazing. Pangolagrass is less productive at low nitrogen levels than bahiagrass or carpetgrass. One of the best management practices is to allow it to attain a height of 30 to 45 cm (12 to 18 in) before grazing and then graze it short. It recovers quickly under a rotational grazing scheme. Cattle should be rotated every week during the summer and every 2 to 3 weeks at other times in the grazing season to maintain a good stand and high forage quality. Pangolagrass makes excellent hay and silage. Hay is usually harvested in late fall, whereas silage is harvested in July and August (33).

Pangolagrass responds well to fertilizer applications and has a high nitrogen efficiency. Levels of fertilization above 225 kg N/ha (200 lb/A) annually should be avoided in areas where winter kill is a factor. Pangolagrass is very susceptible to copper (cu) deficiency and benefits from applications of 0.05 to 0.8 units of copper per hectare (52, 25).

CENTIPEDEGRASS

Centipedegrass, *Eremochloa ophiuroides* (Munro) Hack., is a low-growing, sod-forming, warm-season perennial grass native to southeast Asia. It was introduced to the United States in 1919 from China. It spreads by stolons and is very hard to eradicate. It is adapted to a wide range of soil types. It is not winter-hardy in the upper regions of the south. It is often used as a lawngrass.

The forage is of poor quality. It has a lower protein content and dry matter yield than bermudagrass, carpetgrass, or bahiagrass. It is very difficult to grow legumes with it because of its dense sod snd competitive nature. Due to its low quality, it is used primarily for lawns and erosion control. It is not recommended for pasture use anywhere in the south or the southwest (5).

CARPETGRASS

Carpetgrass, *Axonopus affinis* Chase, is a low-growing, sod-forming, warm-season perennial grass. It is native to the West Indies and Central America. It is adapted in the coastal sections of the United States, from eastern Texas and Louisiana eastward to the coastal region of South Carolina. It is best adapted to sandy soils and soils with a high water table, but will not tolerate water-logged soils. It is adapted to soils of low fertility. It spreads by stolons and seeds, and forms a very dense sod.

Dry matter yields of carpetgrass are lower than those of other perennial warm-season grasses; its nutritive value is also lower. It will produce only about 30% as much beef per hectare as common bermudagrass. It is not recommended as a pasture crop. Most producers or growers try to eliminate carpetgrass if it becomes established in a bermudagrass pasture. It can be controlled by spraying with Dowpon to reduce the competition from the carpetgrass, thus permitting the bermudagrass to increase. Carpetgrass is used primarily as a lawngrass in the south and will withstand moderate shading. It is often found in unimproved pastures throughout the Gulf Coast area (47).

OTHER WARM-SEASON GRASSES

Bluestems

There are several strains of both native and introduced ("old world") bluestems that are common in the southwest. The "old world" bluestems include var. Plains bluestem, *Bothriochloa ischaemum* var. *ischaemum;* King Ranch, *B. ischaemum* var. *songariea;* and Caucasian, *B. caucasius,* all of which are warm-season perennial bunch grasses. The Plains bluestem is best adapted to upland soils and occurs primarily on fine-textured soils. It is winter-hardy throughout Oklahoma and Texas. The Plains bluestem has a more upright growth than King Ranch bluestem, and is more productive and more disease-resistant. Plains is more palatable and digestible than the Caucasian bluestem, but produces less dry matter. Plains bluestem will produce approximately the same amount of dry matter as bermudagrass or weeping lovegrass on the upland soils of the southwest. The King Ranch bluestem is adapted to a wide range of soil types but grows best on fine-textured, rocky limestone soils. It is only fairly winter-hardy and is therefore limited in its northern region of adaptation to the southern two-thirds of Oklahoma.

The seedling vigor of King Ranch bluestem is low, but once it becomes established, it is very aggressive. It has a very fluffy seed and is very difficult to plant. It is drought-resistant and will withstand close grazing. The Caucasian bluestem is a leafier, finer-stemmed strain of bluestem and is more palatable than King Ranch bluestem, but is less palatable than the best native bluestems. It is more resistant to leaf diseases and produces slightly more dry matter than most of the common bluestems (42).

Big bluestem, *A. gerardi* Vitman, is a native, tall, warm-season perennial grass best adapted to the eastern Great Plains, running from North Dakota south to northeastern Texas. Big bluestem grows to a height of 1.8 m (6 ft) at maturity and is a very palatable grass when harvested as a hay, especially if harvested before the seedheads emerge. If it is grazed lightly during the early portion of the season, it will maintain vigorous stands which may later tolerate close grazing (22a, 42).

Little bluestem, *Schizachyrium scoparius* Michx., is a native warm-season perennial which does not grow as tall as big bluestem. At maturity it reaches a height of 0.6 to 1.2 m (2 to 4 ft). This bunch grass possesses great drought resistance. Little bluestem is found in southern Nebraska and the eastern half of Kansas, southward to the Gulf Coast. It grows well on deep, shallow, sandy, fine-textured, and rocky soils. It is one of the most important grasses of the tall-grass prairie. Cattle have been shipped for many years from the south and southwest to the Kansas Flint Hills and the Osage Hills of Oklahoma to be fattened on native range predominately composed of little bluestem.

Grama Grasses

There are several warm-season grama grasses found in the Great Plains. Sideoats grama, *Bouteloua curtipendula* Michx., is a bunch-forming perennial prairie grass most abundant in the central Great Plains. It is the most widely distributed of the grama grasses. Sideoats grama is found throughout the area east of the Rocky Mountains. It is rather low-growing species, reaching heights of approximately 1 m (3.3 ft). It too is very palatable and is readily eaten by all classes of livestock. Its feeding value is about the same as that of big and little bluestem (44).

Bluegrama, *B. gracilis* (H.B.K.) Lag. ex Steud., is a smaller, more drought-resistant native grass than sideoats. It is usually found growing with buffalograss in the short-grass prairie area. Bluegrama has no stolons, but does have a very dense root system. The plant attains a height of 15 to 30 cm (6 to 12 in). It is a very palatable grass and remains nutritious into the winter months (42).

Buffalograss

Buffalograss, *Buchloe dactyloides* (Nutt.) Engelm., is a low-growing perennial found in the lower rainfall portion of the Great Plains. It spreads by stolons and seeds. Plant height is approximately 5 to 15 cm (2 to 6 in). It is generally a very drought-resistant species, but if exposed to long periods of drought or continually overgrazed it may be killed. It maintains itself by dormant seeds that remain in the soil. It is a very palatable, nutritious species while growing and maintains its feed value even into the winter months. It will withstand very heavy grazing (4). Buffalograss is becoming a popular low-maintenance turf or lawn grass in the southern Great Plains.

Lovegrasses

There are several types of lovegrass popular in the central and southern Great Plains. The two most popular ones are sand lovegrass, *Eragrostis trichodes* (Nutt.) Wood, and weeping lovegrass, *E. curvula* (Schrad.) Nees. Both of these grasses are perennial and have a bunch-type growth, reaching heights of about 1.2m (4 ft). Sand lovegrass is more palatable than weeping lovegrass. Sand lovegrass produces good yields from April to late October. Weeping lovegrass is more adapted to the southwest and used more for erosion control, or as a forage on low-fertility soils. It has a shallower and more fibrous root system than sand lovegrass (18).

Switchgrass

Switchgrass, *Panicum virgatum* L., is a tall-growing, sod-forming warm-season perennial, native grass that is grown throughout the bluestem belt of the eastern and central Great Plains. There are two distinct strains of switchgrass, "Upland" and "Bottomland." The Upland type is a coarser-stemmed, broader-leaved plant that reaches heights of 1 to 1.5 m (3 to 4.5 ft). The bottomland type is coarser-stemmed and 0.3 to 1 m (1 to 3 ft) taller than Upland. It is less desirable for grazing or for hay. Switchgrass spreads by short rhizomes. It is used primarily for hay, summer pasture, and erosion control. It should be harvested when plants begin to joint (2a). It is best adapted to the more fertile and moist soils, but does perform much better on droughty and less fertile, erosible soils than do most introduced grass species (2).

Another species resembling switchgrass is blue panicgrass, *P. antidotale* Retz. This species is a tall, sod-forming warm-season perennial grass which spreads by rhizomes and seeds. It is a coarse and woody plant at maturity, and will branch freely at the nodes. The rhizomes are a short, bulbous type. It is adapted to clay soils but not to the more sandy soils. Blue panicgrass is not very winter-hardy and is usually recommended for use in the lower Rio Grande Valley and eastward. This species will not withstand close, continuous grazing and should be rotated if used for pasture. It does make high-quality hay when cut early. The more mature strands become very coarse and unpalatable. Yields generally decline after the first year even when sufficient nitrogen fertilization is added (42).

SUMMARY

There are numerous advantages and disadvantages to each warm-season grass. Most of these considerations were discussed in the section dealing with each species; that is palatability, digestibility, productivity, regions of adaptation, and resistance or susceptibility to various diseases, drought, or flooding. Warm-season grasses play a very vital role, especially in the south, southwest, and Great Plains areas. Most of these native grasses are fairly drought-resistant, disease-resistant, and dependable. In the south and southwest, under natural conditions, warm-season grasses will generally outyield most of the introduced grasses. Many of the warm-season grasses play a very important role in soil-erosion control, especially in the low rainfall areas.

QUESTIONS

1 Describe the general climatic conditions and soil types on regions where most of the warm-season grasses are adapted.

2 Select a particular area or region in the United States and name the various warm-season grasses best adapted to this area. Why did you select these species?

3 List some of the difficulties one may encounter in establishing and growing a warm-season grass. Ecologically speaking, why would the Gulf Coast area offer many more opportunities for warm-season grasses than the upper midwest?

4 Several of the species discussed in this chapter may be considered either a forage crop or a noxious weed. Describe the criteria for such distinctions.

5 Describe, in detail, why bermudagrass is such an important grass in the humid south.

REFERENCES

1 Alexander, J. R., and J. Q. Lynd. Response of four strains of bermudagrass to different rates of nitrogen fertilizer. *Okla. Agr. Exp. Sta. Processed Series P-485,* 1964.

2 Atkins, M. D. Permanent waterways. *Crops and Soils.* 10:2 and 14–15, 1957.

2 a Anderson, Bruce, and A. G. Matches. Forage yield, quality, and persistence of switchgrass and caucasian bluestem. *Agron. J.* 75:119–124, 1983.

3 Beaty, E. R., R. L. Stanley, and J. Powell. Effect of height of cut on yield of Pensacola bahiagrass. *Agron. J.* 60:356–359, 1968.

4 Beetle, A. A. Buffalograss—native of the short grasses. *Wyo. Agr. Exp. Sta. Bull. 293,* 1950.

4 a Belesky, D. P. and S. R. Wilkinson. Response of 'Tifton 44' and 'Coastal' bermudagrass to soil pH, K, and N source. *Agron. J.* 75:1–4, 1983.

5 Bennett, H. W. Johnsongrass, dallisgrass, and other grasses for the humid south. In *Forages,* Iowa State Univ. Press, Ames, IA, 3d ed., pp. 333–343, 1973.

6 Brooks, O. L., W. J. Miller, E. R. Beaty, and C. M. Clifton. Pelleted Coastal bermudagrass—comprehensive investigations. *Ga. Agr. Exp. Sta. Res. Bull. 27,* 1968.

7 Burns, J. C., D. H. Timothy, R. D. Mochrie, D. S. Chamblee, and L. A. Nelson. Animal preference, nutritive attributes, and yield of *Pennisetum flaccidum* and *P. orientale, Agron. J.* 70:451–456, 1978.

8 Burton, G. W. Bermudagrass. In *Forages,* Iowa State Univ. Press, Ames, IA, 3d ed., pp. 321–332, 1973.

9 Burton, G. W. Coastal bermudagrass. *Ga. Agr. Exp. Sta. Bull. N.S. 2,* 1954.

10 Burton, G. W. Registration of Coastcross-1 bermudagrass. *Crop Sci.* 12:125, 1972.

11 Burton, G. W., and J. E. Jackson. Effect of rate and frequency of applying six nitrogen sources on Coastal bermudagrass. *Agron. J* 54:40–43, 1962.

12 Burton, G. W., and J. E. Jackson. Single vs. split potassium application for Coastal bermudagrass. *Agron. J.* 54:13–14, 1962.

13 Burton, G. W., J. E. Jackson, and R. H. Hart. Effects of cutting frequency and nitrogen on yield, *in vitro* digestibility, and protein, fiber, and carotene content of Coastal bermudagrass. *Agron. J.* 55:500–502, 1963.

14 Burton, G. W., and W. G. Monson. Inheritance of dry matter digestibility in bermudagrass, *Cynodon dactylon. Crop Sci.* 12:375–378, 1972.

15 Burton, G. W., G. M. Prine, and J. E. Jackson. Studies of drouth tolerance and water use of several southern grasses. *Agron. J.* 49:498–503, 1957.

16 Burton, G. W., W. S. Wilkinson, and R. L. Carter. Effect of nitrogen, phosphorus and potassium levels and clipping frequency of forage yield and protein, carotene, and xanthophyll content of Coastal bermudagrass. *Agron. J.* 61:60–63, 1969.

17 Chase, A. The North American species of *Paspalum*. *U.S. Nat. Herb.* 28(1):310, 1929.

18 Crider, F. U. Three introduced lovegrasses for soil conservation. *USDA Circ. 330,* 1945.

19 Denman, C. E., W. C. Elder, and V. G. Heller. Performance of weeping lovegrass under different management practices. *Okla. Agr. Exp. Sta. Tech. Bull. T-48,* 1953.

20 Ellzey, H. D. Annual progress report. *Southeast Louisiana Dairy and Pasture Exp. Sta.* pp. 30–68, 1967.

20 a Evers, G. W. Weed control on warm-season perennial grass pastures and clover. *Crop Sci.* 23:170–171, 1983.

21 Fribourg, H. A., K. M. Barth, J. B. McLaren, L. A. Carver, J. T. Connell, and J. M. Bryan. Seasonal trends of *in vitro* dry matter digestibility of N-fertilized bermudagrass and of orchardgrass-ladino pastures. *Agron. J.* 71:117–120, 1979.

22 Fribourg, H. A., and J. R. Overton. Forage production on bermudagrass sods overseeded with tall fescue and winter annual grasses. *Agron. J.* 65:295–298, 1973.

22 a Hall, K. E., J. R. George, and R. R. Riedl. Herbage dry matter yields of switchgrass, big bluestem, and indiangrass with fertilization. *Agron. J.* 74:47–51, 1982.

23 Hart, R. H., and G. W. Burton. Prostrate vs. common dallisgrass under different clipping frequencies and fertility levels. *Agron. J.* 58:521–522, 1966.

24 Hauser, E. W. and A. H. Fred. Johnsongrass as a weed. *USDA Farmers' Bull. 1537,* 1958.

24 a Henderson, M. S., and D. L. Robinson. Enviornmental influences on yield and *in-vitro* true digestibility of warm-season perennial grasses and the relationships to fiber components. *Agron. J.* 74:943–946, 1982.

25 Hodges, E. M., G. G. Killinger, J. E. McCaleb, O. C. Ruelke, R. J. Allen Jr., S. C. Shank, and A. E. Kretschmer Jr. Pangolagrass. *Fla. Agr. Exp. Sta. Bull. 718,* 1967.

26 Hogg, P. G., and J. C. Collins. Clover and Coastal bermudagrass. *Miss. Farm. Res. 28(5):5,* 1965.

27 Holt, E. C. Dallisgrass. *Tex. Agr. Exp. Sta. Bull. 829,* 1956.

28 Holt, E. C., and H. C. Houston. The establishment of dallisgrass. *Tex. Agr. Exp. Sta. Progr. Rep. 622,* 1954.

29 Hoveland, C. S. Bahiagrass for forage in Alabama. *Ala. Agr. Exp. Sta. Circ. 140,* 1968.

30 Hoveland, C. S., W. B. Anthony, J. A. McGuire, and J. G. Starling. Beef cow-calf performance on Coastal bermudagrass overseeded with winter annual clovers and grasses. *Agron. J.* 70:418–420, 1978.

31 Jones, D. W. Bahiagrass in Florida. *Fla. Exp. Serv. Circ. 321A,* 1971.

32 Knox, J. W. Coastal bermudagrass for grazing and hay production. *Annual Res. Rep.* Red River Valley Agr. Exp. Sta., Bossier City, LA, 1976.

33 Kretschmer, A. E. Jr. The effect of nitrogen fertilization of mature pangolagrass just prior to utilization in winter on yields, dry matter, and crude protein contents. *Agron. J.* 57:529–534, 1965.

34 Lovvorn, R. L. The effect of defoliation, soil fertility, temperature, and length of day on the growth of some perennial grasses. *J. Amer. Soc. Agron.* 37:570–582, 1945.

35 McCormick, W. C., D. W. Beardsley, and B. L. Southwell. Coastal bermudagrass pellets for fattening beef steers. *Ga. Agr. Exp. Sta. Bull. N.C. 132,* 1965.

36 Monroe, W. E. Pastures pay with bahiagrass. *La. Agr. Ext. Serv. Pub. 1462,* 1967.

37 Monson, W. G., G. W. Burton, E. J. Williams, and J. L. Butler. Effects of burning on soil temperatures and yield of Coastal bermudagrass. *Agron. J.* 66:212–214, 1974.

38 Nestel, B. L., and M. J. Creek. Pangolagrass. *Herb. Abstr.* 32:265–271, 1962.

39 Oliver, W. M. Grazing stocker cattle on Coastal bermudagrass—a progress report. *La. Agr. Exp. Sta. Hill Farm Facts, Beef Cattle No. 18,* 1976.

40 Prine, G. M., and G. W. Burton. The effect of nitrogen rate and clipping frequency upon the yield, protein content and certain morphological characteristics of Coastal bermudagrass, *Cynodon dactylon* (L.) Pers. *Agron. J.* 48:296–301, 1956.

41 Purd, W. A., and P. G. Hogg. Management and utilization of irrigated Coastal bermudagrass with high level of nitrogen fertilization vs. white clover on Sharkey clay. *Miss. Agr. Exp. Sta. Information Sheet 1062,* 1969.

42 Schwendiman, J. L., and V. B. Hawk. Other grasses for the North and West. In *Forages,* Iowa State Univ. Press, Ames, IA, 3d ed., pp. 231–242, 1973.

43 Semple, J. A. The preparation and feeding value of pangolagrass silage. *Trop. Agr.* 43:251–255, 1966.

44 Smika, D. E., and L. E. Newell. Irrigation and fertilization practices for seed production from established stands of side oats grama. *Nebr. Agr. Exp. Sta. Res. Bull. 218,* 1965.

45 Tabor, P. Early history of dallisgrass in the United States. *Crop Sci.* 3:449–450, 1963.

46 Thurman, C. W., and C. Y. Ward. Johnsongrass management in Mississippi. *Miss. Agr. Exp. Sta. Information Sheet 1046,* 1968.

47 Ward, C. Y., ard V. H. Watson. Bahiagrass and Carpetgrass. In *Forages,* Iowa State Univ. Press, Ames, IA, 3d ed., pp. 314–320, 1973.

48 Watson, V. H., C. Y. Ward, and W. Thurman. Response of johnsongrass swards to various levels of nutrients and management. *Proc. Assoc. Southern Agr. Workers* 67:51, 1970.

49 Wilkinson, S. R., L. F. Welch, G. A. Hillsman, and W. A. Jackson. Compatibility of tall fescue and Coastal bermudagrass as affected by nitrogen fertilization and height of clip. *Agron. J.* 60:359–362, 1968.

CHAPTER 31

SUMMER-ANNUAL GRASSES

A very important objective in forage production is the maintenance of high-quality forage throughout the entire year, especially during the summer months, when drought periods may occur. Following severe winterkilling, or death caused by disease of perennial forage species, emergency or summer-annual forages play an important role in the overall forage program for a given year. They may be used to fill in between major crop production; for example, a program of small grains, fall-seeded and harvested in June, can be supplemented by planting a summer-annual forage between the June harvest and the reseeding time for the small grain in the fall. Some of the characteristics of summer-annual grasses are as follows: (1) rapid establishment; (2) high-yielding capacity in a very short period of time; and (3) production of a leafy, high-quality, nutritious, palatable forage that may be grazed or stored as silage or hay. Within the summer-annual group, the sorghum family and millets probably play the most important role in the emergency summer-annual forage program.

Summer annuals are expensive to grow. When comparing animal performance on summer annuals to their performance on summer perennials, it is difficult to justify the expense of the summer annuals except (1) as a harvested forage where corn is risky, (2) for emergency grazing, and (3) for dairy, where grain is also fed.

SORGHUMS

The sorghum genus, *Sorghum bicolor* (L.) Moench., includes several strains or types: (1) forage sorghums; (2) grain sorghums; (3) sudangrass; (4) sweet sorghums; and (5) sorghum-sudangrass hybrids (12). Sorghums are adapted to approximately the same regions of the United States as corn, particularly the warmer and drier areas where the annual precipitation is lower than it is in corn-growing regions. Although adapted to droughty conditions, the sorghums grow much better where there is sufficient moisture or where irrigation may be applied (11). They have the ability to go dormant during a droughty period and then resume growth when the drought is over. Sorghums also withstand flooding better than does corn.

Botanically speaking, sorghums are perennial but, due to their lack of winterhardiness, they are, agronomically—in practical applications—an annual in temperate regions. The most favorable temperature for growth ranges between 25 and 30°C (70 and 86°F), while the minimum temperature they can tolerate is about 15°C (59°F) (11).

The sorghums prefer a well-drained soil. Sorghums will withstand a pH of 5.7 without effectively reducing production. Young sorghum seedlings withstand very low temperatures of 5 to 10°C (41 to 50°F) (11). However, the soil temperature should probably be 15°C (60°F) before planting.

Plant Description

The sorghum plant grows erect, ranging in height from 45 cm to over 5 m (17 in to 16.5 ft) (2) (Figs. 31-1 and 31-2). The shorter cultivars, or strains, are the grain types, while the tall-growing strains are the forage or sorgo types. The leaf

FIGURE 31-1
Dwarf and normal sorghum. *(Courtesy University of Illinois)*

FIGURE 31-2
Sudangrass plant. *(Courtesy University of Illinois)*

blade is very similar to that of corn, but generally, the corn leaf is shorter and somewhat wider. The leaf blades of sorghum are glabrous and generally have a covering of a waxy material called cuticle. There are fewer stomata on the sorghum leaves than on corn. The leaf sheaths encompass the culm and have overlapping margins. The culm, which is solid, varies in thickness from 5 mm to over 3 cm. Sorghums produce a considerable number of tillers, which are more numerous following a harvest or after some physiological damage to the plant. The root system is very dense. The plant tissue of sorghums contains a higher concentration of sugar, which generally makes it a juicier plant when harvested as either a silage or hay crop (11).

The inflorescence of sorghum is a panicle that ranges from a very compact type in the grain sorghums to a very open, lax-type structure in sudangrass (Figs. 31-3 and 31-4). The sorghums are generally self-pollinated, although a considerable amount of cross-pollination does occur, as hybrid seed is produced for most sorghums in the United States through the use of male sterile lines. Sudangrass seeds are generally much flatter, lighter in weight, and lower in bulk density than grain or forage sorghum seeds. Generally grain sorghums have the largest grain. The forage sorghum seeds are intermediate in size and weight in sorghum-sudan crosses. The seed color varies from white to yellow to red-brown, and even to a deep purple in the case of some sudangrasses. Most of the darker seeds, whether red-brown or purple, contain varying amounts of tannin (2, 11).

FIGURE 31-3
Close-up of a compact forage sorghum
inflorescence. *(Courtesy University of Illinois)*

Uses of Sorghums

Each type of sorghum is best adapted to a different use. Forage sorghums are used as a silage crop; sudangrasses or sudan-sorghum hybrids are used for green chop, silage, and grazing; and the grain sorghums are used mainly as a grain

FIGURE 31-4
Close-up of an open, lax-type sudangrass
inflorescence. *(Courtesy University of Illinois)*

crop, although in some areas, they are used as a silage crop, to a certain extent. The very succulent and juicy stems of sorghums limit their use as a hay crop. High moisture content makes it very difficult to safely dry the hay for storage, but if hay is to be made, the use of a crimper is advised (2).

Approximately 45 days after the seeding of sudangrass, it can be harvested, either for grazing or silage. Repeated harvests, every 30 to 35 days, can occur if a sufficient amount of N is added, accompanied by adequate moisture, until the time of a killing frost (Fig. 31-5). Generally two or three harvests are obtained in a growing season, although in the far southwest, providing irrigation is available, five harvests per season can be taken (11). The sorghum silage yields are generally higher than those of corn silage when the two are planted under similar conditions, but, generally speaking, corn silage may have a higher energy level than sorghum silage. However, because of the high amount of sugar in the forage sorghum, its energy value is not much lower than that of good-quality corn silage (1, 6, 26). Some growers will mix or interseed corn and forage sorghum within the same row when planning for a silage harvest. This will increase both the dry matter yield and the energy level of the final silage product. Quality silage is associated with the nutrient level of the soil, especially N (1, 16, 20, 26). Sorghum has approximately the same nutrient requirements as corn.

Establishment and Management of Sorghums

Generally speaking, sorghums require a slightly higher soil temperature than corn for germination. The optimal soil temperature is between 20 and 30°C (68 and 86°F) (11). A well-prepared, firm, moist seedbed is very important, especially for the establishment of the smaller-seeded sorghums. The seeding depth should be approximately 1.5 to 5 cm (0.6 to 2 in) (11). Most of the forage

FIGURE 31-5
Properly managed sudangrass will provide quality forage during the summer months. *(Courtesy University of Tennessee)*

FIGURE 31-6
Summer annual grasses seeded in rows will provide a high-quality silage crop in a short growing period. *(Courtesy University of Tennessee)*

sorghums and grain sorghums grown for silage are sown in rows and harvested with machines that require row establishment (Fig. 31-6). Press wheels are very important in the establishment of sorghums. This practice ensures good soil-to-seed contact and rapid establishment.

Seeding

Sudangrass is usually sown in rather narrow rows or drilled into the soil. Seeding rates vary widely from the southwest to the more humid midwest. As the row width narrows, the seeding rate must be increased. If the seeding rate for drilling is between 10 and 20 kg/ha (9 and 18 lb/A), the stem size is increased by at least a third over that which would result from the normal seeding rate. The normal seeding rate is 20 to 30 kg/ha (18 to 27 lb/A). If one wishes to reduce the stem size, a higher seeding rate of 50 kg/ha (45 lb/A) will increase the first harvest yields and result in much thinner culms (7, 14).

There has been considerable research in an effort to determine the proper row spacing or row width for sorghum seeding. Overall yields are essentially the same and remain uninfluenced by broadcasting or seeding in various drill widths (7). The total production depends primarily on the number of harvests that are taken (11). Even though sorghums have the ability to tiller, the tillering effect depends on the row spacing or row width; total production, however, is about the same regardless of the spacing (7, 11).

The amount of tillering depends on the size of the stem or culm (Fig. 31-7); the smaller the culm, the greater its ability to produce tillers and promote faster

FIGURE 31-7
Rapid regrowth of sudangrass, 9 days after cutting.

regrowth. There are numerous axillary and basal stems present in sorghums that enhance regrowth (Fig. 31-8). As the cutting height is reduced in sorghums, initiation of the basal or axillary tillers increases (7, 10).

Fertilization

The sorghum family responds to adequate fertilization. It is very important to maintain a high fertility level to ensure high yields. The soil pH should be 5.7 or higher. As compared with pearlmillet, the sorghum family is more sensitive to, and more adversely affected by, a lower pH and lower P and K levels than the millet. The general recommendation is to fertilize the sorghums with the same levels used for corn. Sorghums respond to nitrogen fertilization, up to at least 200 kg/ha (178 lb/A). In the warmer, more humid climates of the Gulf Coast areas, as much as 400 kg/ha (356 lb/A) of nitrogen is recommended. At this nitrogen level, the nitrate content within the plant tissue must be monitored much more closely. For sudangrass strains, one would apply approximately one-half the total amount of nitrogen needed for the first harvest. The remaining one-half would be split for the second two harvests. For forage sorghums, the application of nitrogen should be approximately 40 to 50 kg/ha (35 to 45 lb/A) at the time of seeding, with the remaining amount of nitrogen applied as a

FIGURE 31-8
Regrowth of unevenly grazed sudangrass.

sidedress, as the stand density dictates (11). Depending on the level of P and K in the soil, fertilizer is recommended at the rate of 30 to 60 kg/ha (27 to 54 lb/A) of P, and 60 to 120 kg/ha (56 to 108 lb/A) of K for rapid growth and good yield (20a). The P level should be high enough to ensure rapid development of the root system and early emergence. It has been found that high-yielding sorghums will remove from 30 to 50 kg/ha of P (27 to 45 lb/A), and 150 to 200 kg/ha (134 to 178 lb/A) of K from the soil (11, 16, 23).

In summary, nitrogen fertilization will increase the total nitrogen, prussic acid, and nitrate content in sudangrass. High levels of soil, P, K, and Mg will have very little effect on the prussic acid content and the nitrate concentration of sudangrass (11).

The quality of forage is directly related to the time of harvest. The digestibility of sudangrass at the vegetative-growth stage is approximately 82%, compared with 75% at the boot stage, 65% at the early-head stage, and only 22% at the seed stage (1, 19, 24, 27).

Prussic acid and nitrate poisoning may be problems when grazing sudangrass, as well as with other sorghums. The prussic acid potential of sudangrass is higher in the early stages of growth, but decreases steadily until the fall or a frost. Pearlmillet, on the other hand, does not produce prussic acid, but may cause nitrate poisoning under high nitrogen fertilization (11).

Prussic acid will remain at levels dangerously high to livestock after a frost until all parts of the sorghum plant have completely dried, which takes from 1 to 6 days, depending on the moisture and the humidity present. The young tillers on the sorghum plant can be extremely high in prussic acid (11). Therefore, grazing should be deferred until the plants are 50 to 60 cm (20 to 24 in) tall to help reduce the risk of prussic acid poisoning. Do not graze slow-growing plants until they are at least 60 cm (24 in) tall; then they can be grazed down to approximately 25 to 30 cm (10 to 12 in) in a 10- to 14-day period; afterward, the cattle should be rotated to another sudangrass pasture (11). Sudangrass is usually harvested when in the boot stage, or when it is around 1 m (3 ft) in height, whichever comes first.

Most growers have found that sudangrass can be harvested approximately 45 days after seeding. The prussic acid content is lower in the true sudans or sudan-sudan hybrids than in the sorghum-sudan hybrids (11). It is recommended that, after one grazes down the sudangrass and cattle are rotated to the next pasture, the stubble be clipped to a uniform height to ensure uniform regrowth. It is extremely important to remember that the first growth or regrowth of sudangrass should not be grazed until the plants are approximately 60 cm (24 in) tall, since the young plants are very high in prussic acid. This will ensure a low prussic acid intake by cattle.

Sorghums, sudangrass, and pearlmillets are high in digestibility if they are grazed in the vegetative stage of growth. The percentage of leaves and stem diameter will indirectly or directly affect the fiber content and lignin content of the forage material (22, 27). High lignin content is usually related to advanced maturity and decreases the animal intake and digestibility (1, 19a, 22, 27). The young, tender leaves of the sudangrasses are generally 75% digestible, while the older leaves may be only 50 to 60% digestible. Generally, the stems or culms will be as little as 30 to 50% digestible. Therefore, it is very important to maintain a leafy forage in the sudangraasses when grazing them (1, 16, 19, 27).

The forage sorghums which are generally harvested only once a year have very few problems with high prussic acid content and, therefore, potential animal poisoning. The danger is when one grazes the late regrowth after silage harvest. Silage yields are generally from 33 to over 50 t/ha (15 to over 25 T/A). The forage sorghums will usually yield more than corn silage grown under similar conditions (11).

Forage sorghum should be harvested just prior to the maturity of the plant. This ensures a highly digestible forage. If one harvests prior to the milk or soft-dough stage, the energy level will be greatly reduced (6, 27). Many of the new forage sorghum cultivars are low in grain content; thus, the energy level may also be reduced. Sorghum silage is generally considered to be slightly less palatable than corn silage (24, 27). Some growers are interplanting corn and sorghum in the same row in an attempt to increase the energy level of the silage with the corn additive and overall bulk of dry matter added by the sorghum. Thus, a very favorable silage can be obtained.

Most of the sorghums, except for the sudangrasses, are harvested for silage. Growers use sudangrass as a grazing crop or as an emergency crop. Some grain sorghums grown in the United States are also harvested for silage. There are varying recommendations as to optimal time to harvest grain sorghum as silage. Some researchers suggest that grain sorghum be harvested from the milk stage to the soft-dough stage for dairy animals, and from the soft-dough to the hard-dough stage for beef animals (6, 24, 27). Researchers have found that dry matter intake increases with maturity but makes no difference in milk production (1, 24). If one waits until the hard-dough stage, there is a greater tendency for the grain to pass through the digestive tract intact (27).

SUMMARY

The sorghum family has a great potential for increasing forage production in the United States. This particular family is tolerant of alkaline and salty soils. It responds to irrigation and nitrogen fertilization. The sorghum family, when compared with corn, is very tolerant of drought, and in unpublished studies the sorghums were found to be more tolerant of flooded conditions than corn. Several reasons that sorghums are more drought-tolerant than corn may be: (1) there are more roots per unit of growth area; (2) the sorghums have a thicker cuticle; (3) they have fewer stomata; (4) they have a higher sugar content, which increases uptake of water; and (5) the sorghum family has the genetic ability to go dormant during a droughty period and then recover when moisture is present.

The forage sorghums will generally outyield corn when grown under similar conditions. They have a slightly lower energy level than corn silage. If harvesting the forage sorghum silage for dairy, it should be harvested from the milk to the soft-dough stage, while if harvesting forage sorghum silage for beef, harvest should be from the soft-dough to hard-dough stage.

Sudangrass is generally harvested two or three times per year. Sudangrass is much more leafy than forage sorghums. It is important to be alert to the potential of prussic acid poisoning when grazing sudangrass, but when it is allowed to grow to 60 cm (2 ft) or higher, the prussic acid is generally insignificant. The common sudangrass varieties are lower in yield and prussic acid content, while the sorghum-sudan hybrids are higher in yield and in prussic acid. The most promising type of sudangrass would be sudangrass-sudangrass hybrids. These are higher-yielding strains, with greater leaf content and lower prussic acid levels.

PEARLMILLET

Pearlmillet, *Pennisetum typhoides* (Burm.) Stapf. and C. E. Hubb., is a summer-annual grass that will also withstand drought conditions. It is primarily a cross-pollinated species that is erect in growth habit, reaching 2 to 5 m (6 to 16 ft) in height. The stems are pithy and the plant will tiller at each node on the stem. It is a rather leafy species in which the leaf margins are finely serrated. It is less winter-hardy than sudangrasses or sorghums, especially in the germination or seedling stage. It is more resistant to diseases and high humidity than the sorghum family (11).

Pearlmillet is best adapted to lighter-textured, well-drained soils. It is not as well adapted to alkaline conditions or high pH as sorghums, nor to the heavy black soils of the southeast. It is important to remember that proper liming, which increases the pH of very acid soils, is important in maintaining high forage yield and quality (11).

Pearlmillet should be planted in the late spring or when soil temperatures are above 15°C (60°F), at approximately a 10-cm (4-in) depth. Best germination and growth occur when the air temperatures are between 18 to 29°C (65 to 85°F). Therefore, seeding generally begins in early May and ends by late June. Seeding rates range from 22 to 33 kg/ha (20 to 30 lb/A), again depending on soil fertility and the expected use of the forage (11). Most of the production will occur in the first 6 to 8 weeks of growth. Grazing can begin when plants are around 25 to 38 cm (10 to 15 in) in height; they should be grazed down to a 7.5 to 10 cm (3 to 4 in) stubble before rotating the cattle to another pasture. It is important to maintain a certain stubble height since regrowth comes from axillary buds.

There is no prussic acid problem with pearlmillet pasture, as there is in the sorghum family. The main concern is that under high nitrogen fertilization, nitrate poisoning can become a problem in pearlmillet, as it can with any grass. Most pearlmillet is used as a grazing crop or silage.

Management of pearlmillet will depend on its use. High moisture content and thick stems limit pearlmillet's use as a hay crop. Pearlmillet should not be cut too frequently; in fact, research has indicated that when pearlmillet was cut three times it yielded more than when harvested four times per year, but quality was sacrificed.

The nitrate content of pearlmillet is generally slightly higher than that of common sudangrass or hybrid sudangrass. As the potassium content and soil acidity increase, the nitrate content will increase. Phosphorus and magnesium have no effect on the nitrate content of either pearlmillet or sudangrass (23). Pearlmillet is a leafier forage than sudan-sorghum hybrids. Pearlmillet responds well to good pasture management, but should not be overgrazed.

The fertility requirements for pearlmillet are approximately the same as those for sudangrass or forage sorghum.

Most pearlmillet is used as a pasture crop or made into silage. Some growers will use pearlmillet as green-chopped material. This requires a daily harvest and increases the utilization of the overall forage produced. Green-chop pearlmillet may have a laxative effect on cows and should be used in a limited manner to determine the effect on an individual herd (11). Pearlmillet has proved to be an excellent pasture or emergency pasture for both dairy and beef cattle. Cattle ranchers select it because it is leafier than sudangrass throughout the growing season and there is no danger from prussic acid (Fig. 31-9).

SUMMARY

Pearlmillet may be limited in its production from northern Nebraska eastward to Pennsylvania, but it does have great potential as an emergency forage crop. It requires higher temperatures for growth, but produces a leafier material that is equally digestible as sudangrass or forage sorghum. One of the main advantages of pearlmillet is that there is no danger of prussic acid poisoning, particularly as

FIGURE 31-9
Comparing growth of sudangrass, 'Piper,' with pearlmillet, 'Gahi-1.' *(Courtesy University of Tennessee)*

compared with sudangrass. The main concern is that on highly nitrogen-fertilized stands, nitrate poisoning may potentially become a problem. Pearlmillet does not have quite the drought tolerance or regrowth of sudangrass, nor will it withstand as heavy grazing pressure as sudangrass. In recent years pearlmillet has been grown further north in the United States as a pasture crop. Recently, a few producers have grown soybeans with pearlmillet to increase the protein content of the forage harvested as a silage crop.

The recent release of 'Tif leaf' pearlmillet is significant since it has shown an increase in yield and quality compared with older cultivars. There is considerably more animal gain with the 'Tif leaf' cultivar.

QUESTIONS

1 Compare the adaptation and sensitivity of the sorghum family with these characteristics of pearlmillet in relation to soil type and region of adaptation.
2 Compare the grazing and management of sudangrass and pearlmillet in terms of a green-chop crop, a grazing crop, and a silage crop.
3 Why might a producer consider sudangrass over pearlmillet?
4 What are the different management considerations of sudangrass and pearlmillet?
5 Discuss the features which may make the sorghum family more drought-resistant than other grass species.

REFERENCES

1 Ademosum, A. A., B. R. Baumgardt, and J. M. Scholl. Evaluation of a sorghum-sudangrass hybrid at varying stages of maturity on the basis of intake, digestibility and chemical composition. *J. Anim. Sci.* 27:818–823, 1968.
2 Artschwager, E. Anatomy and morphology of the vegetative organs of *Sorghum vulgare. USDA Tech. Bull. 957,* 1948.

3 Begg, J. E., and G. W. Burton. Comparative study of fine genotypes of pearl millet under a range of photoperiods and temperatures. *Crop Sci.* 11:803–805, 1971.

4 Bernstein, L. Salt tolerance of plants. *USDA Agr. Info. Bull. 283,* 1964.

5 Beuerlein, J. E., H. A. Fribourg, and F. F. Bell. Effects of environment and cutting on the regrowth of a sorghum-sudangrass hybrid. *Crop Sci.* 8:152–155, 1968.

6 Browning, C. B., and J. W. Lusk. Effect of stage of maturity at harvest on nutritive value of combine-type grain sorghum silage. *J. Dairy Sci.* 50:81–85, 1967.

7 Burger, A. W., and W. F. Campbell. Effect of rates and methods of seeding on the original stand, tillering, stem diameter, leaf-stem ratio, and yield of sudangrass. *Agron. J.* 53:289–291, 1961.

8 Burger, A. W., C. N. Hittle, and D. W. Graffis. Effect of variety and rate of seeding on the drying rate of sudangrass herbage for hay. *Agron. J.* 53:198–201, 1961.

9 Burns, J. C., R. F. Barnes, W. F. Weding, C. L. Rhykard, and C. H. Noller. Nutritional characteristics of forage sorghum and sudangrass after frost. *Agron. J.* 62:348–350, 1970.

10 Clapp, J. G., Jr. and D. S. Chamblee. Influence of different defoliation systems on the regrowth of pearl millet, hybrid sudangrass, and two sorghum-sudangrass hybrids from terminal, axillary, and basal buds. *Crop Sci.* 10:345–349, 1970.

11 Fribourg, H. A. Summer annual grasses and cereals for forage. In *Forages,* Iowa State Univ. Press, Ames, IA, 3d ed., pp. 344–357, 1973.

12 Harlan, J. R., and J. M. J. deWet. A simplified classification of cultivated sorghum. *Crop. Sci.* 12:172–176, 1972.

13 Hart, R. H. Digestibility, morphology, and chemical composition of pearl millet. *Crop Sci.* 7:581–584, 1967.

14 Holt, E. C. Relationship of hybrid sudangrass plant populations to plant growth characteristics. *Agron. J.* 62:494–496, 1970.

15 Hoveland, C. S., R. R. Harris, J. K. Boseck, and W. B. Webster. Supplementation of steers grazing sorghum-sudan pasture. *Ala. Agr. Exp. Sta. Circ. 188,* 1971.

16 Jung, G. A., and R. L. Reid. Sudangrass—Studies on its yield, management, chemical composition and nutritive value. *W. Va. Agr. Exp. Sta. Bull. 524T,* 1966.

17 Koller, H. R., and J. M. Scholl. Effect of row spacing and seeding rate on forage production and chemical composition of two sorghum cultivars harvested at two cutting frequencies. *Agron. J.* 60:456–459, 1968.

18 McCartor, M. M., and F. M. Rouquette Jr. Grazing pressures and animal performance from pearl millet. *Agron. J.* 69:983–987, 1977.

19 Moore, J. E., M. E. McCullough, and M. J. Montgomery. Composition and digestibility of southern forages. *Fla. Agr. Exp. Sta. Southern Coop. Ser. Bull. 165,* 1971.

19 a Pederson, J. F., H. J. Gorz, F. A. Haskins, and W. M. Ross. Variability for quality and agronomic traits in forage sorghum hybrids. *Crop Sci.* 22:853–856, 1982.

20 Quinby, J. R., and P. T. Marion. Production and feeding of forage sorghum in Texas. *Tex. Agr. Exp. Sta. Bull. 965,* 1960.

20 a Reneau, R. B. Jr., G. D. Jones, and J. B. Friedericks. Effect of P and K on yield and chemical composition of forage sorghum. *Agron. J.* 75:5–8, 1983.

21 Rouquette, F. M. Jr., T. C. Keisling, B. J. Camp, and K. L. Smith. Characteristics of the occurrence and some factors associated with reduced palatability of pearl millet. *Agron. J.* 72:173–174, 1980.

22 Rusoff, L. L., A. S. Achacoso, C. L. Mondart Jr., and F. L. Bonner. Relationship of lignin to other chemical constituents in sudan and millet forages. *La. Agr. Exp. Sta. Bull. 542,* 1961.

23 Schneider, B. A., N. A. Clark, R. W. Hemken, and J. H. Vandersoll. Relationship of pearl millet to milk fat depression in dairy cows. II. Forage organic acids as influenced by soil nutrients. *J. Dairy Sci.* 53:305–310, 1970.

24 Spahr, S. L., E. E. Ormiston, and R. G. Peterson. Sorghum-sudan hybrid SX-11, Piper sudangrass, and alfalfa-orchardgrass for dairy pastures. *J. Dairy Sci.* 50:1925–1934, 1967.

25 Sumner, D. C., H. S. Etchegaray, J. E. Gregory, and W. Lusk. Summer pasture and greenchop from sudangrass, hybrid sudangrass, and sorghum × sudangrass crosses. *Calif. Agr. Exp. Sta. Circ. 547,* 1968.

26 Thurman, R. L., O. T. Stallcup, and C. E. Reames. Quality factors of sorgo as a silage crop. *Ark. Agr. Exp. Sta. Bull. 632,* 1960.

26 a Touchton, J. T., W. A. Gardner, W. F. Hargrove, and R. R. Duncan. Reseeding crimson clover as a N source for no-tillage grain sorghum production. *Agron. J.* 74:283–287, 1982.

27 Wedin, W. F. Digestible dry matter, crude protein, and dry matter yields of grazing-type sorghum cultivars as affected by harvest frequency. *Agron. J.* 62:359–363, 1970.

GLOSSARY*

acceptability, animal Readiness with which animals select and ingest a forage; sometimes used interchangeably to mean either palatability or voluntary intake.

acid detergent fiber (ADF) Insoluble residue following extraction with acid detergent (van Soest); cell wall constituents minus hemicellulose.

acid detergent fiber digestibility The digestibility of acid detergent fiber (ADF), determined as the difference in ADF in a forage before and after in vitro or in vivo digestion.

acid detergent lignin (ADL) Lignin on residue determined following extraction with acid detergent.

acid pepsin Used in second stage in in vitro forage digestion, 2 g of 1:10,000 pepsin in 1 l of 0.1 N HCL.

ad libitum feeding Daily feed offerings in excess of consumption, generally 115% of consumption.

aflatoxin $C_{17}H_{10}O_6$ A polynuclear substance derived from molds; a known carcinogen. Produced by a fungus occurring on groundnuts.

aftermath Residue and/or regrowth of plants used for grazing after harvesting of a crop.

air-dry weight The weight of a substance after it has been allowed to dry to equilibrium with the atmosphere.

animal day One day's tenure upon pasture by one animal. Not synonymous with animal-unit day.

animal performance Production per animal (weight change or animal products) per unit of time.

animal unit (AU) Considered to be one mature (454 kg or 1,000 lb) cow or the equivalent based on average daily forage consumption of 11.8 kg (26 lb) dry matter per day.

animal unit-month (AUM) The amount of feed or forage required by an animal unit for 1 month; tenure of one animal unit for a period of 1 month. Not synonymous with animal month.

---

*Definitions and descriptions are adaptations from the American Society of Agronomy, Madison, Wisconsin.

antioxidant   An organic compound that accepts free radicals and thus prevents autoxidation of fats and oils; at very low concentrations in food, not only acts in retarding rancidity but also in protecting the nutritional value or minimizing the breakdown of vitamins and essential fatty acids.

antiquality constituents   Constituents that have negative value or that produce negative responses in animals consuming the produce containing the constituent.

apparent dry matter digestibility   See DIGESTIBILITY, APPARENT.

aroma   A characteristic odor, as of a plant, feed, or food.

ash   The residue remaining after complete burning of combustible matter.

atomic absorption spectroscopy   Observation by means of an optical device (spectroscope) of the wavelength and intensity of electromagnetic radiation (light) absorbed by various materials. Particular elements absorb well-defined wavelengths on an atomic level. The wavelengths absorbed are in the visible and infrared regions. Theoretical interpretation of the spectra obtained leads to knowledge of atomic and molecular structure.

backgrounding   Intensive management of young cattle, postweaning, so as to facilitate maximum performance.

bioassay   The use of living organisms to quantitatively estimate the amount of biologically active substances present in a sample.

biomass   Weight of living organisms (plants and animals) in a ecosystem, at a given point in time, expressed either as fresh weight or dry weight.

bird resistance   A genotype or individual plant that is avoided by birds until other food sources are exhausted, or until weathered.

black-layer formation   A cork layer formed at the hilum region of physiologically mature cereal grains at time of cessation of dry matter deposition in the caryopsis.

bloat   Excessive accumulation of gases in the rumen of animals.

bloom, bloomless (sorghum)   Presence or absence, respectively, of a white, waxy, or pruinose covering on the leaves and stems.

bloom, early   Initial flowering (anthesis) in the uppermost portion of the inflorescence.

bloom, full   Essentially all florets in the inflorescence in anthesis.

bloom, late   Most florets in the inflorescence have completed anthesis.

blooming   refers to anthesis in the grass family, or to the period during which florets are open and anthers are extended.

body weight, empty   Conceptually, empty body weight is weight of animal tissue and is equal to live weight minus gut contents; usually estimated from hot carcass weight and appropriate regression equations. (Calculation often desirable to determine gain in short time period where differences in gut fill may exist.)

body weight, fat-free   Conceptually, empty body weight minus body fat; in practice, generally determined from carcass specific gravity to body specific gravity.

body weight, shrunk   Body weight after a period of fast (feed and/or water, usually overnight or 24 hours), to reduce variation in gut-fill contribution to body weight.

bomb calorimetry   Process whereby a substance is completely oxidized in 25 to 30 atmospheres of oxygen to determine gross energy (GE) content.

boot stage   Growth stage when a grass inflorescence is enclosed by the sheath of the uppermost leaf.

bract   a modified or reduced leaf subtending a flower or inflorescence.

bran   Pericarp of grain.

brown midrib   In maize (br) and sorghum (bmr), a single recessive gene character resulting in the dark brown coloration of the back side of the leaf midrib and under the leaf sheaths; associated with reduced lignin content of the plant.

browning　Refers to the reaction between reducing sugars and free amino groups in proteins to form a complex that undergoes a series of reactions to produce brown polymers usually referred to as melanoidins. Higher temperatures and basic pH favor the reaction.

browning reaction　See BROWNING.

browse　(n) Small stems, leaves, flowers, and fruits of shrubs, trees, or woody vines available for grazing; (v) to consume browse.

bundle sheath　A sheath of one or more layers of parenchymatous or of sclerenchymatous cells surrounding a vascular bundle.

calorie (gram calorie)　The amount of heat required to raise the temperature of one gram of water one degree Celsius. One kilocalorie (kcal) = 1,000 calories; one megacalorie (Mcal) = 1,000,000 calories.

canopy　The vertical projection downward of the aerial portion of plants, usually expressed as percent of ground so occupied.

carbohydrates, nonstructural　Soluble carbohydrates found in the cell contents, as contrasted with structural carbohydrates in the cell walls.

carbohydrates, structural　Carbohydrates found in the cell walls, for example, hemicellulose, cellulose.

carrying capacity　Number of animals a given pasture will support at a specified level of animal gain or production of milk, wool, etc., for a given period of time. See GRAZING CAPACITY.

cellulose　A carbohydrate formed from glucose that is a major constituent of plant cell walls. A colorless solid; insoluble in water.

cell wall constituents　Compounds that make up or constitute the cell wall, including cellulose, hemicellulose, lignin, and minerals (ash).

cell wall content　That proportion of plant material made up of cell walls as opposed to cell contents.

chaff　Glumes, hulls, and small fragments of straw separated from grain or seed in the threshing process.

chromic oxide　A completely indigestible chemical (Cr_2O_3) used as an indicator to estimate forage intake.

colorimetry　The process of measuring the concentration of a known solution constituent through comparison of its color with colors of standard solutions of that constituent.

companion crop　A crop sown with another crop, used particularly with small grains with which forage crops are sown. Preferred to the term "nurse crop."

concentrate　All feed, low in fiber and high in total digestible nutrients, that supplies primary nutrients (protein, carbohydrate, and fat), for example, grains, cottonseed meal, wheat bran.

consumer acceptance　The state of consumers' using the genotype, phenotype, or product of their own choice.

continuous grazing　The grazing of a specific unit by livestock throughout a year, or for that part of a year during which grazing is feasible without removing the livestock from the pasture.

cored hay samples　Samples taken from stored hay with a hollow cylinder that removes the core.

coumarin　A white, crystalline compound with a vanillalike odor that gives sweet clover its characteristic odor. An antiquality component of sweet clovers.

coumestrol　Estrogenic factor occurring naturally in forage crops, especially in ladino clover, strawberry clover, and alfalfa.

cover (1) The combined aerial parts of plants and mulch; (2) shelter and protection for animals and birds.

crimped Rolled with corrugated rollers.

crop residue Portion of plants remaining after seed harvest; said mainly of grain crops such as corn stover or of small-grain straw and stubble.

crude fiber Coarse, fibrous portions of plants, such as cellulose; partially digestible and relatively low in nutritional value. In chemical analysis it is the residue obtained after boiling plant material with dilute acid and then with dilute alkali.

cubing Process of forming hay into high-density cubes to facilitate transportation, storage, and feeding.

cuticle A waxy layer secreted by epidermal cells on the outer surface of plants.

cutin A waxy, somewhat waterproof outer covering of plants.

cyanogenesis The release of hydrocyanic acid (HCN) in the process of chemical change. Cyanogenetic is the adjective form.

days per ha, animal Total tenure of animals on pasture, expressed as animal days per ha, usually per unit of time (month, year, etc.)

deferment Delay or discontinuance of livestock grazing on an area for an adequate period of time to provide for plant reproduction, establishment of new plants, or restoration of vigor of existing plants.

deferred grazing See GRAZING, DEFERRED.

deferred rotation Any grazing system having a stocking density index greater than 1 and less than 2, which provides for a systematic rotation of the deferment among pastures.

defoliation Removal of the leaves (tops) from a plant by cutting or grazing; causing the leaves of a plant to fall off, especially by the use of a chemical spray or dust.

density (1) The number of individuals per unit area; (2) the relative closeness of individuals to one another.

diet The usual food regularly offered to or consumed by a person or animal.

digestibility, apparent Digestibility determined by animal feeding trials, calculated as feed consumption minus excretion (feces), expressed as percent; does not account for endogenous excretions in the feces.

digestibility, true Actual digestibility or availability of feed, forage, or nutrient as represented by the balance between intake and fecal loss of the same ingested material with endogenous excretions in feces accounted for; in vitro digestibility without adjustment to in vivo base.

digestible dry matter (DDM) Feed intake minus feces, expressed as a percent of feed dry matter consumed.

digestible energy (DE) Feed-intake gross energy minus fecal energy, expressed as calories per unit feed dry matter consumed.

digestible energy intake Feed consumption expressed as units of digestible energy.

digestible nutrients Portion of nutrients consumed that is digested and taken into the animal body. This may be either apparent or true digestibility; generally applied to energy and protein.

digestible protein Feed protein minus feces protein (N × 6.25), expressed as a percentage of amount in feed.

digestion The conversion of complex, generally insoluble foods to simpler substances which are soluble in water.

dough stage Seed development stage at which endosperm development is pliable, like dough (for example, soft, medium, hard), usually used when 50% of seeds on an inflorescence are in this stage of development.

dry matter The substance in a plant remaining after oven drying to a constant weight at a temperature slightly above the boiling point of water.

dry matter accumulation Total (aboveground) plant development up to some point in time, including both new and old growth.

dry matter disappearance (DMD) (1) Pasture: Forage present at the beginning of a grazing period plus growth during the period, minus forage present at the end of the period. (2) Digestibility: Loss in dry weight of forage exposed to in vitro digestion.

dry matter intake, daily Amount of dry matter ingested by an animal on a daily basis.

dye-binding capacity The amount of monosulfonic azo dye (acid orange 12) bound by the basic amino acids during the Udy Dye Binding Procedure; the more dye-binding capacity, the more the basic amino acid content.

early leaf Term used to describe a stage in the development of a plant; requires a reference point to be useful.

embryo percent The amount of embryo compared to endosperm and other seed parts. The percent of embryo in the whole seed.

enclosure An area enclosed by fence or wall to confine animals (range); a caged or fenced area within a pasture to confine grazing.

ensilage Silage.

enzymatic degradation The chemical breakdown of a given substance by the specific enzyme catalyst for that particular chemical reaction of a biological process.

enzyme An organic catalyst containing protein that speeds up a specific reaction.

epinasty Increased growth on upper surface of a plant organ or part (especially the leaf) causing it to bend downward.

ether extract Fats, waxes, oils, and similar plant components which are extracted with warm ether in chemical analysis.

exotic plant An introduced plant that is not fully naturalized or acclimated.

extruded Pushed through orifices of a die under pressure; refers to feed or food.

fat-free body weight See BODY WEIGHT, FAT-FREE.

fecal index Indirect method of estimating indigestibility of dry matter by determining concentration of an indicator in feces.

feeding value Characteristics that make feed valuable to animals as a source of nutrients; the combination of chemical, biochemical, physical, and organoleptic characteristics of forage that determine its potentials to produce animal meat, milk, wool, or work. Considered by some as synonymous with nutritive value.

fermentation Anaerobic chemical transformation induced by the activity of the enzyme systems of micro-organisms such as yeast enzymes that produce carbon dioxide and alcohol from sugar.

fescue foot Red and swollen skin at junction of hoof in animals grazing tall fescues, followed by sloughing off of hoofs, tail tips, and ear tips in advanced stages, along with loss of appetite and emaciation. Problem is most severe in cold, rainy, or overcast weather.

fescue toxicity Toxicity to animals grazing tall fescue; may take one or all of three forms: fescue foot, fat necrosis, and poor animal performance (also called summer slump or summer syndrome).

fiber A unit of matter characterized by a length at least 100 times its diameter or width.

fibrous Being finely lined in appearance or composed of fibers.

fill (rumen content) Amount of ingested feed or water present in the rumen.

fines Material that passes through a screen whose openings are smaller than the specified minimum size of the product being processed.

fistula   A surgical opening, duct, or passage from a cavity, or hollow organ of the body.

flag leaf   The uppermost leaf on a fruiting (fertile) culm; the leaf immediately below the inflorescence or seed head.

flaked   Prepared by a method involving the use of high temperatures, tempering, and rollers set close together.

floury   Refers to an endosperm characteristic of cereals. The floury endosperm is in the center of the cereal kernel and has many voids between the starch granule and the protein matrix. These voids cause light to be refracted, and hence it appears as a light area when the kernel is cut in half and viewed with reflected light. The void will appear opaque to transmitted light. For example, high lysine corn has a floury or opaque endosperm characteristic. This is mentioned in the description of chalky grain.

fluorescence   The emission of electromagnetic radiation, especially of visible light, resulting from the absorption of incident radiation, and persisting only as long as the stimulating radiation is continued.

flowering stage   The physiological stage of a grass plant in which anthesis (blooming) occurs, or in which flowers are visible in nongrass plants.

fodder   Coarse grasses such as corn and sorghum harvested with the seed and leaves green or alive, cured and fed in their entirety as forage.

foliage   The green or live leaves of growing plants; plant leaves collectively, often used in reference to aboveground development of forage plants.

forage   Herbaceous plants or plant parts consumed by domestic animals (generally, the term refers to such material as pasturage, hay, silage, dehy, and green chop in contrast to less digestible plant material known as "roughage").

forage crop   Plants grown primarily for livestock feed and either used for grazing or harvested for green-chop feeding, silage, or hay.

forage nutritive value   See NUTRITIVE VALUE.

forage quality   Characteristics that make forage valuable to animals as a source of nutrients; the combination of chemical and biocharacteristics of forage that determines its potential to produce meat, milk, wool, or work. Considered by some as synonymous with feeding value and nutritive value.

fractionation   To separate into components, as by distillation, crystallization, or physical separation.

fruiting period   The period during which heading, pollination, and seed maturation occur.

gelatinized   Ruptured by a combination of moisture, heat, and pressure. Refers to starch granules of feed.

gel filtration   A type of fractionation procedure in which molecules are separated from one another according to differences in size and shape; the action is similar to that of molecular sieves. Dextran gels (three-dimensional networks of polysaccharide chains) are usually used in this method.

germ   Biology: A small organic structure or cell from which a new organism may develop. Seed: Refers to the embryo.

gluten   A mixture of plant proteins occurring in cereal grains, chiefly corn and wheat; substance in wheat flour that gives cohesiveness to dough.

gossypol   A phenolic pigment in cottonseed that is toxic to some animals.

grain grade   Market standard established to describe the amount of contamination, grain damage, immaturity, test weight, and marketable traits.

grain maturity   When no further dry matter is accumulated in the grain.

grain, percent

$$\text{Grain, percent} = \frac{\text{threshed grain weight}}{\text{threshed grain weight} + \text{stalk weight}} \times 100$$

gram/stalk ratio

$$\text{Grain/stalk ratio} = \frac{\text{threshed grain weight}}{\text{stalk weight}}$$

grassland Any plant community in which grasses and/or legumes compose the dominant vegetation.

grass tetany (hypomagnesemia) Condition of cattle and sheep marked by tetanic staggers, convulsions, coma, and frequently death, characterized by a low level of blood magnesium.

grazer Animal that grazes in situ grass as herbage. Animals on experimental pastures which may or may not remain on specified pasture treatment for the entire grazing period or season, but which are of a kind or physiological condition not necessarily represented on all pasture treatments for the entire grazing period or season.

grazing capacity Number of animals a given pasture will support at a specified level of animal gain or of production of milk, wool, etc., for a given period of time. See CARRYING CAPACITY.

grazing, continuous Grazing an area over a period of time without interruption, usually throughout the pasture season.

grazing, deferred Delay or discontinuance of livestock grazing on an area for a period of time adequate to provide for plant reproduction, establishment of new plants, or restoration of vigor.

grazing, high-intensity–low-frequency A grazing system in which the forage on individual pastures is removed by grazing in a relatively short period (high-intensity) and the pasture is not grazed again for a relatively long period (low-frequency).

grazing management The manipulation of livestock grazing to accomplish a desired result.

grazing pressure The amount of forage allowed per animal of specified kind and physiological condition at a specific time, or, conversely, the number of animals per unit available forage.

grazing, rotational Grazing of two or more pastures in succession, with the grazing cycle usually repeated throughout the pasture season.

gross energy (GE) The amount of heat that is released when a substance is completely oxidized in a bomb calorimeter containing 25 to 30 atmospheres of oxygen.

groat The caryopsis (kernel) of oats after the husk has been removed.

haylage Product resulting from ensiling forage with about 45% moisture in the absence of oxygen.

head components Components of the inflorescence of grains and grass crops; generally, grain versus vegetative structures, but may include individual vegetative structures such as rachis, peduncle, and pedicel.

heading The stage of development of a grass plant from initial emergence of the inflorescence from the boot, until the inflorescence is fully exerted.

hemicellulose Polysaccharides that accompany cellulose and lignin in the cell walls of green plants; differs from cellulose in that it is soluble in alkali, and, with acid hydrolysis, gives rise to uronic acid, xylose, galactose, and other carbohydrates, as well as glucose.

high-quality protein A protein containing the appropriate proportions of amino acids for a particular dietary usage.

histochemistry The chemistry of cells and tissues.

introduced species A species not a part of the original fauna or flora of the area in question.

in vitro digestible dry matter (IVDDM) See IN VITRO DRY MATTER (DIGESTIBILITY) DISAPPEARANCE (IVDMD); the procedures and the numerical values are the same for IVDDM and IVDMD.

in vitro dry matter (digestibility) disappearance A gravimetric measurement of the amount of dry matter lost upon filtration following the incubation of forage in test tubes with rumen microflora, usually expressed as a percentage:

$$\frac{\text{weight dry matter sample} - \text{weight residue}}{\text{weight dry matter sample}}$$

in vivo In a living organism, such as in the animal or in the plant.

in vivo nylon bag technique System of determining dry matter disappearance of forage placed in fine-mesh nylon bags, either placed in the rumen or suspended in the rumen from a cannula cover of a fistulated animal.

in vitro organic matter (digestibility) disappearance (IVOMD) Similar to IVDMD, except expressed on an organic matter basis:

$$\frac{(\text{weight dry matter sample} - \text{weight sample ash}) - (\text{weight residue} - \text{weight residue ash})}{\text{weight dry matter sample} - \text{weight sample ash}}$$

kernel A mature ovule which has the ovary wall fused to it. Same as caryopsis.

ketosis A pathological accumulation of ketone bodies in an organism.

lesions A wound or injury; a circumscribed pathological alteration of tissue.

ley In the United States, "ley" is interpreted as biennial or perennial hay or pasture portion of a rotation, including cultivated crops.

lignify To make woody. The thickening, hardening, and strengthening of plant cells by the disposition of lignin on and in the walls of plant cells.

lignin An organic chemical which strengthens and hardens the walls of cells, especially wood cells.

liquid chromotography An analytical method based on the separation of the components of a mixture in solution by selective absorption. All systems contain a moving solvent; a means of producing solvent motion (gravity or a pump); a means of sample introduction; a fractioning column; and a detector. Innovations in functional systems provide the analytical capability for operating in three separation modes: (1) Liquid/Liquid: Partitition in which separations depend on relative solubilities of sample components in two immiscible solvents (one of which is usually water); (2) Liquid/Solid: Absorption where the differences in polarities of sample components and their relative absorption on an active surface determine the degree of separation; (3) Molecular size separations: Depend on the effective molecular size of sample components in solution. Common solvents used in liquid chromotography: isooctane, methyl ethyl ketone, acetone chloroform, tetrahydrofuran, hexane, and toluene.

Common packing materials: silica gel, alumina, glass beads, polystyrene gel, and ion exchange resins.

lodging, root   Stalk fall without stalk breakage, because of weak root system, root damage, or soil condition.

lodging, stalk   Stalk breakage above the ground level.

maillard browning reaction   See BROWNING.

mesophyll   The leaf cells which contain chloroplasts and are located between the upper and lower epidermis.

metabolic body weight ($W^{0.75}$)   Basal metabolic rate (energy expenditure per unit body weight per unit time; that is, kcal heat/weight/day) varies as a function of fraction power of body weight, usually determined to be body weight raised to the 0.75 power. Loss of protein from body also varies by a similar fractional power of weight. Thus the metabolic mass to body weight is presumed to be related by the same exponential power of body weight.

metabolizable energy (ME)   Digestible energy (DE), less the energy lost as methane and lost in urine by ruminant animals.

milk stage   In grain (seed), the stage of development following pollination in which the endosperm appears as a whitish liquid somewhat like milk.

middlings   A byproduct of flour milling that contains varying proportions of endosperm, bran, and germ.

milling, dry   Refers to processes in which grains and other commodities are subjected to grinding followed by sifting, sizing, or other separation techniques to produce more refined products for use in human or animal feeds. The process usually involves only a small amount of water to temper the grain prior to the milling operation. For example, in the milling of wheat, the grain is tempered to approximately 15 to 17% moisture, which facilitates the separation of the endosperm, the pericarp, and germ. Another example is in corn dry milling where the grain is tempered to 20% or slightly higher levels to ensure the separation of the germ from the endosperm.

milling, wet   Refers to the counter-current steeping of grain in water for an extended period of time until the grain reaches the 40 to 50% moisture level. Then the starch is separated from the other constituents of grain. In contrast to this, dry milling only separates the grain into its major anatomical portions, for example, germ (embryo), bran (pericarp), and endosperm (flour or grits).

moisture equilibrium   The condition reached by a sample when it no longer takes up moisture from, or gives up moisture to, the surrounding atmosphere.

moisture regain   The amount of moisture in a material determined under prescribed conditions and expressed as a percentage of the weight of the moisture-free specimen.

mycotoxin   A toxin or toxic substance produced by a fungus.

native pasture   A term referring to native, predominately herbaceous, vegetation used for grazing in untilled areas. The term "tame" or "introduced" is used instead of "native" for pastures that include mainly species that are not native.

native species   A species indigenous to an area; not introduced from another environment or area.

naturalized pasture   Plants introduced from other countries which have become established in, and more or less adapted to, a given region by long continued growth there. The name is appropriate for pastures made up of plants such as white clover, bluegrass, bermudagrass.

naturalized species   See NATURALIZED PASTURE.

net energy (NE)   Metabolizable energy minus the energy lost in the heat increment.

neutral detergent fiber neutral detergent insoluble residue, synonymous with cell wall constituents.

neutral detergent fiber digestibility The digestibility of neutral detergent fiber determined as the difference in NDF in a forage before and after in vivo or in vitro digestion.

nitrate poisoning See NITRATE TOXICITY.

nitrate toxicity Conditions in animals resulting from ingestion of feed high in nitrate; the toxicity actually results when nitrate is reduced to nitrite in the rumen.

nitrogen-free extract (NFE) The unanalyzed portion of a plant, consisting mostly of carbohydrates, that remains after the protein, ash, crude fiber, ether extracts, and moisture content have been determined.

NMR oil content The estimation of the amount of oil in a whole seed sample by use of the nuclear magnetic reasonance spectroscopy technique.

nonenzymatic browning See BROWNING.

nonnutritive fiber That portion of fiber in a feed which is not digestible and hence is of no nutritive value.

nuclear magnetic resonance (NMR) A type of radio frequency or microwave spectroscopy, based on the magnetic field generated by the spinning of the electrically charged nucleus of certain atoms. This nuclear magnetic field is caused to interact with a very large magnetic field of the instrument magnet. Each nuclear species requires a different instrumental setting to obtain reasonance of the magnetic field frequency of the instrument with that of the nucleus. The setting also varies with the electron distribution around each atom, so it is possible, for instance, to distinguish between the three kinds of hydrogen atoms in ethyl alcohol. The direct relation of NMR spectral bank height to the concentration of chemical compound allows NMR to be used for quantitative analysis and the following of rates of reaction and equilibrium of chemicals in solution.

nutrient, animal Food constituent or group of food constituents of the same general chemical composition required for support of animal life.

nutritive value Relative capacity of a given feed to furnish nutrition for animals; may be prefixed by low, high, moderate, etc.

nutritive value index (NVI) Daily digestible amount of forage per unit of metabolic body size relative to a standard forage.

opaque-2 An endosperm mutant of maize associated with suppressed prolamine production in the endosperm, resulting in increased lysine content of the protein fraction.

orts Rejected feedstuffs left under conditions of ad libitum stall feeding.

overgrazing The grazing of a number of animals on a given area that, if continued to the end of the planned grazing period, will result in less than satisfactory animal performance and/or less than satisfactory pasture forage production.

overstocking The placing of a number of animals on a given area that will result in overuse if continued to the end of the planned grazing period. Not to be confused with overgrazing, as an area may be overstocked for a short period, but the animals be removed before the area is overused. However, continued overstocking will lead to overgrazing.

paddock Small fenced field used for grazing purposes.

palatability Plant characteristics eliciting a choice between two or more forages or parts of the same forage, conditioned by the animal and environmental factors that stimulate a selective intake response.

pasturage Vegetation on which animals graze, including grasses or grasslike plants, legumes, forbs, and shrubs.

pasture A fenced area of land covered with grass or other herbaceous forage plants and used for grazing animals.

pasture carrying capacity See CARRYING CAPACITY.

pasture, native See NATIVE PASTURE.

pasture, permanent Composed of perennial or self-seeding annual plants kept for grazing indefinitely.

pasture, rotation Used for a few seasons and then plowed for other crops.

pasture, semipermanent Pasture of perennial or reseeding annual species that is grazed for only a limited number of years.

pasture stage Refers to the vegetative stage, usually prior to any appreciable heading or flowering.

pasture, supplemental A crop used to provide grazing for supplemental use, usually during periods of low pasture production.

pasture, tame Grazing lands, planted with introduced or domesticated forage species, which may receive periodic cultural treatments such as renovation, fertilization, and weed control.

pasture, temporary A field of crop or forage plants grazed for only a short period, usually not more than one crop season.

pearled Reduced by machine brushing to smaller, smooth particles. Refers to dehulled grains.

perennial Persisting for several years, usually with new growth from a perenniating part.

pericarp The ripened and variously modified walls of a plant ovary, especially in cereal caryopsis.

perloline A plant alkaloid, found to interfere with cellulose digestion by rumen micro-organisms; commonly associated with tall fescue.

phloem The conducting tissue present in vascular plants, chiefly concerned with the transport of elaborated food materials in the plant. When fully developed, the phloem consists of sieve tubes and parenchyma, with companion cells present or absent.

photosensitization A noncontagious disease resulting from the abnormal reaction of light-colored skin to sunlight after a photodynamic agent has been absorbed through the animal's system. Grazing certain kinds of vegetation or ingesting certain molds under specific conditions causes photosensitization.

pith A usually continuous central strand of spongy tissue in the stems of most vascular plants which probably functions chiefly in storage.

prebloom The stage or period immediately preceding blooming.

preservative An additive used to protect against decay, discoloration, or spoilage.

protein acceptability See PROTEIN QUALITY.

protein, crude An estimate of protein content based on a determination of total nitrogen content times a constant (6.25).

protein fraction Refers to solubility of proteins as originated by Osborne, who designed a sequential extraction scheme based on water, dilute salt, 70% alcohol, and dilute acid or alkali.

protein matrix It is one of the major structural forms of protein inside the cereal caryopsis. It surrounds the starch granules as a continuous phase in the outer portion of the endosperm. However, in the floury endosperm portion, the protein matrix is discontinuous and sometimes difficult to see. The proteins composing the matrix are mainly alkali-soluble, although some protein bodies which are composed of primarily alcohol-soluble proteins are involved in the matrix. This is especially true for corn and sorghum.

protein quality   Refers to the balance of essential amino acids in the protein, as well as the biological availability of the protein. In general, for cereals the first limiting amino acid and the one with the greatest effect on protein quality is lysine.

proximate analysis   Analytical system that includes the determination of ash, crude fiber, crude protein, ether extract, moisture (dry matter), and nitrogen-free extract.

prussic acid   A poison, produced as a glucoside by several plant species, especially sorghums. Also called hydrocyanic acid.

pubescence   A general term for hairs or trichomes.

pubescent   Covered with fine, soft, short hairs or trichomes.

pulses   The edible seeds of various leguminous crops (such as peas, beans, lentils).

put-and-take animal   See GRAZER.

range   Embraces rangelands, and many forest lands which support an understory or periodic cover of herbaceous or shrubby vegetation amenable to certain range management principles or practices.

rangeland   Land on which the native vegetation (climax or natural potential) is predominantly grasses, grasslike plants, forbs, or shrubs, suitable for grazing or browing use. Rangelands include natural grasslands, savannahs, shrublands, most deserts, tundra, alpine communities, coastal marshes, and wetland meadows.

range management   The science of maintaining maximum range forage production without jeopardy to other resources or uses of the land.

ration   The total amount of feed (diet) allotted to one animal for a 24-hour period.

reducing sugars   Sugars that have the ability to donate electrons to copper cations to produce copper metal. Some common reducing sugars are glucose, fructose, and maltose, as opposed to nonreducing sugars such as sucrose, raffinose, melibiose, and stachyose.

residue biomass   The biomass that remains following removal or utilization of part of the biomass by grazing, harvesting, burning, etc.

resins   Sticky to brittle plant products from essential oils, and sometimes possessing marked odors. Used in medicines, varnishes, etc.

resistance   (1) The ability of a plant or crop to grow and produce even though heavily inoculated or actually infected or infested with a pest; (2) The ability of a plant to survive a period of stress such as drought, cold or heat.

rind   The epidermis and sclerenchyma tissue on the outer surface of stems of corn, sorghum, and other grass plants.

ripe   Fully grown and developed; mature.

rotation grazing   See GRAZING, ROTATIONAL.

roughage   Animal feeds that are relatively high in crude fiber and low in total digestible nutrients and protein.

rumen   First compartment of the stomach of a ruminant or cud-chewing animal.

ruminant   Of or relating to a suborder of mammals having a complex multichambered stomach.

saponin   Any of various plant glucosides that form soapy colloidal solutions when mixed and agitated with water.

sclerenchyma   Strengthening tissue made up of heavy lignified cell walls; supports and protects the softer tissues of the plant.

seed   (n) Ripened (mature) ovule consisting of an embryo, a seedcoat, and a supply of food that, in some species, is stored in the endosperm; (v) to sow, as to broadcast or drill small-seeded grasses and legumes or other crops.

seed size   Usually expressed as weight per unit number of seed.

senescence   A slowing in growth rate of a plant or plant organ, usually due to old age.

shrub   A perennial woody plant smaller than a tree and having several stems arising at a point near the ground.

silage   Forage preserved in a succulent condition by partial anaerobic acid fermentation.

silage, additive   Material added to forage at the time of ensiling to enhance favorable fermentation process.

silage preservative   See SILAGE, ADDITIVE.

single-kernel analysis   Use of one seed or a portion of one seed to obtain estimates of chemical composition and seed quality factors.

spectroscopy   Observation by means of an optical device (spectroscope) of the wavelength and intensity of electromagnetic radiation (light) absorbed or emitted by various materials. Theoretical interpretation of well-defined wavelengths of elements (often minute quantities) in obtained spectra leads to knowledge of atomic and molecular structure.

stalk diameter   The diameter of a stalk, usually at a designated node or internode.

stalk girdle   Ring made by insect around the stem or stalk which may result in lodging of the stalk. Girdle may be either internal or external.

stalk tunnels   Longitudinal tunnels in plant stalks produced by insects. Examples of stalk-tunneling insects are the sugarcane borer and lesser corn stalk borer.

starch fractions   Refers to amylose and amylopectin. In most cereal starches, there is usually 20 to 30% amylose and 70 to 80% amylopectin in the starch. However, waxy cereals contain 100% amylopectin, (branch-chain starch), whereas high-amylose corn varieties have been developed that contain 80% amylose (linear-chain starch) and only 20% amylopectin in the starch.

starch granules   The fundamental unit in which starch is deposited in the storage tissue of many higher plants. It is paracrystalline cold-water-insoluble and has a characteristic size and shape depending on the species which produced it.

stocker   Young cattle, post weaning, subjected to pasture grazing situations.

stocking density   The relationship between number of animals and area of land at any instant of time. It may be expressed as animal units per acre, animal units per section, or AUM/ac.

stocking density index   The reciprocal of the fraction: land available to the animals at any one time/land available to the animals for the entire grazeable period.

stocking plan   The number and kind of livestock assigned to one or more given management areas or units for a special period.

stocking pressure   See GRAZING PRESSURE.

stocking rate   The area of land allotted to each animal or animal unit of specified kind and physiological condition, or, conversely, the number of animals per unit land.

stockpiled   Accumulated growth of forage for later use.

stover   The matured, cured stalks of such crops as corn or sorghum, from which the grain has been removed. A type of roughage.

stubble   The basal portion of the stems of herbaceous plants left standing after harvest.

substrate   (1) A substance that is acted upon in a chemical reaction; (2) A culture medium.

supplement Nutritional additive (salt, protein, phosphorus, etc.) intended to improve nutrition balance and remedy deficiencies of the diet.

supplemental feeding Supplying concentrates or harvested feed to correct deficiencies of the pasture diet. Often erroneously used to mean emergency feeding. Compare maintenance feeding.

surfactants Compounds that are active at the interface between nonpolar (oil) and polar (water) molecules, that is, soaps, detergents.

sustained yield The continuation of desired forage or animal production.

sward The ground cover of a stand of grass, mixtures of grasses, or mixtures of grass and legumes.

terminal Of or relating to an end or extremity; growing at the end of a branch or stem.

tester animals Animals of like kind and similar physiological condition used in grazing experiments to measure animal performance or pasture quality; usually assigned to a treatment for the duration of the graing season, versus "grazer" animals, which may be assigned temporarily.

test weight Weight of a volume of seed, for example, weight per bushel, bulk density.

texture The surface, appearance, and hand of a textile material.

total digestible nutrients (TDN) Sum total of the digestibility of the organic components of plant material and/or seed, for example, crude protein + NFE + crude fiber + fat.

total nonstructural carbohydrates (TNC) See CARBOHYDRATES, NONSTRUCTURAL.

toxicity Injury, impairment, or death resulting from poison or toxin.

toxin A substance that is a specific product of the metabolic activities of a living organism and is usually very unstable, notably toxic when introduced into the tissues, and typically capable of inducing antibody formation.

toxoid Toxin that has been treated to be rendered nontoxic, but which will still induce the formation of antibodies.

trichome A filamentous outgrowth; an epidermal hair structure on a plant.

true digestibility See DIGESTIBILITY, TRUE.

undergrazing Utilizing pasture forage with grazing animals at a rate less than that required for optimum animal performance and/or forage production.

vascular bundle An elongated strand containing xylem and phloem, the conducting tissues.

vascular tissue Conducting tissue with vessels or ducts.

vegetative A descriptive term referring to stem and leaf development in contrast to flower and seed development. Commonly used as a synonym of "nonsexual," in contrast to sexual type of development and reproduction in plants.

vegetative cover A soil cover of plants, irrespective of species.

vegetative propagation See VEGETATIVE REPRODUCTION.

vegetative reproduction (1) In seed plants, reproduction by means other than seeds; (2) In lower forms, reproduction by vegetation spores, fragmentation, or division of the plant body.

vegetative state Stage prior to the appearance of fruiting structures.

vigor Indicative of active growth, relative absence of disease or other stresses.

vitality The capacity to live and develop; power of enduring or continuing.

voluntary intake Ad libitum intake achieved when an animal is offered an excess of a single feed or forage.

weather deterioration The loss of quality in a crop due to the effects of weather on the product or process.

whole-kernel analysis Chemical analysis of a sample made up of ground whole kernels.

whorl stage Stage in the development of a grass plant prior to the emergence of the inflorescence, synonymous with the vegetative stage in the grass family.

xylem The portion of the conducting tissue which is specialized for the conduction of water and minerals.

INDEX